设施园艺技术进展

——2013第三届中国·寿光国际设施园艺高层学术论坛论文集

Technology Advances in Protected Horticulture

—— Proceedings of 2013 the 3rd High-level International Forum on Protected Horticulture（Shouguang•China）

杨其长 Toyoki Kozai（日本） Gerard P.A. Bot（荷兰） Nicolas Castilla（西班牙） 主编
Edited by Yang Qichang Toyoki Kozai（Japan） Gerard P.A. Bot（the Netherlands） Nicolas Castilla（Spain）

中国农业科学技术出版社

图书在版编目（CIP）数据

设施园艺技术进展——2013 第三届中国·寿光国际设施园艺高层学术论坛论文集/杨其长等主编．—北京：中国农业科学技术出版社，2013.4
ISBN 978－7－5116－1239－7

Ⅰ.①设…　Ⅱ.①杨…　Ⅲ.①园艺—保护地栽培—文集　Ⅳ.①S62－53

中国版本图书馆 CIP 数据核字（2013）第 051991 号

责任编辑　张孝安
责任校对　贾晓红

出 版 者　中国农业科学技术出版社
北京市中关村南大街 12 号　邮编：100081
电　　话　(010)82109708(编辑室)　(010)82109704(发行部)
(010)82109709(读者服务部)
传　　真　(010)82109708
网　　址　http://www.castp.cn
经 销 者　各地新华书店
印 刷 者　北京科信印刷有限公司
开　　本　787 mm×1 092 mm　1/16
印　　张　21.5
字　　数　518 千字
版　　次　2013 年 4 月第 1 版　2013 年 8 月第 2 次印刷
定　　价　86.00 元

2013 第三届中国·寿光国际设施园艺高层学术论坛组织委员会名单

组委会主席：

吴孔明　中国工程院院士　中国农业科学院副院长

朱兰玺　山东省寿光市人民政府市长

组委会副主席：

杨德峰　山东省寿光市人大常委会主任

梅旭荣　中国农业科学院农业环境与可持续发展研究所所长　研究员

组委会成员：

王惠玲　中共寿光市委常委　寿光市人民政府副市长

王小虎　中国农业科学院科技局局长

袁龙江　中国农业科学院成果转化局局长

栗金池　中国农业科学院农业环境与可持续发展研究所党委书记

张燕卿　中国农业科学院农业环境与可持续发展研究所副所长

杨其长　中国农业科学院农业环境与可持续发展研究所研究员

孙德华　山东省寿光市蔬菜高科技示范园管理处党支部书记

王启龙　中国（寿光）国际蔬菜科技博览会组委会办公室主任

隋申利　山东省寿光市蔬菜高科技示范园管理处副主任

陈青云　中国农业大学教授　中国农业工程学会设施园艺工程专委会主任

何启伟　山东省园艺学会理事长　山东省农业专家顾问团蔬菜分团团长

王秀峰　山东农业大学园艺科学与工程学院院长

王乐义　山东省寿光市三元朱村书记

2013 第三届中国·寿光国际设施园艺高层学术论坛学术委员会名单

学术委员会主席：

Kozai Toyoki 教授　日本千叶大学
杨其长　研究员　中国农业科学院农业环境与可持续发展研究所
Gerard P. A. Bot 教授　荷兰瓦赫宁根大学
Nicolás Castilla 教授　西班牙格拉纳达大学

学术委员会成员（按姓氏字母顺序）：

艾希珍　教授	山东农业大学
白义奎　教授	沈阳农业大学
别之龙　教授/主任	华中农业大学
Constantinos Kittas 教授	希腊塞萨洛尼基大学
崔　瑾　副教授	南京农业大学
陈　超　教授	北京工业大学
陈青云　教授/副院长	中国农业大学
Egbert Heuvelink 教授	瓦赫宁根大学
Eiji Goto 教授	日本千叶大学
方　炜　教授	台湾大学
郭世荣　教授/主任	南京农业大学
Haruhiko Murase 教授	日本大阪府立大学
郝秀敏　教授	加拿大温室与加工作物研究中心
Kotaro Takayama 教授	日本爱媛大学
李天来　教授/副校长	沈阳农业大学
李亚灵　教授	山西农业大学
刘　红　教授/俄罗斯自然科学院外籍院士	北京航空航天大学
刘士哲　教授/副院长	华南农业大学
李建设　教授/处长	宁夏大学
罗卫红　教授	南京农业大学
Meir Teitel 教授	以色列 Volcani 农业研究组织
毛罕平　教授/院长	江苏大学
马承伟　教授/主任	中国农业大学
乔晓军　教授	北京市农林科学院
宋卫堂　教授	中国农业大学
Jung Eek Son 教授	韩国首尔大学

Organizing Committee of 2013 the 3rd High-level International Forum on Protected Horticulture

Chairman of the Organizing Committee

Kongming Wu (Academician of Chinese Agricultural Academy of Sciences)

Lanxi Zhu (Mayor of the People's Government of Shouguang City)

Vice-Chairman of the Organizing Committee

Defeng Yang (Director of the Standing Committee of Shouguang People's Congress)

Xurong Mei (Head of Institute of Environment and Sustainable Development in Agricultural)

Members of the organizing committee

Huiling Wang (Vice-Mayor of the People's Government of Shouguang City)

Xiaohu Wang (Director General of Ministry of Science and Technology Bureau, Chinese Academy of Agricultural Sciences)

Longjiang Yuan (Director of Chinese Academy of Agricultural Sciences Results into the Bureau)

Jinchi Li (Party secretary of Institute of Environment and Sustainable Development in Agricultural, Chinese Academy of Agricultural Sciences)

Yanqing Zhang (Vice-Director of Institute of Environment and Sustainable Development in Agricultural, Chinese Academy of Agricultural Sciences)

Qichang Yang (Professor of Institute of Environment and Sustainable Development in Agricultural, Chinese Academy of Agricultural Sciences)

Dehuan Sun (Party secretary of Vegetable Hi-Tech Demonstration Park Office in Shouguang, Shandong)

Qilong Wang (Director of International Vegetable Science and Technology Expo Organizing Committee Office in Shouguang, Shandong)

Shenli Duo (Vice-Director of Vegetable Hi-Tech Demonstration Park Office in Shouguang, Shandong)

Qingyun Chen (Professor of China Agricultural University, Chinese Society of Agricultural Engineering Horticultural Engineering Director of Special Committees)

Qiwei He (Chairman of Horticultural Society in Shandong Province, Expert Advisory of Shandong Agricultural Vegetables Regiment)

Xiufeng Wang (Dean of the College of Horticulture Science and Engineering, Shandong Agricultural University)

Leyi Wang (Secretary of Sanyuanzhu Village in Shouguang City, Shandong Province)

Scientific Committee of 2013 the 3rd High-level International Forum on Protected Horticulture

Chairman of the Scientific Committee

Kozai Toyoki, Professor Chiba University in Japan

Yang Qichang, Professor Institute of Environment and Sustainable Development in Agricultural, CAAS

Gerard P. A. Bot, Professor Wageningen University

Nicolás Castilla, Professor University of Granada in Spain

Member of Scientific Committee

Xizhen Ai, Professor	Shandong Agricultural University
Yikui Bai, Professor	Shenyang Agricultural University
Zhilong Bie, Professor/Director	Huazhong Agricultural University
Constantinos Kittas, Professor	University of Thessaloniki
Jin Cui, Professor	Nanjing Agricultural University
Chao Chen, Professor	Beijing University of Technology
Qingyun Chen, Professor/Vice president	China Agricultural University
Egbert Heuvelink, Professor	Wageningen University
Eiji Goto, Professor	Chiba University, Japan
Wei Fang, Professor	National Taiwan University
Shirong Guo, Professor/ Director	Nanjing Agricultural University
Haruhiko Murase, Professor	Japan Osaka Prefecture University
Xiuming Hao, Professor	Canada Greenhouse and Processing Crops Research Center
Kotaro Takayama, Professor	Japan Ehime University
Tianlai Li, Professor/Vice-President	Shenyang Agricultural University
Yaling Li, Professor	Shanxi Agricultural University
Hong Liu, Professor	Beijing University of Aeronautics and Astronautics
Shizhe Liu, Professor /Vice president	South China Agricultural University
Jianshe Li, Professor/Director	Ningxia University
Weihong Luo, Professor	Nanjing Agricultural University
Meir Teitel, Professor	Volcani Institute-Agricultural Research Organization, Israel
Hanping Mao Professor/Dean	Jiangsu University
Chengwei Ma, Professor/Director	China Agricultural University

Xiaojun Qiao, Professor	Beijing Academy of Agriculture and Forestry
Weitang Song, Professor	Chinese Agricultural University
Jung Eek Son, Professor	Seoul University Korea
Zhongfu Sun, Professor	IEDA, Chinese Academy of Agricultural Sciences
Zhiqiang Sun, Professor/Director	Henan Agricultural University
Guohong Tong, Professor	Shenyang Agricultural University
Xiufeng Wang, Professor/Dean	Shandong Agricultural University
Min Wei, Professor	Shandong Agricultural University
Xiangzhen Wen, Professor	Shanxi Agricultural University
Hui Xu, Professor/Vice president	Shenyang Agricultural University
Yulan Xiao, Professor	Capital Normal University
Youngshik Kim, Professor	Samgmyung University
Xianchang Yu, Professor	Institute of Vegetables and Flowers, CAAS
Zhibin Zhang, Professor	Institute of Vegetables and Flowers, CAAS
Zhenxian Zhang, Professor	China Agricultural University
Changji Zhou, Professor/Vice director	Chinese Academy of Agricultural Engineering
Zhirong Zou, Professor / Dean	Northwest Agriculture and Forestry University of Science and Technology
Weijun Zhou, Professor	Zhejiang University

前　　言

自2009年以来，借助中国（寿光）国际蔬菜科技博览会平台，中国农业科学院和寿光市人民政府先后于2009年、2011年在博览会期间连续举办了两届“中国·寿光国际设施园艺高层学术论坛”（High-level International Forum on Protected Horticulture，HIFPH）。论坛分别以“设施园艺与现代科技”、“节能高效，绿色安全”为主题，汇聚了国内外数十位知名设施园艺专家以及数百位参会代表，就设施园艺科技发展、节能与新能源利用、环境优化控制、高效栽培、植物工厂技术等专题进行深入研讨，论坛取得了圆满成功。作为论坛的成果之一，两届论坛还汇集与会专家的70余篇论文，正式编辑出版了两本论文专集，受到业内的广泛关注。

近年来，设施园艺产业发展迅速，中国设施栽培面积已经突破350万公顷。设施园艺产业的快速发展为改善城乡居民的生活质量、增加农民收入作出了重要贡献，但随着耕地的不断减少、化石能源的日益紧缺、劳动力成本的持续上升以及人们对食品安全的高度关注，设施园艺产业也面临着诸多亟待解决的难题。如何利用现代科技成果解决设施园艺生产中面临的资源、环境和可持续发展问题，是摆在世界设施园艺专家面前的重大课题。为此，本届论坛选择以“低碳、节能、高效、安全”为主题，邀请了40余位国内外知名专家作大会主题报告和专题报告，并围绕设施环境调控、高效栽培、节能与新能源利用、新型材料与装备、绿色安全生产、物联网技术以及植物工厂等热点内容进行交流与研讨，探讨实现设施园艺节能、绿色、安全、高效生产的技术途径，并汇集与会专家的45篇论文，正式编辑出版。

在论坛组织过程中，得到了中国农业科学院、山东省寿光市人民政府、荷兰瓦赫宁根大学（Wageningen UR）、日本千叶大学（Chiba University）、山东农业大学、山东农业科学院、中国园艺学会设施园艺分会、中国农业工程学会设施园艺工程专业委员会等单位的大力支持，在此表示衷心感谢。

由于时间仓促，论文集中难免会有错漏之处，恳请各位同仁和读者批评指正。

编　者

2013年4月

目 录

综 述

设施园艺工程技术

设施栽培理论技术

综
述

Role of Plant Factory with Artificial Light in Urban Life

Toyoki Kozai

(*Japan Plant Factory Association*, *c/o Chiba University*, *Kashiwa city*, *Chiba* 277-0882, *Japan*)

Abstract: Role of plant factory with artificial light (PF) in urban areas is discussed. Importance of fresh food production in urban areas is emphasized with respect to resource saving, environment conservation, losses of quality and quantity of produce during transportation and community communication. Concepts of closed plant production system and resource use efficiency are introduced to discuss the resource saving characteristics of PF. Annual productivity per land of PF is compared with that in the open field. PF can be a key technology of urban agriculture in the forthcoming decades. A trial project of networks of PF is introduced.

Key words: Closed plant production system; Resource use efficiency; Urban agriculture

1 Introduction

In 2013, over 50% of world population (over 7 billions) is living in large cities or urban areas, and this percentage as well as the world population is estimated to continue increasing up to around 70% until 2050, while the world agricultural population and the percentage of farmers/growers younger than 60 years old have been decreasing and will continue to decrease in the forthcoming decades.

Furthermore, the world is facing with issues on unusual weather, environmental pollution, shortages of water, fossil fuel, arable land area and plant biomass resources, and fears over food safety, security and the stability of food supplies. In the face of such issues, the world is turning increasingly to urban agriculture or controlled environment agriculture including plant factories with artificial light (called PF hereafter) (Giacomelli and Munday, 2013; Kozai, 2013a; 2013b) and vertical farming (Despommier, 2010).

PF is briefly defined in this paper as a closed plant production system with artificial light where the environment of the growing plants is controlled within the room enabling the year-round planned production of plants such as vegetables. According to this definition, PF neither requires automation nor a large scale operation.

This paper addresses a potential contribution of controlled environment agriculture to solving a part of the above-mentioned issues from an aspect of fresh food production in urban areas, with special attention to PF.

2 Issues in Urban Areas

By producing fresh foods in a large city, can we reduce the total amounts of resource consumption and emission of environmental pollutants per unit weight of fresh food in a fresh food supply chain from production to processing of food waste? As is shown in Figure 1, huge amounts of food, water, fossil fuel, electricity and various kinds of products are transported every day into a large city from remote areas. In return, huge amounts of heat, CO_2, waste water and garbage/waste are generated and emitted in the large city. In order to collect, transport and process these environmental pollutants, considerable amounts of energy and materials are consumed. In addition, large numbers of vehicles and personnel are going in and out of the large cities to transport and handle such resources and environmental pollutants. Furthermore, there are other issues in a large city regarding relatively high percentages of unemployment, empty rooms and buildings, traffic jams, social security, crime, etc., resulting in lowering quality of life.

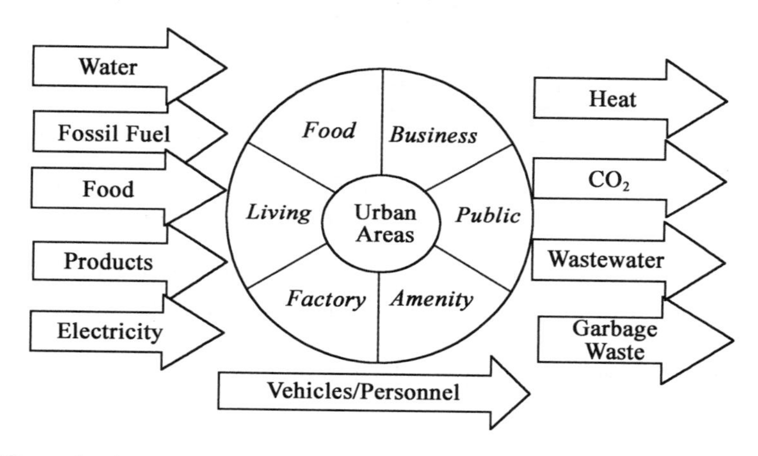

Figure 1 A scheme showing resource inflows and outflows in urban areas

3 Crops Which Can be Commercially Grown in Urban Areas

Foods originated from crops can be roughly classified into staple or calorie foods and functional foods. Staple foods such as wheat, corn, rice, potato, etc. are primarily taken for calories, almost all of which are grown in the open fields. Staple food plants are not suitable for commercial production in PF, because its economic value perkg (dry mass) is generally much lower than that of fruit and leaf vegetables, etc.

Functional foods such as tomatoes and lettuces are primarily taken for their functional components such as vitamins containing minimum calories. Some of fruit vegetables such as tomatoes and leaf vegetables such as lettuces are grown in the greenhouse. Calories, protein and vitamin A contained in wheat are, for example, respectively, 14.2MJ/kg, 0.137kg/kg, and 0μg RAE (retinol equivalent), while calories, protein and vitamin A contained in tomato are, for example, respectively, 0.754MJ/kg, 0.009kg/kg, 42μg RAE.

Land area necessary for growing these crops are the largest for staple foods and the smallest for fruit and leaf vegetables (Figure 2). On the other hand, annual yield and sales volume per land area are the highest in the leaf and fruit vegetables if grown all year round using hydroponics under controlled environment in the greenhouse. It should be noticed that annual productivity per land area or floor are of PF with 10 tiers is around 100-fold compared with that in the open field (Table 1; Kozai. 2013b).

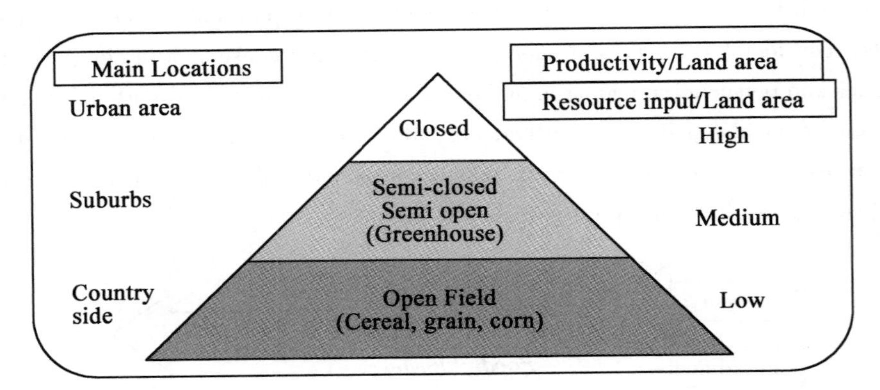

Figure 2　Three types of food production sysems with respectto the degree of closeness of corps with their environments

Table 1　Estimated relative annual production capacity and sales volume of PF per unit land area by its components in comparison with those in the open field.

No.	Magnification by PF compared with the open field	Component factor	Multiplied factor
1	10-fold by use of 10 tiers	10	10
2	2-fold by optimal environment control for shortening the culture period from transplanting to harvesting by half	2	20
3	2-fold by extending the annual duration of cultivation to year-round production with virtually no time loss between harvest and next transplanting	2	40
4	1. 5-fold by increased planting density per cultivation area	1. 5	60
5	1. 5-fold per cropping by no damage due to abnormal weather such as typhoon, heavy rain and drought, and outbreak of pest insects	1. 5	90
6	1. 3-fold sales price due to improved quality and less loss of produce after harvest	1. 3	117

4　Benefits of Fresh Vegetable Production in Urban Areas

It would be beneficial to produce fresh vegetables under controlled environments in urban areas, because most fresh vegetables with water content of around 90% are perishable and tend to lose their quality and quantity during transportation. In order to minimize the losses, significant amounts of energy and materials are additionally needed for cooling and delicate packing.

Leaf and fruit vegetables can be grown on the rooftop and other small open space in urban areas

(Giacomelli and Munday, 2013). Basement and other empty rooms in the building can be used to grow leaf vegetables commercially if artificial light source such as LED (light emitting diodes) is installed. Then, we are requested to answer a question - which is more resource saving, more land area- and room space saving, higher quality, more economically profitable and more comfortable for citizens-urban area or remote area?

5 Why PF in Urban Areas?

Use of PF is limited to the production of leaf vegetables, other short plants in height for use in herbal medicines and supplements for health, and seedlings or any kind of transplants. Oh the other hand, it should be noticed that PF can make a significant contribution to solving the following growing concern and/or demands listed in Table 2.

Table 2 Benefits of PF to meet growing concerns and/or demands in urban areas

No.	Concerns and/or demands
1	safety, security, consistency of supply and price stability of fresh vegetables, particularly the rising demand for consistency in purchasing for the catering (restaurants etc.) and home-meal replacement industries (prepared food and *bento* retailers).
2	highly functional fresh vegetables arising from concerns with health and improved quality of life.
3	Water saving methods of culture in regions with insufficient or saline irrigation water. PF requires around a fiftieth only of the irrigation water required by greenhouses.
4	consistent all-year-round production of plants using relatively low electricity consumption in extremely cold, hot and arid regions.
5	greater local self-sufficiency in vegetables for local consumption of local production and reduction in food mileage.
6	more employment opportunities for the aged, disabled and unemployed (PF provides safe and pleasant work environments and enable year-round human work).
7	lifestyle for inner city PF in conjunction with convenience stores, supermarkets, restaurants, hospitals and social welfare facilities, apartment flats, etc.
8	improvements in the cost-performance of lighting, air conditioning, thermal insulation and information equipment.
9	development of new businesses from the electrical, information, construction, health care and food industries.
10	the efficient use of vacant land, unused storage spaces, shaded areas, rooftops and basements etc.
11	high quality transplants for uses in horticulture, agriculture, reforestation, landscaping and dessert rehabilitation.

6 Current Situation of Commercial PF in Japan

It is estimated that the number of PF in commercial production exceeded 150 in December 2012, and this number is predicted to increase further in 2013. The largest PF in Japan, Spread Co., Ltd., can produce over 20 000 leaf lettuce heads per day or over 7million heads per year. It is roughly estimated that 20% of PFs are making profit, 60% are breaking even, and 20% are losing money. The number and percentage of PF that are profitable have been increasing steadily since 2009.

The depreciation cost accounts for roughly 30%, labor cost for 25%, and electricity cost for 20% of the total production cost (Kozai, 2013[A1] a). Electricity price per kWh is roughly 0.17 EURO. In 2012, the total production cost per lettuce head including depreciation per head was about 0.6 EURO and its sales price was 0.7 ~ 0.8 EURO. It is estimated that half the costs of setting up a PF unit are used by the construction of the outer structure, while about half are needed for equipping the utility. The total initial investment of a PF with 10 tiers is said to be around 4 000 EURO per square meter. Then, it will take 5 ~ 7 years to recover the initial investment.

7 Plants Suitable for Production in PF

Plants suitable for commercial production using PF are those that grow well at relatively low light intensity, thrive at a high planting density and for which most parts (leaves, stems and roots) are edible or saleable at a high price. Also, the plants need to be shorter than around 30cm in height because the distance between tiers is around 40cm for maximum use of the air space. These include leaf vegetables, herbs, root crops, medicinal plants and others with a height of 30cm or shorter.

In addition to the PF for production of leaf vegetables, smaller PF units with a floor area of 15 ~ 100m^2 have been used for commercial production of seedlings at about 150 different locations in Japan as of October 2012. Using this type of PF, grafted and non-grafted seedlings of tomato, cucumber, eggplant, etc., seedlings for hydroponic spinach, lettuce, etc. and seedlings/cuttings of ornamental plants and bedding plants are produced.

8 Increasing Demands in Food and Health Care Industry

Currently, less than half of leaf vegetables and herbs produced in PF are sold at supermarkets or grocery stores, and the rest is sold to food service industry including the home-meal- replacement industry. In such industries, the cost for hygiene processing is considerably reduced by using PF-produced leaf vegetables, which, without washing, do not contain pesticide, contaminants and insects.

PF-produced leaf vegetables are also used to produce paste for baby food and food for fragile elderly and sick persons. R & D is under way to use safe and high quality PF-produced vegetables and medicinal herbs for use as raw materials of pickles, frozen foods, food and drink additives, sauce, traditional medicine, supplements, cosmetics, aromatic essential oils, herb tea and more.

[A1] Is this referring to Kozai, 2013a or 2013b (see references)

9 Concept of Cpps or PF

PF is a form of closed plant production system (CPPS) which is an important concept for resource saving and environmental conservation with high yield of high quality plants. The CPPS consists of 6 principal structural elements: ①Thermally insulated warehouse-like structure, with opaque walls; ②From four to fifteen racks or tiers (40cm between racks vertically) equipped with lighting devices; ③Heat pumps (also known as air conditioners) with fans for air circulation, principally used for cooling to eliminate heat generated by lamps); ④a CO_2 delivery unit; ⑤a nutrient solution delivery unit; ⑥an environmental control unit.

The CPPS is designed and operated to meet the goals: ①the material and energy balance is controlled to produce a maximum possible amount of value-added plants using minimum amount of resources, and ② maximum possible percentages of all the resources added as raw materials are converted into product, meaning that resource utilization efficiencies are highest possible, and as a result ③ minimum costs for essential resources with virtually no release of pollutants into the environment. The concept of CPPS was proposed by Kozai *et al.* (2000), and a CPPS for transplant production was commercialized in 2003 (Kozai, 2005, 2007; Kozai *et al.*, 2005, 2006). The concept and methods of CPPS are now considered one of basic concepts in PF.

10 Resource Saving Characteristics of CPPS or PF

Essential resources for growing plants in PF are light energy, water, CO_2 and inorganic fertilizers. However, in PF, electricity is needed to provide the light energy and to control the air temperature, air current speed and nutrient solution cycling and control. Lighting, air conditioning and nutrient solution control account for approximately 80, 15 and 5% of respective total electricity consumption in the cultivation room (Kozai, 2007; Kozai, 2013[A2]b).

PF which is thermally well-insulated is always cooled by using air conditioners even during the nights of the coldest season. This is because at least 30% ~40% of the lamps are always turned on at night to minimize the daily maximum power consumption and minimize lighting and cooling costs. A significant amount of heat energy generated from lamps must thus be removed by air conditioners.

While cooling the air, these conditioning units condense about 95% of the water evaporated from leaves, which is returned to the nutrient solution tank for recycling. The rest of the irrigated water (about 5%) is either kept in the plants or released to the outside through the air gaps within walls (Kozai, 2007).

The PF must also be airtight because CO_2 is enriched at $1\,600 \sim 2\,000 \times 10^{-6}$ ($4 \sim 5$ times higher than that of outside air) to promote the photosynthesis and plant growth. The PF is sealed also

[A2] Is this referring to Kozai, 2013a or 2013b (see references)

to prevent insects and dusts from entering PF. Then, enriched CO_2 use efficiency is generally around 90% or higher (This efficiency is generally around 50% or lower in the greenhouse). Use efficiency of inorganic fertilizer of PF is over 90% because the culture bed is isolated from the soil and nutrient solution is continuously re-circulated, while use efficiency of inorganic/organic fertilizer in soil culture is often lower than 5% due to leaching, runoff and accumulation in the soil.

Light energy use efficiency of plants in PF is a few times higher than can be obtained in the greenhouse. Yet it is believed that this efficiency can still be improved further (Kozai, 2011).

11 Other Characteristics of CPPS

In addition to the features mentioned above, CPPS has the following features. ①Light quality and intensity, temperature, relative humidity or water vapor deficit, CO_2 concentration, air flow speed, nutrient concentrations and so forth are all kept as desired, regardless of weather conditions, with plant flowering/leaf emergence timing and so forth being controllable. It is also possible to use environmental controls to retard growth and vary quality in response to different needs and unanticipated changes of plan. Controllability of relative humidity or water vapor deficit is an advantage for improving the quality of products. ②By enhancing the horizontal air flow within plant community, photosynthesis and transpiration of plants are promoted and the relative humidity within the plant community can be reduced which prevents elongation of hypocotyls, stems or shoots even at high planting density. ③The absence of ventilation and gaps of CPPS prevents the entry of insects, pathogens and dangerous substances, and also prevents damage to and theft from the structure. ④The working environment is comfortable, because the cultivation space is maintained at an air temperature of 20℃ to 25℃, 60% to 70% relative humidity, air flow speed of $0.5 m \cdot s^{-1}$. The workers can experience the plants growing at a relatively high speed. ⑤There are systems established in society to recycle the principal structural components - lamps, air conditioner, thermal insulation boards and racks and so forth-after they have reached the ends of their service lives. ⑥The relationship between the environment and plant growth and the plant's functional components is relatively straightforward. It is easy to devise and implement effective methods of controlling the environment, and systems of cultivation can proceed sustainably. These features enhance people's will to work and make the work pleasurable. ⑦It can be built in a shaded area, a building, a basement, etc. ⑧No heating cost is necessary even in cold regions and cooling cost is minimum because the plant factory is covered with thermally well-insulated walls and minimum ventilation.

12 Visualizing Rate Variables in PF

The growing process in plants is complex, and there are differences in physiological characteristics among plat species and cultivars. In the closed system, this growing process can be evaluated and controlled by measuring various kinds of state and rate variables. State variables show quantities that do not include the time dimension in units, and rate variables show flows that include the time dimension in units. Examples of state variables are the air and leaf temperatures, water content of culture bed, pH and EC of nutrient solution, chlorophyll phosphorescence in plant leaves, leaf area, plant height and plant community architecture. Those of rate variables are net photosynthetic rate, dark respiration rate, transpiration rate, and supply rates of CO_2 and irrigation water.

Efficient PF production intended to achieve both high quality/high yield and resource-saving/environmental protection requires the visualization of both state and rate variables. Active research and development is in progress into the intact measurement of state variables such as temperature, stomatal conductance and chlorophyll phosphorescence in plant leaf surfaces (e. g. , Ohmasa and Takayama, 2003).

On the other hand, there has been little research into the visualization of the rate variables such as rates of net photosynthesis, transpiration, and supply rates of CO_2, irrigation water, fertilizer components, and petrochemical fuels in the plant factory, and their application to environmental controls.

If such rate variables are measured over time, the resource utilization efficiencies can be calculated online. Then, set points of environmental factors can be determined by considering the utilization efficiencies, rates of photosynthesis, dark respiration and transpiration, the costs of resources and CO_2 emission rate. For that reason, the visualization of rate variables has become an important R & D subject. The methods of measurement and estimation of rates of net photosynthesis, dark respiration, transpiration of plants in PF for rational cropping management and environmental controls elsewhere. Has been proposed by Li *et al.* (2012a, b and c; Kozai, 2013c).

13 Possibilities of PF Network (Kozai, 2013b)

Individual small owners of the small PF won't be left alone, as internet technologies now enable to link such small PFs to a large network of users. This community of PF growers will have access to the latest information on cultivation methods and plant varieties, downloadable from cloud servers. They will also have access to FAQ page and social network service (SNS) functions where they will be able to exchange tips about cultivation and food preparation. The performance of plants in the units will be checked automatically and remotely to assess how plants are doing in real time using in-built cameras and specialized software. These will be useful to ascertain the optimization of environmental conditions and eventually adjust these settings. PF growers will have the capacity to upload their own cultivation instructions, which they will upload onto cloud servers, making them available to others. Social media tools such as Twitter and Facebook will also be used.

14 A Trial Project of PF Network (Kozai, 2013b)

A trial project along these lines using well-designed small PFs was launched in 2012 in the Kashiwa-no-ha district of Kashiwa City in Chiba Prefecture, Japan (Figure 3). This project is being conducted by the Working Group of "Town PF Consortium Kashiwa-no-ha", a consortium made up of Chiba University, Mitsui Real Estate Co. , Ltd. , Panasonic Corporation, and Mirai Co. , Ltd. , with the collaboration of local residents acting as end-users for the project (http: //

www. miraibatake. net/).

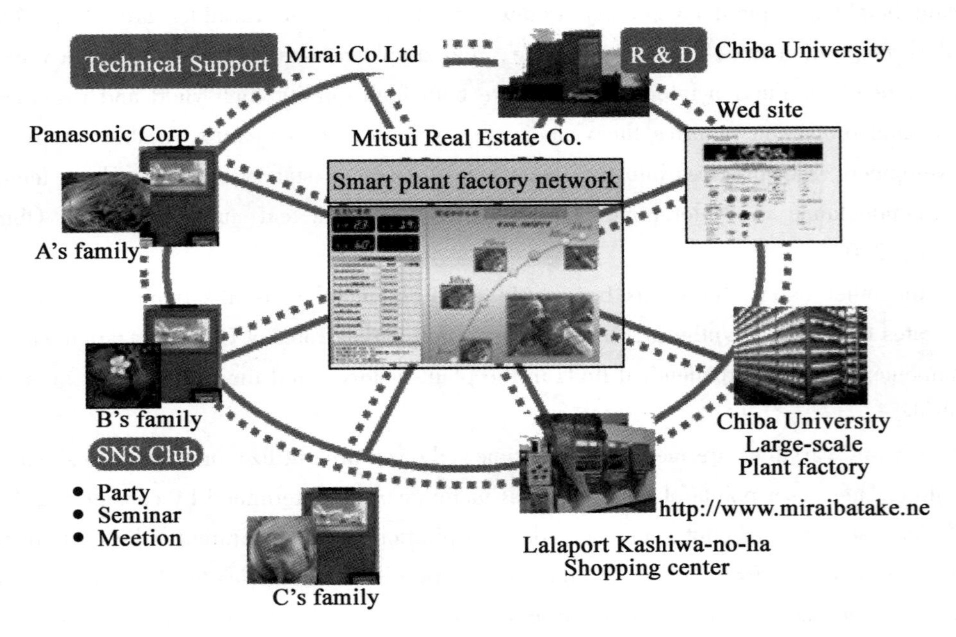

Figure 3　A scheme of the household plantfactory network being conducted in 2012 at Kashiw－no－ha area in Kashiwa city, Japan (Kozai, 2013b)

14.1　Objectives of the trial

The trial project was intended to examine: ① how people can grow vegetables in household spaces; ② the effectiveness of the supply of cultivation advice and the operability of the factory equipment; and ③ the value added from creating a network. It also assessed the usefulness and commercial feasibility of an exclusively web-based service where PF growers can ask experts questions on cultivation management, share information on their own circumstances and experiences, and arrange to exchange the vegetables they have grown. This project is an integral part of making Kashiwa-no-ha in Kashiwa City a "smart, streamlined" community, keeping with its designation as an "Environmental-Symbiotic City" that is sensitive to preserving green spaces and maintaining the harmony between cities and agriculture.

14.2　Further applications

This project is assessing the implementation of a network linking people's homes, the experiences and results, but this project can be extended to networks that link schools, local communities, hobby grower groups, restaurants, hospitals, and hotels (Figure 4). Then, so called 'big data' can be collected via Internet for 'data mining'.

One can also foresee that similar networks could be established for fish farming, raising tropical fish and aquatic plants and creating biotopes integrating plants, mushrooms, insects, and small animals. The small-scale experiences and results could also be applied to large-scale commercial and industrial PFs. Lastly by being connected to household and local community energy management sys-

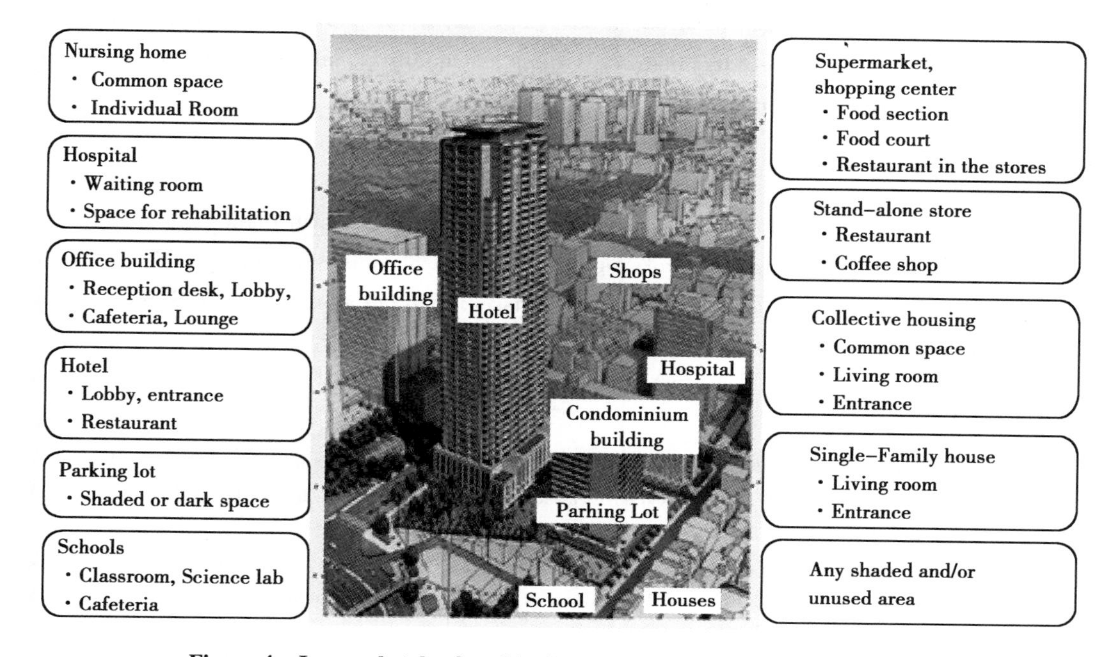

Figure 4 Image sketch of a ubiquitous plant factory in urban areas in the near auture (Kozai, 2013b)

tems, these networks can form their backbone.

14.3 Virtual PF connected with real PF

With a virtual system incorporating an e-learning function, and a real system involving the actual cultivation of plants, these networks have a dual aspect. In the virtual system, growers can operate a PF using a simulator that incorporates a plant growth, an environment and a business model; after attaining a certain level of proficiency in the simulator, they can transfer their newly obtained knowledge to the real PF. They can obviously partake both approaches in parallel.

15 Conclusion

Interest in the community in plant factories (PF) has grown over recent years, especially in urban areas, with the unusual weather, the uncontrolled movements in the oil price, and greater community concern for health, safety and security. On the other hand, PFs have attracted much criticism and concern. We should accept these criticisms earnestly, analyze their reasons calmly, study the issues from all sides and search for solutions to them. This paper addresses the criticism that PF is a system that makes heavy use of oil, and that they have high production costs and are not financially viable' . This paper shows that, through proper selection of amounts of resource inputs to yield, production volumes and unit value generated, PF can contribute to provide value-added plants in urban areas at high yield with minimum resource consumption and environmental pollution.

Residents/users living in urban areas and having little chance to grow plants in the open field may enjoy using a household PF. It is suggested that such a PF and its network have a potential to contribute to a better life of people in urban areas, and that it will provide an educational perspective to the people about science, technology, virtual community, plant growing, origin of food, global ecosystem and global productivity.

16 Acknowledgement

The author would like to thank the Working Group members of "Town Plant Factory Consortium Kashiwa-no-ha", Prof. M. Takagaki, Prof. and Prof. H. Hara of Chiba University, Mr. J. Kawai of Mitsui Real Estate Co., Ltd., Mr. M. Miyaki of Panasonic Corporation, and Mr. S. Shimamura of Mirai Co., Ltd. for their collaboration.

Literature Cited

[1] Despommier, D. 2010. The Vertical Farm: Feeding the world in the 21st century. St. Martin's Press, LLC., USA. 213p

[2] Giacomelli and Munday (eds.). 2013. Controlled environment agriculture-Finding paths to feed the world-. Resource. Special Issue March/April. ASABE. 30 pp

[3] Kozai, T., Kubota, C., Chun, C., Afreen, F., and Ohyama, K. 2000. Necessity and concept of the closed transplant production system, p. 3 ~ 19, In: C. Kubota and C. Chun (eds.) Transplant Production in the 21st Century. Kluwer Academic Publishers, The Netherlands

[4] Kozai, T., 2005. Closed systems for high quality transplants using minimum resources (In: Plant Tissue Culture Engineering, SBN: 1-4020-3594-2, (eds. Gupta, S. and Y. Ibaraki, 480pp.), Springer, Berlin. 275 ~ 312

[5] Kozai, T., F. Afreen and S. M. A. Zobayed (eds.). 2005. Photoautotrophic (sugar-free medium) micropropagation as a new micropropagation and transplant production system, Springer, Dordrecht, The Netherlands, 315pp

[6] Kozai, T., K. Ohyama and C. Chun, 2006. Commercialized closed systems with artificial lighting for plant production, Acta Hort. 711: 61 ~ 70

[7] Kozai, T. 2007. Propagation, grafting, and transplant production in closed systems with artificial lighting for commercialization in Japan, J. Ornamental Plants. 7 (3): 145 ~ 149

[8] Kozai, T. 2011. Improving light energy utilization efficiency for a sustainable plant factory with artificial light. Proc. of Green Lighting Shanghai Forum 2011, p. 375 ~ 383

[9] Kozai, T. 2013a. Innovation in agriculture: plant factories with artificial light. APO News. Jan. -Feb. Issue, p. 2 ~ 3. (http: //npoplantfactory. org/file/APO%20News. pdf)

[10] Kozai, T. 2013b. Plant Factory in Japan – Current Situation and Perspectives. Chronica Horticulturae, June Issue. (in press)

[11] Kozai, T. 2013c. Resource use efficiency of closed plant production system: concepts, estimation and application to plant factory. (submitted)

[12] Li, M., Kozai, T., Niu, G. and Takagaki, M. 2012a. Estimating the air exchange rate using water vapour as a tracer gas in a semi-closed growth chamber. Biosystems Engineering, 113: 94 ~ 101

[13] Li, M., Kozai, T., Ohyama, K., Shimamura, S., Gonda, K. and Sekiyama, S. 2012b. Estimation of hourly CO_2 assimilation rate of lettuce plants in a closed system with artificial lighting for commercial production. Eco-engineering, 24 (3): 77 ~ 83

[14] Li, M., Kozai, T., Ohyama, K., Shimamura, D., Gonda, K. and Sekiyama, T. 2012c. CO_2 balance of a commercial closed system with artificial lighting for producing lettuce plants. HortScience, 47 (9): 1 257 ~ 1 260

[15] Ohmasa, K. and K. Takayama. 2003. Simultaneous measurement of stomatal conductance, non-photochemical quenching, and photochemical yield of photosystem II in intact leaves by thermal and chlorophyll fluorescence imaging. Plant and Cell Physiology, 44 (12): 1 290 ~ 1 300

Greenhouse Technology for Cultivation in Arid and Semi-Arid Regions

E. J. Baeza, N. Castilla, IFAPA Centro La Mojonera, IFAPA Centro Camino de Purchil, Almería. Spain, Granada. Spain

(*estebanj. baeza@juntadeandalucia. es*)

Abstract: Greenhouses have expanded from Northern areas to very different climatic areas in the World, including arid and semi-arid regions. The main limiting factors for greenhouse cultivation in such areas are the high ambient temperatures and low humidity, typical of these regions during almost all the year, and the availability of good quality water for irrigation and cooling. Therefore, the technological options to grow in these areas are quite dependent on the average maximum temperatures along the year. If the area has average maximum temperatures not higher than 30 ~ 35℃ during the autumn, winter and spring months, a good choice is the combination of a plastic covered greenhouse with high natural ventilation capacity, selective shading (NIR reflection) and evaporative cooling with a fog system, provided that there is water available. During the summer months, cultivation can be moved to highlands, with cooler temperatures, using screenhouses combined with fogging (low or high pressure, depending on the water quality). If temperatures are higher than 30 ~ 35℃ during the whole year, pad and fan combined with selective shading is possibly the best option, avoiding cultivation during the summer months. Other more sophisticated technologies have been proposed for these regions, like the seawater greenhouse (limited to greenhouses close to the coast). If large greenhouse clusters are to be built near the coast, closed greenhouses could be cooled using deep-sea cold water, but technology to prevent corrosion in the collection, pumping and distribution systems must be developed.

Key words: natural ventilation; humidity; selective shading; evaporative cooling; desalination

1 Introduction

During the last decades, greenhouses have expanded from the cold zones of the globe, where they mostly developed until the plastic film technology appeared, to many other climate areas of the globe. Arid and semi-arid areas are not an exception to this trend. One of the most paradigmatic examples of growing interest in greenhouse cultivation in the world is the Arabian Peninsula. The climate in this region is characterized by an arid climate with brackish water resources. Summer season is hot and long (the ambient temperature exceeding 45℃ at around noon in summer), high solar radiation flux (the daily solar radiation integral reaches to 30 MJm^2), dusty and dry weather (relative humidity of the ambient air drop below 10% at around noon), and water resources being scarce and brackish (salty) (Abdel-Ghany *et al.*, 2012).

Cultivating horticultural crops under these conditions is not easy, and that is why most of the some of the sunniest countries in the world have become net importers of food (FAOSTAT, 2004). This situation has food security implications (Qasam, 1990). Besides, the transportation of fresh products consumes a lot of energy and it is not always sustainable. It is therefore important to consider how to create cool environments for cultivation in hot climates. Greenhouses are generally regarded as necessary only to provide a warm environment in cold climates, it has been shown that with properly designed greenhouse structure and cooling system/s, it is possible to improve plant growing conditions under extensively hot conditions. Protected cultivation also has the potential benefit of substantially increasing the water use efficiency, which is essential in desert areas. However, a discussion for adapting an adequate cooling technique that can be used for greenhouses in these areas is still missing.

The present paper aims to review the different technological alternatives (including greenhouse structure and its cover plus the equipment for climate control) available nowadays that can be used in arid areas, like Saudi Arabia, to successfully (both economically and environmentally) grow horticultural crops.

2 Strategy A: Combination of Classical Cooling Techniques

Starting from the principle that conventional refrigeration is generally just too expensive to install and run in greenhouses, compared to the cost of importing the food produce from cooler regions and according to Abdel-Ghany *et al.* (2012), three main categories of cooling techniques are commercially used to cool the greenhouse under high radiation conditions: ventilation, evaporation and heat prevention. In arid regions, it is not possible to successfully grow along the year without a combination of the previous methods. Let's analyze the possible alternatives:

Methods combining ventilation and water evaporation.

a) *Natural ventilation + fog system*: natural ventilation is cheap as it does not rely on fuel energy but it must be properly designed to make it as efficient as possible and it must be combined with some type of efficient fog system. The main problem of this combination of techniques is that under low wind conditions (the most limiting from the temperature point of view), in environments like the Arabian Peninsula, it would not be possible to grow during the summer months, also because of the high night temperatures, that limit the quality of the produce (May to September).

Obviously the greenhouse structure providing maximum ventilation capacity for arid and semi-arid areas, where rain is almost inexistent throughout the year are the screenhouses, where roof and sidewalls are covered with a screen material that must decrease the radiation intercepted by the crop, protect it from strong winds and from the pests, while providing maximum ventilation capacity. Tanny (2013) has reviewed all research concerning screenhouses and the microclimate generated inside them. The choice of the screen is the most critical decision. In desert areas, radiation levels are very high and thus, the screen used must have an important shading effect (via absorption or reflection of the radiation). However, and according to Hemming (2011) fruits exposed to high levels of direct radiation may suffer sunburn which significantly reduces the marketable yield but, on the other hand, extreme shading may, as well, deteriorate production and yield due to light shortage. In the market, a large catalogue of screens can be found, and usually information provided by the manufacturer is incomplete. The best would be to know the colour, the shading percentage, the porosity and the thread diameter, to allow for an estimation of the effect on the airflow. If pest pressure in the area is very high, insect proof screens (that have lower porosity) can be used, at the expense of losing ventilation capacity. If pest pressure is not so high, it is better to use a shading screen. Dust accumulation will increase the shading effect but strongly decrease the air-exchange, strongly limiting the use of screenhouses unless the screen is cleaned during the growing cycle. For more detailed information, the review by Tanny (2013) is extremely comprehensive and provides all the information required to better understand the microclimate generated by screenhouses.

If a greenhouse is to be built, with screened openings in the roof and in the sidewalls, regardless of the covering material (glass, flexible film, etc.), recent research works, assisted by the use of computational fluid dynamics, have provided some of the key factors required for building a highly efficient natural ventilation design (Figure 1):

* Maximum greenhouse volume: which is achieved by increasing the width of the spans (commercial greenhouses of 12.60m span width are offered nowadays by different manufacturers) as well as the gutter and ridge height (ridge heights of up to 8 ~ 9m are available). The higher the greenhouse volume above the canopy the higher is the vertical temperature gradient and the cooler the canopy volume. Besides, the higher is the distance between the sidewall vents and the roof vents; the airflow created by the buoyancy effect (when wind speed is low) becomes larger.

* Maximum ventilator area: the literature recommends a percentage between 15% ~ 30% of ventilator area in relation to the covered area (White & Aldrich, 1975). However, these values do not consider the low porosity insect proof screens, of inevitable use in arid climates, to prevent pest invasions. Such screens strongly affect the air exchange (Teitel, 2007). Therefore, the ventilator area must be maximized, by combining large sidewall vents (if possible 3m high in the four greenhouse sides) with double roof vents in every span. The sidewall vents must open from below to top, to enhance the airflow caused by temperature difference (buoyancy flow). To avoid the direct impact of the incoming airflow on the plants, a simple baffle device can be installed all along the sidewall vent inside the greenhouse (Figure 2). The double roof vents must be both opened at zero or very low wind velocities ($<2m \cdot s^{-1}$) to maximize thermally driven ventilation, and then managed by the climate controller to ensure the optimum airflow for each wind speed and direction. The roof vents must be large. According to Baeza (2007), vent widths of more than 2m only provide very small increases in the ventilation rate. In order to obtain maximum synergy between roof and sidewall vents the width of the greenhouse must not exceed 60 ~ 70m. For wider greenhouses, spots of very high temperature will appear in the centre of the greenhouse.

* Increase the slope of the span to values around 30° strongly increases the air exchange and improves the homogeneity of the air movement inside the greenhouse (Baeza, 2007).

* The screened area must be maximized with different possible solutions (Figure 3), a quite recommendable practice for arid climates. In the volume between the screen and the sidewall vent, extra lines of foggers can be installed to cool the air entering the greenhouse through the sidewall

vent.

The main problem for using high-pressure fog systems is the brackish water, which causes a quick clogging of the orifice of the nozzles making maintenance a nightmare. An alternative is to use compressed air-water fog systems, which are of equal efficiency as high-pressure fog systems and water quality is not as limiting as the orifices in the nozzles are larger (Montero *et al.*, 1990). On the other hand, these systems are high energy consuming. According to Toida *et al.* (2006) it is possible to enhance the evaporation ratio of a fog system, supplying upward air with small fans, moderating the downward air stream caused by evaporation of fog under indoor conditions. They found that fog evaporation with fans was 1.6 times greater than that without fans, and that the upright nozzle with fans distributed the fog over a wider area, and resulted in a more uniform and lower air temperatures than did upright and horizontal nozzles without fans under similar environmental conditions.

b) *Wet pad & fan*: it is probably the most commonly used method to cool greenhouses in arid and semi-arid regions (Al-Helal *et al.*, 2004; Ganguly & Ghosh, 2007; Al-Helal, 2007), being possible to decrease temperature up to 10 ~ 12℃ and increase humidity in the greenhouse up to 30% with pure fresh water available (Al-Helal & Abdel-Ghany, 2011). A well known limitation of wet-pad & fan systems is that the number of fans and their power must be well calculated and their location in the greenhouse well designed to obtain the most efficient cooling airflow in the canopy area. To avoid large temperature and humidity gradients, the distance between wet pads and the fans must not exceed 40m; the rest of recommendations for the design of such systems can be found in ASAE (2000). The advantage of this system in relation to the use of combined natural ventilation and fogging is that, if properly designed, it can provide a certain cooling capacity regardless of the external wind and temperature conditions, but at the expense of a high energy use. Its use is thus justified only when the price of the energy is low. Despite of the previous, the main problem associated to the efficient use of these systems is the water quality. For instance, in the Arabian Peninsula water resources are scarce and the salinity of the water is often very high. This causes a fast deterioration in the cooling performance of the wet-pad fan systems due to clogging of the pad by the salt build up on the pad surfaces that also restricts the airflow. According to Al-Helal *et al.* (2004) clogged pads were found to reduce the cooling system performance significantly, increase the electric energy consumed by the fan motors by about 22%, increase the inside greenhouse air temperature up to 55℃, and reduce the relative humidity below 10%. A method has been proposed by Soaud *et al.* (2010) based on introducing acidified water to remove and prevent future build-up of scaling deposits in greenhouse cooling pads. The acid water was maintained by utilizing sulphur by-product from gas production plant using sulphur-burning equipment, the average pH of the acid water generated by this technique was 6.5. After several trials, preliminary results indicated the acidified water was effective in removing scale and salt accumulation from the cooling pads. The visual observations in (Figure. 4) showed clear evidence that acidified water effectively removed the deposits form the cooling pads.

Methods combining selective shading with ventilation and water evaporation.

According to the previous, evaporative cooling can be quite efficient in arid climates, but good quality water is required. In case good quality water is available (i. e. seawater desalinated water) the problem is solved, otherwise the grower needs to produce good quality water from the brackish/sea water. In the scientific literature, different systems have been proposed to cope with common requirement of arid areas. It will be revised later.

A third type of cooling technique may be used in combination with any of the two previous cooling alternatives analyzed: selective shading. A deep and recent review on this subject can be found in Abdel-Ghany *et al.* (2012). The main conclusions we can derive is that the option of using a LRF (liquid radiation filter) flowing on the greenhouse roof, even if it is selective (low NIR transmission), for economic reasons among others. The other options are therefore, to use a permanent cover material with a NIR filter (NIR reflection is preferable to NIR absorption for obvious reasons), a movable screen with NIR reflection or a selective whitewash. In regions like the Arabian Peninsula, the most popular material due to its lower price and simplicity of the structure is LDPE film. Therefore, using a NIR-reflecting plastic film for covering greenhouse permanently in the Arabian Peninsula is preferable to non-permanent filters because the winter season is relatively short and the daytime ambient temperature and solar radiation flux are relatively high. Maximum drops in air temperature achieved under an NIR-reflecting greenhouse cover with the best of the tested materials are 5℃, not enough to cool the greenhouse all along the year if such a cover is not combined with evaporative cooling. The perfect NIR-filtering plastic film cover that is suitable for an arid climate with high solar radiation flux is not yet available. More research is needed to develop such kinds of NIR-reflecting PE film covers. Besides, it is technically possible to concentrate the NIR reflected in PV cells that can generate electricity that can be used, for instance to distil water for the evaporative cooling system. In any case, studies like Alhamdan & Al-Helal, (2009) show that the high radiation and temperature together with dusty winds, characteristics of arid areas greatly shorten the shelf life of PE films used as greenhouse covers. In relation to mechanical stress, the best is to use multi-layer film (i. e. three layers, PE + EVA + PE) instead of monolayer.

3 Strategy B: Composite Systems

3.1 Earth to air heat exchange systems and similar systems

Since at certain depth, the temperature of the ground remains more or less constant year round (26 ~ 28℃), this property can also be used for cooling the greenhouses. According to Sethi & Sharma (2007) if the hot greenhouse air is circulated through buried pipes (2 ~ 4m depth) heat is dissipated to the underground soil. However, the literature review shows limited studies exclusively related to exploring the cooling potential of EAHES for agricultural greenhouses (Levit *et al.*, 1989; Santamouris *et al.*, 1994, 1995). To summarize, such systems are not suitable to cool the greenhouses in very harsh climates like in the Arabian Peninsula year round. They can only lower the inside air temperature by 3 ~ 5℃ as compared to the control greenhouses. Another major disadvantage of using EAHES is the cost of digging the soil and laying the pipes p to 2 ~ 4m depth. Horizontal laying and fitting of pipe network at this depth is not easy. Condition of the pipes, and leakage from the joints once laid cannot be checked. The corrosion of metallic pipes and/or deterioration of plastic pipes under the pressure of the soil also make this system little convenient for long duration projects. Similar can be said about aquifer coupled cavity flow heat exchanger systems, as proposed by Sethi and Sharma (2007), which besides is dependent on the ability of deep underground water and does not provide good enough conditions in summer period.

3.2 The seawater greenhouse concept: linking water solar desalination with greenhouse production

The seawater greenhouse concept is very interesting and simple and was easily explained by Raoueche & Bailey (1997): the function of the seawater greenhouse is to produce normal greenhouse crops and also to provide water for irrigation. It is intended for use in hot arid coastal regions, which have steady winds from a predominant direction to create airflow through the greenhouse for ventilation. Several prototypes of this type of greenhouse have been designed and built along the last decade in different locations (Tenerife in Spain, Oman, etc.). The most recent concept can be explained as follows (Goosen *et al.*, 2003): Surface seawater trickles down the front wall evaporator, through which air is drawn into the greenhouse. Dust, salt spray, pollen and insects are trapped and filtered out leaving the air pure, humidified and cool. Sunlight is selectively filtered by the roof elements to remove radiation that does not contribute to photosynthesis. This helps to keep the greenhouse cool while allowing the crops o grow in high light conditions. Air passes through second seawater evaporator and is further humidified to saturation point. Saturated air passes through the condenser, which is cooled using cold deep seawater. Pure distilled water condenses and is piped to storage. Fans raw the air through the greenhouse and into a shaded house area (Figure 5). The concept is promising and different simulation works have been done, using CFD to optimize the performance of this system (Davies & Paton, 2005). In the trials performed, crop production in terms of quality and quantity was commercially acceptable, with the greenhouse supplying the water required for irrigation in excess. The different projects have enabled for an extensive validation of a thermodynamic simulation model (references), so given appropriate meteorological data, the model accurately predicts and quantifies how the Seawater Greenhouse will perform in other parts of the world.

3.3 A greenhouse with a solar desalination module integrated in the roof

Chaibi (2003) proposed a greenhouse concept with integrated water desalination considered for small-scale applications at remote locations in areas where only saline water is available. In such a greenhouse the roof light transmission is reduced as solar radiation is absorbed by a layer of flowing water on a glass covered by a top glass. Fresh water is evaporated, condensed on the top glass and collected at the roof eaves (Figure 6). Results indicate that considerably less extreme climate conditions were registered in an experimental greenhouse with roof desalination compared to a conventional greenhouse. With the system integrated in 50% of the roof area of a widespan greenhouse, the capacity to cover the annual demand for a low canopy crop was covered. A similar capacity for a high canopy crop required asymmetric roof design and the desalination system in the whole roof area. Simulated yield reductions for these cases were 25 and 18 % and seasonal fresh water storage was required. Lower yield reductions could be achieved with application of more light selective glass materials in the roof absorber.

3.4 The Watergy greenhouse concept

The Watergy prototype (Zaragoza *et al.*, 2007) proposes new concept of solar collector based on a humid air circuit powered by thermal solar energy. The collector is formed by a greenhouse connected with a solar chimney, which in the prototype built in Almería, was located in the centre of the greenhouse. Inside of the tower a cooling duct contained a simple air-to-water heat exchanger connected to a heat accumulator. The buoyancy forced the air heated by the sun and humid due to evapotranspiration from the plants to go up the tower. The aim is to have very hot and humid air at the top of the solar tower. Cold water in the heat exchanger cools the air. On the surface of the heat exchanger, the cooling of the humid air creates condensation, releasing additional thermal energy and distilled water. The cold and dry air falls back o the greenhouse, where it is heated and humidified starting the cycle again. The hot water accumulated can be used to heat during the night (if required) by reversing the flows. Main drawbacks of this design is that natural buoyancy could not create enough air circulation to ensure that temperatures were kept within an acceptable range for most of the typical crops of Almería during he spring summer and early autumn seasons. The position of the tower in the centre of the greenhouse shadowed the north of the greenhouse all along the year. Therefore, although an interesting concept for arid areas, it needs further re-design to become commercially successful.

4 Strategy C: Cooling Large Greenhouse Clusters

4.1 Cooling of greenhouse districts using deep seawater

In many countries with significant greenhouse area (Spain, Netherlands, Italy, Greece, etc) greenhouses are located near coastal regions. In addition, deep cold ocean and seawater is considered a valuable natural resource that can be used for energy production, cooling, desalination, aquaculture and agriculture. The most economically viable use of this deep water is to air condition buildings with cooling requirements through a District Cooling System using Sea Water as heat source or sink. In general, District Cooling is the technology for providing energy from a central plant to multiple users. Despite the obvious advantages of District Cooling Technologies mainly by using seawater, they have never been implemented till now in greenhouse industry, even if the concentration of greenhouse companies near coastal regions indicates that greenhouse industry could be an ideal sector for applying these technologies. The main components of such a system are (Figure 7):

- Seawater uptake and outfall system.
- Cooling station and heat exchanger for seawater.
- Network of pre-insulated pipes to distribute chilled water.

· Heat Exchangers and distribution systems for applying chilled water inside the greenhouse.

· Management tools to operate seawater pumping, cooling station and delivery of chilled water.

Along many ocean coastlines and lake shorelines, there is reasonable access to naturally cold water that is as cold as or colder than the water used in conventional air conditioning systems. If this water can be tapped, then the significant power for operating mechanical chillers to keep the chilled water cold can be eliminated. These projects are large and involve research to develop new materials to avoid corrosion by seawater, but could be a very promising technology for countries with many km of coast, like in the Arabian Peninsula.

4.2 Solar cooling

Basically, the idea would be to produce hot water using high efficiency solar collectors (i. e. CPC's) and link them to absorption chillers, which would provide the cold water required for cooling the greenhouses by means of air-water heat exchangers. Again, in order to make this technology sustainable from an economic point of view, a large area of greenhouse is required, in order to centralize the production of hot and cold water, making it cheaper.

4.3 Figures (1 ~ 7)

Figure 1　A new prototype of greenhouse of high efficiency in natural ventilation for arid and semi-arid areas

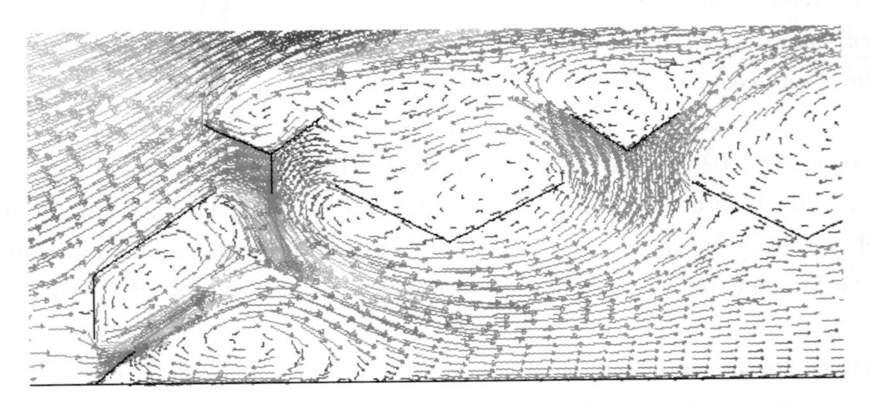

Figure 2　CFD air velocity vector field showing the effect of simple baffles on the incoming ventilation airflow

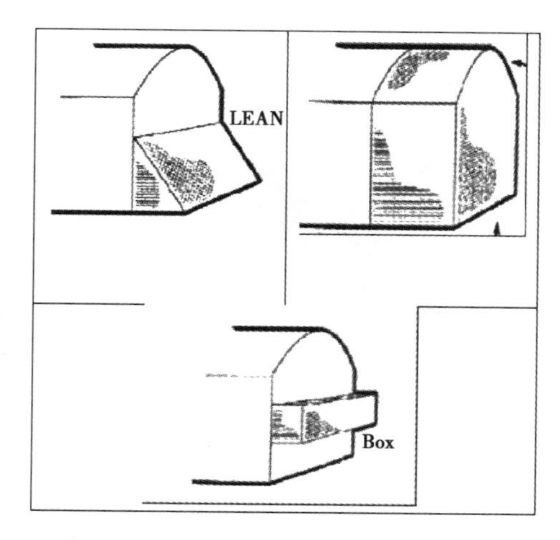

Figure 3　Different possible solutions to increase the screened area in the sidewall vents

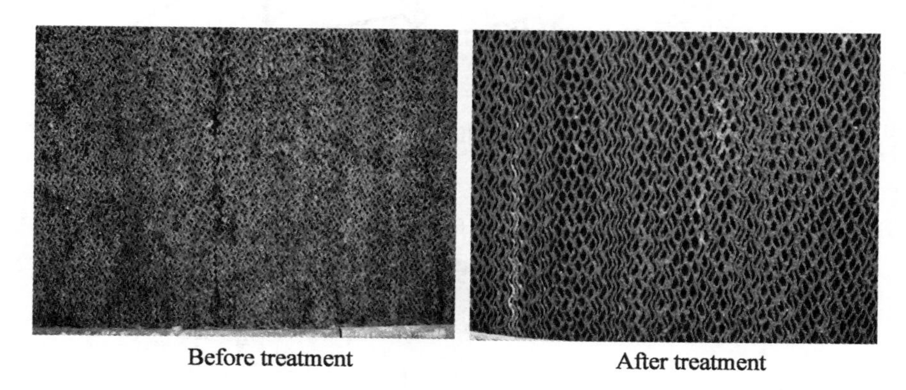

Figure 4　The effect of acidified water in removing scaling and salt deposits from cooling pads

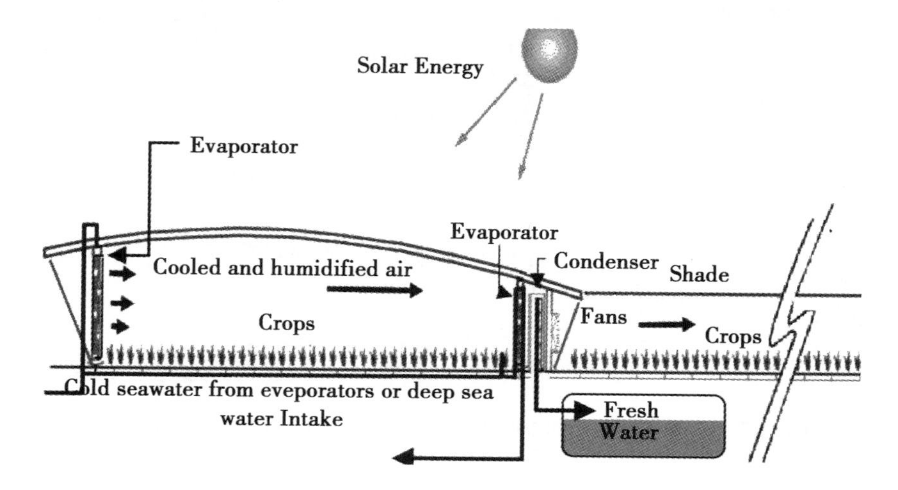

Figure 5　Seawater Greenhouse scheme

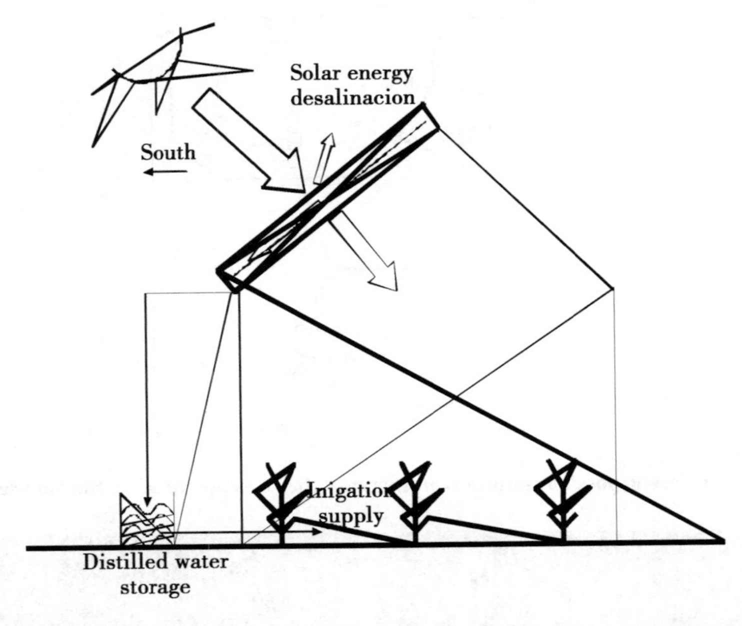

Figure 6　Principle of a water desalination system integrated in a greenhouse roof

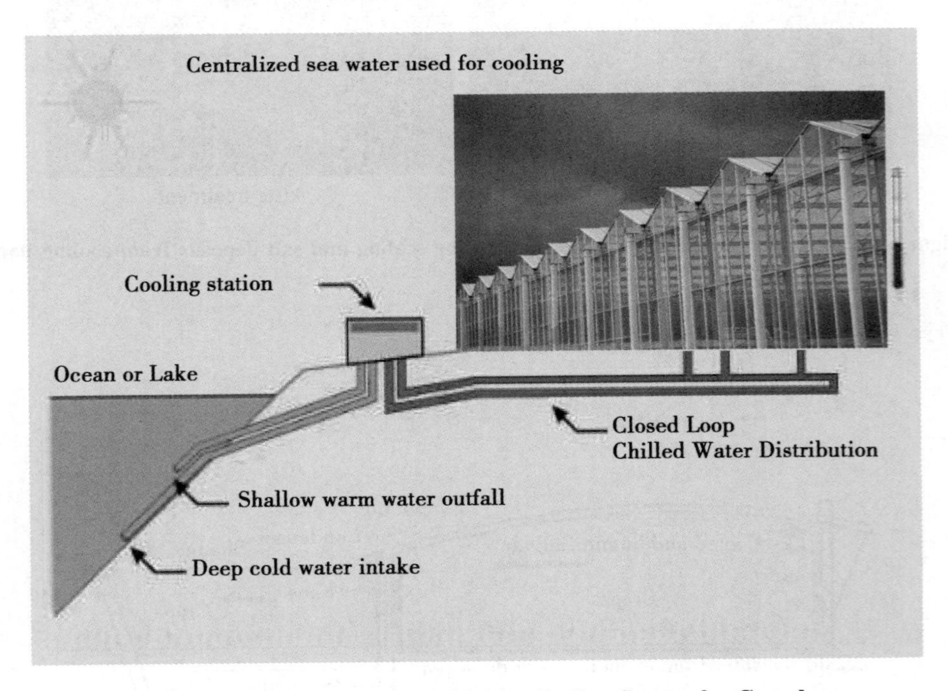

Figure 7　Sketch of a Centralized District Cooling System for Greenhouses

Literature Cited

[1] Abdel-Ghany, A. M. , I. M. Al-Helal, S. M. Alzahrani, A. A. Alsadon, I. M. Ali and R. M. Elleithy, "CoveringMaterials Incorporating Radiation-Preventing Techniques to Meet Greenhouse Cooling Challenges in Arid Regions: A Review", The ScientificWorld Journal Volume 2012, 11 pages

[2] Al-Helal I. M. and A. M. Abdel-Ghany, "Energy partition and conversion of solar and thermal radiation into sensible and latent heat in a greenhouse under arid conditions," Energy and Buildings, vol. 43, no. 7, pp. 1740 ~ 1747, 2011

[3] Al-Helal, I. M. "Effects of ventilation rate on the environment of a fan-pad evaporatively cooled, shaded greenhouse in extreme arid climates," Applied Engineering in Agriculture, vol. 23, no. 2, pp. 221 ~ 230, 2007

[4] Al-Helal, I. , M. Al-Abbadi, and A. Al-Ibrahim, "A study of fan & pad performance for a photovoltaic powered greenhouse in Saudi Arabian summer," International Agricultural Engineering Journal, vol. 13, no. 4, pp. 113 ~ 124, 2004

[5] Al-Helal, I. , N. Al-Abbadi, and A. Al-Ibrahim, "A study of fanpad performance for a photovoltaic powered greenhouse in Saudi Arabian summer," International Agricultural Engineering Journal, vol. 13, no. 4, pp. 113 ~ 124, 2004

[6] Alhamdan, A. M. And Ibrahim M. Al-Helal, "Mechanical deterioration of polyethylene greenhouses covering under arid conditions" Journal of Materials Processing Technology 2009, 63 ~ 69

[7] American Society for Agricultural Engineers: ANSI/ASAE EP406. 3 MAR98, Heating, Venting and Cooling Greenhouses, pp. 675 ~ 682, 2000

[8] Baeza, E. J. "Optimizaci6n del dise? de los sistemas de ventilaci6n en invernaderos tipo parral" . PhD dissertation, Escuela Politécnica Superior, Departamento de Ingeniería Rural, Universidad de Almería, 2007

[9] buildings coupled with earth-to-air-heat-exchangers. Solar Energy 54: 375 ~ 380

[10] Chaibi, M. T. 2003. Greenhouse systems with integrated water desalination for arid áreas based on solar energy. Doctor's dissertation. ISSN 1401 ~ 6249, ISBN 91-576-6438-2

[11] Davies, P. A. and C. Paton, "The Seawater Greenhouse in the United Arab Emirates: thermal modelling and evaluation of design options"

[12] FAOSTAT, 2004. Food and Agriculture Organization Statistical Database, http: //faostat. fao. org

[13] Ganguly A. and Ghosh, S. "Modeling and analysis of a fan-pad ventilated floricultural greenhouse," Energy and Buildings, vol. 39, no. 10, pp. 1 092 ~ 1 097, 2007

[14] Goosen, M. F. A. S. S. Sablani, C. Paton, J. Perret, A. Al-Nuaimi, I. Haffar, H. Al-Hinai, W. H. Shayya, "Solar energy desalination for arid coastal regions: development of a humidification¨Cdehumidification seawater greenhouse", Solar Energy 75 (2003) 413 ~ 419

[15] Hemming, S. "Use of natural and artificial light in horticulture e interaction of plant and technology" . Acta Horticulturae, 907, 25 ~ 35, 2011

[16] Levit, H. J. , Gaspar, R. , Piacentini, R. D. , 1989. Simulation of greenhousemicroclimate by earth-tube-heat-exchangers. Agriculture and Forest Meteorology 47, 31 ~ 47

[17] Montero, J. I. , A. Anton, C. Biel, A. Franquet, "Cooling of greenhouse with compressed air fogging nozzles", Acta Horticulturae, 281, 199 ~ 209, 1990

[18] Qasam, S. , 1990. Issues of food security in the Arab countries. Chapter 10. In: Salman, A. (Ed.), Agriculture in the Middle East. Paragon House, ISBN 0-943852-80-3

[19] Raoueche A. and B. J. Bailey, "Performance aspects of a seawater greenhouse", 23rd WEDC Conference, 23rd WEDC Conference, Durban, South Africa, 1997

[20] Santamouris, M. , Mihalakakou, G. , Balaras, C. A. , Argiriou, A. , Asimakopoulos, D. , Vallindras, M. , 1995a. On the performance of

[21] Santamouris,M. , Mihalakakou, G. , Balaras, C. A. , Argiriou, A. , Asimakopoulos, D. , Vallindras, M. , 1995b. Use of buried pipes for energy conservation in cooling of agricultural greenhouses. Solar Energy, 55 (2): 111 ~ 124

[22] Sethi, V. P. & S. K. Sharma, "Survey of cooling technologies for worldwide agricultural greenhouse applications", Solar En-

ergy 81 （2007）1 447 ~ 1 459

[23] Soaud A. A. and Fareed H. Al Darweesh，“Increasing Water Efficiency in Greenhouse Cooling system in Arid Regions Using Sulfur Burning Technology”，19th World Congress of Soil Science，Soil Solutions for a Changing World，pp 56 ~ 59，2010

[24] Tanny，J. “Microclimate and evapotranspiration of crops covered byagricultural screens：A review”，Biosystems engineering 114，pp 26 ~ 43，2013

[25] Teitel，M. “The effect of screened openings on greenhouse microclimate”，Agricultural and Forest Meteorology 143 （2007） 159 ~ 175

[26] Toida，H.，T. Kozai，K. Ohyama，Handarto，“Enhancing Fog Evaporation Rate using an Upward Air Stream to improve Greenhouse Cooling Performance”，Biosystems Engineering，93 （2）：205 ~ 211，2006

[27] Zaragoza，G.，M. Buchholz，P. Jochum，and J. Perez-Parra. Watergy project：Towards a rational use of water in greenhouse agriculture and sustainable architecture. Desalination 211：296 ~ 303，2007

Greenhouse Ventilation and Cooling

Meir Teitel

(*Institute of Agricultural Engineering*, *Agricultural Research Organization*, *the Volcani Center*, *P. O. B.* 6, *Bet Dagan*, 50250)

(*Israel. grteitel@ agri. gov. il*)

Abstract: The manipulation of the greenhouse environment to achieve optimum growing conditions that will produce sustainable outcomes, in both economic and environmental terms, is still a significant challenge. In warm climates the main challenge remains the reduction of excessive temperature and humidity during the hot season. Improved ventilation of naturally ventilated greenhouses has been an important technological development that contributed to sustainable production. The utilization of air exchange theory into the design of greenhouse ventilation systems has resulted in greenhouse structures that can effectively ventilate under natural buoyancy and wind driven conditions. Still, however, in warm climates the use of insect-proof screens on the openings, to exclude insects, resulted in a warmer and more humid microclimate; a problem that needs to be solved. When ventilation does not sufficiently reduce the crop temperature, evaporative cooling is applied, a method that is now widely used. Techniques include sprinkling, fan and pad system and fogging. The hot dry conditions of some greenhouse regions are well sited to the combined use of fogging systems in conjunction with naturally ventilated greenhouse structures, an approach that is suitable to very large structures. This paper reviews few technologies that are related to the alleviation of heat load from greenhouses and discusses the applications of each technology.

Key words: Fan; Roof vent; Side vent; Air exchange; Ventilation efficiency; Evaporative; Cooling; Refrigeration

1 Introduction

The primary purpose of greenhouse ventilation is to prevent excessive rise of temperature and humidity. This is achieved by replacing the hot and humid air of the greenhouse with ambient cooler and dryer air. In some cases, it is applied to prevent CO_2 depletion due to the crop photosynthesis and non adequate air exchange between the greenhouse and the environment. At the same time, ventilation can also reduce the concentration of pollutant gases (e. g., toxic gases generated by incomplete combustion in a heating system). Furthermore, ventilation is important since it generates air movement within the greenhouse and thus reduces the boundary layer thickness near the leaves. This improves sensible and latent heat transfers from the crop to the greenhouse air and enhances CO_2 transfer to the leaves. Two types of ventilation can be distinguished: *natural* and *forced*. Natural ventilation is driven by two mechanisms, namely the pressure field induced by the wind around the greenhouse and the buoyancy force induced by the warmer and more humid air in the greenhouse. Forced ventilation is accomplished by fans that are usually installed on sidewalls and are capable to move large quantities of air at relatively low pressure drop. As was pointed out by (Sase, 2006) the phenomenon of natural ventilation is complex and its design is generally considered more difficult than forced ventilation. When ventilation is not sufficient to lower the crop temperature to the desired values, active cooling has to be applied. In warm climates such as the Mediterranean region, greenhouse cooling in the hot season is often vital to reduce crop damage.

2 Forced Ventilation

The basic requirements of a fan for greenhouse ventilation are that it should be capable of moving large quantities of air at relatively low pressure drops. Of the different types of fan available the most suitable is the axial fan which consists of direct-driven or belt-driven impellers. The blades are made from galvanized sheet steel formed into shape so that the air is propelled axially through the fan. Propeller fans are limited to low pressure difference applications and function to move large quantities of air rather than generating significant total pressure differences. Blade tip clearance is an important factor in the performance of such fans, with a small, uniform clearance being more desirable. Small tip clearances prevent air from flowing back around the propeller, which would be a short circuit. The air leaves the blades in a circular discharge pattern and anything on the discharge side of the fan which impedes the swirling motion acts to degrade the fan's performance. Fans must be guarded both on the inlet and outlet sides for safety. To prevent inflow of cold air during winter and entrance of insects during the whole year (when the fans are not operating), the exhaust side of each fan should be fitted with shutters which open under air pressure when the fans start. Several mechanisms for opening the shutters at the initial stage of rotation are available (e. g. , based on centrifugal forces, electromagnet). The shutters are usually made of galvanized steel, UV stabilized PVC or aluminum. The fans are usually mounted on the sidewalls because of structure limitation and to minimize shading. They should be located so that the general airflow is in the same direction as the external summer prevailing winds. That is, air inlets should be located on windward side. If fans must be located on windward sides, their capacity should be increased by about 10%. Wind speed and direction determine how much the ventilators should be opened. A total air inlet area of about 10% ~20% of floor area is recommended. To ensure reasonably uniform airflow fans should not be spaced too far apart. The ASABE standard EP406. 4 (2007) recommends that exhaust fans should be spaced no more than 7. 6m apart along the walls or ends of the greenhouse. Ventilation fans should be sized to deliver the required flow rate at 0. 015 kPa static pressure when all guard and louvers are in place. The fans should provide a higher static pressure when insect-proof screens are to be used. Thermostats, humidistats and other sensing devices used to control ventilation should be located at least 3m away from an outside wall. The total rate of ventilation must be divided into increments from economic reasons. First stage should be 10% ~25% of total rate. For large structures two or more groups of fans are economical. Teitel *et al.* (2004) proposed to use variable speed fans to control the speed of rotation of the fans according to the heat load on the greenhouse. They showed that with such control it is possible to reduce electricity consumption compared with ON-OFF operation by an amount which depends on the weather. In their study, the average energy consumption with a variable speed system over a period of one month was about 0. 64 of that with an ON-OFF system. A minimum velocity of 0. 2m/sec is recommended for preventing temperature, humidity or CO_2 stratification. It is recommended to maintain a distance of 4 ~ 5 fan diameters between the fan discharge and any nearby obstruction.

3 Natural Ventilation

Natural ventilation can be achieved by opening windows at the top of the greenhouse and/or at the sidewalls. The number and size of the windows and mechanisms for window opening vary. Many different arrangements of opening ventilators have been used in glass and plastic covered houses. Ridge openings can be classified under the type headings of continuous or non-continuous and they are usually on both sides of the ridge though hoses with openings on one side only are also constructed. Roof vents are either fixed or fully automatic (movable roof vents). A fixed overlapping vent on gable ridge provides ventilation while preventing penetration of rain and hail. Movable roof vents are formed by e. g. film roll-up from gutter to ridge, ridge hinged arched vents, vertical opening at the center of the arc which run the entire length of the roof, vertical roof opening that starts at the gutters and extends to a height of about 1m, vertical opening at the center of the arched roof which run the entire length of the roof. The position and hinging of the vent at the ridge are the basis of a better evacuation of the hot and humid air which builds up at the top of the greenhouse. In Venlo greenhouses the ventilators in most of the houses are hinged from the ridge and extend half way to the gutter or as far as the gutter. The idea is to provide a large opening area especially in warm and humid areas. Recent greenhouse designs provide retractable roofs. The new designs are traditional A-frame greenhouses with articulating roofs that either hinge at the gutters and open at the peak or hinge at one gutter and the peak while opening at the opposite gutter and moving across the greenhouse bay. Side ventilation is usually achieved by rolling up curtains with central mechanism operated manually or by an electric motor. Mechanisms that open the side vents from bottom to top or vice versa are available, where the most common operate from bottom to top. Side openings with flaps that are hinged from the top are also used however they are more common in glasshouses than in plastic covered houses. The theory of greenhouse ventilation was summarized in many scientific papers (e. g. , Roy *et al.* , 2002; Boulard *et al.* , 2002; Bailey *et al.* , 2003) and is therefore not given here.

3. 1 Ventilation and Air Exchange Efficiencies

Bournet and Boulard (2010) indicate that heterogeneity of the flow patterns and temperature and humidity distributions should also be considered in greenhouse management. Therefore, the maximization of the ventilation rate of the greenhouse is not the only criterion to be considered to evaluate the ventilation performance. One can imagine a situation in which the airflow rate through the greenhouse is high but most of the air does not flow through the crop. Consequently, the aerodynamic resistance of the crop will be high, resulting in poor exchange of vapor and CO_2 between the crop and the air and thus negatively influencing photosynthesis, transpiration and crop water use efficiency.

The concept of age of air was used by Etheridge and Sandberg (1996) to define a measure of air exchange efficiency ε_a, for the entire space of a room:

$$\varepsilon_a = \frac{\tau_{\min}}{2[\bar{\tau}]} \tag{1}$$

Where $\tau_{\min} = V/Q$ is the minimum possible time to complete one air exchange (i. e. the time taken if ideal 'piston' or 'displacement' flow occurred) and $[\tau]$ is the mean age of air in the room as whole. The volume of the room is V and the ventilation flow rate is Q.

The mean age of air in the room as whole can be calculated from:

$$[\bar{\tau}] = \frac{\int_0^\infty tC_e(t)\,dt}{\int_0^\infty C_e(t)\,dt} \tag{2}$$

Where $C_e(t)$ is the concentration at the exhaust at time t. The air exchange efficiency ε_a that is given in Eq. (1) is a measure of effectiveness of ventilation air delivery to the room (Awbi, 2003). It is the ratio between the shortest possible time it takes to replace the air and the actual average time it takes to replace the air in the room.

The purpose of ventilation is to limit contaminant exposure from indoor pollutant sources. The definition of ventilation efficiency should therefore reflect the performance of a system with regard to this effect. Thus, ventilation efficiency for an entire space that is in steady state is given by:

$$\varepsilon_v = \frac{C_e - C_s}{[C] - C_s} \tag{3}$$

Where C_e and C_s denote the concentrations of the contaminant at the extract vent and at the supply vent, respectively, and $[C]$ is the average concentration within the space.

The issue of greenhouse ventilation efficiency was raised by several authors. Molina-Aiz *et al.* (2012) investigated experimentally the ventilation efficiency of an Almeria-type greenhouse in which a mature tomato crop grew. Using the temperature data that they have measured and an equation similar to Eq. (3) they have calculated ventilation efficiency values in the range of 0.68 ~ 0.87 for wind speeds of 3.3 to 8.4m/s.

3.2 Effect of Insect-Proof Screens

The use of screens to reduce insect entry into greenhouses has become a common practice in many countries. The screens act as a mechanical barrier that prevents migratory insects from reaching the plants, and thus reduce the incidence of direct crop damage and of insect-transmitted virus diseases. As a consequence, the need for pesticide application is reduced; growers can follow international mandatory regulations, and can use biological control agents as well as insect pollinators. The exclusion of very small insects is achieved by installing fine mesh screens across the greenhouse openings. Since the porosity (ratio between open and total areas) of these screens is usually low, they impede ventilation. Pérez Parra *et al.* (2004), suggested a simple expression for the ratio between ventilation rates of a greenhouse with and without insects on the vents. Based on a correlation derived from the results of four studies, including their own they have suggested the ratio:

$$\frac{N_{sw}}{N_w} = \varepsilon(2 - \varepsilon) \quad (4)$$

Where, ε is the porosity of the screen. They further indicated that the reciprocal of this expression could be used to determine the increase in the opening area necessary to offset the effect of introducing an insect-proof screen.

Teitel (2007) by using correlations which relate the temperature difference between the greenhouse interior and the outside, with and without a screen, arrived at the following correlation:

$$\Delta T_{sw} = \Delta T_w(5 - 4\varepsilon) \quad (5)$$

The agreement between the data set obtained using Eq. (5) and the data from which the equation was established (a set of eleven values) is significant ($P < 0.05$), resulting in a coefficient of determination $r^2 = 0.5$. Nevertheless, it should be kept in mind that Eq. (5) probably provides only a rough estimate of ΔT_{sw}, since the relationship between the temperature differences with and without a screen are dependent on the type of greenhouse, the crop, the weather and the location where the inside air temperature was measured. Yet it should be mentioned that Eq. (5) is based on data of studies made in different greenhouses with a crop and that the interior temperature was nearly always measured at the center of the greenhouses.

4 Greenhouse Cooling

When greenhouse ventilation is not sufficient for cooling the greenhouse to a desired temperature other cooling techniques have to be applied. The large quantity of heat that has to be removed from greenhouses, especially in warm regions, precludes the possibility of using mechanical refrigeration. Therefore, all commercial greenhouses that utilize a cooling system use evaporative cooling which is much cheaper. Evaporative cooling is based on the conversion of sensible heat into latent heat. When non-saturated air comes in contact with free moisture and the two are thermally isolated from outside heat source, there is transfer of mass and heat. Because the vapour pressure of the free water is higher than that of the unsaturated air, water transfers in response to the difference in vapour pressure. The transfer involves a change of state from liquid to vapour, requiring heat for vaporization. This heat comes from the sensible heat of the air and the water resulting in a drop of temperature in both of them. Since no outside heat is added during this process, it can be assumed that the total heat (enthalpy) of the air does not change. The lowest temperature the air can get is the wet bulb temperature (when the air is completely saturated). Under practical greenhouse conditions, however, the air does not become completely saturated. In this process the wet bulb of the temperature is remained constant, the dry bulb temperature is lowered and the relative humidity is increased. The efficiency of the cooling unit is the ratio between the change in saturation actually achieved to the potential change in saturation.

4.1 Evaporative Cooling

The advantages of evaporative cooling are: no need for sophisticated machinery; relatively low cost of operation; effective mainly in regions with low humidity. The following lists several systems, which can satisfactorily provide cooling by evaporating water:

- Sprinkling
- Pad-and-fan
- Fogging

4.2 Sprinkling

Spraying water onto a surface of the roof and the canopy using sprinklers enlarges the free surface of the water and hence increases the evaporation rate. The evaporation process causes cooling of the canopy and of the air in the immediate vicinity, in accordance with the local microclimate. The advantage lies in the low cost. The main disadvantages of this method are its poor cooling effect compared with pad-and-fan and fogging systems, and the creation of conditions favorable to the development of fungal diseases. Also, sprinkling usually results in scalding and deposition of precipitates on the surfaces of the leaves and the fruits, especially when water quality is poor. Therefore, sprinkling is inferior in this respect to the pad-and-fan and fog systems.

4.3 Pad-and-fan

The currently accepted method is based on placing fans in one wall and the wet pad in the opposite one (usually the northern one to prevent shading). Outside air is sucked into the greenhouse through the wet pad, and is thus humidified and cooled. From the wet pad, this air flows through the greenhouse, absorbing heat and water vapor, and is removed by the fans at the opposite end. However, it is not able to treat the accumulated excessive heat (absorbed solar energy) in the greenhouse. The advantages of this method lie in its simplicity of operation and control and also in that it does not entail any risk of wetting the foliage. The main disadvantages are: high cost; lack of uniformity of the climatic conditions, which are characterized by rising temperature and falling humidity along the length of the structure and in the airflow direction; electric power failure transforms the greenhouse into a heat trap; low cooling effect compared with a fogging system; general diminution in efficiency with increasing humidity; and waste of water-to prevent blockage of the wet pad, water-bleed is necessary. Ganguly and Ghosh (2011) emphasized the point that a pad and fan system requires uninterrupted electric supply to drive the fans and water pump which becomes a major constraint for system application in rural areas, especially for developing countries like India, where a considerable number of villages do not have access to uninterrupted electricity supply.

Two types of pads are accepted, the vertical and the horizontal pads. Dripping water onto the upper edge can wet the porous vertically mounted pads. A drip collector and return gutter are mounted at the bottom of the pad and are used to re-circulate the water. In the horizontal pad, loose straw or small pieces of wood are distributed over horizontally supported wire mesh netting. The water is sprayed on the entire surface of the pad. Horizontal pads are more effective than vertical pads regarding pad plugging in dusty areas. The outside air is generally cooled by evaporation to about 1.5℃ of the wet bulb temperature. In regions with very low humidity the outside air temperature can be reduced by as much as 10 ~ 25℃ cooler than ambient temperature. The efficiency of wet pads is usually around 80% ~ 90%. Temperature differences ranging from 3 ~ 7℃ between fan and pad are quite common. When the system is operating, overhead ventilators and other openings must be closed. Pads must be continuous to avoid warm areas in the greenhouse. Dry spots or holes in the pad permit warm air to enter the greenhouse. Air velocity at the face of the pad should be in the range of 0.8 ~ 2m/sec. For corrugated cellulose 100mm thick the velocity should be about 1.3m/sec. When the thickness is 150mm the velocity should increase to about 1.8m/sec. The most common pads are made of aspen wood excelsior, corrugated cellulose, honeycomb paper and PVC. Air inlets should be constructed so that they may be easily covered during the winter to prevent heat loss.

4.4 Fogging

This method is based on supplying water in the form of the smallest possible drops (in the fog range-diameter 2 ~ 60μm so as to enhance the heat and mass exchange between the water and the air. This is because (for a given quantity of water) the surface area of the water in contact with the air increases in a direct relationship to the diminution in the diameter of the drops. Also characteristic of drops in this size range is that the frictional forces arising from movement of the drops through the air are relatively large, so that the terminal velocity of the falling drops is low, which results in a long residence time, allowing complete evaporation of the drops. Furthermore, because of their small size these drops are properly carried by the airflow. These combined characteristics ensure highly efficient evaporation of the water, while keeping the foliage dry. The high efficiency is because, in addition to the evaporation of water to cool the air that enters the greenhouse (similarly to the wet pad), it is possible to evaporate water in quantities sufficient to match the energy absorbed in the greenhouse (under normal conditions about $700g \cdot m^{-2} \cdot h^{-1}$).

Most fogging systems are based on high-pressure nozzles, characterized by low cost and high cooling effect relative to other systems, as discussed by Arbel *et al.* (1999). In the light of these considerations, the following scheme is recommended, comprising uniform roof openings, fans in all walls and nozzles uniformly distributed at the height of the structure. The air that enters the greenhouse through the roof openings carries the water drops with it, and the water evaporates within the flow. As a result, the air is cooled (by water evaporation), both on its entry into the greenhouse and in the course of its passage among the canopy, and absorbs excess heat.

Arbel *et al.* (2003) focused on the operational characterization of a fogging system in combination with forced ventilation following the above scheme. The results obtained revealed that inside the greenhouse air temperature and relative humidity of 28°C and 80% respectively, were maintained during the summer at midday. Furthermore, the results obtained revealed generally high uniformity of the climatic conditions, within the greenhouse. Such an arrangement may lead to: uniform climatic conditions within the desired range, even in the summer months; good control over the climatic conditions through reduction of the influence of the wind; operation of the greenhouse by means of a relatively simple control system; the establishment of greenhouses in large units and, consequently, better exploitation of the land area and significant reduction in the cost per unit area of the structure.

4.5 Refrigeration

Two main systems are available: vapor compression cycle and the absorption cycle. Refrigeration systems enable quite accurate control of temperature and humidity and it is possible to reduce temperature to low levels. However, the equipment, installation and operation costs are very high. Refrigeration is therefore mainly used in small research greenhouses.

List of references

[1] Arbel, A., Yekutieli, O. Barak, M. 1999. Performance of a fog system for cooling greenhouses. J. Agricultural Engineering Res., 72 (2): 129 ~ 136

[2] Arbel, A., M. Barak and A. Shklyar, 2003. Combination of forced ventilation and fogging systems for cooling greenhouses. Biosystems Engineering, 84 (1): 45 ~ 55

[3] ASABE standard 2007. Heating, ventilating and cooling greenhouses. ANSI/ASAE EP406.4, American Society of Agricultural and Biological Engineers. MI, USA, pp 787 ~ 795

[4] Awbi, H. B. 2003. Ventilation of buildings. Spon Press, New York, NY, USA, 522p

[5] Bailey, B. J., Montero, J. I., Pérez Parra, J., Robertson, A. P., Baeza, E. and Kamaruddin, R. 2003. Airflow resistance of greenhouse ventilators with and without insect screens. Biosyst. Eng. 86 (2): 217 ~ 229

[6] Boulard, T., Kittas, C., Roy, J. C., Wang, S. 2002. Convective and ventilation transfers in greenhouses, part 2: determination of the distributed greenhouse climate. Biosystems Engineering, 83: 129 ~ 147

[7] Bournet, P. E. and Boulard, T. 2010. Effect of ventilator configuration on the distributed climate of greenhouses: a review of experimental and CFD studies. Computers and Electronics in Agriculture, 74: 195 ~ 217

[8] Ethridge, D. and Sandberg, M. 1996. Building ventilation: theory and measurement. John Wiley & Sons, New York, NY, USA, 724p

[9] Ganguly, A. and Ghosh, S. 2011. A review of ventilation and cooling technologies in agricultural greenhouse application. Iranica Journal of Energy & Environment, 2 (1): 32 ~ 46

[10] Molina-Aiz, F. D., Valera, D. L., López, A., Álvarez, A. J. 2012. Analysis of cooling ventilation efficiency in a naturally ventilated Almeria-type greenhouse with insect screens. Acta Horticulturae, 927: 551 ~ 558

[11] Pérez Parra, J., Baeza, E., Montero, J. I. and Bailey, B. J. 2004. Natural ventilation of Parral Greenhouses. Biosyst. Eng. 87 (3): 355 ~ 366

[12] Roy, J. C., Boulard, T., Kittas, C., Wang, S. 2002. Convective and ventilation transfers in greenhouses, part 1: the greenhouse considered as a perfectly stirred tank. *Biosystems Engineering*, 83: 1 ~ 20

[13] Sase, S. 2006. Air movement and climate uniformity in ventilated greenhouses. Acta Horticulturae, 719: 313 ~ 323

[14] Teitel, M. (2007). The effect of screened openings on greenhouse microclimate. *J. Agricultural and Forest Meteorology*, 143: 159 ~ 175

[15] Teitel, M., Zhao, Y., Barak, M., Bar-lev, E., Shmuel, D. 2004. Effect on energy use and greenhouse microclimate through fan motor control by variable frequency drives, Energy Conversion & Management, 45: 209 ~ 223

Protected Cultivation in Europe

Esteban J. Baeza[1], Cecilia Stanghellini[2], Nicolas Castilla[3*]

(1. *Invited researcher Vegetable Prod. Dpt. University of Almeria Almeria*, *Spain*;
2. *Wageningen UR Greenhouse Horticulture Droevendaalsesteeg* 1 6708 *PB Wageningen*, *The Netherlands*; 3. *IFAPA- Camino de Purchil Apartado* 2027 18080 *Granada*, *Spain*)

Abstract: During the last decades, the horticultural production in Europe has gradually moved from the northern countries towards the Mediterranean basin. Two well-differentiated agrosystems can be distinguished as predominant in each area. The northern agrosystems (the most paradigmatic example is The Netherlands), sophisticated and oriented towards high yields and highly dependent on energy and expensive technology, and the southern agrosystems (the South-east of Spain is the best example), based on low investments, low yields and low dependency on energy. The main trends in the northern agrosystems are to develop innovations that reduce costs (mainly by saving energy) and increasing production, through improved optical and thermic properties of the cover material (diffusive and antireflective glass, NIR filters, etc.) and improved climate and crop management (semi-closed greenhouses, robotics, de-humidification strategies, etc.). The trend in the southern agrosystem is to improve yield with moderate investments: improve the greenhouse structures (enhancing the light transmission in winter, optimizing the natural ventilation systems, collecting condensation, increasing tightness, etc.), using better covering materials (anti-pest plastic films, more diffusive films, etc.) and improving equipment (combination of passive techniques like double covers with cheap heating systems, or biomass heating, implementation of CO_2 enrichment systems, fog or mist systems, etc.). Another trend is to move summer production to cooler areas using simple screenhouse structures to complement the year-round supply of vegetables. Diversification of crops would also be desirable, because until now diversification has mostly been limited to different presentations of the same products. Marketing is dominated by a market driven approach, with consumers demanding high quality and environmentally safe products, forcing the growers to set up integral quality management systems and in the case of Mediterranean countries, as stated before, to find the balance between investing in new technology and maintaining the benefits, having to choose between different available technological packages.

Key words: Agrosystem; Screens; Semi-closed greenhose; Cooling; Heating; Marketing; Environment; LCA

1 Introduction: Protected Cultivation In Europe

During Imperial Rome, small mobile structures were used to grow cucumber plants, which were taken outdoors when the weather was favourable or covered otherwise, constituting the first documents of protected cultivation in the world. Mica and alabaster sheets were used as the covering material. Much time later, during the Renaissance period (15th to 16th centuries) the precursors of the actual greenhouses appeared. In the 19th century, the first greenhouses with a gabled cover were built, expanding quickly from Europe to America and Asia, being located near the larger cities. In

* E-mail: nicolas. castilla@ juntadeandalucia. es

the 20th century, economic development, especially after the Second World War, spurred the building of glasshouses, especially in the cold countries.

However, the appearance of plastics, to be used as cladding, enabled the enormous expansion of the surface area of greenhouses towards the south, to milder climates, especially in Mediterranean countries. Italy, which counts nowadays with 65 800hm^2 of protected crops, out of which 28 300hm^2 are greenhouses, and Spain with the largest greenhouse area in Europe with more than 50 000hm^2 are the two leading countries in protected cultivation in the area (Sigrimis *et al.*, 2010). The Orient (mainly China and Japan) has the largest greenhouse area of the world. Huge differences in protected cultivation area exist between the Mediterranean and central Europe countries. Only The Netherlands, with an almost constant glasshouse area of around 10 000hm^2 during the last decades competes with the Mediterranean horticulture. Recently, protected cultivation is greatly expanding also in other Mediterranean countries like Turkey, with a total area of high tunnels + greenhouses of 27 864hm^2 (TUIK, 2009) or Morocco, with a greenhouse area of 16 500hm^2 in 2006 (Hanafi, 2007).

2 Objectives of Protected Cultivation

The general goal of protected cultivation is to modify the natural environment by several techniques in order to achieve all or at least some of the following objectives (Castilla, 2007; Wittwer and Castilla, 1995):

a. Increase yield, improve product quality, and preserve resources.

b. Extend the production areas and crop cycles.

c. Stabilize the supply of high-quality products for horticultural markets.

d. Reduce water needs.

e. Protect crops from low temperatures.

f. Reduce wind damage.

g. Reduce solar radiation.

h. Protect from heavy rainfall.

i. Limit the impact of arid and desert climates.

j. Reduce damage from insects, pests, diseases, nematodes, weeds, birds, and other predators.

Normally, achieving all these goals requires a greater investment than in conventional open field cultivation, as well as more inputs per unit surface area, though not necessarily per unit yield and not necessarily involving greater environmental impact, if properly managed.

3 Types of Protected Cultivation in Europe

As stated by Castilla (2007), a screen or protective shelter placed next to the plant changes the environmental conditions affecting the whole plant or part of it, altering the energy balance of the surroundings. Depending on how this screen is placed in relation to the crop, we can talk of a different type of protection. In this sense, if we only use a protection device (screen, plastic, hedge, etc.) on the laterals of the plot, we talk about windbreaks. If we only cover the soil, then we have a mulch (i. e. enarenado-sand mulch- in Spain). When the screen is placed over the plants, we have a third group of protections, which may range from floating covers (or direct covers, directly laid over the plants without any supporting structure), tunnels (which can be low, normally less than 1m high, or high, also known as walk-in tunnels, in which workers can walk inside, and crops are grown to a certain height) and greenhouses (much more solid, higher and wider than high tunnels, allows for cultivating of even fruit trees). In any case, distinguishing between high tunnels and greenhouses is not obvious and usually high tunnels are considered a subgroup of greenhouses.

4 European Greenhouse Agrosystems

In the last decades, the greenhouse industry has spread widely around the world, where we can identify two basic and more extended "greenhouse agrosystems" . One type is the sophisticated with high climate control "Northern greenhouse agrosystem", typical of the high latitude areas of Europe and North America, that developed first in these colder climates. The other type of agrosystem only provides a very limited climate control, enabling the plants to adapt to a sub-optimum environment, but able to produce an economical yield with a much lower investment (Enoch, 1986). Obviously, we can find intermediate technology levels in between these two well-known extremes, especially in the southern countries (in northern countries the level of sophistication must be high in order to make profitable crops) but we can consider that the "Mediterranean greenhouse agrosystem" is of the low cost type.

5 The Northern Greenhouse Agrosystem

5.1 Greenhouse structures and covering materials: State of the art and trends

The "northern" greenhouses achieve profit by delivering very high yield of first-quality products, in a reliable production pattern. This is achieved by an elevated degree of "decoupling" from external conditions, that is a climate management based on heating, continuous control of ventilation, high carbon dioxide concentrations and sometimes even artificial light and/or cooling. Nevertheless, profitability has been under pressure the past few years, due to high prices of resources (energy and labour) and low prices of product (Vermeulen, 2011).

One of the main characteristics of Central European (Dutch) greenhouse design, is the high-degree of standardization. The Venlo design (multi-span, with fixed span width and glass cover with slope of about 25° degree and fully automated roof ventilation) is the most common by far. Indeed, Vanthoor *et al.*, 2012 have shown that both the glass cover and the ventilation area as usually applied are "optimal" for the Dutch conditions of both climate and market. Innovations deal with reducing costs (mainly by saving energy) and increasing production, through improved optical and thermic properties of the cover material and improved climate and crop management.

Innovative (spectral selective) cover materials have the potential of reducing pest pressure, saving energy and improving summer growing conditions, although unfortunately very often the requirements of a high PAR (Photosynthetically Active Radiation) transmittance, spectral selectivity and high thermal insulation are conflicting.

The environmental and economic potential of combinations of NIR-blocking (NearInfraRed) plastic additives and of glass coatings was evaluated through a dynamic greenhouse climate and production simulator (Kempkes *et al.*, 2011). The results are disappointing about the potential for NIR-filters in the greenhouse cover material (Stanghellini *et al.*, 2011). A filter can work two ways: reflecting or absorbing. Evolution has already endowed leaves with a high (0% ~ 50%) NIR-reflectance, so that NIR-reflection will lead to multiple reflection between crop and cover and a fraction of crop reflection will not escape the greenhouse. Therefore only NIR-filters with very high reflectance will have some consequence. On the other hand, NIR-absorption will warm up the cover, so that a fraction of the withheld energy will end up in the greenhouse at longer wavelengths and through convection. In addition, as the contribution of NIR to heating the greenhouse may be welcome often enough, a permanent NIR filter may backfire, in terms of either a worse winter climate or higher energy requirement in un-heated and heated greenhouses, respectively.

In the climatic conditions of Central-Northern Europe, high light transmissivity coupled to high insulation are the conflicting requirements for greenhouse covers. Modern anti-reflection (AR) coatings allow for panes of double glass to have the total transmittance of a single standard pane. The obvious potential for energy saving can be increased further by a low-emission coating (Hemming *et al.*, 2011) as demonstrated experimentally by Arkenstein *et al.* (2012).

Light diffusion has been proven to increase photosynthetic efficiency of crops. Dueck *et al.*, 2009 have shown that the productivity of cucumber in The Netherlands could be increased by 9.2% by a highly diffusive cover (haze 70%), in spite of an overall reduction in transmission of 3%. Recent developments (such as AR coatings on diffusing panes) are able to reduce the [inevitable] light loss caused by diffusion to almost nothing. Indeed, an increase of production of tomatoes of 10% has been observed in a recent research (Dueck *et al.*, 2012). In regions where the fraction of direct radiation in a year is larger than the 30% typical of the Netherlands, the potential for increasing productivity through diffusing cover materials must be even larger.

5.2 Equipment and management: State of the art and trends

All North-Central European greenhouses are fitted with automatic climate control (heating and ventilation) and carbon dioxide fertilisation. However, natural ventilation to discharge excess heat and vapour from the greenhouse environment has serious drawbacks. Pests and diseases find their way through the openings; carbon dioxide fertilisation becomes inefficient and the inescapable coupling of heat and vapour release results often in sub-optimal conditions for either temperature or humidity. The present trend, therefore, is to reduce ventilation as much as possible (Heuvelink *et al.*, 2008). This relies obviously on improved means for diminishing the heat load and proper use of cooling equipment, which might be a financially reasonable option only when coupled to thermal storage, usually in shallow natural aquifers. Such a design could combine the benefits of cooling the greenhouse air with serious energy conservation. However, opposite to the clear benefits there are also serious investments associated with active cooling of greenhouses.

Therefore, present wisdom is that greenhouses should be semi-closed (De Zwart, 2012), that is the cooling capacity should be enough to reduce significantly the need for ventilation, but natural ventilation is the cheapest mean by far to discharge peak amounts of excess energy, so that there is an economic optimum to be struck. In The Netherlands, Qiang *et al.* (2011) achieved an increase in tomato production by 14% in a closed greenhouse (700 W/m^2 cooling capacity) and of 10% and 6% in semi-closed greenhouses fitted with 350 and 150 W/m^2 cooling capacity, respectively.

Given a certain technical infrastructure of the greenhouse, energy consumption can be further reduced by energy efficient climate control and crop management. Essential elements are to allow more fluctuation in air temperature than growers are used to, lower crop transpiration, allow higher humidities, make efficient use of light and create fluent transitions in set points. That higher humidity and lower transpiration do not affect yield at all was already shown by Li *et al.* (2001) and explained by Li *et al.* (2004). Stanghellini *et al.* (2012) have shown that cooling may affect the vertical distribution of [reduced] transpiration in a tall crop such as tomatoes. However, Dieleman *et al.* (2012) have demonstrated that there is no consequence for plant growth and yield.

Indeed, humidity control has been shown to account for some 20% of energy consumption in traditional Dutch greenhouses (Bakker *et al.*, 2008). The main reason is that the need for natural (roof-top) ventilation conflicts with the use of thermal screens. Therefore the newest greenhouses are fitted with side-wall fans that are automatically operated for humidity control, as first proposed by Campen (2008). Even more energy can be saved whenever the incoming air-flow gains heat from the out-going air, in a cross-flow heat -exchanger, although of course the latent heat cannot be recovered (De Zwart, 2012). Multiple, dedicated screens (aluminized thermal screen, transparent PE screen for daytime use, shadow screen) are increasingly used and have been shown to significantly reduce energy requirement with respect to the one fits all (usually a shadow screen with aluminium stripes (De Zwart *et al.*, 2010).

5.3 Protected crops

It is a fact that, even with due attention to cost reduction, greenhouse production in North-Central Europe relies on very expensive production means. Profitability, therefore, is to be sought in high-value crops. Indeed, "bulk" products have been in fact displaced by cheaper products from Southern Europe or even North Africa. The trend is towards niche products such as high-lycopene tomatoes or fancy peppers and, obviously, ornamental plants and cut flowers.

Actually, the product that has kept Dutch greenhouse horticulture in the black for the past few years has been the electricity produced by co-generator engines, whose waste heat and CO_2 are used in the greenhouse. Indeed, the total installed electric capacity in the Dutch greenhouse horticulture has triplicated between 2005 and 2008, to reach a total of 3 GW that is more than 10% of the national electricity use (Centraal Bureau voor de Statistiek, 2009).

6 The Mediterranean Greenhouse Agrosystem

Greenhouse structures and covering materials: State of the art and trends.

In the Mediterranean greenhouse regions, protection of the crops is achieved by means of simple structures which often have been empirically developed in each area (using local available materials, such as wood for the supports), covered with plastic films and with almost no climate control, which is limited to opening and closing of the greenhouse vents (with very scarce automation of this operation) and shading using whitening during the times of the growing cycle when radiation levels are very high.

6.1 Greenhouse Structure

As stated by Castilla and Montero (2008) the two major types of greenhouse constructions found in the Mediterranean area are the artisan and the industrial. The artisan type is the prevalent, for instance, in the Spanish industry and in other large greenhouse areas as Sicily (Italy). The case of Spain is typically represented by the *parral* type greenhouse, made by a vertical structure of rigid pillars (of wood, iron, or steel) on which a double wire grid is placed in the roof and sidewalls, to attach the plastic film in between (i. e. in a "sandwich" way). As in other Mediterranean zones, the availability and price of local materials as well as installation expertise (which in the case of the *parral* came from previous structures used for decades to grow the table grapes) have been fundamental in the greenhouse expansion. These type of greenhouses, despite of their popularity, have several problems that prevent the growers from achieving higher productions of better quality along the growing cycle, among which we could highlight: the lack of tightness (easier penetration of pests and low efficiency of climate control systems due to the high infiltration), the lack of natural ventilation (caused mainly by the low ventilation area, the use of inefficient ventilator designs, the use of low porosity insect-proof screens and the small distance left often between adjacent greenhouses) as well as the low values of light transmission in winter time (Baeza, 2007).

The arch-roof multi-span (multitunnel) is the dominant design among the industrial type greenhouse, which is primarily covered with plastic film or, in some cases, with rigid or semirigid plastic panels (typically polycarbonate). The supporting frame of the arc-shaped multi-tunnels is normally made of galvanized steel and the ornamental growers and nurseries prefer them. Multi-tunnels are more airtight than the *parral* type greenhouses and easier to equip. The glasshouse area is very scarce, due to the high initial investment costs associated both to the structure and its equipment.

Considerable enhancement of greenhouse light transmission has been achieved by increasing the roof slope from nearly zero, typical from areas like the Mediterranean with low rainfall, to roof angles values close to 30°. This has had a direct effect on crop response in the winter time (Soriano *et al.*, 2004). The increase in slope also aims at improving the collection of the condensation during the winter time, going from semi-circular arch type multi-span greenhouse to the so-called gothic type arch. There is no scientific evidence of how much improvement is achieved (how much more water is collected) in these new types of arch-roof multi-span types.

Some recent attempts have been made to develop closed greenhouses in Mediterranean areas, such as the Watergy project (Buchholz *et al.*, 2006). It is the combined benefits, not only of conserving fossil fuel but also the requirement of saving water for irrigation, and the limitation of pest entrance in the greenhouse through the vents, which makes the closed-greenhouse concept attractive. Despite of this, the need to store the hot/cold water on a day night basis, together with the absence of easily accessible aquifers (which is the case in The Netherlands, where closed greenhouses have gained more popularity) as well as the high radiation load during part of the year (which forces to cool the water during the night using energy, by means of a heat pump and/or a cooling tower) makes this systems very little competitive for Mediterranean climate.

Therefore, the prevailing "Mediterranean approach" for developing future greenhouse systems is based on the development of more efficient passive greenhouses (better natural ventilation designs, higher light transmission in winter, collection of the condensation from the roof), for those areas where heating may be convenient but is not essentially needed. This approach normally leads to attain lower production levels, but also achieved at lower investment and operating costs and therefore more economically feasible.

6.2 Greenhouse Cladding Materials

The first step of change in most countries is to use long life and thermal polyethylene (PE) films. Multi-layer films that combine the desirable characteristics of various materials (anti-drip, diffusing direct radiation, high PAR transmission, anti-dust, UV absorption, high IR absorption) are developing fast as a second step. As stated by Heuvelink and González Real, (2008) different types of photoselective plastic films are now commercial: their effect ranges from altering the ratio "red-far red" (R/FR) of incoming light, thus affecting plant morphogenesis, or reducing crop diseases (B*otrytis*), or modifying the behavior of insects (influencing their vision) by blocking certain wavebands of the solar radiation spectrum (UV blocking), or limiting the sun energy load (NIR, near infra-red, blocking).

In the case of the UV blocking film, there is enough scientific evidence of their positive effect in decreasing the incidence of the most important pests affecting Mediterranean greenhouses (*Bemisia tabaci*, *Frankliniela occidentalis*, and different species of aphids) (Diaz and Fereres, 2007). The possible harmful effect on the pollinators widely used in the Mediterranean greenhouses (bumblebees and bees) has prevented among the growers a higher expansion of this new development, although most of scientific works evidences an adaptation (at least of bumblebees, as studies show that bees are more sensitive) to the lack of UV light after several days, as well as the importance of opening roof vents to provide enough UV light for the pollinators (Pérez *et al.*, 2009). No negative effect of these types of plastics has been observed on the beneficial insects used in biological control in different greenhouse crops (Soler *et al.*, 2009). Other special covering materials, like fluorescent and colored plastic films, can be of interest assuming that they do not reduce PAR transmission, but their use is limited.

There is a general trend to use covering materials which diffuse the incoming light, without relevant reduction of transmission values, therefore contributing to improve crop production and radiation use efficiency (Heuvelink and Gonzalez-Real, 2008; Hemming *et al.*, 2008; Magan *et al.*, 2011). In Mediterranean climate, recent works have developed models that can be used to study the effect of different roof geometries (on mono modular or multi modular greenhouses), given the optical properties of the cover, on the distribution of diffuse light inside the greenhouse (Cabrera *et al.*, 2009). In this sense new highly diffusive materials are being tested at the moment in the Experimental Station of the Cajamar Foundation in Almería (Spain), and considering the increases in yield achieved in The Netherlands of up to 10% (Hemming *et al.*, 2008) with highly diffusive glass, promising results can be expected given the much larger number of clear days in the Mediterranean area in relation to The Netherlands.

As stated before, the NIR-selective filters that are commercially available can be applied in three different procedures: ① as permanent additives or coatings to the cover; ② as seasonal "whitewash"; and ③ as movable flexible film screens. Whereas a permanent filter may have a useful application in tropical environments, it seems that in Mediterranean climates there is probably potential for either movable screens, seasonal filters or filters whose optical properties vary with temperature, presently under investigation (Montero *et al.*, 2008). Until now, no commercial NIR-blocking films are available in which the PAR range is not affected as well, causing small yield decreases that do not compensate for the money saved in whitewashing the greenhouse.

As stated by Tanny *et al.* (2006), covering the crops with flexible porous screens (nets) is becoming a common practice in many Mediterranean areas. The so called "screenhouses" are effective and economical structures to provide an "oasis effect" in dry or semi-dry climate areas: shading the crops, protecting them from wind and hail, improving the temperature and humidity regimes, saving irrigation water and excluding insects (pests) and birds (Tanny *et al.*, 2006). As stated by Castilla and Montero (2008) in the south of Spain, the lack of greenhouse production in coastal areas during the high radiation summer season is being replaced with vegetable produce from screenhouses in the cooler highlands, enabling year-round supply to the market. In addition, the highlands where these screenhouses are developing are economically depressed areas with high agricultural unemployment.

A large number of studies have been devoted during the last years to better understand the interaction of the different screen types available in the market (materials, colors, porosity, optical properties of the threads, etc) with the meteorological factors, mainly with the wind and radiation. In areas where pest pressure is high, the need to prevent the entrance of very small insects (i. e. white fly or thrips) forces to use screens with very low porosity, but today we know much more on how these structures ventilate (Tanny *et al.*, 2003) and the microclimate generated inside (Moller *et al.*, 2004; Tanny *et al.*, 2006). Recently, different works have also dealt with the radiation transmission through these types of screens (Möller *et al.*, 2010; Al-Helal & Abdel-Ghani, 2011). More attention must be paid in the future to characterize better the diffuse light transmission and distribution inside the screenhouses, and the quality of the light under the color-nets that have proved to be a very attractive solution

Greenhouse Equipment and management: State of the art and trends.

6.3 Greenhouse Equipment

The grower can choose between a large number of technology packages (greenhouse and its equipment) depending on factors such as the local climate conditions, the crop, the growing cycle, etc. The equipment used is also quite dependent on the type of greenhouse (i. e. the artisan type greenhouses are not very tight, thus the use of heating may be quite inefficient). Therefore, the technology used must be oriented to optimize the performance of the crop, but not just from a technical point of view but also keeping economic and environmental soundness.

Heating. The problem of low air temperatures during winter can be solved by some heat supply to the greenhouse during the critical time periods. The problem is not technical, as it is easy to heat an enclosure, but economical, as the investment and the running cost are relatively high (Bartzanas *et al.*, 2005). In the Almeria province of Spain, less than 1% of the greenhouses are heated (FIAPA, 2001). Heating is installed mostly in high-technological level greenhouses (multitunnels), or for frost protection within inexpensive structures using pulsed hot air heating systems. Low-temperature (40 ~ 50℃) water-heating systems that use plastic tubes for heat distribution are preferred to high-temperature systems that require steel tubes, since they are cheaper and heating set points maintained are low. If the latter are installed, the heating tubes provide the rails for mechanization. The use of movable internal thermal screens is not common, as they are difficult to install in low-cost greenhouses, where double covers during the winter is preferred. However, inflated double PE film covers are very rarely constructed.

Air heaters are generally used as the primary heating sources in greenhouses where the heating needs are minimal. Their main advantages are the promptly response to control changes in air temperature and their lower capital cost, while the disadvantages are their lower energy efficiency (Bartzanas *et al.*, 2005). There are other technical solutions of lower energy use, which could be and are indeed applied, to improve the thermal and hygrometric conditions during the winter time in Mediterranean greenhouses, but which have been little studied until now (Bailey, 1988). The use of passive climatization during the cold period, although non-optimum conditions are achieved, would improve the actual conditions (Baille *et al.*, 2006), with a lower environmental and economic impact than traditional systems. First method is natural ventilation, but this system will be treated extensively in the next section. In unheated Mediterranean greenhouses the use of plastic mulch can improve the soil and air temperature during the first stages of development of the crop that start at the end of autumn or during the winter, but the available information is scarce, as the majority of the work on mulching in greenhouses have studied the soil solarization in summer (Stapleton, 2000). Another passive system used to improve the greenhouse microclimate during the cold period is the fix double roof. This technique, which consists on placing above the crop a very thin transparent thermal plastic film, horizontally, with very little slope, is widely used by Mediterranean growers that produce during the winter. The double roof limits the exit of thermal radiation increasing slightly the air temperature (Montero *et al.*, 1985) and, especially, avoiding the fall of condensation water over the crop, which decreases the risk of diseases.

There is a big opportunity of biomass heating, to become a system used by the growers to increase the quantity and quality of their yields during the winter, especially where cheap biomass is available and subsides are available for the growers (i. e. In Spain up to 40% of the investment on the biomass heating system currently can be subsided by the government). Biomass consumption as fuel for greenhouse heating is related to both the greenhouse surface and the specific energy needs of crops It is obvious, that taxes and regulations (i. e. subsides for buying biomass boilers) of each country may have big effect in pushing growers to use these systems.

Natural Ventilation. In the Mediterranean, achieving optimum levels of ventilation should be sufficient to maintain the greenhouses located near the coast under almost optimum temperature and humidity conditions during a large part of the typical growing cycles (Montero, 2006). Unfortunately, most of the greenhouses in Spain suffer from inefficient and insufficient ventilation. There are four reasons to explain this: the limited venting area (12% of floor area as an average in the Almeria area), which are covered in their majority with low porosity insect-proof nets (to prevent the entrance of small pest insects). Another reason is the inefficient ventilator design (small roof vents are preferred by growers due to fear of sudden winds since automation is almost inexistent) and the greenhouses are "packed" together with almost no distance between them, making sidewall ventilation almost completely inefficient in many cases. Therefore, more attention must be given to improving ventilation in these coastal greenhouses before contemplating investments in more complex technology such as evaporative cooling systems.

Natural ventilation is preferable to fan ventilation as it uses significantly less energy (Sase, 2006). During the last decade, a significant improvement in the knowledge of natural ventilation systems and their improved design has been obtained from computational fluid dynamic methods or CFD.

CFD analysis indicates that windward ventilation provides higher ventilation rates than leeward ventilation, especially in narrow greenhouses, although leeward ventilation provides a slightly more homogenous airflow. Since ventilation systems must be designed for "worst case scenarios", new greenhouse constructions have larger openings oriented towards the prevailing winds, especially in the first spans facing the wind. Good ventilation efficiency involves not just a high ventilation rate, but also a good mixing of the incoming air with the internal air, also in the canopy area. To correct designs in which this is not achieved, the use of deflectors to conduct the entering air through the crop area is strongly recommended (Baeza, 2007). However, if a new greenhouse is to be built, the CFD simulations have demonstrated that the greenhouse roof slope has a significant effect on ventilation rate and also on the homogeneity of air movement in the greenhouse. Traditional horizontal or almost horizontal (10 ~ 12℃ roof angles) greenhouses should be replaced with symmetrical or asymmetrical greenhouses with near 30° roof angles, because CFD simulations show that above 30° of roof angle no significant increase in ventilation has been identified (Baeza, 2007). Montero *et al.*, (2008), in view of the important harmful effect that neighbor greenhouses have on ventilation efficiency, is studying the effect of the surrounding greenhouses or other structures on the ventilation rate and the uniformity of air movement inside the greenhouse. In any case, the combination of roof and side ventilation in not to wide greenhouse modules is the most efficient design to obtain enough ventilation regardless of the time of the year in Mediterranean greenhouses, both under zero wind and windy conditions and for increasing number of spans (Kacira *et al.*, 2004; Baeza *et al.*, 2009).

Efficient roof-opening greenhouse designs, for improving both ventilation and solar radiation transmission were developed in the Mediterranean region but they were not implemented, due to their high cost (Brun and Lagier, 1985). Some growers have adapted simple systems for changing the greenhouse cover, from a plastic film to a screen and on the opposite, in order to improve the inside microclimate. Prototypes of this interchangeable cover mechanism have been developed, as in other areas (Sase *et al.*, 2002).

Other Cooling Methods. At certain moments, natural ventilation is unable to provide the right conditions for optimum crop growth and development, thus plant suffer stress due to excessive temperatures, and the grower has to decide and use cooling techniques that are available: mechanical ventilation, shading nets (inside or outside the greenhouse), shading by whitewashing, evaporative cooling (fan-and-pan or fogging, high or low pressure) or a combination of them.

The simplest and more extensively used method in the Mediterranean area is to shade, decreasing the solar radiation transmission and, hence, the heat load and air temperature. Unfortunately, most of shading methods cause a decrease of PAR, and thus decrease crop productivity is to be expected (Gonzalez-Real and Baille, 2006). Fix shading screens are much less efficient than external mobile shading, because of the ability to control the shade depending on the incident radiation and temperature, and recent studies have shown that if control settings are precise enough the system can be as efficient decreasing temperatures as a fog system (at least in areas of moderate humidity levels), without water use. Main drawback is that it is expensive to install and cannot be used under high wind conditions (García *et al.*, 2011). Cover whitening is an inexpensive and popular method to reduce the heat load (decreasing VPD and temperature) during the high solar radiation season and may have a positive effect in diffusing the light. Whitening does not affect the greenhouse ventilation, while internal shading nets do, although new models of internal screens with better porosity are being commercialized.

Misting nozzles are inexpensive and can be useful for cooling low cost greenhouses or screenhouses (trying to design the network so that we avoid wetting the plants), but a simple VPD-based controller, capable of providing intermittent misting for relative short time intervals, should control the system (Montero, 2006). Comparative studies on the use of the various cooling methods are scarce, although works such as Gazquez *et al.* (2006) estate that cover whitening is the most profitable cooling treatment as compared with high pressure fogging or fan ventilation, when growing soilless pepper during the summer season. Fan ventilation should be discarded due to high investment and running costs, and also for higher risk of pest infestation under high pest pressure conditions (Gazquez *et al.*, 2006). Summarizing, most Mediterranean growers rely on the combination of natural ventilation and shading, with a few using evaporative cooling during the first stages of development of early summer crops.

Carbon dioxide (CO_2). The fact that many greenhouses in the Mediterranean area have poor ventilation, or the use of techniques that improve the thermal performance of the greenhouse, such as fixed double covers, may produce carbon dioxide (CO_2) depletion inside the greenhouse, thereby limiting photosynthesis and reducing yield. As stated by Stanghellini *et al.* (2008) under mild winter conditions, poor ventilation may be the ultimate limiting factor hampering higher production, through its effect on internal carbon dioxide concentration and high humidity that enhances the risk of diseases. Despite of this, only high-technological-level greenhouses are equipped for CO_2 enrichment (0.2 % of total greenhouses in the Almeria area) and the need to ventilate during the central hours of the day, avoids a much more efficient use of the applied CO_2.

6.4 Protected crops

As stated by Castilla (2002), the greenhouse industry of the Mediterranean area is in its majority devoted to grow edible vegetable crops, accounting for 84% of the greenhouse area in Italy, near 90% in Spain, and 92% in Greece (Castilla, 2002), remaining the ornamental sector in a much more weak position, often unable to compete with this strong sector in The Nethelands.

Castilla and Hernandez (2005) in the case of Spain identified the main vegetable crops as tomato, sweet pepper, cucumber, green beans, strawberry, melon, watermelon, eggplant, squash and lettuce, in descending order. Production of flowers (mainly carnation and rose) and ornamentals constitute only around 5% of the greenhouse area whereas. Banana is the major tree crop with an increasing area of table grape under temporary plastic cover in the south east of the country (Castilla and Hernandez, 2005)

6.5 Diversification

Diversification of the greenhouse production would be desirable to diversify the risks, and until now this has been achieved by growers, not by introducing new crops, which require from research and marketing programs for the consumers, but developing and growing new presentations of different species (tomato is a good example of this, with cherry and cluster tomatoes representing 60% of the greenhouse tomato production in Italy) (Castilla, 2002; La Malfa and Leonardi, 2001). New presentations of well-established crops include variations of the colour, shape or size, recovery of old varieties with better taste (i. e. marmande RAF tomato in Spain) as well as quality labels. (La Malfa and Leonardi, 2001) and of "baby" vegetables (Castilla, 2002). The use of transgenic cultivars is still too controversial in Europe (Castilla, 2002).

7 Marketing: State of The Art and Trends

As stated by van Uffelen *et al.* (2000) the European consumer is demanding high quality products which must be grown taking into account the environment, the safety for the workers in an hygienic growing environment, and the grower to be controlled to prove this. The traceability of the produce has become the tool used by consumers to ensure the demanded security measures on the production method.

The greenhouse industry has evolved to a highly focused market-driven approach to production (Sullivan *et al.*, 1999), which has derived in the translating of the specifications of the market into production standards. It has become a standard practice, to include in the contract detailed protocols to be followed by growers for the sustainable production of safe and nutritious vegetables. In this sense, there is a wide use of good agricultural practices (GAP) both for farm and post-harvest management processes.

We can state that the "Mediterranean greenhouse agrosystem" is characterized by the high priority in setting up integral quality management systems, which prove the quality of the product in relation to other production strategies. Hence, consumers are more and more faithful to labeled, quality certified products.

8 Environmental Aspects

Quantitative environmental assessments have become an important tool to prove that greenhouse production does not necessary involve a high environmental impact. An example of this is the Life Cycle Assessment (LCA) methodology, which was used by Muñoz *et al.*, (2008) to compare the environmental impacts of greenhouse versus open-field tomato production in the Mediterranean region determining that greenhouse production, if properly managed, has a smaller environmental impact than open-field crops in most of the evaluation categories. If we take into account that the greenhouse may reduce water consumption almost by half for a tomato crop, in relation to open filed, greenhouse proves to be a good alternative to open field in semi-arid regions. It is obvious that in some aspects the greenhouse has an impact, such as the visual impact in coastal touristic areas of the Mediterranean basin (for which the use of vegetal barriers around the greenhouses would be a good solutions) or the contribution to global warming (Muñoz *et al.*, 2008) due to the energy and gas emissions during the manufacture of the steel and the concrete (the search for alternative more environmentally friendly and equally resistant materials could be matter of interesting research).

Good agricultural practices (GAP) especially regarding irrigation and fertilization programmers (the use of nitrogen could be decreased in Almería greenhouses without effect on yield as stated by Thompson *et al.*, 2007 and Gallardo *et al.*, 2009) are highly important to reduce emissions. However, waste management by composting the plant biomass (or alternative methods as producing pellets or brickets for biomass heating or feeding animals, etc) and recycling of operational materials (especially substrates) is a must for future sustainable greenhouses. Plastic films are a good example of good waste management, as for instance, in the Almeria area of Spain, practically all the greenhouse plastic cover residues are recycled, as well as the crop residues.

The sustainability of the greenhouse produce in Spain is proved by recent studies, when compared to the Dutch industry, showing that primary fuel consumption for cultivation and transport purposes per kilogram of tomato, sweet pepper and cucumber is estimated to be 13, 14 ~ 17 and 9 times greater, respectively, in the Netherlands (Van der Velden *et al.*, 2004).

9 Greenhouse Production Strategies: State of the Art and Trends

The growers that want to meet the market demand of producing high quality products on a year round basis in the Mediterranean area face during the summer time the need to choose between two opposite strategies: A. -Investing on high technology greenhouses. B. - Growing in two, or more, different locations, whose harvesting periods are complementary, enabling a continuous and coordinated year-round supply to the markets (Castilla and Hernandez, 2007).

In the Mediterranean area, alternative A is in most cases (except for some growers with better marketing channels) not economical, in most cases, basically due to the competitive open-air and low-cost greenhouses produce from nearby areas, whose harvesting periods sometimes overlap. Producing in two different locations (alternative B), usually with different greenhouse technological levels, is becoming an increasingly adopted strategy. In a similar way, some large growers in Spain are shifting their high labor-demanding greenhouse crop production to developing countries, where labor costs are lower, and coordinate the marketing of the produce with other commodities grown in Spain.

Probable the best example of the general adoption of GAP is the massive implementation of integrated pest management (IPM) in greenhouse crops, at least in some crops like pepper, in which is a major step forward to guarantee the sustainability of greenhouse production (Van der Blom, 2009).

10 Socio-economical Aspects

The money spent on the operations of the greenhouse and also the quality achieved are quite dependent on the specific climate conditions and latitude (which should determine the design of the greenhouse) of the greenhouse location, which in its turn determines the cost of transportation and the "technological package" (Castilla and Hernandez, 2007), which includes both the greenhouse structure and internal equipment for climate control. Other aspects to take into account before selecting the greenhouse location are water and electricity supply, communications, labor availability, etc. which may affect to a great extent production costs how competitive the farm is.

In the passive greenhouses, the markets are forcing the growers to invest in technology to remain competitive, to maintain quality during the whole year, through better climate control. As stated by Castilla *et al.* (2004) for each situation, the grower must find a compromise between the higher costs associated to some of the equipments that need to be implemented in the greenhouses for good agricultural performances in order to produce commodities of good quality at competitive levels (Castilla *et al.*, 2004). Obviously, the transportation costs associated to distant export markets limit the competitiveness of the produce with production areas that are closer (i. e. Almería or Sicily and The Netherlands, in relation to Germany, the largest vegetable consumer in Europe).

11 Greenhouse Technological Packages

The passive greenhouse represents the lowest level of technology, in which no climate control equipment is used, and the fully equipped greenhouses are in the top level. In the Northern greenhouses, the top option is the only that ensures profits, but in the Mediterranean area, as stated by Castilla and Hernández (2007) and according to the local microclimate conditions and the chosen greenhouse production strategy, different "greenhouse technological packages" are available. Tables 1 and 2 summarize these packages for the Spanish and Dutch case, respectively.

Growing on low cost and no active climate control greenhouses (Table 1) always involves limiting high quality production during certain periods of the year, but the investment costs of this level in mild winter climate are around 10% of the investing cost of a standard glasshouse in the Netherlands. Some growers are rising the quality of their crops and extending their growing calendars by using a better equipped option, with air heating system (level 2, Table 1), as a first step. Further improvements require usually a special marketing channel to payback for the higher investments of a fully equipped plastichouse in the Mediterranean area (Table 1).

As stated by Castilla *et al.* (2004) the production costs are very dependent not just on the fixed costs (related with the greenhouse package), but also on the variable costs, very influenced by the yield performances and the costs of energy, specially due to heating.

Literature Cited

[1] Al-Helal, I. M and Abdel-Ghany, A. M. 2011. Measuring and evaluating solar radiative properties of plastic shading nets. Solar Energy Materials and Solar Cells, 95: 677 ~ 683

[2] Arkesteijn, M., Janse, J., and Kempkes, F. L. K. 2012. Met zo weinig energie zulke goede tomaten: Tomaten met 50% minder energie in Venlow Energy Kas (tomatoes with 50% less energy in the Venlow greenhouses). Onder Glas 9 (3): 32 ~ 33

[3] Baeza, E. J. 2007. Optimización del diseño de los sistemas de ventilación en invernadero tipo parral. Tesis Doctoral. Universidad de Almería. Escuela Politécnica Superior

[4] Esteban J. Baeza, Jerónimo J. Pérez-Parra, Juan I. Montero, Bernard J. Bailey, Juan C. López and Juan C. Gázquez. 2009. Analysis of the role of sidewall vents on buoyancy-driven natural ventilation in parral-type greenhouses with and without insect screens using computational fluid dynamics, Biosystems Engineering, 104 (1): 86 ~ 96

[5] Bailey, B. J., 1988. Energy conservation and renewable energies for greenhouse heating. FAO-REUR Technical Series, No. 3, pp. 17 ~ 39

[6] Baille A., López J. C., Bonachela S., González- Real M. M. and Montero J. I., 2006. Night energy balance in a heated low-cost plastic greenhouse. Agriculture Forest Meteorology, 137: 107 ~ 118

[7] Bakker, J. C., Adams, S. R., Boulard, T. and Montero, J. I. 2008. Innovative technologies for an efficient use of energy. Acta Horticulturae, 801: 49 ~ 62

[8] Bartzanas, T., Tchamitchian, M., and Kittas, C. 2005. Influence of the heating method on greenhouse microclimate and energy consumption. Biosystems Engineering, 91 (4): 487 ~ 499

[9] Brun, R and Lagier, J., 1985. A new greenhouse structure adapted to Mediterranean growing conditions. Acta Horticulturae, 170: 37 ~ 46

[10] Buchholz, M.; Buchholz, P.; Jochum, P.; Zaragoza, G. and Pérez-Parra, J. 2006. Temperature and humidity control in the Watergy greenhouse. Acta Horticulturae. 719: 401 ~ 407

[11] Cabrera, F. J., Baille, A., López, J. C. González-Real, and M. M. Pérez-Parra, J. 2009. Effects of cover diffusive properties on the components of greenhouse solar radiation. Biosystems Engineering, 103: 344 ~ 356

[12] Campen, J. B. 2008. Vapour Removal from the Greenhouse Using Forced Ventilation when Applying a Thermal Screen. Acta Horticulturae, 801: 863 ~ 868

[13] Castilla, N. 2002. Current situation and future prospects of protected crops in the Mediterranean region. Acta Horticulturae,

582：135～147

[14] Castilla，N. 2007. Invernaderos de plástico：Tecnología y manejo. 2ª Edición corregida y ampliada. Mundi-Prensa Libros. Madrid. 462 pp

[15] Castilla，N. 2009. Sistemas productivos en horticultura protegida. Conferencia invitada：XVIII Congreso Internacional de Plásticos en Agricultura（CIPA-2009）. Almería. España. pp. 23～25

[16] Castilla，N. and Hernández，J. 2005. The plastic greenhouse industry of Spain. Chronica Horticulturae，45（3）：15～20

[17] Castilla，N. and Hernandez，J. 2007. Greenhouse technological packages for high-quality crop production. Acta Horticulturae，761：285～297

[18] Castilla，N.，Hernandez，J.，and Abou-Hadid，A. F. 2004. Strategic crop and greenhouse management in mild winter climate areas. Acta Horticulturae，633：183～196

[19] Castilla，N.，and Montero，J. I. 2008. Environmental control and crop production in Mediterranean greenhouses. Acta Horticulturae，797：25～36

[20] Castilla，N.，Soriano，T.，and Hernández，J. 2010. Protected Cultivation. In："Current Topics in Agriculture". Gonzalez-Fontes，A.，Garate，A. and Bonilla I.（Eds.）. Studium Press，Houston，Texas，USA. 327～348

[21] Centraal Bureau voor de Statistiek，2009. Explosieve groei warmtekrachtvermogen in glastuinbouw zet door. Webmagazine，maandag 2 november 2009 9：30. http：//www. cbs. nl/nlNL/menu/themas/industrieenergie/publicaties/artikelen/archief/2009/2009-2947-wm. htm Comparison of climate and production in closed，semi-closed and open greenhouses

[22] Diaz，B. M.，and Fereres，A. 2007. Ultraviolet-Blocking Materials as a Physical Barrier to Control Insect Pests and Plant Pathogens in Protected Crops. Pest Technology 1（2）：85～95

[23] De Zwart，H. F.（2011）. Lessons learned from experiments with semi-closed greenhouses. Acta Horticulturae（GreenSys 2011，in press）.

[24] De Zwart，H. F.；Stanghellini，C.；Knaap，L. P. M. van der（2010）. Hoog isolerende en lichtdoorlatende schermconfiguraties. Wageningen UR Greenhouse Horticulture，Bleijswijk，Nota

[25] Dieleman，J. A.，Gelder，A. De，Qian，T.，Elings，A.，Marcelis，L. F. M. 2011. Crop physiology in semi-closed greenhouses. Acta Horticulturae（GreenSys 2011，in press）.

[26] Dueck，T. A.，Poudel，D.，Janse，J. and Hemming，S.，2009. Diffuus licht - wat is de optimale lichtverstrooiing? Wageningen UR Glastuinbouw，Rapport 308

[27] Dueck，T. J. Janse，J.，Kempkes，F. L. K.，Li，T.，Elings，A. and Hemming，S.，2012. Diffuus licht bij tomaat. Wageningen：Wageningen UR Greenhouse Horticulture，Bleiswijk，Report in press

[28] Gallardo，M.，Thompson，R. B.，Rodríguez，F. and Fernández，M. D. 2009. Simulation of transpiration，drainage，N uptake，nitrate leaching，and N uptake concentration in tomato grown in open substrate. Agricultural Water Management，96：1 773～1 784

[29] García，M. L.，Medrano，E.，Sánchez-Guerrero，M. C.，Lorenzo，P. 2011. Climatic effects of two cooling systems in greenhouses in the Mediterranean área：External mobile shading and fog system. Biosystems Engineering，108：133～143

[30] Gazquez，J. C.，J. C. Lopez，E. Baeza，M. Saez，M. C Sanchez-Guerrero，E. Medrano and P. Lorenzo. 2006. Yield response of a sweet pepper crop to different methods of greenhouse cooling. Acta Horticulturae，719：507～514

[31] Gonzalez] Real，M. M. and Baille，A.，2006. Plant response to greenhouse cooling. Acta Horticulturae，719：427～438

[32] Hemming，S.，MohammadkhaniV. and Dueck，T. 2008. Diffuse Greenhouse Covering Materials -Material Technology，Measurements and Evaluation of Optical Properties. Acta Horticulturae，797：469～476

[33] Hemming，S.；Kempkes，F. L. K.；Mohammadkhani，V. 2011. New glass coatings for high insulating greenhouses without light losses - energy saving crop production and economic potentials. Acta Horticulturae，893：217～226

[34] Heuvelink，E.，Gonzalez-Real，M. M. 2008. Innovation in plant-greenhouse interactions and crop management. Acta Horticulturae，801：63～74

[35] Heuvelink，E.；Bakker，M. J.；Marcelis，L. F. M.；Raaphorst，M. 2008. Climate and Yield in a closed greenhouse. Acta Horticulturae 801：1 083～1 092

[36] Kacira, M. , S. Sase, and L. Okushima. 2004. Effects of side vents and span numbers on wind-induced natural ventilation of a gothic multi-span greenhouse. JARQ 38 (4): 227 ~ 233

[37] La Malfa, G. and Leonardi, C. 2001. Crop practices and techniques: Trends and needs. International I. S. H. S. Symposium on protected cultivation in mild winter climates. Acta Horticulturae, 559: 31 ~ 42

[38] Li, Y. L. , C. Stanghellini and H. Challa, 2001. Effect of electrical conductivity and transpiration on production of greenhouse tomato. Scientia Horticulturae, 88: 11 ~ 29

[39] Li, Y. L. , L. F. M. Marcelis and C. Stanghellini, 2004. Plant water relations as affected by osmotic potential of the nutrient solution and potential transpiration in tomato (*Lycopersicon esculentum* L.). The Journal of Horticultural Science and Biotechnology, 79 (2): 211 ~ 218

[40] Magan, J. J. , Lopez, J. C. , Granados, R. , Perez-Parra, J. , Soriano, T. , Romero-Gamez, M. and Castilla, N. 2011. Global radiation differences under a glasshouse and a plastic greenhouse in Almeria (Spain): Preliminary report. Acta Horticulturae, 907: 125 ~ 130

[41] Meneses, J. F. and Castilla, N. 2009. Protected Cultivation in Iberian Horticulture. Chronica Horticulturae, 49 (4): 37 ~ 39

[42] Möller, M. , Cohen, S. , Pirkner, M. , Israeli, Y. and Tanny, J. 2010. Transmission of short-wave radiation by agricultural screens. Biosystems Engineering, 107: 317 ~ 327

[43] Möller, M. , Tanny, J. , Li, Y. and Cohen, S. , 2004. Measuring and predicting evapotranspiration in an insect-proof screenhouse. Agriculture Forest Meteorology, 127: 35 ~ 51

[44] Montero, J. I. 2006. Evaporative cooling in greenhouses: Effects on microclimate, water use efficiency and plant response. Acta Horticulturae, 719: 373 ~ 383

[45] Montero, J. I. ; Castilla, N. ; Gutiérrez de Ravé, E. ; Bretones, F. 1985. Climate under plastic in the Almería area. Acta Horticulturae, 170: 227 ~ 234

[46] Montero, J. I. ; Stanghellini, C. and Castilla, N. 2008. Invernadero para la producción sostenible en áreas de clima de invierno suaves, Horticultura internacional, 65: 12 ~ 26

[47] Muñoz, P. , Antón, A. , Paranjpe, A. , Ariño, J. and Montero, J. I. 2008. High decrease in nitrate leaching by lower N input without reducing greenhouse tomato yield. Agronomie and Sustainable Development, 28: 489 ~ 495

[48] Pérez, C. , López, J. C. , Gázquez, J. C. , Marín, A. y Bermúdez, M. S. , 2009. Experiencias con plásticos antiplagas en cultivos de tomate y sandía. VI Congreso Ibérico de Ciencias Hortícolas (SECH). Logroño. Acta de Horticultura, 54: 204 ~ 205

[49] Qian, T. ; Dieleman, J. A. ; Elings, A. ; Gelder, A. de; Marcelis, L. F. M. ; and Kooten, O. van 2011. Comparison of climate and production in closed, semi-closed and open greenhouses. Acta Horticulturae, 893: 807 ~ 814

[50] Thompson R. B. , Martínez-Gaitán C. , Gallardo M. , Jiménez C. and Fernández M. D. , 2007. Identification of irrigation and management practices that con- tribute to nitrate leaching loss from an intensive vegetable production system by use of a comprehensive survey. Agricultural Water Management, 89 (3): 261 ~ 274

[51] Sase, S. 2006. Air movement and climate uniformity in ventilated greenhouses. Acta Horticulturae, 719: 313 ~ 324

[52] Stapleton, J. J. 2000. Soil solarization in various agricultural production systems. Crop Protection, 19 : 837 ~ 841

[53] Stanghellini C. , Incrocci L. , Gázquez J. C. and Dimauro B. 2008. Carbon Dioxide Concentration in Mediterranean Greenhouses: How Much Lost Production? Acta Horticulturae, 801: 1 541 ~ 1 548

[54] Stanghellini, C. ; Dieleman, J. A. ; Driever, S. M. and Marcelis, L. F. M. 2011. Modeling the vertical profile of trasnpiration in greenhouse. Acta Horticulturae (GreenSys 2011, in press).

[55] Stanghellini, C. ; Jianfeng, D. and Kempkes, F. L. K. 2011. Effect of near-infrared-radiation reflective screen materials on ventilation requirement, crop transpiration and water use efficiency of a greenhouse rose crop. Biosystems Engineering 110 (3): 261 ~ 271

[56] Soler, A. , Parra, A. , Arevalo, A. B. , Pérez, C. and López, J. C. 2009: Efecto de los plásticos antiplagas en el éxito del control biológico en cultivos hortícolas intensivos del sureste español. Acta de XVIII Congreso Internacional de CIPA y XI Congreso CIDAPA. Almería

[57] Soriano, T., Hernández, J., Morales, M. I., Escobar, I. and Castilla, N. 2004. Radiation transmission differences in east-west oriented plastic greenhouses. Acta Horticulturae, 663: 91 ~ 97

[58] Sullivan, R. 2009. Corporate Responses to Climate Change: Achieving Emissions Reductions Through Regulation; Self Regulation and Economic Incentives. Sheffield, UK: Greenleaf

[59] Tanny, J. and Cohen, S. 2003. The effect of a small shade net on the properties of wind and selected boundary layer parameters above and inside a citrus orchard. Biosystems Engineering, 84: 57 ~ 67

[60] Tanny, J., Haijun, L. and Cohen, S. 2006. Airflow characteristics, energy balance and eddy covariance measurements in a banana screenhouse. Agricultural and Forest Meteorology, 139: 105 ~ 118

[61] Van der Blom, J., 2009. Control Biológico en cultivos en invernaderos de Almería. Seminario sobre control Biológico en Producción Agroalimentaria. Ministerio de Medio Ambiente y Medio Rural y Marino. Murcia. 6-8 de mayo

[62] Van der Velden, N. J. A., Jansen, J., Kaarsemaker, R. C., Maaswinkel, R. H. M.. 2004. Sustainability of greenhouse fruit vegetables: Spain versus the Netherlands; Development of a monitoring system. Acta Horticulturae, 655: 275 ~ 281

[63] Van Uffelen, R. L. M., van der Mass, A. A., Vermeulen, P. C. M. and Ammerlaan, J. C. J. 2000. T. Q. M. applied to the Dutch glasshouse industry: State of the art in 2000. Acta Horticulturae, 536: 679 ~ 686

[64] Vanthoor, B. H. E.; Stigter, J. D.; Henten, E. J. van; Stanghellini, C.; Visser, P. H. B. de; and Hemming, S. 2012. A methodology for model-based greenhouse design: Part 5, greenhouse design optimisation for southern-Spanish and Dutch conditions. Biosystems Engineering, 111 (4): 350 ~ 368

[65] Vermeulen, P. C. M. (2010). Kwantitatieve Informatie voor de Glastuinbouw 2010 : Kengetallen voor Groenten - Snijbloemen - Potplanten teelten. 21st edition. Wageningen : Wageningen UR Greenhouse Horticulture, Bleiswijk, Report 1037

Tables (1 ~ 2)

Table 1　Average investment costs in the south of Spain (in 2006) for three different levels of greenhouse technological packets (structure and equipment). Prices of the land and climate and fertigation computers are not included. Costs calculated for one ha size minimum area. Polyethylene (PE) film cladding is included (As stated by Castilla and Hernández, 2007)

			Euros m^{-2}
Level 1	Low roof slope (parral type) greenhouse		8. 5
	Drip irrigation system		2. 2
		Total	10. 7
Level 2	High slope (parral type) greenhouse		11. 3
	Drip irrigation system		2. 2
	Hot air heating system		3. 1
		Total	16. 6
Level 3	Arched multispan greenhouse		18. 0
	Drip irrigation system		2. 2
	Heating system (steel tubes)		13. 6
	Fans (air mixing)		1. 2
	Misting system (low pressure)		2. 0
	Shading-thermal screen		5. 0
		Total	42. 0

Source: Dr. I. Escobar (personal communication)

Simple soilless system (without drip) costs 1. 5 euros m^{-2}.

Greenhouses with motorised side and roof vents (simple system).

Table 2 Average investment costs in the Netherlands (in 2005) for two glasshouse technological packets (structure and equipment). Price of the land is not included. Costs calculated for one ha size minimum area. Standard Venlo includes structure, wide single glass, energy screen, rain and condensation water collection, water heating system (steel pipes), CO_2 injection, heat storage tank, drain water disinfection and soilless recirculation systems. (As stated by Castilla and Hernández, 2007).

			Euros m^{-2}
Standard	Standard Venlo type greenhouse and equipment		96.0
		Total	96.0
High Level	Standard Venlo type greenhouse and equipment		96.0
	Lighting system (600 W high pressure sodium lamps; one lamp per 12.5m^2)		22.2
		Total	118.2

Sources: Van Woerden (2005); E. Van Os (personal communication).

Figures (1)

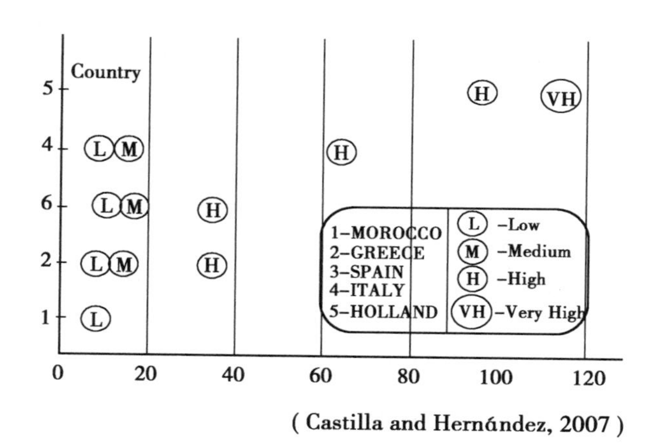

Figure 1 "Greenhouse technological packages" costs according to their level (Euros/m^2) (Castilla and Hernández, 2007)

New trends and innovation technologies for sustainable greenhouses

C. Kittas and N. Katsoulas[1], T. Bartzanas[2]

(1. *University of Thessaly, Sch. of Agr. Sc., Dept. of Agriculture Crop Prod. & Rural Env., Lab. of Agricultural Constructions and Environmental Control, Fytokou str*, 38446, *N. Ionia, Magnesia, Greece, ckittas@ uth. gr, n. katsoulas@ uth. gr*; 2. *National Center for Research and Technology, Institute of Research and Technology Thessaly*, 1st *Industrial Area*, 38500, *Volos, Greece, bartzanas@ cereteth. gr*)

Abstract: In recent years several factors (climate change, unpredictably weather conditions, water shortage, energy crisis and environmental pollution) lead to an increase of agriculture production under protected environments, such as greenhouses. Nowadays, the main challenge of greenhouse industry is its sustainability, i. e to increase the production efficiency and to reduce its environmental impact. Construction improvements and development of new greenhouse types and several techniques and methods, such as use of computers, sensors and models for climate and crop control, higher levels of CO_2 dosing etc. can increase the sustainability index of greenhouse industry. An overview of the recent scientific and technical advances on issues concerning the greenhouse microclimate, control and management is presented and discussed with focus on Mediterranean greenhouses. The main greenhouse climate control systems presented and discussed are the heating, dehumidification, ventilation, shading and cooling systems. These systems are the most widely used for greenhouse climate control especially in the Mediterranean area. The prospective and needs for future research and development in greenhouse climate control are presented in the conclusion.

Key words: ventilation; cooling; shading; heating; dehumidification

Main Challenges For Greenhouse Industry

Half of the European Union's land is farmed. This fact alone highlights the importance of farming for the EU's economy, employment, energy use and environment. The globalisation of markets has increased the competitiveness whereas consumers' needs for healthy, safe and locally produced products highlight the need for high quality production. According to Food Agriculture Organisation (FAO) in the next 20 years world food production must increase by 50% and 80% of this increase must come from intensification. In recent years several factors lead to an increase of agriculture production under controlled environments. Some of the reasons that lead to this continuous increase are: (i) extreme and unpredictable outside climate conditions as a result of climate change, (ii) water shortage, which is critical especially in Mediterranean countries (according to FAO, water use efficiency is 3 ~ 5 times higher in greenhouses compared to open field crops), (iii) environmental pollution & food security problems & (iv) ability of greenhouses to provide high-quality product all-year round.

Concerning agricultural production under controlled environments greenhouse cultivations constitute the most productive form of primary agricultural production. Greenhouses have been extremely successful in providing abundant, cheap and high-quality produce, by using resources (water, minerals, pesticides) with a very high economic efficiency. After Eastern Asia, Europe is the second largest region in the world with greenhouse cultivations (Figure 1). Greenhouses were steadily increased because they can provide high-quality product all-year round with an efficient use of resources, such as water, fertilisers, pesticides and handlabour. Moreover, the enhanced awareness of environmental pollution provoked by agriculture, the increasing demand of healthy foods and the shortage of resources like water and energy together with the new reformed CAP and the climate change, are forcing the growers to seek for alternative dynamic cultivations and to introduce more sustainable growing techniques. Today greenhouse industry in Europe is a leading industry in horticulture industry with more than 170.000 ha covered by greenhouses and 200.000 active SMEs which employ 700.000 workers and have a turnover of about 80 billion euro.

A report from FAO pointed out that "Greenhouse cultivations seem to be one of the most efficient mean of agricultural production, especially in terms of water usage. However in order to more widely used and adopted greenhouse cultivations should become more sustainable". Therefore, increase of production efficiency and reduction of environmental impact are the major challenges greenhouse industry is facing today. However, the problems and challenges for greenhouse sustainability are not the same in all countries in Europe. Different climate and market conditions in Europe have had a strong influence on the technological and economic development of protected horticulture. For example, with the mild winters in southern Europe, thin plastic film is used as the covering for practically all the greenhouse area, and most greenhouses are unheated and use simple technology due to the favourable climate conditions. In contrast, in northern cold-winter climates, the majority of greenhouses are covered with glass, and efficient heating and advanced technology is needed for high productivity. This can be better illustrated using the climograph of selected areas.

The climograph of some Mediterranean and North Europe regions is shown in Figure 2. It shows that in temperate climates e.g. in the Netherlands, heating and ventilation enables the temperature to be controlled over the whole year, however, at lower latitudes, e.g. Almeria, Ierapetra and Volos, the daytime temperatures are too high for ventilation to provide sufficient cooling during the summer. The attainment of suitable temperatures then requires positive cooling.

Heating and Energy

A severe bottleneck for sustainable energy production is the energy used, especially for heating during the winter period of the year, which for the northern regions exceeds six or even seven months. Only Dutch greenhouse (Bot, 2001) industry takes about 12.5% of the annual national gas consumption. In light of EU policies for sustainable reduction of greenhouses gases, according to Kyoto protocol, CO_2 emissions should also significantly be decreased in each industrial sector, including greenhouse industry. It is estimated that almost 3 kg of CO_2 were emitted in the atmosphere for every litre of oil burned for greenhouse heating. In a north country of Europe it is estimated that 50 m^3 of gas are used in order to heat 1 m^2 of a high equipped greenhouse. That means that the heating of one hectare of greenhouse results in the emission of about 1100 tons of CO_2. Consequently energy consumption has to be reduced in a sustainable way. To do so, greenhouse systems should be improved.

To decrease fossil fuels and natural gas consumption for greenhouse heating, several alternative techniques were developed. Fundamental investigations of effective methods of thermal energy storage have been significantly intensified since the 1973 - 1974 energy crisis, leading to the development of passive solar heating systems which can satisfy 30% ~60% of the annual heating requirements and achieve temperatures 3 ~ 10℃ higher than the minimum outdoor temperature (Santamouris et al. , 1994) . However these systems are still on improvement (Ozgener and Hepbasl, 2007) . Microwave heating systems, which are capable of heating plants in preference to the surrounding air, can save 45% of energy, were early developed by Teitel et al. (2000) as an effort for more energy saving. Another effort to the same direction realised nowadays with the alternative energy sources (Sethi and Sharma, 2007) and biofules (Wright, 2006) use for greenhouse heating and energy supply proposes.

During the last decade another significant contributions to energy saving, was made by the application of thermal screens, since about 40% saving in heat supply can be achieved in this way (Bot, 2001) and by using double cover materials for the greenhouse covering. Double cover materials reduce winter heat losses, but invariably cause a 65% reduction in light transmission (Critten, 1990), thereby reducing crop growth rate. Capital costs are also increased. Although progress during the last decade is limited mainly to experimental studies of the behaviour of individual sheets of cover materials, looking at a variety of properties

Ventilation, Shading and Cooling

Sufficient ventilation is very important for optimal plant growth, especially in the case of high outside temperatures and solar radiation, which are common conditions during summer in Mediterranean countries. In order to study the variables, determining the greenhouse air temperature and to decide about the necessary measurements for greenhouse air temperature control, a simplified version of the greenhouse energy balance is formulated. According to Kittas et al. (2005), the greenhouse energy balance can be simplified to:

$$Va = \frac{0.0003\tau Rs,o - \max}{\Delta T} \quad (1)$$

where:

V_a is the ratio Q/A_g, Q is the ventilation flow rate, in m^3 [air] s^{-1} and A_g is the greenhouse ground surface area, in m^2

τ is the greenhouse transmission coefficient to solar radiation

$R_{s,o\text{-}max}$ is the maximum outside solar radiation W m^{-2},

ΔT is the temperature difference between greenhouse and outside air, in ? C.

Using eqn. (1), it is easy to calculate the ventilation needs for several values of $R_{s,o\text{-}max}$ and

ΔT. Figure 3 presents the variation of V_a for several values of $R_{s,o\text{-}max}$, τ and ΔT.

From Figure 3 it can be seen that for the area of Magnesia, Greece, where during the critical summer period, values of outside solar radiation exceed the value of 900 W m^{-2} (Kittas et al., 2005), a ventilation rate of about 0.06 m^3 s^{-1} m^{-2} (which corresponds, for a greenhouse with a mean height of 3 m, to an air- exchange of 60 h^{-1}) is needed in order to maintain a ΔT of about 4℃.

The necessary ventilation rate can be obtained by natural or by forced ventilation. For effective ventilation, ventilators should, if possible, be located at the ridge, on the side walls and the gable. Total ventilator area equivalent to 15 - 30% of floor area. Above 30%, the effect of additional ventilation area on the temperature difference was very small.

Systems like exhaust fan; blower, etc. can supply high air exchange rates whenever needed. These are simple and robust systems and significantly increase the air transfer rate from the greenhouse and allow maintaining inside temperature to a level slightly higher than the outside temperature by increasing the number of air changes (Figure 3).

The principle of forced ventilation is to create an air flow through the house. Fans suck air out on the one side, and openings on the other side let air in. Forced ventilation by fans is the most effective way to ventilate a greenhouse, but consumes electricity. It is estimated that the needs for electrical energy for greenhouse ventilation, for a greenhouse located in the Mediterranean are about 100000 kWh per greenhouse ha.

Many researchers also studied the effects of insect-proof screens in roof openings on greenhouse microclimate. However, the obstruction offered by fine mesh screens to flow through the openings results in air velocity reduction (Figure 4) and higher temperature and humidity as well as increased of thermal gradients within the greenhouse (Bartzanas et al., 2005).

The entry of unwanted radiation (or light) can be controlled by the use of shading or reflection (Kittas et al., 1999). Shading can be done by various methods such as by the use of paints, external shade cloths, use of nets (of various colors), partially reflective shade screens and water film over the roof and liquid foams between the greenhouse walls. Shading is the ultimate solution to be used for cooling greenhouses, because it affects the productivity. However, in some cases, a better quality can be obtained from shading. One of the most used methods adopted by growers due to the low cost is white painting, or whitening, the cover material. The use of screens has progressively been accepted by growers and has gained, through the last decade, a renewed interest as shown by the increasing area of field crops cultivated under screenhouses (Cohen et al., 2005), while roof whitening, due to its low cost, is a current practice in the Mediterranean Basin.

Baille et al. (2001) reported that whitening applied onto a glass material enhanced slightly the PAR proportion of the incoming solar irradiance, thus reducing the solar infrared fraction entering the greenhouse. This characteristic of whitening could represent an advantage with respect to other shading devices, especially in warm countries with high radiation load during summer. Another advantage of whitening is that it does not affect the greenhouse ventilation, while internal shading nets affect negatively the performance of the roof ventilation. Whitening also significantly increases the fraction of diffuse irradiance, which is known to enhance the radiation use efficiency.

One of the most efficient solutions for alleviating the climatic conditions is to use evaporative cooling systems (Giacomelli et al., 1985; Montero, 2006; Kittas et al., 2001) based on the conversion of sensible heat into latent heat by means of evaporation of water supplied directly into the greenhouse atmosphere (mist or fog system, sprinklers) or through evaporative pads (wet pads). Evaporative cooling allows simultaneous lowering of temperature and vapour pressure deficit and its efficiency is higher in dry environments. The efficiency of fog systems is often limited by insufficient natural air convection, in the absence of wind, and by the risk of wetting the plants when water droplet evaporation is not complete. The main disadvantage of fan and pad cooling systems is the creation of large temperature gradients inside the greenhouse, from pads on one side to extracting fans on the opposite side (Figure 5. Bartzanas et al., 2012). In order to overcome these large thermal gradients from pads to fans, growers usually combine pad and fan evaporative cooling systems with partial shading.

Several studies were concerned with the effects of fog cooling on greenhouse microclimate (e. g. Baille et al., 1994; Ishii et al., 2006), the fog cooling efficiency of fog systems (e. g. Abdel-Ghany and Kozai, 2006; Perdigones et al., 2008) or the impact of fog cooling on crop performance (e. g. Baille et al., 2006; Katsoulas et al., 2007). Many studies have already shown that reducing transpiration by modifying the microclimate (e. g., reducing vapour pressure deficit and incoming solar radiation, direct wetting of the leaves) inside the greenhouse improves the physiological adaptation of plants to stress conditions such as salinity or unfavourable external climatic conditions (Katsoulas et al. 2001; 2002). However, undesired effects sometimes arise from the response of crops to cooling, such as the appearance of blossom-end rot (Meca et al., 2006), which is often associated with the reduction in air vapour pressure deficit (Lorenzo et al., 2004).

According to Gates et al. (1991) and Arbel et al. (2003), increased cooling efficiency in relation to water consumption can be expected if fogging is combined with low ventilation rate. Li et al. (2006) concluded that fog cooling efficiency increases with spraying rate and decreases with ventilation rate. Sase et al. (2006) concluded that control of the fog system by temperature set point (on/off control) and of vent openings by relative humidity set point (proportional control) resulted in 21% fog water saving compared to no control of vent openings (maximum opening). Perdigones et al. (2008) found that pulse width modulation strategy applied to fog system control reduced fog water consumption by 8% ~15% compared to a simple on/off control with fixed fog cycles. Finally, Toida et al. (2006) suggested incorporating small fans to fogging nozzles to increase (about 1.5 times) evaporation ratio of sprayed water and cooling system efficiency. The above reports indicate that fog system efficiency could easily increase, resulting in reduced greenhouse water consumption.

Dehumidification

Humidity is one of the key factors in greenhouse climate. It usually tends to be high due to crop transpiration. Energy saving measures in greenhouses can result in high levels of humidity which can lead to yield loss and have detrimental effects on product quality. Without humidity control, high relative humidity levels may lead to loss of crop quality due to fungal diseases, leaf necrosis, calcium deficiencies and soft and thin leaves.

Condensation refers to the formation of liquid drops of water from water vapor. Condensation occurs when warm, moist air in a greenhouse comes in contact with a cold surface such as glass, fiberglass, plastic or structural members. The air in contact with the cold surface is cooled to the temperature of the surface. If the surface temperature is below the dew point temperature of the air, the water vapor in the air will condense onto the surface. Condensation will form heaviest in greenhouses during the period from sundown to several hours after sunrise. During the daylight hours, there is sufficient heating in the greenhouse from solar radiation to minimize or eliminate condensation from occurring except on very cold, cloudy days. The time when greenhouses are most likely to experience heavy condensation is sunrise or shortly before. Condensation is a symptom of high humidity and can lead to significant problems, including germination of fungal pathogen spores, including Botrytis and powdery mildew. Condensation can be a major problem and it is unfortunately one which, at least at certain times of the year, is almost impossible to avoid entirely.

There are several ways and technologies for reducing the high humidity levels in greenhouse and the most used ones are:

Combined used of heating and ventilation systems

By opening the windows, moist greenhouse air is replaced by relatively dry outside air. This is common practice to dehumidify a greenhouse. This method does not consume any energy when excess heat is available in the greenhouse and ventilation is needed to reduce the greenhouse temperature. Though, when the need for ventilation to reduce the temperature is less than the ventilation needed to remove moisture from the air, dehumidification consumes energy. The warm greenhouse air is replaced by cold dry outside air, lowering the temperature in the greenhouse. The type of ventilators is critical in these systems since it governs the whole process (Kittas and Bartzanas, 2007, Fig 6).

Absorption using a hygroscopic material

The research on the application of hygroscopic dehumidification in greenhouses is minimal because the installation is too complex and the use of chemicals is not favourable in greenhouses. During the process, moist greenhouse air is in contact with the hygroscopic material releasing the latent heat of vaporisation as water vapour is absorbed. The hygroscopic material has to be regenerated at a higher temperature level. A maximum of 90% of the energy supplied to the material for regeneration can be returned to the greenhouse air with a sophisticated system involving several heat exchange processes including condensation of the vapour produced in the regeneration process.

Condensation on cold surfaces

With this system wet humid air is forced to a clod surface which is located inside the greenhouse (Figure 7), different than the covering material. In the cold surface condensation occurs, the condensate water is collected and can be re-used and absolute humidity of the wet greenhouse air is reduced.

One metre of finned pipe used in their study at a temperature of 5°C can remove 54 g of vapour per hour from air at a temperature of 20°C and 80% relative humidity.

Forced ventilation usually with the combined use of a heat exchanger

Mechanical ventilation is applied to exchange dry outside air with moist greenhouse air, exchanging heat between the two airflows.

Based on the results of Campen et al. (2003) a ventilator capacity of 0.01 $m^3 s^{-1}$ is sufficient for all crops. The energy needed to operate the ventilators is not considered because the experimental study by Speetjens (2001) showed that the energy consumption by the ventilators is less than 1% of the energy saved.

Anti-drop covering materials

The use of anti-drop covering materials is an alternative technology for greenhouse dehumidification. The "anti-dripping" films that contain special additives which eliminate droplets and form instead a continuous thin layer of water running down the sides. The search for anti-drip cover materials has been mainly focused on the optical properties of the cover materials.

Prospective and Needs for Future Research

Modern on-demand agricultural production requires a high degree of control of the whole process with respect to timing, quantity and quality of production. Greenhouse horticulture is the sector where such control is possible, in view of the even more high tech means for environmental management, the use of innovative techniques and the increasing application of ICT. Full advantage, of these tools, however, can be gained only by improving, on present knowledge about plant processes and its environment (aerial and root) . However it is obvious that many proposed methods or practices can give a solution to a problem but in the same time they rise another one. This is because they faced the problems independently and for this reason the used techniques and methods can not be any more efficient. However there is the consideration that more energy efficient and accordingly grater profit may be achieved, simply by globalize the problems in the greenhouse, considering that greenhouse environment, crop, energy inputs (pesticides, fuel, water fertilizers, work), quality of product, profit of producer and effluents on the environment, constitute parameters in the same system which is susceptible of one optimal solution. This means that the investment in new technology should be also accompanied from a strategy of management that would guarantee the optimal use of equipment. Under this consumption technology can play a more essential role in using energy more rationally. This project aims to the development of a globalize system for the management of greenhouse ecosystem for sustainable production of high quality products, using, combining and improving all the gathered knowledge concerning greenhouse industry.

References

[1] Abdel-Ghany, A. M., and T. Kozai. 2006. Cooling efficiency of fogging systems for greenhouses. Biosyst. Eng., 94 (1):

97 ~ 109

[2] Arbel A; Yekutieli O; Barak M 1999. Performance of a fog system for cooling greenhouses. Journal of Agricultural Engineering Research, 72, 129 ~ 136

[3] Bartzanas, T., T. Boulard, C. Kittas, 2002. Numerical simulation of airflow and temperature patterns in a greenhouse equipped with insect-proof screen. Computers and Electronics in Agriculture, 34, 207 ~ 221

[4] Bartzanas, T., N. Katsoulas, C. Kittas, T. Boulard, M. Mermier, 2005. Effect of Vent Configuration and Insect Screen on Greenhouse Microclimate. International Journal of Ventilation, 4 (3), 193 ~ 202

[5] Bartzanas, T., D. Fidaros, C. Baxevanou, C. Kittas, 2012. Climate distribution in a fan and pad evaporative cooled greenhouse: A CFD approach, AgEng-CIGR2012, Valencia, Spain

[6] Baille, A., Leonardi, C., 2001. Influence of misting on temperature and heat storage of greenhouse grown tomato fruits during summer conditions. Acta Hort., 559, 271 ~ 278

[7] Baille, A., M. M. Gonzalez-Real, J. C. Gazquez, J. C. Lopez, J. J. Perez-Parra, and E. Rodriguez. 2006. Effects of different cooling strategies on the transpiration rate and conductance of greenhouse sweet pepper crops. Acta Hort., 719: 463 ~ 470

[8] Baille, M., Baille, A., Delmon, D., 1994. Microclimate and transpiration of greenhouse rose crops. Agric. For. Meteorol., 71, 83 ~ 97

[9] Bakker, J. C. 1991. Analysis of humidity effects on growth and production of glasshouse fruit vegetables. Ph. D. thesis, Wageningen Agricultural University, Wageningen, p. 155

[10] Bot, G. P. A. 2001. Developments in indoor sustainable plant production with emphasis on energy saving Computers and Electronics in Agriculture, 30: 151 ~ 165

[11] Campen, J. B., Bot, G. P. A., de Zwart, H. F. 2003. Dehumidification of Greenhouses at Northern Latitudes. Biosystems Engineering, 86 (4): 487 ~ 493

[12] Cohen S, Raveh E, Li Y, Grava A, Goldschmidh EE (2005) Physiological response of leaves, tree growth and fruit yield of grapefrui trees under reflective shading screens. Science Horticulturae, 107, 15 ~ 35

[13] Critten, D. L. 1990. Simplified Model of Light Loss due to Structural Members in Multispan Greenhouse Roofs. J. Agric. Engng Res., 47, 197 ~ 205

[14] Ishii, M., Sase, S., Moriyama, H., Kubota, C., Kurata, K., Hayashi, M., Ikeguchi, A., Sabeh, N., Romero, P., Giacomelli, G. A., 2006. The effect of evaporative fog cooling in a naturally ventilated greenhouse on air and leaf temperature, relative humidity and water use in a semiarid climate. Acta Hort., 719, 491 ~ 498

[15] Katsoulas, N., A. Baille, and C. Kittas. 2001. Effect of misting on transpiration and conductances of a greenhouse rose canopy. Agric. Forest Meteorol., 106 (3): 233 ~ 247

[16] Katsoulas, N., Baille, A., Kittas, C., 2002. Influence of leaf area index on canopy energy partitioning and greenhouse cooling requirements. Biosyst. Engin., 83, 349 ~ 359

[17] Katsoulas, N., Kittas, C., Tsirogiannis, I. L., Kitta, E., Savvas D., 2007. Greenhouse microclimate and soilless pepper crop production and quality as affected by a fog evaporative cooling system. Transactions of the ASABE, 50 (5): 1 831 ~ 1 840

[18] Kacira M. 2011. Greenhouse Production in US: Status, Challenges, and Opportunities. Presented at CIGR 2011 conference on Sustainable Bioproduction WEF 2011, September 19 ~ 23, 2011 Tower Hall Funabori, Tokyo, Japan

[19] Kittas C; Bartzanas T; Jaffrin A 2001. Greenhouse evaporative cooling: measurement and data analysis. Transactions of the America Society of Agricultural Engineering, 44, 683 ~ 689

[20] Kittas, C., Baille, A., Giaglaras, P., 1999. Influence of covering material and shading on the spectral distribution of light in greenhouses. J. Agric. Engng. Res., 73, 341 ~ 351

[21] Kittas, C., Bartzanas, T., 2007. Greenhouse Microclimate and Dehumidification Effectiveness under Different Ventilators Configuration. Energy and Buildings: 42 (10): 3 774 ~ 3 784

[22] Li, S., Willits, D. H., Yunker, C. A., 2006. Experimental study of a high pressure fogging system in naturally ventilated

greenhouses. Acta Hort. , 719, 393 ~ 400

[23] Lorenzo, P. , Sanchez-Guerrero, M. C. , Medrano, E. , Garcia, M. L. , Caparros, I. , Coelho, G. , Gimenez. M. , 2004. Climate control in the summer season: a comparative study of external mobile shading and fog system. Acta Hort. , 659, 189 ~ 194

[24] Meca, D. , Lopez, J. C. , Gazquez, J. C. , Baeza, E. , Pérez-Parra, J. , Zaragoza, G. , 2006. Evaluation of two cooling systems in parral type greenhouses with pepper crops: Low-pressure fog system verses whitening. Acta Hort. , 719, 515 ~ 519

[25] Montero, J. I. , 2006. Evaporative Cooling in Greenhouses: Effects on Microclimate, Water Use Efficiency and Plant Response. Acta Hort. , 719: 373 ~ 383

[26] Ozgener, O. and Hepbasl, A, 2007. A parametrical study on the energetic and exergetic assessment of a solar-assisted vertical ground-source heat pump system used for heating a greenhouse. Building and Environment, 42, 11 ~ 24

[27] Perdigones, A. , Garcia J. L. , Romero, A. , Rodr? guez, A. , Luna, L. , Raposo, C. , de la Plaza S. , 2008. Cooling strategies for greenhouses in summer: Control of fogging by pulse width modulation. Biosyst. Engin. , 99, 573 ~ 586

[28] Sase, S. , Ishii, M. , Moriyama, H. , Kubota, C. , Kurata, K. , Hayashi, M. , Sabeh, N. , Romero, P. , Giacomelli, G. A. , 2006. Effect of natural ventilation rate on relative humidity and water use for fog cooling in a semiarid greenhouse. Acta Hort. , 719, 385 ~ 392

[29] Santamouris, M. , Balaras, C. A. , Daskalaki, E and Vallindras M. 1994. Passive solar agricultural greenhouses: A worldwide classification and evaluation of technologies and systems used for heating purposes. Solar Energy, 53, (5), 411 ~ 426

[30] Speetjens, S. L. 2001. Warmteterugwinning uit ventilatielucht. [Heat recovery from ventilation.] Report Nota V 2001-86, IMAG, Wageningen, The Netherlands

[31] Teitel, M. , Dikhtyer, V. , Elad, Y. , Shklyar, A. , Jerby, E. , 2000. Development of a microwave system for greenhouse heating. Acta Horticulturae, 534, 189 ~ 195

[32] Toida, H. , Kozai, T. , Ohyama, K. , Handarto, 2006. Enhancing fog evaporation rate using an upward air stream to improve greenhouse cooling performance. Biosyst. Engin. , 93, 205 ~ 211

[33] Wright, L. , 2006. Worldwide commercial development of bioenergy with a focus on energy crop-based projects. Biomass and Bioenergy, 30, 706 ~ 714

Figures

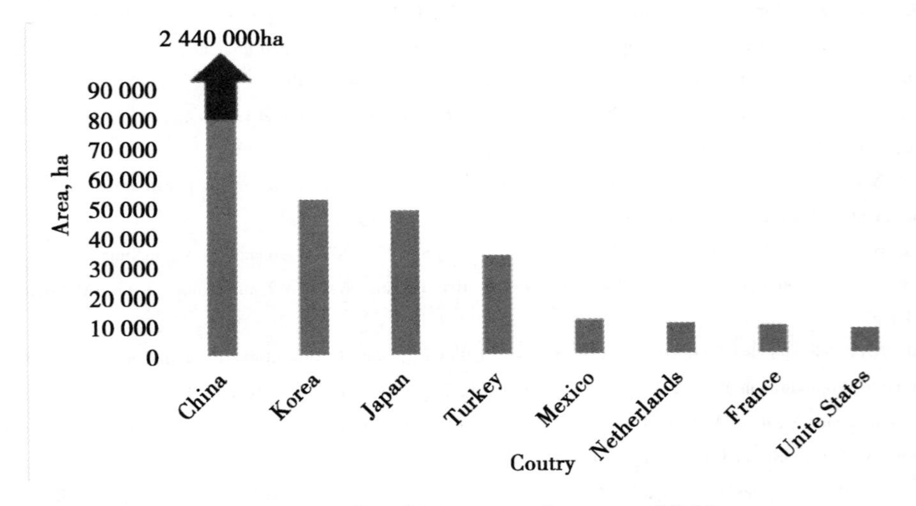

Figure 1　Greenhouse covered areas worldwide

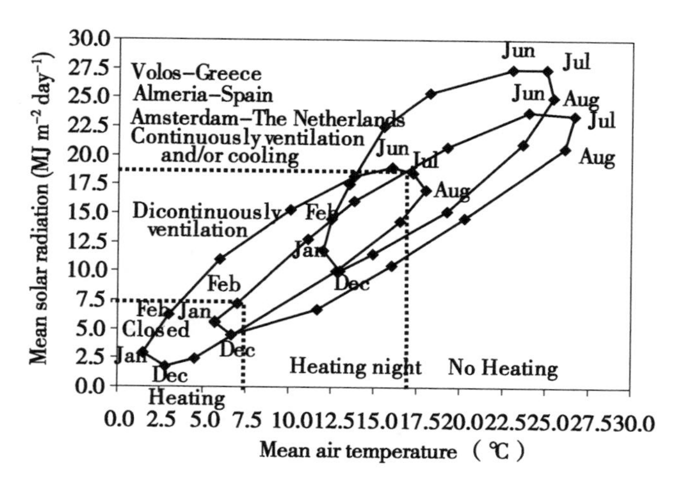

Figure 2 Mean solar radiation vs. mean air temperature for several locations around Europe: The climograph. Dotted lines indicate limits for different control actions

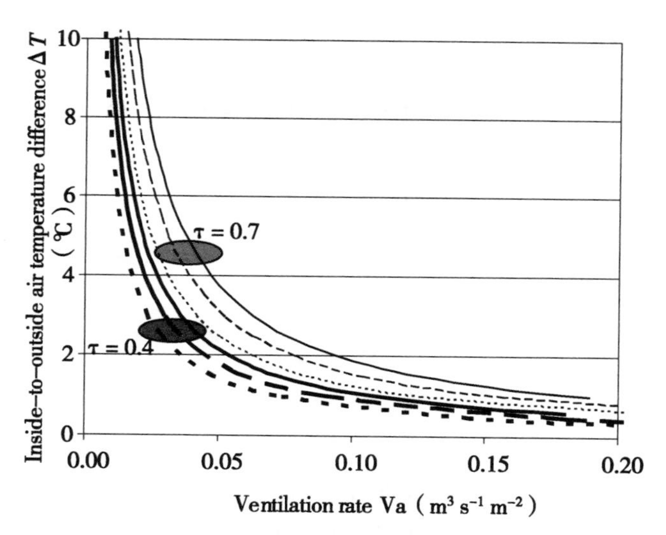

Figure 3 Inside-to-outside greenhouse air temperature difference ΔT as a function of ventilation rate V_a as calculated using eqn. (3), for a greenhouse with a regularly transpiring rose crop. Thin lines correspond to a greenhouse cover transmission to solar radiation τ of 0.7. Heavy lines correspond to a τ of 0.4. (—): $R_{s,o}$ = 900 W m^{-2}, (— —): $R_{s,o}$ = 750 W m^{-2}, (- - -): $R_{s,o}$ = 600 W m^{-2}. (Kittas et al., 2005

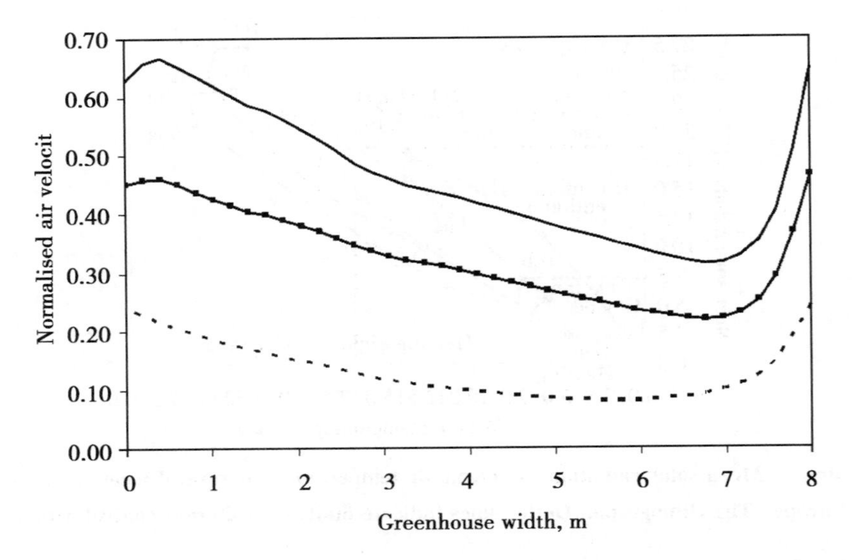

Figure 4 Normalised air velocity along the greenhouse width at a height of 1. 1m above ground, for a tunnel greenhouse with side vents without screen (——), with anti-aphid insect screen (— — —) and with anti-bemisia insect screen (—■—■—)

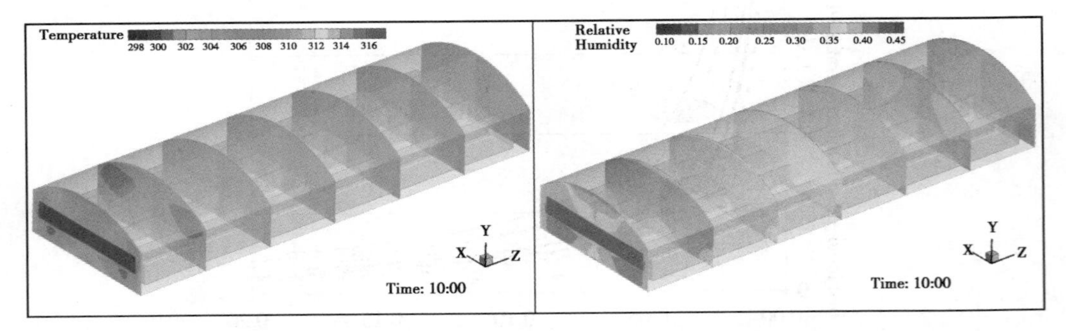

Figure 5 Traverse sections of the temperature (left) and relative humidity (right) contours in a fan and pad evaporative cooled greenhouse (Bartzanas et al. 2012)

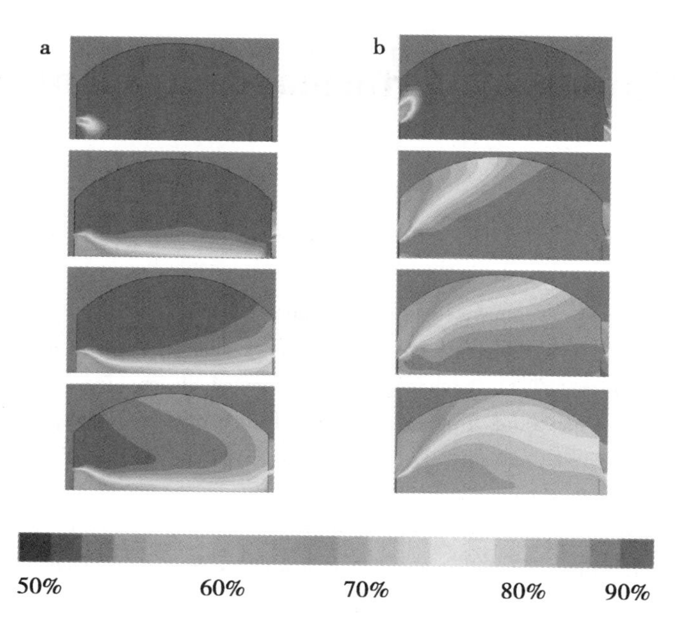

Figure 6 Unsteady distribution of air relative humidity in an arc type tunnel greenhouse during the dehumidification process with simultaneously heating and ventilation, (a) roll-up type vents and (b) pivoting door type vents

Figure 7 Photograph of the greenhouse where finned tubes are placed under the gutter and an enlargement of a finned tube with condensation on the surface (Campen & Bot, 2002)

Quantification of Performance in Plant Factory

Fang Wei

(*Dept. of Bio-Industrial Mechatronics Engineering National Taiwan University*)

Abstract: A quantitative method was introduced for the evaluation of overall performance of lamps and plants in plant factory using artificial light only (PF_ AL). A method to calculate the overall total light integral per plant (oTLI) and total power integral per plant (oTPI) throughout the overall growth stages (from seeding to harvest) were developed. With the given average harvesting fresh weight under specific oTLI and oTPI, quantum yield (QY) and power yields (PY) can be derived. Converting factors of oTLI vs. oTPI and QY vs. PY, in term of mole per Joule, represent the converting efficiency of light sources. With the price of the light sourceavailable, the cost-performance value of lampscan be derived for comparison. The method developed can also be used in comparing the profitability of various types of plants grown in PF_ AL. Some examples on how to use methods developed were introduced using the hardware and settings of PF_ AL in operation in Taiwan.

Key words: Plant Factory; Quantification of Performance; Quantum Yield; Power Yield

植物工廠效能評估之量化

方　煒*

（生物產業機電工程學系，台灣大學生物資源暨農學院）

摘要：本研究建立植物工廠效能的量化評估方法，可用來評估栽培全程（由播種到收穫）的燈具利用與作物成長效能。首先計算栽培全程單株的光子數與能耗，分別以 oTLI 與 oTPI 表示。當收穫時的鮮重為已知，可以計算光子產能（QY）與電能產能（PY），兩者的比例可做為評估燈具效率的指標。以上方法還可以用來評估在相同狀況下栽培不同作物的結果，可據此選出在該條件下最具生長效率的作物。當燈具成本為已知，也可以計算性價比以利於在建廠時評估選用何種燈具。本文最後使用台灣順利運作中兩家植物工廠的數據，來解說如何使用本研究所建立的方法來計算上數關鍵指標以評估植物工廠的效能。

關鍵詞：植物工廠；效能評估；量化；光子產能；電能產能

1 Introduction

Plant factory (PF) / Vertical farming gained its attention recently due to several 'GloNaCal' reasons such as food safety, food security, resource (oil and fresh water) shortage, global warming, extreme weather, and steady supply. GloNaCal is a new term describing issues related not only locally, but nationwide as well as globally. In spite of its high investment cost and operating cost, features of PF such as capable of controlling quantity, quality of crops and risk and time required to grow, attract people from all sectors.

* 作者联系方式：E-mail：weifang@ ntu. edu. tw

The growth of crops in PF is normally separated into several stages with various growing density in order to make better use of resources. In between stages, the transplanting can be done manually or automatically. The environmental settings and cultural practices in different stage might be different to suit different purposes such as maximize the growth, minimize the cost, enhance the quality, etc. Those settings can be categorized into 4 different aspects: ①Air: temperature, humidity, wind velocity, and CO_2 content. ②Crop: variety of crop, duration and growing density. ③Light: intensity, spectral quality, and duration. ④Solution: pH, EC, DO, and temperature. The performance of growth is the integrated outcome of above mentioned governing factors.

Many researches were conducted to investigate on various combination of above parameters aiming at different targets. However, lack of systematic means to describe various settings in various stages limit the spread of valuable research outcome. Also make it difficult to duplicate the experiment and to exchange the experience.

Environmental factors of crops grown in open fields and/or greenhouse are time and site specific. That is why phytotron and growth chamber (Ph & GC) were developed for research purposes. 'PF with artificial light only' (PF_ AL) is the facility similar with Ph&GC just that it is not for research but for mass production of quality crops. Due to the capability of controlling most of the parameters mentioned above, a systematic mean to describe various settings in various stages will be beneficial for knowledge exchange as well as for the finalize of a protocol for the production of target crop.

In PF_ AL, cost on electricity occupied more than 30% in operating costs and the cost on lighting equipment (light source, power supply, reflector, etc.) also occupied a great deal in total fixed costs. In general, artificial light source, air conditioner and pump are three major equipment required electricity in a PF_ AL. When the COP of the air conditioner nears 3, the electricity consumption of light source is roughly 70% ~75% of the total. How to reach high efficiency in using all kind of resource is of great concern in the research of PF_ AL (Kozai, 2011).

There are various types of artificial light such as fluorescent lamp, LED, CCFL, HEFL, etc. Same type of lamp from different sources might have various spectral qualities. Means to evaluate the overall performance of lamps in term of lighting efficiency and quantum / power yield are valuable for the selection of proper light source and crop.

The concept of LUE_d (Light energy Use Efficiency) and PUE_p (Photosynthetic photon Use Efficiency) are not new. LUE and PUE are calculated based on the dry weight of the biomass produced and the photosynthetic rate of the plant, respectively. Goto (2011) suggested that by adding one mole of photons lead to an increase of 0.4 g of dry mass for lettuce and the chemical energy of one unit of dry mass is around $20 kJ \cdot g^{-1}$. Due to the average energy of 1mole of photons (visible light) is about 250 kJ, the LUE_d is around $0.4 \times 20/250$ that is 0.032 (Kozai, 2012; Fang, 2012). The approach is purely theoretical and if the water content of the plant and energy converting efficiency of the light source/power supply are unknown, one can't link it to the real world situation: revenue: fresh weight information and cost: kW. hr consumed. The PUE approach only deal with the instance of measurement not for the entire growth period.

The method introduced in this study deals with not only the complete growth stages but also use the fresh weight of biomass Instead of dry weight as the base for calculation. It is more practical because no PF_ AL in the world estimate their production and their revenue use the dry weight data. Also, when the cost in using the artificial light is of interests, the power consumed in term of kW. hr (degree) is of great concern. One mole of photons (visible light) is about 250 kJ is important to know. However, it take how many degree (kW · h = 3 600kJ) of power to provide this 250 kJ is a practical question that need to be answered.

Methods

Daily light integral (DLI, in mol · $m^{-2}day^{-1}$) of various stages can be calculated using values of PPF (in μmol · $m^{-2}s^{-1}$) and L (duration of light period per day, in hr · day^{-1}) as shown in eq. 1a. Total light integral of each stage (i, from 1 to 3 or more), denoted TLI (i) (in mol · m^{-2}), can be calculated as shown in eq. 2a, where D (i) is the duration of days in stage i. The summation of TLI of all stages is the overall TLI (OTLI, in mol · m^{-2}) as shown in eq. 3a.

$$DLI(i) = PPF(i) \times L(i) \times 3.6/1\,000 \quad (1a)$$

$$TLI(i) = DLI(i) \times D(i) \quad (2a)$$

$$OTLI = TLI(i) \quad (3a)$$

DLI, TLI and OTLI are area specific parameters. They are values associate with unit area, i. e. square meter. However, in different stages, crop grown in different density, denoted as d (i), in term of number of plants per square meter (plt · m^{-2}). Even in the same stage, different crop might have different density as well. The OTLI value shown in eq. 3a is somehow misleading due to different number of plants among stages. The term oTLI (in mol · plt^{-1}) was defined as shown in eq. 4a converting the per area value into per plant value.

$$oTLI = TLI(i)/d(i) \quad (4a)$$

Similar approach can be used to calculate power consumption throughout entire growth stages both per area basis (eq. 1b ~ 3b) and per plant basis (eq. 4b). As shown in eq. 1b, W (i) is the power consumption (in Watts) of artificial light source per unit growth area and A (i) is the value of unit growth area in square meters. Unit of DPI (daily power integral) is deg · $m^{-2}day^{-1}$. Unit for TPI and OTPI is deg · m^{-2}. Unit for oTPI is deg · plt^{-1}.

$$DPI(i) = W(i) \times L(i) / (A(i) \times 1\,000) \quad (1b)$$

$$TPI(i) = DPI(i) \times D(i) \quad (2b)$$

$$OTPI = TPI(i) \quad (3b)$$

$$oTPI = TPI(i)/d(i) \quad (4b)$$

In some cases, the calculation of above values based on per unit area is not preferred, but instead based on per growth unit such as per layer or per bench is more straight forward. Thus, equations 1b to 4b can be modified as shown below, where W' (i) is the stage-wise total watts per growth unit and d' (i) is the stage-wise no. of plants grown per growth unit. The values derived in eq. 4b and 4c should be the same.

$$DPI'(i) = W'(i) \times L(i)/1\,000 \tag{1c}$$

$$TPI'(i) = DPI'(i) \times D(i) \tag{2c}$$

$$OTPI' = TPI'(i) \tag{3c}$$

$$oTPI' = TPI'(i)/d'(i) \tag{4c}$$

Quantum-yield (QY, in $g \cdot mol^{-1}$) and power-yield (PY, in $g \cdot deg.^{-1}$) in term of fresh weight (FW, in $g \cdot plt^{-1}$) per oTLI (in $mol \cdot plt^{-1}$) and per oTPI (in $deg \cdot plt^{-1}$) was defined, as shown in eq. 5a and 5b, respectively. Both terms can be used to describe the performance of crop in converting light energy to biomass under given conditions.

$$QY = FW/oTLI \tag{5a}$$

$$PY = FW/oTPI \tag{5b}$$

Equation 5 provides a straightforward method in finding the best treatmentamong complicated situations when fresh weight of produces is of great concern. Some applications are:

1.1 Values of converting factor among oTLI and oTPI reveal effectiveness of light source converting energy to quantum. The best light source most suitable to grow specific crop has the highest value among treatments.

1.2 Outcomes of QY (PY) for same crop under various light sources with same PPF (W) treatments can be used to identify the best light source (same type with various spectra or various types) for the specific crop.

1.3 Outcomes of QY (PY) for same crop under same light sources with various PPF (W) treatments can be used to identify the best light intensity for the specific crop and light source.

1.4 Outcomes of QY (PY) for various crops under same conditions can be used to identify the best crop for the given condition.

1.5 Outcomes of PY can be converted to dollar basis when utility fee (UF) per degree is given. Thus, the electricity cost of providing light per g of fresh weight and how much fresh weight can be produced per unit cost can be determined.

2 Case examples

Growth conditions of cases in different PFs of Taiwan were used as examples to introduce the steps in calculating the QY and PY. As shown in Tables 1 and 2 are steps in calculating oTLI and oTPI. The ratio (converting factor) of these two values indicates the overall conversion efficiency of the light source throughout growth period. The unit of the value of oTLI vs. oTPI is mol per degree which indicates the number of 'mol' of quantum provided by every degree of power consumption by the light source. In this specific case, T5 fluorescent lamps were used and the value is 1.617/0.366 equals 4.414mol $\cdot$ deg^{-1} as shown in the 5th row of Table 3.

Table 3 shows more in depth information with two more given data, including the utility fee (3 NT $ per degree) and the average fresh weight of the leafy greens (50g $\cdot$ plt^{-1}) produced using protocols shown in Tables 1 and 2. The plants were grown for 28 days, sum of D (i), with low light intensity (150 ~ 180 PPF) as shown in Table 1.

The QY value calculated based on eq. 5, shown in the last row of Table 3, represents how efficient the photons were used by the crop. For this specific crop, almost 31 g of fresh weight can be produced per mole of quantum provided. The QY values of various crops under same growth conditions can be compared and the type with highest QY value is the most efficient crop.

The PY value can be link to cost if the utility fee (UF) is given. As shown in the 3rd row from the bottom of Table 3, PY value is 136 g per degree of power consumption and value of PY/UF shows that each NT $ spent in providing light can produce 45 g of fresh produce.

Data shown in Tables 4 and 5 were compiled from another PF company in Taiwan. oTPI and oTLI related data were shown in Table 4. As shown in Table 4, differences in stage 1 and 2 are on the number of LED lamps used and the duration of the stage. The differences among 3 treatments were mainly on intensity of light (number of lamps used) and duration of light period per day in stage 3.

Table 5 shows experimental and calculated results for further explanation on methods developed. Data under treatments A, B, C are from Table 4 and data under treatment X isfrom Tables 1 to 3. In spite of great differences among settings of the experimental conditions, as shown in Table 5, effects of treatments can be compared easily in terms of values of oTLI, oTPI, QY, PY, PY/UF and converting factors. These terms are generic and can be used in various combinations of conditions, even for various types of crops. The last row shows converting factors ranging from 4.42 to 4.99 for 4 treatments. If there are no loss from converting electricity to quantum, the ideal value should be 14.4 (1 degree equals 3 600 kJ and 3 600/250 equals 14.4).

As shown in Table 6, PPF values of three panels measured in 15cm distance are 136, 345 and 154 in average and the area for crop production under the panel is 0.12m^2. LED2 is the one provides the largest PPF and also the most expensive as shown in the 2nd right column. The converting efficiency is listed under the columnwith 'mol · J^{-1}' as the heading, LED2 is the worst and LED1 is the best among 3 panels.

However, if market price is considered, the cost-performance (CP) comparison on 3 LED panels as shown in the last column reveals that LED2 has the least CP value and it might be the best choice among 3 options. It will be very interested to know the QY and PY of the three if the fresh weight of the produces harvested using these panels are available. The values imply the bulk effect of spectrum and other factors not mentioned.

3 Conclusion

A systematic quantitative method was developed for the evaluation of various treatments in the overall stage-wise production of crops in plant factory with artificial light only. With the given conditions of light source and cultural condition in various stages, overall performance of values in quantum basis such as oTLI and QY, and values in power basis such as oTPI and PY can be derived. Values in power basis can be further converted to cost basis with given utility fee information. Converting factor among oTLI and oTPI reveals effectiveness of light source converting energy to quantum.

The beauty of PF_ AL is to allow us to play with various hardware, software and crops. Systematic approach in finding the best protocol (combination of environmental settings) is of vital importance. The usage of the method developed can be very generic and can serve various purposes depend on the treatments designed. By comparing QY or PY values among treatments, the best protocol related to treatments can be identified no matter the interests is in hardware such as type of light source, spectral of light, etc., in software such as settings of parameters, in types of crops to grow, or in combinations of above.

Reference

[1] Goto, E. 2011. Environmental control to get the best out of plant. Proceedings of OSAKA International forum, Japan Bio-Environmental Engineering Society. 25 ~ 36. (in Japanese)

[2] Fang, W. 2012. Chapter 8. Utilization efficiency of light and electric energy. In 'Plant factory with artificial light'. Co-published by Ohmsha, Ltd. and Harvest Farm Magazine, Ltd. (in Traditional Chinese). ISBN 978-957-9157-55-1

[3] Kozai, T. 2011. Improving utilization efficiencies of electricity, light energy, water and CO2 of a plant factory with artificial light. Proceedings of 2011 the 2nd high-level International Forum on protected horticulture (Shouguan, Sandong Province, China, April 19-22, 2011), 2 ~ 8

[4] Kozai, T. 2012. Chapter 8. Utilization efficiency of light and electric energy. In 'Plant factory with artificial light'. Ohmsha, Ltd. (in Japanese)

Table 1 Example in calculating oTLI based on eqs. 1a to 4a

i	μmol · $m^{-2}s^{-1}$	hr · day^{-1}	days	mol · $m^{-2}day^{-1}$	mol · m^{-2}	plt · m^{-2}	mol · plt^{-1}
	PPF (i)	L (i)	D (i)	DLI (i)	TLI (i)	d (i)	oTLI (i)
Stage 1	180	24	10	15.55	155.52	750	0.207
Stage 2	150	12	4	6.48	25.92	533	0.049
Stage 3	150	12	14	6.48	90.72	67	1.361

oTLI = 1.617

Table 2 Example in calculating oTPI based on eqs. 1c to 4c

i	No.	W	W/layer	deg. /layer	plts/layer	deg · plt^{-1}
	lamps/layer	power/lamp	W' (i)	TPI' (i)	d' (i)	oTPI' (i)
Stage 1	3	30	90	21.6	540	0.040
Stage 2	3	30	90	4.32	384	0.011
Stage 3	3	30	90	15.12	48	0.315

oTPI = 0.366

Table 3 Example in calculating quantum yield and power yield and related cost

Avg. Fresh Weight	FW	50	g · plt^{-1}
Utility Fee	UF	3	NT \$. $deg.^{-1}$
Overall total light integral	oTLI	1.617	mol · plt^{-1}
Overall total power integral	oTPI	0.366	deg · plt^{-1}
Converting factor	oTLI/oTPI or PY/QY	4.414	mol · $deg.^{-1}$

(continued)

Avg. Fresh Weight	FW	50	g · plt^{-1}
Cost per mole of quantum	UF * oTPI / oTLI	0.68	NT $. mol^{-1}
Power yield*	PY = FW/oTPI	136.5	g · deg.$^{-1}$
Fresh Yield per NT $	PY/UF	45.51	g · NT $$^{-1}$
Quantum yield	QY = FW/oTLI	30.93	g · mol^{-1}

* Fresh Yield per degree of power consumption

Table 4a Basic information of parameters related to oTPI calculation with three treatments in stage 3

i	Treatments	hrs · day^{-1} L (i)	days D (i)	m^2 · layer^{-1} A (i)	plts · layer^{-1} d' (i)	lamps · layer^{-1} n (i)	W/LED tube	W/layer W' (i)
Stage 1	A, B, C	24	6	0.84	600	3	20	60
Stage 2	A, B, C	24	7	0.84	600	5	20	100
Stage 3a	A	18	21	1.4	49	6	20	120
Stage 3b	B	16	21	1.4	49	8	20	160
Stage 3c	C	16	21	1.4	49	10	20	200

Table 4b Basic information of parameters related to oTLI calculation with three treatments in stage 3

i	Treat-ments	hrs · day^{-1} L (i)	days D (i)	μmol · $m^{-2}s^{-1}$ PPF (i)	mol · m^{-2} DLI (i)	μmol · m^{-2} TLI (i)	plt · m^{-2} d (i)	mol · plt^{-1} oTLI (i)	Stage1 ~ 3 oTLI
Stage 1	A, B, C	24	6	90	7.78	46.7	714	0.065	—
Stage 2	A, B, C	24	7	110	9.5	66.5	714	0.093	—
Stage 3a	A	18	21	120	7.78	163.3	35	4.67	4.83
Stage 3b	B	16	21	160	9.22	193.5	35	5.53	5.69
Stage 3c	C	16	21	200	11.52	241.9	35	6.91	7.07

Table 5 Quantitative comparison on various treatments in growing same crop

Treatments	A	B	C	X*
Days from seeding to harvest	34	34	34	28
FW (in g · plt^{-1})	64.3	86.8	98	50
oTPI (in deg · plt^{-1})	0.97	1.14	1.41	0.366
PY (in g · deg.$^{-1}$)	66.42	76.17	69.32	136.5
PY/UF (in g · NT $$^{-1}$)	22.14	25.39	23.11	45.51
oTLI (in mol · plt^{-1})	4.83	5.69	7.07	1.617
QY (in g · mol^{-1})	13.3	15.3	13.9	30.9
Converting factor (in mol · deg.$^{-1}$)	4.99	4.98	4.99	4.42

Note 1: Utility Fee (UF) is 3 NT $ · deg.$^{-1}$

*: Values in the last column were from Tables 1 and 3.

Table 6 Cost-performance of LED panels available in Taiwan

	PPF*	V	A	W	m^2	$\mu mol \cdot s^{-1}$	$mol \cdot J^{-1}$	NT\$ · $panel^{-1}$	NT\$/$\mu mol \cdot s^{-1}$
LED1	135.9	48	0.45	21.6	0.12	16.31	0.76	1200	73.6
LED2	345.3	48	1.34	64.32	0.12	41.44	0.64	1750	42.2
LED3	153.7	48	0.55	26.4	0.12	18.44	0.70	1350	73.2

* PPF in $\mu mol \cdot m^{-2}s^{-1}$, measured in 15cm distance

提高设施农业增施 CO_2 利用效率的方法探讨*

全宇欣[1,2**]，杨其长[1,2***]

（1. 中国农业科学院农业环境与可持续发展研究所，北京 100081；
2. 农业部设施农业节能与废气物处理重点实验室，北京 100081）

摘要：为了提高设施农业增施 CO_2 的利用效率和经济效益，减少 CO_2 的逸散损失，降低成本，本文以增施 CO_2 利用效率计算方法为基础，分析了增施 CO_2 利用效率的可能影响因素，并探讨了提高 CO_2 施肥经济效益的有效技术与方法。分析结果表明，影响增施 CO_2 利用效率的主要因素为设施的换气次数、设施内外 CO_2 浓度差和植物的光合能力。因此，在进行 CO_2 施肥时，应综合考虑植物种类、生育阶段、栽培条件及其他环境要素等条件，选择适宜的 CO_2 增施方法、施肥浓度和施肥时间。

关键词：CO_2 浓度，换气次数，CO_2 施肥，利用效率，设施农业

Improvement of the Enriched CO_2 Utilization Efficiency in Protected Agriculture

Tong Yuxin[1,2]，Yang Qichang[1,2]

（1. *Institute of Environment and Sustainable in Agriculture*，*Chinese Academy of Agricultural Science*，*Beijing* 100081，*China*；2. *Key Lab. for Energy Saving and Waste Disposal of Protected Agriculture*，*Ministry of Agriculture*，*Beijing* 100081，*China*）

Abstract：To improve the utilization efficiency and economic benefits of enriched CO_2 and reduce the CO_2 leaked outside and operation cost of protected agriculture，in this paper，the possible effect factors on the enriched CO_2 utilization efficiency were analyzed based on its calculations. The effective methods and technologies to improve the enriched CO_2 utilization efficiency were discussed. This paper show that the main effect factors of the enriched CO_2 utilization efficiency are the air exchange rate，CO_2 concentration difference between inside and outside facilities and the plant photosynthetic capacity. To maximize the enriched CO_2 utilization efficiency，the optimum CO_2 enriched method，CO_2 enriched concentration and enriched time should be decided by considering plant species，growth stage，cultivation conditions，other environmental factors，etc. while the CO_2 enrichment is conducted.

Key words：CO_2 concentration，air exchange rate，CO_2 enrichment，utilization efficiency，protected agriculture

1 引言

近年来，随着设施工程技术和栽培育种技术的不断进步，设施农业作物产量也得到了极

---

* 基金项目：863 计划资助项目（（2013AA103007））；农业科技成果转化资金项目（2011GB23260011）

** 作者简介：全宇欣（1982），女，山东，助理研究员，博士，主要从事设施园艺环境工程、植物工厂等研究。E-mail：yxtong07@ hotmail. com

*** 通讯作者：杨其长（1963），男，安徽，博士，研究员，博士生导师，主要从事设施农业、植物工厂、LED 农业应用、生物环境工程等研究。E-mail：yangq@ ieda. org. cn

大的提高。但是，由于不适生长环境因子（如光照强度、温度、CO_2 浓度等）及病虫害等的影响，设施农业作物实际的产量只发挥了其生产潜力的 24% 左右，其中不适生长环境已成为作物产量最重要的限制因素[1]。与设施农业环境控制技术发达的荷兰和日本相比，我国设施农业作物的平均产量分别为其平均产量的 1/6 和 1/3 左右[2]。其主要原因为，我国的农业设施主要以大棚和日光温室为主，室内环境调控能力较差。另外，许多研究学者和生产者较注重温室光透性与温度控制问题，而对其他环境因子，如 CO_2 浓度，则未给予相应重视。

CO_2 是作物进行光合作用的原料之一，因此，CO_2 浓度是影响作物生长的一个很重要的环境因子。在设施农业生产中，作物进行光合作用会消耗大量的 CO_2，若室内 CO_2 得不到及时补充，CO_2 浓度会迅速下降。在不通风情况下，CO_2 浓度会降低到作物 CO_2 补偿点以下，即使在通风的情况下，室内 CO_2 浓度也可能低于室外 CO_2 浓度[3]。因此，过低的 CO_2 浓度已成为设施作物光合的主要限制因素，制约了作物生长发育，降低了作物产量和品质[3]。

众所周知，在作物 CO_2 饱和点以下，光合作用随着 CO_2 浓度的升高而增强。因此，提高设施内 CO_2 浓度是提高设施农业作物产量和改善品质的有效措施。研究表明[4~9]，增施 CO_2 可以：①增加产量达 20% ~60%；②改善品质。外观品质好，个大，瓜粗，畸形果少，色泽鲜艳，果肉厚实，耐贮运；③提早成熟上市。收获期或采摘期一般可提前 3 ~7d，价格高，效益好；④增强抗病抗旱能力。植株健壮，抗病虫害的能力增强。

鉴于此，19 世纪 80 年代以来，许多研究学者致力于 CO_2 施肥技术的研究，特别是一些欧洲和美国学者[9]。但是在研究初期，由于对 CO_2 施肥技术认识不足（如室内 CO_2 浓度不能得到很好控制，CO_2 增施的浓度往往达到作物 CO_2 饱和点以上），试验方法和条件的限制（如采用同时会产生其他空气污染物的燃烧法生成 CO_2 进行施肥，对照试验利用土耕栽培并施用可放出大量 CO_2 的有机肥），试验效果并不明显，有些试验结果甚至表明作物产量和品质反而会低于对照[10]，增施 CO_2 的利用效率也较低。致使一些研究学者对 CO_2 施肥技术的研究热情一度低迷，在研究初期的四五十年内，CO_2 施肥技术并未得到很大提高。近年来，随着对 CO_2 施肥技术认识的增加及设施农业环境控制技术的提高，一些国家，尤其是设施农业技术发达的国家，配备 CO_2 施肥装置的设施面积正不断增加。据 2011 年日本农林水产省报道[11]，2009 年配备 CO_2 施肥装置的温室面积比 2001 年增加了 56%。荷兰也有 70% 以上的温室配备了 CO_2 施肥装置。而在我国，配备 CO_2 施肥装置的温室较少，且普遍缺乏 CO_2 调控装置及监测系统，给 CO_2 施肥带来更多技术上的困难，而一些技术措施还缺乏适当的理论指导，致使 CO_2 施肥的经济效益低，限制了 CO_2 施肥技术的推广应用[12]。

因此，为了提高设施增施 CO_2 的经济效益，本文在参考大量文献基础上，从设施内增施 CO_2 利用效率的计算方法入手，分析了增施 CO_2 利用效率和经济效益的可能影响因素，探讨了提高 CO_2 施肥经济效益的有效技术与方法。

2 设施内增施 CO_2 利用效率计算方法

设施内增施 CO_2 的利用效率（*ECUE*，enriched CO_2 utilization efficiency）是 CO_2 施肥经济效益评价的最重要指标。其计算公式为[13]：

$$ECUE = 1 - \frac{L_{co_2} + Dc/dt}{S} \tag{1}$$

其中，L_{co_2}为 CO_2 通过设施缝隙逸散到设施外的损失率（$kg \cdot m^{-2} \cdot h^{-1}$），$S$ 为设施内 CO_2 的增施速率（$kg \cdot m^{-2} \cdot h^{-1}$），$dC/dt$ 为设施内 CO_2 浓度变化速率（$kg \cdot m^{-2} \cdot h^{-1}$）。

L_{co_2}可由下式求得[13]：

$$L_{co_2} = \frac{K_{co_2} \cdot N \cdot V \ (C_{out} - C_{in})}{A} \tag{2}$$

其中，K_{CO_2}为标准状态下 CO_2 的密度（$1.96kg \cdot m^{-3}$），N 为设施的换气次数（h^{-1}），C_{out}为设施外 CO_2 浓度（$\mu mol \cdot mol^{-1}$），C_{in}为设施内 CO_2 浓度（$\mu mol \cdot mol^{-1}$），V 为设施体积（m^3），A 为栽培面积（m^2）。

dC/dt 可由下式求得[13]：

$$\frac{dC}{dt} = R_{co_2} + S - P_n - L_{co_2} \tag{3}$$

其中，R_{CO_2}为土壤释放出 CO_2 速率（$kg \cdot m^{-2} \cdot h^{-1}$），若土壤表面有覆盖膜，$R_{CO_2}$一般可忽略不计，$P_n$为植物净光合速率（$kg \cdot m^{-2} \cdot h^{-1}$）。

3　增施 CO_2 利用效率的影响因素

结合（1）、（2）和（3）式可以看出，增施 CO_2 的利用效率主要受设施的换气次数、设施内外 CO_2 浓度差和植物光合能力的影响。日本学者横井等[14]假设设施面积为 $1\ 000m^2$，体积为 $3\ 000m^3$，设施内 CO_2 浓度为 $1\ 000\mu mol \cdot mol^{-1}$，设施外 CO_2 浓度为 $350\mu mol \cdot mol^{-1}$，换气次数分别为：$0.1h^{-1}$、$0.5h^{-1}$、$1.0h^{-1}$、$2.0h^{-1}$、$10h^{-1}$，设施外气温为 27°C，由土壤微生物呼吸所放出的 CO_2 忽略不计，得出增施 CO_2 的利用效率与植物的净光合速率和设施换气次数的关系图（图 1）。

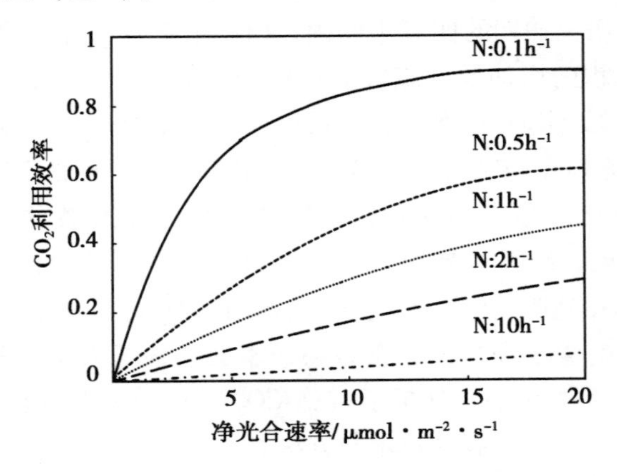

图 1　增施 CO_2 的利用效率与净光合速率和换气次数的关系

Figure 1　Schematic diagram of the responses of the net photosynthetic rate and air exchange rate to the enriched CO_2 utilization efficiency

3.1 设施换气次数对增施 CO_2 利用效率的影响

温室在密闭状态下的换气次数一般在0.3~0.5 h^{-1}。若室内 CO_2 增施到高浓度（例如，1 000μmol·mol^{-1}），增施 CO_2 的利用效率最高可达到0.6~0.8（图1）。而对于密闭性较好的植物生产设施，如人工光型植物工厂，其换气次数一般在0.01~0.04 h^{-1}，因此，即使室内 CO_2 浓度增施到1 000μmol·mol^{-1}或以上，增施 CO_2 的利用效率也可达到0.9以上[15]。

温室在通风状态下的换气次数一般在10 h^{-1}以上，若室内 CO_2 仍增施到高浓度，那么增施的 $CO_2$90%以上将会逸散到室外，不但造成经济损失，而且还会增加产生温室效应的 CO_2 排出量。试验表明[3,13]，即使在通风状态下，由于植物光合作用，温室内 CO_2 浓度有时也会低于大气 CO_2 水平50~100μmol·mol^{-1}。而在低 CO_2 浓度，植物净光合速率与 CO_2 浓度几乎成直线增长关系。Allen 等[16]研究发现当 CO_2 浓度从330μmol·mol^{-1}逐渐升高到800μmol·mol^{-1}时，大豆的净光合速率随 CO_2 浓度升高而成直线增加。综合上述分析可知，增施 CO_2 到室外浓度水平可以在很大程度上提高植物的净光合速率。由式（2）可知，若增施 CO_2 浓度到室外浓度水平，即使设施换气次数很大，或者说，即使设施在充分通风的情况下，也不会有 CO_2 逸散到室外。那么，增施 CO_2 的利用效率约为1。

3.2 植物光合能力对增施 CO_2 利用效率的影响

植物光合能力也是影响增施 CO_2 利用效率的一个很重要的因素。因此，任何影响植物光合能力的因素，同样也会影响增施 CO_2 的利用效率。植物光合能力主要受到植物种类及其生长环境因素的影响。

3.2.1 不同植物种类对增施 CO_2 利用效率的影响

植物根据其光合碳同化途径不同，可分为 C_3，C_4 和 CAM 植物。其中，C_3 植物占95%左右，而设施农业作物基本上属于 C_3 植物[17,18]。试验证明[19]，与 C_3 植物相比，C_4 植物的 PEP 羧化酶的活性强，对 CO_2 的亲和力大。因此，C_4 植物的 CO_2 补偿点比 C_3 植物低，在大气 CO_2 浓度水平，C_4 植物利用低浓度 CO_2 进行光合的能力比 C_3 植物高（图2）[17]。C_3 植物的光合能力随 CO_2 浓度升高而增加的幅度较大[20,21]。Havelka 等[22]发现在1390μmol·mol^{-1} CO_2 浓度下生长的小麦净光合速率比332μmol·$mol^{-1}$$CO_2$ 浓度下增加了50%。Ziska 和 Teramura[23]，Kimball[24]等的研究也都表明了增施 CO_2 对 C_3 植物有明显的增产效果。增施 CO_2 对 C_4 植物的增产效果则不太明显[25]。而 CAM 植物的气孔一般会在傍晚或夜晚开放，进行 CO_2 的吸收与固定，因此，在傍晚或夜晚进行 CO_2 施肥可能会促进其生长[9]。

3.2.2 不同环境因子对增施 CO_2 利用效率的影响

植物利用 CO_2 进行光合的能力还受到不同环境因子的影响。CO_2 浓度水平与其他环境因子（光照强度、叶温、湿度、气流速度和水肥吸收速度等）在对植物进行光合的影响上存在着既相辅相成又相互制约的关系[3,26~29]。

如图3所示[15]，增施 CO_2 可以降低光强不足对植物光合的影响，而提高光照强度可以降低植物 CO_2 补偿点同时提高 CO_2 饱和点，补偿 CO_2 浓度不足对植物光合的影响[9,12]。从图3还可以看出，在较强光照强度下进行 CO_2 施肥，增施 CO_2 的利用效率较高。因此，根据光强大小增施 CO_2 至不同浓度，效果更显著。

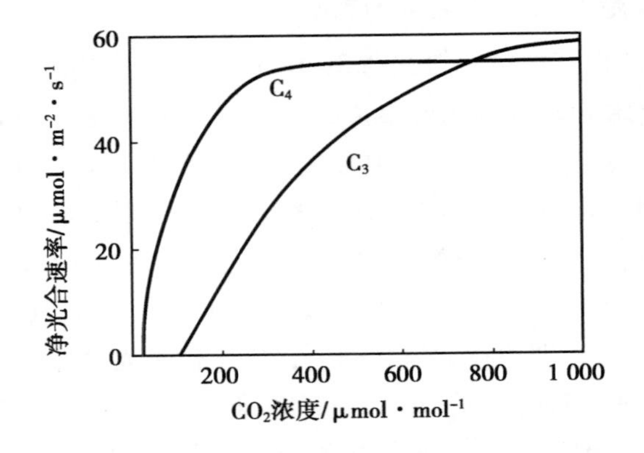

图2 增施 CO_2 对 C_3 和 C_4 植物净光合速率的影响

Figure 2 Schematic diagram of the response of CO_2 enrichment to the net photosynthesis of C_3 and C_4 plants

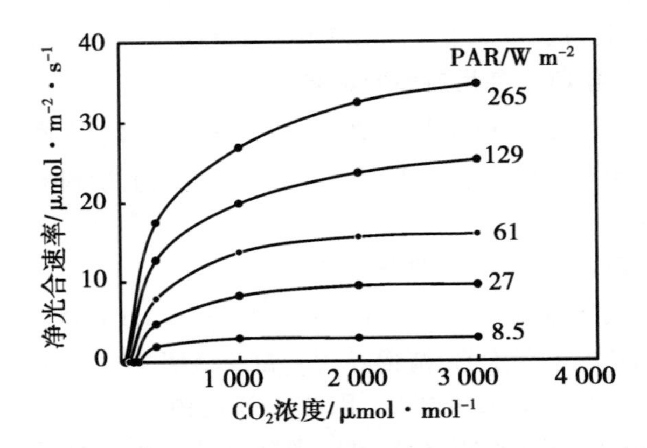

图3 净光合速率与 CO_2 浓度和光合有效辐射（PAR）的关系

Figure 3 Schematic diagram of the responses of CO_2 concentration and photosynthetic active radiation to net photosynthetic rate

增施 CO_2 的效果还会受到叶温的影响（图4）[15]。Eamus 等[30]对桉树进行 CO_2 施肥试验证明，当 CO_2 浓度从大气浓度水平提高到 700μmol·mol^{-1} 后，桉树生长的最适叶温从 29°C 升高到了 34°C。Idso 等[31]在不同温度下对一些植物品种进行了 CO_2 施肥试验，结果表明，在高 CO_2 浓度下，植物生长的最适叶温也相应提高了 2～5°C，若在低温下（如低于 18°C）进行 CO_2 施肥，其效果不明显，有时反而会抑制植物生长。因此，在进行设施 CO_2 施肥时，可适当调高设施开窗通风的设定温度，延长 CO_2 施肥时间，使其达到更好的效果。

设施内空气相对湿度和风速间接影响了增施 CO_2 的效果（图5）[15]。过低或过高的空气相对湿度都会降低植物叶片的气孔导度[3,19]，从而增加 CO_2 进入叶片的阻力，降低蒸腾速率，尤其是在低水肥供给条件下，会导致植物水分、营养匮乏，降低增施 CO_2 的效果[32～34]。

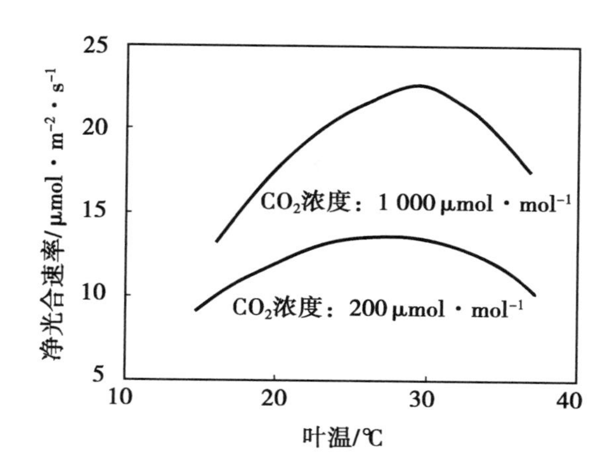

图 4　净光合速率与叶温和 CO_2 浓度的关系

Figure 4　Schematic diagram of the responses of the leaf temperature and CO_2 concentration to net photosynthetic rate

风速大小会影响植物冠层与群落内部 CO_2 的均匀分布，从而影响增施 CO_2 的效果。植物进行光合作用消耗大量的 CO_2，若风速较小，会使 CO_2 的扩散速率减慢，造成植物群落内部 CO_2 得不到及时补充，从而会降低植物的光合速率。一般当风速在 0.3 ~ 1m · s^{-1} 时，随着风速增大，植物叶片的边界层阻力减少，气孔导度增大，增施 CO_2 的效果增加[3]。若风速超过 1m · s^{-1}，尤其是在较低的相对湿度环境下，再增大风速，会导致植物的部分气孔关闭，气孔导度降低，增施 CO_2 的效果降低[13]。

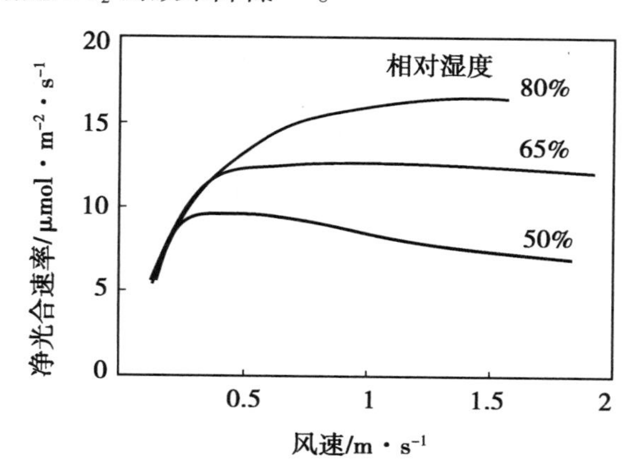

图 5　净光合速率与风速和相对湿度的关系

Figure 5　Schematic diagram of the response of the wind speed and relative humidity to net photosynthetic rate

植物水肥供给条件会影响增施 CO_2 的效果，而 CO_2 施肥也会影响植物的水肥吸收速度。Kanemoto 等[32]在 332 ~ 910μmol · mol^{-1} 之间的 6 个 CO_2 浓度水平上进行大豆试验表明，在水分充足的情况下，虽然高浓度 CO_2 增大了大豆的叶面积，但同时也降低了大豆叶片的气孔导度，从而减少了每株大豆的水分蒸发量。在水分胁迫情况下，高 CO_2 浓度下大豆表现出较强的抗旱性。Sánchez-Guerrero 等[7]研究也发现在水胁迫下增施 CO_2 可以降低设施黄瓜的蒸腾速率，提高水分利用效率。Reddy 等[35]在 350～900μmol·mol^{-1}之间的 5 个 CO_2 水平上对棉花进行试验表明，棉花对水的利用效率随着增施 CO_2 浓度的升高而增大，其主要原因为增施 CO_2 提高了植物冠层光合能力，减少了其蒸腾速率。由上可知，CO_2 施肥会降低植物的蒸腾速率，从而减少水肥的吸收，因此，若在进行 CO_2 施肥的同时提高水肥的供给水平，则可以在很大程度上提高增施 CO_2 的效果，从而提高植物产量和品质[36]。

综上所述，环境因子对植物利用 CO_2 进行光合的影响存在着既相互制约又相互促进的关系。因此，为提高增施 CO_2 的利用效率应注意环境因子的综合调控，提高其经济效益。

4　提高增施 CO_2 利用效率的技术

大量的试验研究表明，不当的 CO_2 施肥方法会对植株造成伤害，如出现徒长、营养缺乏、加速老化，有时甚至会造成减产。因此，采用适当的 CO_2 施肥方法是提高其利用效率和经济效益的关键。

4.1　增施 CO_2 的方法

如表所示，CO_2 的增施方法主要包括：通风换气、液态 CO_2（或干冰）[37]、碳水化合物燃烧[38]、化学反应、发酵、利用动植物产生的 CO_2 等[12,39]。目前，在中国，由于大部分设施结构简单，综合环境控制技术差，增施的 CO_2 主要采用成本较低的方法获得，例如，通风换气、化学反应和自然降解法等。在荷兰、日本等设施农业技术较高的国家，主要采用 CO_2 浓度可精确控制或可进行多目的应用的方法，例如，纯 CO_2（液体 CO_2 或工业副产品）和燃烧法（在冬季利用较多）[40]。表 1 中对各种常用方法的优缺点做了相关介绍。在进行方法选择时，应充分考虑设施栽培条件、栽培植物、环境控制条件、经济条件等因素，以取材方便、操作简单、安全可靠、无污染物影响植物生长和便于自动控制等为原则，合理选择一种或几种可以协同利用的方法，提高增施 CO_2 的利用效率。

4.2　CO_2 施肥浓度

对于植物而言，并非 CO_2 浓度越高越好。如图 2 所示，在低 CO_2 浓度时，植物的光合速率随 CO_2 浓度升高几乎呈直线增加，越接近 CO_2 饱和点，光合速率增加的越慢。若 CO_2 浓度增施到饱和点以上，光合速率则不再增加。过高的 CO_2 浓度还会减小植物叶片气孔导度，降低植物蒸腾，使植株表现为营养缺乏，落叶，降低 CO_2 的利用效率，造成经济损失。因此，适宜的 CO_2 浓度应根据设施的密闭状况、植物的种类、品种、生育阶段和其他环境因子而定。一般蔬菜的 CO_2 饱和点都在 1 000μmol·mol^{-1}以上，且随着光强增加而升高[12]。实际生产中，在设施密闭性较好、室内光、温等环境条件较为适宜的条件下，增施 CO_2 的浓度，叶菜类蔬菜以 600～1 000μmol·mol^{-1}为宜，果菜类蔬菜以 1 000～1 500μmol·mol^{-1}为宜，生长发育前期和阴天取低限，生长发育后期和晴天取高限。

4.3　CO_2 施肥时间

选择适宜的 CO_2 施肥时间，可以提高 CO_2 的利用效率并增加产量。适宜的 CO_2 施肥时间根据植物不同生育阶段、栽培方式等的不同而有所变化。

同一种植物，在不同的生育阶段或采用不同的栽培方式，其利用CO_2进行光合的能力是存在差别的。多层立体栽培的叶菜类蔬菜或种苗生产，单位土地面积上的叶面积指数大，群落光合能力强，增施CO_2的利用效率高，CO_2施肥可以在整个生育期进行。而在一般的设施果菜类栽培中，植物苗期的叶面积指数小，利用CO_2进行光合的能力较弱，增施CO_2的利用效率低，不宜施用。而在果菜类植物进入开花结果后期，CO_2吸收量增加，增施CO_2可以促进果菜生殖生长，增产效果好。因此，在开花结果期，增施CO_2增产效果较明显。

表　CO_2的增施方法

Table　CO_2 enrichment methods

方法	来源	优点	缺点
通风换气法	利用天窗或侧窗，通过通风换气来补充设施内CO_2，减小室内外CO_2浓度差	操作简单，无成本	只能将设施内CO_2浓度提高到设施外浓度水平，且受到时间限制，比如在冬季为避免设施内设施过低，不宜开窗
液体CO_2法	释放瓶装液体CO_2	操作简单，可精确控制设施内CO_2浓度	成本较高，约8元/kg
固体CO_2法	施入地表或浅埋土中的固体CO_2颗粒气肥，借助光温效应自行潮解释放CO_2	操作简单	CO_2释放速度不易控制，因此设施内CO_2浓度无法进行精确控制
燃烧法	利用燃烧煤、油、天然气、沼气等碳水化合物释放的CO_2	燃烧释放的热量可用于设施内加温	燃烧同时会产生一些大气污染物，如SO_2、NO_x等，未完全燃烧产生的CO会造成人身伤害，成本高
化学反应法	利用碳酸氢铵、碳酸氢钠或碳酸钙与稀硫酸（3∶1）进行化学反应产生CO_2	反应剩余物可做肥料，成本较低	若温度过高导致碳酸氢铵分解，会产生氨中毒。硫酸对人、对物有腐蚀作用
酵解法	将有机物在酵母作用下酵解，或增施有机肥，将秸秆和畜禽粪便混合进行堆肥，利用微生物酵解产生的CO_2	操作简单，成本低，同时可以提高土壤肥力	CO_2释放速度不易控制，堆肥时可能同时会产生一些大气污染物，如SO_2、NO_x等
种植食用菌法	在栽培的空闲空间或在可以进行气体交换的设施中种植食用菌，利用食用菌释放的CO_2	无成本	CO_2释放速度不易控制，设施内CO_2浓度无法进行精确控制
养殖动物法	在设施旁边建设畜（禽）舍，利用畜禽呼吸产生的CO_2	无成本，畜禽的粪便可以作为有机肥料	CO_2释放速度不易控制，设施内CO_2浓度无法进行精确控制

一天当中CO_2施肥的适宜时间取决于室内CO_2浓度和光、温等环境条件。在相对密闭的设施内，由于夜间植物呼吸和土壤有机物经微生物分解释放的CO_2积蓄于室内，日出前，室内CO_2浓度较高，一般可达800$\mu mol \cdot mol^{-1}$。日出后，植物开始进行光合作用吸收大量

的 CO_2，室内 CO_2 浓度迅速下降，因此，CO_2 施肥应当在日出后半个小时左右进行。为避免高温对植物伤害，一般进行通风换气，所以，在通风前半个小时应停止 CO_2 施肥，避免浪费。^{14}C 同位素跟踪试验表明，上午增施的 CO_2 在果实、根中的分配比例较高，而下午增施的 CO_2 在叶内积累较多。而植物全天光合产物的3/4在上午产生，可知，植物的光合作用主要在上午进行，因此 CO_2 施肥也应主要集中在上午。在光照强度较低的阴雨天，可不施或进行 CO_2 低浓度施肥。

CO_2 施肥时间的长短也会影响增施 CO_2 的利用效率。相关研究发现[41~43]，若对一些植物进行长期的 CO_2 施肥，会使光合产物在植物叶片中积累，使叶绿素浓度和光合反应酶的活性下降。一些学者对不同植物进行长期 CO_2 施肥试验表明，在进行 CO_2 施肥的初期，植物的净光合能力普遍增强，但几周后，植物的净光合能力则会下降到对照试验水平或更低。Delucia 等[44]研究发现，棉花在 $350\mu mol \cdot mol^{-1}$、$675\mu mol \cdot mol^{-1}$ 和 $1\ 000\mu mol \cdot mol^{-1}$ 的 CO_2 浓度环境下生长4周后，高 CO_2 浓度增加了棉花的生物量，但由于碳水化合物，如淀粉，在植物叶片中的积累，分解了叶片中部分叶绿素，并降低了光合反应酶的活性，从而使叶绿素浓度和光合能力下降。烟草在 $1\ 000\mu mol \cdot mol^{-1}$ 的 CO_2 浓度下生长几周后也出现了20%净光合能力的下降。高浓度 CO_2 可导致番茄叶片中光合产物积累使叶片光合能力下降。但也有试验表明，对植物进行长期 CO_2 施肥不会降低其光合能力。Eamus 等[30]研究表明，以大气 CO_2 浓度为对照，桉树在 $700\mu mol \cdot mol^{-1}$ CO_2 浓度下分别生长12月、18月和30个月后，虽然叶片叶绿素浓度有所降低，但植物的光合能力却得到增强。Michael 等[45]也有相同的报道。因此，CO_2 施肥时间的长短应因植物品种不同而异。

5　结论

为了提高设施增施 CO_2 的利用效率和经济效益，本文在分析增施 CO_2 利用效率计算方法的基础上，给出了增施 CO_2 利用效率的可能影响因素，探讨了提高 CO_2 施肥经济效益的有效技术与方法。主要结论如下。

第一，从计算公式可知，增施 CO_2 的利用效率主要受设施的换气次数、设施内外 CO_2 浓度差和植物光合能力的影响。

第二，为提高增施 CO_2 的利用效率，设施内增施 CO_2 的浓度应随设施换气次数的增大而降低。

第三，CO_2 施肥对不同植物种类的增产效果不同。其中，对 C_3 植物的增产效果最显著。

第四，植物利用 CO_2 进行光合的能力受其环境因子的影响，因此，为提高增施 CO_2 的利用效率，应注意环境因子的综合调控。

第五，不当的 CO_2 施肥方法会对植物造成伤害，甚至减产。因此，应在综合考虑植物种类、生育阶段、栽培条件及其他环境因子等因素的基础上，选择适宜的 CO_2 的增施方法、施肥浓度和时间。

参考文献

[1] Ikeda T. The optimal environment for agriculture production. Japan-China International Workshop of Horticulture, 2010, Japan, Tokyo

[2] Kato S, Matsuda R, Anjyo K, *et al.* Growth modeling on analyzing the yield difference of greenhouse tomato between Japan and Holland. Japanese Society for Agricultural, Biological, and Environment Engineers and Scientists, 2011. (In Japanese)
[3] Tongbai P, Kozai T, Ohyama K. CO_2 and air circulation effects on photosynthesis and transpiration of tomato seedlings. Scientia Horticulturae, 2010, 126 (3): 338 ~ 344
[4] Baker J T, Allen L H J, Boote K J. Temperature effects on rice at elevated CO_2 concentration. Journal of Environment Botony, 1992, 43 (7): 959 ~ 964
[5] Jaffrin A, Bentounes N, Joan, A M, *et al.* Landfill biogas for heating greenhouses and providing carbon dioxide supplement for plant growth. Biosystems Engineering, 2003, 86 (1): 113 ~ 123
[6] Critten D L, Bailey B J. A review of greenhouse engineering developments during the 1990s. Agricultural and Forest Meteorology, 2002, 112 (1): 1 ~ 22
[7] Sánchez-Guerrero M C, Lorenzo P, Medrano E, *et al.* Effects of EC-based irrigation scheduling and CO_2 enrichment on water use efficiency of a greenhouse cucumber crop. Agriculture Water Management, 2009, 96 (3): 429 ~ 436
[8] Tisserat B, Vaughn S F, Berhow M A. Ultrahigh CO_2 levels enhances cuphea growth and morphogenesis. Industrial Crop and Products, 2008, 27: 133 ~ 135
[9] Mortensen L M. Review: CO_2 enrichment in greenhouses. Crop responses. Scientia Horticulturae, 1987, 33 (1 ~ 2): 1 ~ 25
[10] Peet M M, Huber S C, Patterson D T. Acclimation to high CO_2 in monecious cucumbers. II. Carbon exchange rate, enzyme activities, and starch and nutrient concentrations. Plant Physiology, 1986, 80: 63 ~ 67
[11] Ministry of Agriculture, Forestry and Fisheries: Investigation on Protected Horticulture and plastic used in Agriculture. 2011 http: //www. maff. go. jp/j/tokei/kouhyou/engei/index. html
[12] 周长吉. 现代设施工程. 北京: 化学工业出版社, 2009
Zhou C. Modern greenhouse engineering. Beijing: Chemical Industry Press, 2009
[13] Tong Y. Integrated Greenhouse Environment Control Using Heat Pumps with High Coefficient of Performance. Japan, Chiba University, 2011
[14] Yokoyi M, Kozai T, Nagay G, *et al.* Effects of leaf area index of tomato and air exchange rate on the CO_2 and water use efficiency in closed systems. Japanese Society for Agricultural, Biological, and Environment Engineers and Scientists, 2005, 17 (4): 182 ~ 191
[15] Kozai T. Plant factory with artificial light. Japan: Ohmsha, 2012
[16] Allen L H, Valle J. Mishoe R R, *et al.* Soybean leaf gas exchange responses to CO_2 enrichment. Proceedings - Soil & Crop Science Society of Florida, 1990, 49: 192 ~ 198
[17] Taiz L. Zeiger E. Plant Physiology. The Benjamin/Cummings Pub. Co. , Redwood City, CA. 1991
[18] Tans P. NOAA/ESRL. Trends in atmospheric carbon dioxide-global, http: //www. esrl. noaa. gov/gmd/ccgg/ trends/#global, 2010
[19] Morison J I L, Gifford R M. Stomatal sensitivity to carbon dioxide and humidity-A comparison of two C_3 and two C_4 Grass species. Plant Physiology, 1983, 71 (4): 789 ~ 796
[20] Masafumi O, Makie K, Hiroko T, *et al.* Is yield enhancement by CO_2 enrichment greater in genotypes with a higher capacity for nitrogen fixation Agricultural and Forest Meteorology, 2011, 151: 1 385 ~ 1 393
[21] Remy M, Andreas P, Hans-Joachim W. Effect of free air carbon dioxide enrichment combined with two nitrogen levels on growth, yield and yield quality of sugar beet: Evidence for a sink limitation of beet growth under elevated CO_2. European Journal of Agronomy, 2010, 32: 228 ~ 239
[22] Havelka U D, Wittenbach V A, Boyle M G. CO_2-enrichment effects on wheat yield and physiology. Crop Science, 1984, 24 (6): 1 163 ~ 1 168
[23] Ziska L H, Teramura A H. CO_2 enrichment of growth and photosynthesis in rice (*Oryza Sativa*) . Plant Physiology, 1992, 99: 473 ~ 481
[24] Kimball B A. Carbon dioxide and agricultural yield: an assemblage and analysis of 430 prior observations. Agronomy Joutnal, 1983, 75: 779 ~ 788

[25] Strain B R, Cure J D. Direct effects of increasing carbon dioxide on vegetation. United States, U. S. Dept. of Energy, 1985

[26] Jonghan K, Lajpat A, Bruce K, *et al.* Simulation of free air CO_2 enriched wheat growth and interactions with water, nitrogen, and temperature. Agricultural and Forest Meteorology, 2010, 150: 1 331 ~ 1 346

[27] Renu P, Priya M, Chacko M L, *et al.* Higher than optimum temperature under CO_2 enrichment influences stomata anatomical characters in rose (*Rosa hybrida*). Scientia Horticulturae, 2007, 113 (1): 74 ~ 81

[28] Franzaring J, Weller S, Schmid I, *et al.* Growth, senescence and water use efficiency of spring oilseed rape (*Brassicanapus* L. cv. Mozart) grown in a factorial combination of nitrogen supply and elevated CO_2 Environmental and Experimental Botany, 2011, 72 (2): 284 ~ 296

[29] Zhu Q, Jiang H, Peng C, *et al.* Evaluating the effects of future climate change and elevated CO_2 on the water use efficiency in terrestrial ecosystems of China. Ecological Modeling, 2011, 222 (14): 2 414 ~ 2 429

[30] Eamus D, Duff G A, Berryman C A. Photosynthetic responses to temperature, light flux-density, CO_2 concentration and vapour pressure deficit in *Eucalyptus tetrodonta* grown under CO_2 concentration. Environment Pollution, 1995, 90 (1): 41 ~ 49

[31] Idso S B, Kimball B A, Anderson M G, *et al.* Effects of atmospheric CO_2 enrichment on plant growth: The interactive role of air temperature. Agricultural Ecosystems Environmental, 1987, 20 (1): 1 ~ 10

[32] Kanemoto K, Yamashita Y, Ozawa T, *et al.* Photosynthetic acclimation to elevated CO_2 is dependent on N partitioning and transpiration in soybean. Plant Science, 2009, 177 (5): 398 ~ 403

[33] Zhang X C, Zhang F S, Feng Y X, *et al.* Effect of Nitrogen Nutrition on Photosynthetic Function of Wheat Leaf under Elevated Atmospheric CO_2 Concentration. Acta Agronomica Sinica, 2010, 36 (8): 1 362 ~ 1 370

[34] Li W, Han X, Zhang Y, *et al.* Effects of elevated CO_2 concentration, irrigation and nitrogenous fertilizer application on the growth and yield of spring wheat in semi-arid areas. Agricultural water management, 2007, 87 (1): 106 ~ 114

[35] Reddy V R, Reddy K R, Hodges H F. Carbon dioxide enrichment and temperature effects on cotton canopy photosynthesis, transpiration, and water-use efficiency. Field crops research, 1995, 41 (1): 13 ~ 23

[36] Franzaring J, Holz I, Fangmeier A. Different responses of Molinia caerulea plants from three origins to CO_2 enrichment and nutrient supply. Acta ecological, 2008, 33 (2): 176 ~ 187

[37] Wittwer S, Robb W. Carbon dioxide enrichment of greenhouse atmospheres for food crop production. Economy Botany 1964, 18 (1): 34 ~ 56

[38] Hanan J J. Greenhouses advanced technology for protected horticulture. CRC Press, 1998

[39] Louis-Martin D, Mark L, Vale'rie O. Review of CO_2 recovery methods from the exhaust gas of biomass heating systems for safe enrichment in greenhouses. Biomass and bioenergy, 2011, 35 (8): 3 422 ~ 3 432

[40] Kozai T. Solar light Plant factory. Japan: Ohmsha, 2009

[41] Rowland-Bamford A J, Allen L H, Baker J T, *et al.* Acclimation of rice to changing atmospheric carbon dioxide concentration. Plant Cell Environment, 1991, 14 (6): 577 ~ 583

[42] Besford R T, Hand D W. The effects of CO_2 enrichment and nitrogen oxides on some Calvin cycle enzymes and nitrite reductase in glasshouse lettuce. Journal of Experiment Botany, 1989, 40 (3): 329 ~ 336

[43] Besford R T, Ludwig L J, Withers A C. The greenhouse effect: acclimation of tomato plants grown in high CO_2, photosynthesis and ribulose-l, 5-bisphosphate carboxylase protein. Journal of Experiment Botany, 1990, 41: 925 ~ 931

[44] Delucia E H, Sasek T W, Strain B R. Photosynthetic inhibition after long-term exposure to elevated levels of atmospheric carbon dioxide. Photosynthesis Research, 1985, 7 (2): 175 ~ 184

[45] Michael H, Daniel C M, Anne L. Long-term effects of nutrient and CO_2 enrichment on the temperate coral Astrangia poculata (Ellis and Solander, 1786) Cohen Journal of Experimental Marine Biology and Ecology, 2010, 386 (1 ~ 2): 27 ~ 33

我国设施蔬菜栽培机械装备研究进展*

杨雅婷**，陈永生，胡　桧
（农业部南京农业机械化研究所，南京　210014）

摘要：随着我国设施农业的迅速推广，设施栽培面积居世界第一。设施农业是高投入、高产出的现代农业方式，但设施农业机械化程度低和技术含量低已经成为制约设施农业发展和效益提升的重要原因。本文以作物栽培生长为视角，以农艺为主线，根据栽培时序将设施栽培机械分为耕整地机械、育苗播种机械、环境调控机械、植保机械和收获机械，对我国设施农业机械装备现状作一简要综述。
关键词：设施农业；农业工程；机械；设备

The Research Progress of Machine and Equipment in Protected Agriculture

Yang Yating, Chen Yongsheng, Hu Hui
(*Nanjing Research Institute for Agricultural Mechanization*, *Ministry of Agriculture*, *Nanjing* 210014, *China*)

Abstract: The area of protected agriculture in China is largest all over the world with the rapid development of protected agriculture. Protected agriculture is the way achieving modern agriculture within high input and high yield, but the development and benefit growth of protected agriculture have been restricted by low level of mechanization and low technical content. The protected agriculture equipment in China was reviewed. According the agriculture timing, the protected agriculture equipment was divided into scarification equipment, seeding equipment, environment controlling equipment, plant protection equipment and harvesting equipment.
Key words: Protected agriculture; Agricultural engineering; Machine and equipment

1　引言

设施农业，亦称环境控制农业或工厂化农业，是利用工程技术手段和工业化生产方式为动植物生产提供适宜生长环境，使其在最经济的生长空间内，获得最高的产量、品质以及经济效益的一种高效农业，是现代农业的重要标志之一[1]，可分为设施栽培、设施养殖以及设施水产。近二十年来，我国设施农业发展迅速，尤其是设施蔬菜和园艺作物栽培的面积增长迅猛。截至2009年年底，我国人均占有设施面积已达25m^2[2]。2010年，我国日光温室和塑料大棚的面积分别超过480万亩***和930万亩，设施园艺总面积居世界第一[3]。仅在南京市，设施农业面积接近50万亩，占全市耕地面积的12%[4]。

---

* 基金项目：江苏省农业三新工程项目（SXGC［2012］370）

** 杨雅婷，女，1984年生，甘肃玉门人，硕士，农业部南京农业机械化研究所助理研究员，南京农业大学在职博士生研究生，研究方向为设施农业装备研发，210014，南京市玄武区柳营100号，E-mail：yangyating9@gmail.com。

*** 1亩=667m^2，15亩=1hm^2，全书同

温室生产作业机械及其配套设施是现代农业的重要装备之一，是农机技术与园艺技术相结合的产物[5]。近几年，在需求拉动、资金推动、政策带动等多重刺激的影响下，我国设施栽培生产作业机械得到大力发展，卷帘机、保温被、卷膜器、开窗机、二氧化碳发生器等设施栽培装备已经列入《国家支持推广的农业机械产品目录》。随着农机具的进一步推广和人工费用的不断上涨，设施栽培装备的普及讲有利于打破设施农业发展的瓶颈，显著提高设施栽培整体效益，促进设施农业又快又好地发展。

以往设施栽培装备综述通常采用将各类机械罗列的分类方法[5~6]，或者采用耕作机械、配套栽培机械的分类方法[7~8]，存在杂乱无章和笼统含糊的缺点。本文以农艺为主线，根据“种前——播种——生长——收获”的栽培顺序，将设施栽培机械分为耕整地机械、育苗播种机械、环境调控机械、植保机械和收获机械。其中，由于设施栽培中植保技术的重要性，特将植保机械单独提出。

2　耕整地机械

过去常采用6.0～7.5kW 的微耕机（也称田园管理机）和2.5～4.5kW 动力驱动的手扶耕作机械，虽然具有便于温室大棚边角作业等优点，但由于作业效率低、扶手振动大、机手工作条件差，已逐渐被淘汰。目前，主要采用以18.4～22.1kW 动力驱动的小型作业机具，驱动主机为小四轮拖拉机。通过更换后部作业机具，可以进行犁耕、旋耕、起垄、开沟、施肥作业。农业部南京农机化所和盐城市盐海拖拉机制造有限公司研发的“1GVF-125/140 设施栽培多功能复式作业机”解决了需要更换机具、多次进棚的问题，一次进棚可以完成旋耕、起垄、施肥作业，填补了国内空白。同时，为了和草莓栽培农艺相结合，通过简单拆装，该机型还可以完成一次性一行双垄的作业。

3　育苗播种机械

播种机械有条播机、穴播机、穴盘精量播种机等。由于集中育苗能大大提高种苗的成活率、壮苗系数以及蔬菜品质，穴盘育苗和钵体育苗方式得到极大的发展。

3.1　蔬菜穴盘精量播种机

蔬菜穴盘精量播种机主要有填土、排种、覆土、刮平、浇水等5 个工作环节，其中最主要也是最重要的是排种器。按照排种原理可将排种器分为机械式和气力式，气力式按照结构形式则又可分为针（管）式、板（盘）式和滚筒式。针式精度高，但只适用于圆形小粒、尺寸规则的种子，且生产效率不高；板式精度不高，易造成空穴，但生产效率比针式高；滚筒式效率最高，且可用于异形种子，但精度不高，有种子有一定损伤。浙江一鸣机械设备有限公司生产的蔬菜花卉精量播种流水线已经投放市场，售价20 万元左右，约为韩国大东电器生产的Helper 蔬菜播种流水线售价的一半。

3.2　制钵机

钵体育苗效果明显，成活率高，而且无炼苗期，生长健壮。制钵机解决了人工制钵效率低下和劳动量大的问题，并能明显改善钵体成型质量。由农业部南京农机化所和江苏省滨海县金辉机械厂联合开发的ZB-2500 型双头对冲设施育苗制钵机制钵效率为手工作业的5 倍，操作人员2～3 人，生产率 ≥2 500只/h，成型率≥95%，钵体外紧内松，最大功率1.5kW，移栽成活率≥93.6%。

3.3 蔬菜移栽机

国内蔬菜移栽机可分为钳夹式、导苗管式和吊篮式等，可将穴盘苗和钵体苗定植到土壤中，吊篮式尤其适用于尺寸较大的钵体。我国对蔬菜移栽机的研究起步较晚，目前有沈阳农业大学工学院张为政等研发有悬杯式蔬菜移栽机[9]、黑龙江省畜牧机械化研究所研发的鸭嘴式蔬菜移栽机、现代农装股份有限公司研发的2ZJ-2型多功能移栽机等，但实际生产效果仍需要市场检验。

3.4 蔬菜嫁接机

蔬菜嫁接机常用于茄科作物和瓜类作物，能提高嫁接作业质量，提高嫁接苗愈合成活率，工作效率比人工至少提高一倍以上。日本洋马、井关等农机公司研发的蔬菜嫁接机工作可靠，效率可达800株/h以上，但价格普遍在35万元以上，远远超出一般农户和中小型育苗企业的承受范围。国内最早研发蔬菜嫁接机的是中国农业大学工学院张贴中教授，于1998年研制的2JSZ-600型自动蔬菜嫁接机适用于瓜类，工作效率为600株/h；东北农业大学辜松教授研制了2JC-350型蔬菜插接式嫁接机和2JP-1000型茄科蔬菜嫁接机，分别适用于瓜类和茄科，效率分别可达350株/h和1 000株/h；国家农业智能装备工程技术研究中心研发了TJ-300半自动蔬菜嫁接机和TJ-800型蔬菜嫁接机，适用于瓜类和茄科，前者只需一个人完成上苗和上夹，效率较低，后者需要两个人同时进行上苗操作，效率较高。

4 环境调控装备

设施内的影响作物生长的主要环境因子包括光、温、水、气、肥等，因此也对应了光环境调控设备、温度调控设备、灌溉机械设备、CO_2调控设备、养分调控及综合电气控制设备等六大环境调控装备。

4.1 光环境调控设备

光环境对设施蔬菜生产具有决定性的影响，有遮光、补光两种调控设备。无论是外覆盖还是内覆盖的遮光设备，大多采用电动或者手动齿轮副传动调节方式，可较好地控制光照度和光周期。补光常用光源有荧光灯、高压钠灯、低压钠灯、金属卤化物灯、LED光源等，除荧光灯和LED光源外，其他光源发热量都较大，既不能近距离照射作物，损失能量，也易灼伤作物。LED光源已经成功应用于植物组培、植物工厂等领域，但在温室补光中的研究还刚刚起步。

4.2 温度调控设备

温度是设施栽培的首要环境条件，任何作物生产都必须满足“三基点温度”。温度调控设备主要分为保温与加温设备和降温设备。

4.2.1 卷帘机

在设施屋面覆盖幕帘、无纺布、草帘甚至保温被等覆盖材料，可有效提高室温。我国北方日光温室常采用保温被和草帘等的多层覆盖，其覆盖物厚重，人力拉卷费时费工，雨雪大风天气易造成损失。卷帘机械可在几分钟内完成卷铺作业，大大减轻了劳动强度。卷帘机可按工作原理分为绳拉式和卷轴式，前者适用于较重的保温材料，后者适用于较轻的保温材料。

4.2.2　加温设备

按照热量来源不同，采暖系统可分为煤炭、燃气、电、地热等；按照热媒不同，可分为热风、热水、蒸汽等；按照吸热介质不同，又可分为加热空气、加热土壤（基质）两类。由于加温设备运行成本高、能耗大，因此我国加温温室面积不大。通常多在高档花卉温室、育苗温室及观光温室等产出较高的温室中应用加温设备，常用热风或者热水采暖系统。此外，北方一些地区也采用炉火烟道或者小型暖风炉加热。

4.2.3　降温设备

常用降温手段有通风、蒸发吸热以及二者结合等方式。自然通风简易有效，采用天窗、侧窗或者地裙上部扒缝、顶部扒缝等方式。单独使用蒸发冷却时，降温效果不明显，常与强制通风配合使用，效果良好。如湿帘风机系统、喷雾风机系统等，一般由风机、湿帘、水循环系统和控制装置组成。该系统设备简单、能耗低、效率高，北方夏季降温效果通常可达4～8℃，在南方高温高湿地区降温效果受到限制，但在正午高温时段仍可完成有效降温。

此外，育苗工厂、蔬菜工厂等环境要求严格的设施采用空调作为温湿度调节工具，具有控制准确的优点，但成本高、能耗大，只用于特殊设施环境中。遮阳网除了降低光照度和调节光周期外，也能起到一定降低室温的作用。

4.3　灌溉机械设备

传统设施栽培灌溉常采用沟灌、浸灌等方式，具有需水量大、土壤养分流失严重、劳动强度大、棚内湿度过大、难以有效进行作物需水量控制等缺点。目前，我国设施蔬菜栽培中正在大力推广节水灌溉技术，以喷灌技术和微灌技术为主，其中又以微灌技术应用最为广泛，已经在不少大田粮食作物、果树、蔬菜、花卉、药材、园林、苗圃、绿地等经济作物进行了大面积应用。微灌节水设备，包括滴灌、微喷灌等。滴灌和微喷灌的主要区别在于灌水器的不同，滴灌的灌水器分为滴头或者滴灌管（带）两类，微喷灌喷头则有折射式、离心式和旋转式。微灌节水技术除了具有节水、降湿等优点外，还可以随灌溉施肥，实现水肥合一、施肥均匀，有利于作物对养分的吸收，达到增产、优质的目的。

4.4　CO_2 调控设备

CO_2 是作物光合作用的原料，其浓度高低直接影响光合速率。设施内 CO_2 浓度过低或者分布不均匀，会造成作物光合强度降低或差异，影响作物产量和品质。CO_2 肥源常来源于液态 CO_2、燃料燃烧、颗粒气肥和化学反应，常见设备有以燃料燃烧为肥源、同时可用于加温的 CO_2 发生机、中央锅炉系统和净烟燃煤炉，以及强酸和碳酸盐反应产气的 CO_2 气肥发生器。此外，强制或者自然通风设备也可以迅速补充 CO_2，但浓度升高有限，寒冷季节不适用。

4.5　养分调控设备

养分调控设备常用于无土栽培技术，尤其是营养液栽培。该系统由液源、营养液搅拌设备、水泵、流量调节器、过滤器、输液管道等组成，常和滴灌系统结合使用，实现水肥合一。

5　植保机械

我国设施栽培中使用的传统植保机械，仍大部分采用手动背负式喷雾器。其价格相对便宜，但具有雾化效果差、药业覆盖不均匀、用水量大、劳动强度大、作业人员易中毒等诸多缺点。常温烟雾机、弥雾机、担架式喷雾机以及行走式喷雾机解决了易中毒、劳动强度大等问题，近年来得到大力推广。其中以常温烟雾机是塑料大棚和温室农作物的主要防治机具，其尤其适合对密集型作物进行防治，不受农药品种限制，工作人员在喷雾作业时无须进入棚内，节省农药和大量水资源，不增加温室内湿度。此外，驱虫灯、诱虫灯、臭氧防治机等病虫害防治机械也得到一定应用，但是防治效果还有待于加强。

6 收获机械

我国设施蔬菜栽培中使用的收获机械还较少，主要可分为果菜收获机械和叶菜收获机械。果菜收获机械常通过采摘机器人技术得以实现，中国农业大学工学院以李伟教授为首的研究团队研制了黄瓜采摘机器人，其中“基于多光谱段融合的非结构环境下果实信息获取与感知关键技术”达到国际先进水平[10]。国家农业智能装备工程技术研究中心研制了草莓嫁接机，可以实现对下垂的草莓果实模型的采摘。

7 发展前景展望

综合分析设施农业装备机械发展现状，我国设施农业装备发展还面临行业瓶颈和技术难题。应当加大对设施农业农机装备的投入，加强农机部门与农艺部门之间的互相配合与协调，农机研发和推广部门应积极同农技推广部门做好农机农艺融合工作，有助于设施农业装备的推广和发展。同时，农机研发高校和单位应当更多关注移栽、采收以及叶菜清洗包装等我们尚未拥有自主知识产权和核心技术的设施农业装备，争取尽早实现设施蔬菜栽培全程机械化；也应当加强对 CO_2 气肥发生器、臭氧杀菌消毒等可以明显提高蔬菜产量品质的装备的研究和推广。

参考文献

[1] 杨其长．中国设施农业的现状与发展趋势．农业机械，2002，2（1）：36～37
[2] 杨　雪．五大措施推动我国设施农业科学发展．中国农机化导报，2009，11（30）：1
[3] 王国占．设施农业及装备技术发展成绩显著．中国农机化导报，2010，12（27）：10
[4] 乐农网．南京市年内完成10万亩设施农业建设［EB/OL］http：//www.chinalenong.com/snyw/snyw_detail.aspx?id=3417，2011-04-09
[5] 龙建明，李敏科，朱亮亮等．设施农业机械设备及应用．农机化研究，2006（11）：50～53
[6] 张晓文，王影，邹岚，程存仁等．中国设施农业机械装备的现状及发展前景．农机化研究，2008（5）：229～232
[7] 樊桂菊，李汝莘，杜辉．国外设施农业机械的发展．农业装备技术，2003（2）：47～48
[8] 戴志中．我国设施农业机械的现状及发展趋势分析．中国农机化，2000（4）：10～11
[9] 张为政，王君玲，张祖立．悬杯式蔬菜移栽机的设计．农机化研究，2011（8）：104～106
[10] 冯青春，袁挺，纪超等．黄瓜采摘机器人远近景组合闭环定位方法．农业机械学报，2011，42（2）：154～157

组培苗瓶外生根技术研究进展*

刘　丽[1**]，丁国昌[1,2]，周锦业[3]，马志慧[1]，许珊珊[1,2]，林思祖[1,2***]
（1. 福建农林大学林学院，福州　350002；2. 福建省杉木工程技术研究中心，
福州　350002；3. 福建农林大学园林学院，福州　350002）

摘要：随着植物组织培养技术的迅速发展，近年来试管外生根技术也已经逐步成熟并被广泛应用于工厂化育苗实践中。本文总结了组培苗瓶外生根技术的研究进展，为进一步开展组培苗瓶外生根技术提供参考。
关键词：组培苗；瓶外生根；生根方法；影响因素

Progress of Study on ex vitro rooting of plantlets in vitro

Liu Li[1], Ding Guochang[1,2], Zhou Jinye[3], Ma Zhihui[1], Xu Shanshan[1,2], Lin Sizu[1,2]
(1. *Forestry College*, *Fujian Agriculture and Forestry University*, *Fuzhou* 350002, *China*;
2. *Chinese Fir Engineering Research Center of Fujian*, *Fuzhou* 350002, *China*;
3. *College of Landscape Architecture*, *Fujian Agriculture and Forestry University*, *Fuzhou* 350002, *China*)

Abstract: With the rapid development of plant tissue culture technology in recent years, ex vitro rooting technique is getting mature gradually and put into industrialized breeding practice widely. This paper summarized the progress in ex vitro rooting technology, which provided a reference for the further development in ex vitro rooting technology.
Key words: Plantlets in vitro; Ex vitro rooting; Rooting methods; Influence factors

针对有些植物种类在组培瓶中难以生根或根系发育不良，移栽后成活率低的特点，同时也是为了缩短育苗周期和降低生产成本，1988 年，英国格拉斯哥植物园秋海棠研究中心成功开发了秋海棠瓶外生根技术[1]，并得到越来越广泛的应用，成为近年来组培生根技术的研究热点。

在国外，试管外生根技术广泛被应用到苗木商业化生产中，特别是一些名贵药用植物、观赏性植物以及濒危植物的开发利用[2,3]。研究内容主要包括瓶外生根的方法、瓶外重要的主要影响因素及物形态解剖学上的比较研究等几个方面[4,5]。目前，我国在工厂化育苗方面仍以瓶内生根为主，虽然近几年对瓶外生根技术的研究逐渐增多，也还是处于一个尚未成熟的阶段，虽然部分物种可以通过试管外生根技术来大量育苗，主要集中在药用植物、观赏性植物、果树等植物上[6~8]，但是，大部分物种还没有形成一个相对比较成熟的技术体系。本

* 基金项目：福建省自然科学基金项目（2010j01063）资助
** 作者简介：刘丽（1987—），女，江西南城县人，硕士研究生，研究方向：森林培育理论与技术。E-mail：jiangxiliuli@ qq. com
*** 通讯作者：林思祖（1953—），男，教授，博导，研究方向：森林培育理论与技术。E-mail：Szlin53 @ 126. com

文对有关研究进行了总结，期望给该项技术能很好地运用在大部分物种的工厂化育苗中，也为该技术的推广应用提供借鉴。

1 组培苗瓶外生根的方法

瓶外生根的方法很多，有基质培养法、水培养法、气雾培养法以及滤纸桥法等[9]。国内外大部分研究侧重于使用基质培养法，而其他方法较为少见。

1.1 基质培养法

基质培养法是国内外在研究试管外生根时使用最多的一种方法，大多数都采用微枝扦插到基质中生根的方法，一般使用两种或3种基质按一定比例混合使用，也有只使用一种基质的，一般情况下两种或两种以上混合的基质较单一的好。邓桂秀等[10]以南方高丛蓝浆果（*V. corymbosum hybrids*）增殖培养若干代的组培苗为试验材料，研究发现以泥炭+珍珠岩（v：v=1：1）基质上的生根率大于以苔藓为扦插基质，而且生根苗的长势最好。

1.2 水培养法

董玲等[11]采用瓶外水培生根法进行满天星组培苗生根，得到很好的效果。邢朝斌等[12]研究了4种生根方法（组培法、水培法、扦插法和滤纸桥法）对非洲堇组培苗生根的影响。结果表明，4种方法均能诱导其产生不定根，但水培法较另外3种方法更有利。

1.3 气雾培养法

Isutsa等[13]研究使用气雾培养法对蓝莓组培苗进行生根试验。冯学赞等[14]较为详细的研究了木本植物和草本植物的气培生根法，该方法用于工厂化育苗能显著地提高生产效率，并且能使试管苗的育苗成本下降35%以上。

1.4 滤纸桥法

滤纸桥法是在水培法的基础上进行改进的一种方法，该方法克服了水培法中根系长期浸泡在水中导致根系发育不良的问题。孙仲序等[15]研究使用滤纸桥法对珠美海棠、樱桃colt砧木、弗吉尼亚草莓组培苗进行瓶外生根试验。所用成本明显下降到原来的13.26 %。滤纸桥法简单、易行、低廉、快速，并且已经在生产上推广应用，效果良好。

2 组培苗瓶外生根的主要影响因素

一般来说，影响组培苗瓶外生根的因素主要有苗木的质量、基质类型，环境条件以及植物生长调节剂。林艳等[16]对影响日本花椒组培苗瓶外生根的主要因素进行了研究，结果表明，组培苗质量、生根基质、生根促进剂浓度及环境因素（光照、温度、湿度）对其瓶外生根均会产生显著影响。

2.1 植物种类及苗木质量对组培苗瓶外生根的影响

植物种类不同以及相同植物的不同基因型，产生不定根的难易程度也会有所不同[17]。程淑云[18]对4个蓝莓无性系组培苗进行瓶外生根试验结果表明，不同无性系之间组培苗生根率差异较大。此外，苗木的质量也会影响瓶外生根率。因此，一般在生根前对组培苗进行壮苗培养，生根效果会很好。刘敏等[19]在研究欧美杨107组培苗瓶外生根时得出瓶内复壮后瓶外生根比直接瓶外生根好。在进行瓶外生根之前进行炼苗可提高成活率，在一定范围内，炼苗时间越长组培苗移栽成活率越高[20]。另外，采用松口炼苗比封口炼苗能提高组培苗的瓶外生根率[21]。

2.2　基质类型及含水率对组培苗瓶外生根的影响

随着科学技术的发展，除土壤、河沙外，出现了越来越多的轻质材料基质，像珍珠岩、蛭石、锯末、岩棉、海绵等。还有一些含有丰富养分的基质，比如，草炭土、泥炭土、腐殖土、苔藓等。除此之外，还有用到煤渣、塑料泡沫等，使得资源得到了充分利用，同时也为工厂化育苗节约了成本。基质种类不同，其理化性质也往往不同，主要体现在养分含量、保水能力、透气性以及酸碱度等[17]。黄国辉等[22]以蓝莓组培复壮苗为插条，研究发现不同基质对生根率有较大影响，苔藓上最高，达60%，苗质量好；其次是草炭土上为26.67%，苗质量较好；而2份草炭土：1份珍珠岩和2份草炭土：1份河沙生根情况较差。另外，基质的含水率也是影响生根率的重要因素，基质中水分过少时，会抑制根系的生长，但是过多又会造成缺氧，对根的生根有害，甚至不会正常生根，苗木基部容易腐烂[23]。梁立东等[24]在研究新西伯利亚银白杨微枝试管外生根与微环境因子的关系时，发现基质含水率为50%的微枝生根效果最好。

2.3　环境条件对组培苗瓶外生根的影响

瓶外生根过程中的环境条件影响因素主要体现在湿度、光照、温度3个方面。这里湿度主要指空气湿度，湿度过高会使得扦插苗叶片腐烂，容易滋生菌；但是过低，苗易萎蔫失水。李玉巧等[25]在研究红花刺槐瓶外生根中发现，当湿度为75%时，生根率只有2.6%，当湿度为85%时，生根率为79.8%。光照从光照强度、光质、光周期3个方面影响瓶外生根。组培苗出瓶扦插，由异养变成自养，光照强度自然要适当提高，以促进其光合作用；但是，光照强度又不能太强，过强使得微枝叶绿素遭到破坏，蒸腾作用也会加强，破坏水分平衡。关于培养温度的控制主要要依据植物的生态类型来决定，一般情况下认为在20～28℃下，微枝能正常生根。

2.4　生根促进剂的种类、浓度和处理时间

生长素是促进植物不定根形成的主要激素。其种类、浓度和处理时间都会不同程度地影响微枝不定根的形成[26]。谭巍等[27]以草莓继代组培苗为瓶外生根试验材料，研究结果得出激素浓度对组培苗试管外生根的影响显著，而激素种类对组培苗试管外生根的影响不显著。国内外研究中使用较多的生长素为NAA、IBA、ABT等，有将两种生长素混合使用的，也有只使用一种，大量试验研究证明，在进行试管外生根前使用一定量的生长素对试验材料进行预处理一定时间，可以普遍提高生根率以及产生不定根的数量。Shekafandeh[28]研究了不同植物生长调节剂对伊朗桃金娘微枝瓶外生根的影响，结果表明，不加任何处理和用浓度为1.5mg/LNAA+0.31mg/LIBA浸泡插条24h相比，微枝生根率由0变为91.7%。

3　瓶外生根技术中存在的问题及发展方向

（1）关于瓶外生根的生根机理方面的研究，在国外仅见在栀子花和栗树上的研究POD酶活性与不定根形成过程中的关系有报道[3,29]，今后可加强瓶外生根机理方面的研究，进一步促进瓶外生根技术的发展与应用。

（2）瓶外生根方法上有基质培养法、水培养法、气雾培养法以及滤纸桥法，但目前应用最多的是基质培，在实际应用中应该根据植物的生物学特性来选择合适的方法。

（3）在对植物进行瓶外生根时，首要考虑的应该是明确在生根率达到最高时生长素的种类、浓度以及预处理时间，其次在此基础上结合完善的环境调控设施，才能使其朝有利于生根的方向生长发育。因此，在今后的研究中，环境调控与化学调控的有益结合对建立适宜各类苗木组培苗瓶外生根技术体系形成组培苗瓶外生根的标准化将是一个重要的发展方向。

4 总结

组培苗瓶外生根技术相比传统瓶内生根具有苗木质量高、育苗周期缩短、育苗成本降低、移栽成活率高等优势，在商业化组培苗的生产中具有非常大的发展潜力，是一项很实用并且值得推广的育苗技术。

参考文献

[1] 徐振华，王学勇，李敬川等．试管苗瓶外生根的研究进展．中国农学通报．2002（4）：84～86

[2] Singh A，Reddy M P，Chikara J，*et al.* A simple regeneration protocol from stem explants of Jatropha curcas—A biodiesel plant. Industrial Crops and Products. 2010，31（2）：209～213

[3] Jose B，Satheeshkumar K，Seeni S. A protocol for high frequency regeneration through nodal explant cultures and ex vitro rooting of *Plumbago rosea* L. Pak J Biol Sci.，2007，10（2）：349～355

[4] Hatzilazarou S P，Syros T D，Yupsanis T A，*et al.* Peroxidases，lignin and anatomy during in vitro and ex vitro rooting of gardenia（Gardenia jasminoides Ellis）microshoots. Journal of Plant Physiology. 2006，163（8）：827～836

[5] Borkowska B. Morphological and physiological characteristics of micropropagated strawberry plants rooted in vitro or ex vitro. Scientia Horticulturae. 2001，89（3）：195～206

[6] 江芹，董玲，宁志怨等．百蕊草无性系建立与瓶外生根研究（英文）．Agricultural Science & Technology. 2008（5）：47～49

[7] 叶顶英．三角梅组培苗试管内外生根研究．北方园艺．2011（15）：169～171

[8] 焦淑华．欧李试管苗瓶外生根研究．赤峰学院学报（自然科学版）．2008（1）：31～33

[9] 黄卓忠，严华兵．试管苗瓶外生根技术．广西农业科学．2007（2）：124～126

[10] 邓桂秀，宋鹏飞，姜燕琴等．‘南月’蓝浆果实生优株组培苗瓶外生根研究．北方园艺．2010（13）：132～134

[11] 董玲，陈静娴，聂凡．组培满天星生根途径试验报告．安徽农业科学．1997，25（3）：272～274

[12] 邢朝斌，李占军，李玉彦，等．非洲堇无根苗的生根方法研究．安徽农业科学．2007（35）：11 386～11 412

[13] Isutsa D K，Pritts M P. Rapid Propagation of Blueberry Plants Using ex Vitro Rooting and Controlled Acclimatization of Micropropagules. Hort Science. 1994，29（10）：1 124～1 126

[14] 刘立秋，冯学赞，陈文龙等．组培苗气培生根法的研究．林业科技通讯．1996（10）：18～20

[15] 孙仲序，王玉军，刘静等．植物组培快繁滤纸桥生根新技术．山东农业大学学报（自然科学版）．2002（3）：257～263

[16] 林艳，郭伟珍，毕君等．日本花椒组培苗瓶外生根影响因素研究．林业科技开发．2004（5）：38～39

[17] 沈海龙等．树木组织培养微枝试管外生根育苗技术．北京：中国林业出版社，2009

[18] 程淑云．蓝莓组培苗瓶外生根技术的研究．农业科技通讯．2009（4）：48～50

[19] 刘敏，苏乔，刘纪文．‘欧美杨107’组培苗瓶外生根．植物生理学通讯．2010（10）：1050～1054

[20] 郭琪，杨晖，张军等．桃儿七组织培养苗炼苗移栽技术．林业实用技术．2012（9）：29～31

[21] 何云芳，余有祥，裘丽珍等．金线莲组培苗的试管外生根和大田移栽技术．浙江林业科技．1998（2）：22～25

[22] 黄国辉，姚平．蓝莓组培苗瓶外生根的研究．江苏农业科学．2011（4）：227～228

[23] 杜振宇，马海林，王清华等．不同基质对大樱桃试管苗生长的影响．山东林业科技．2003（2）：3～5

[24] 梁立东，杨玲，沈海龙等．新西伯利亚银白杨微枝试管外生根与微环境因子的关系．安徽农业科学．2009（30）：14 983～14 986

[25] 李玉巧，梁珍海，蒋泽平等. 2种木本植物无根试管苗的移栽技术. 南京林业大学学报（自然科学版）. 2005（2）: 69 ~ 72
[26] 王蒂. 植物组织培养. 北京：中国农业出版社，2004
[27] 谭巍，沙春燕，王娟. 不同处理对草莓试管苗瓶外生根影响研究. 中国林副特产. 2010（2）: 13 ~ 15
[28] Shekafandeh A. Effect of Different Growth Regulators and Source of Carbohydrates on in and ex vitro Rooting of Iranian Myrtle. International Journal of Agricultural Research. 2007，2（2）: 152 ~ 158
[29] Gonçalves J C，Diogo G，Amâncio S. In vitro propagation of chestnut（*Castanea sativa* × *C. crenata*）: Effects of rooting treatments on plant survival，peroxidase activity and anatomical changes during adventitious root formation. Scientia Horticulturae. 1998，72（3 ~ 4）: 265 ~ 275

设施园艺工程技术

Evaluation of Global, Photosynthetically Active Radiation and Diffuse Radiation Transmission of Agricultural Screens

M. Romero-Gámez, E. M. Suárez-Rey, T. Soriano and N. Castilla
(*IFAPA-Centro Camino de Purchil. Apartado* 2027, 18080 *Granada. Spain.*)

Abstract: Transmittance of a material depends on the type of radiation impinging on the material (direct or diffuse), the angle of incidence of the sun's rays (in direct radiation conditions) and the structure and characteristics of the material itself. The aim of this study was to evaluate the performance of nine agricultural screens of different densities, colours, thread diameters and porosities. A simple metal frame was used to quantify global, diffuse and photosynthetically active radiation (PAR) transmission to determine their transmittance values (as a function of the incidence angle of solar rays in direct radiation conditions) for global radiation and PAR, as well as the diffuse radiation transmitted characteristics of the screens (ratio of diffuse radiation to global radiation transmitted by screens). Non-coloured (translucent) screens contributed to a higher and more efficient proportion of diffuse radiation ($D_i/G_i \approx 90\%$). The green screen behaved similarly to the non-coloured 20 × 10 type as far as global radiation was concerned, although the PAR transmittance values were lower (up to 12.4% at 15°). Variation in transmission according to diameter and density of the threads was greatest in the black screens. The non-coloured-thread screens may be the best option for maximising the transmission of diffuse radiation and the 6 × 6 green screen for vegetables cultivation during the summer in inland areas with latitudes close to 36°N.

Key words: Covering materials; Incidence angle; Porosity; Screenhouse; Transmittance

1 Introduction

During recent years growing in screenhouses vegetable crops has taken on considerable socio-economic importance especially for small-growers in semiarid and arid countries. In southern Spain, these systems are complementing the growing and marketing calendars of the plastic greenhouses production in the coastal areas.

Transmittance of a material depends not only upon the type of radiation impinging on the material (direct or diffuse) and the angle of incidence of the sun's rays (in direct radiation conditions), but also upon the structure and characteristics of the material itself (composition, porosity, colour and shape and diameter of the thread) (Soni *et al.*, 2005; Soriano *et al.*, 2006; Sica & Picuno, 2008) and the absorbance and reflectance properties of the screen. As far as direct radiation is concerned, the more perpendicular the incidence of the sunlight (lower incidence angle), the more radiation will be transmitted through the screen, whilst at greater angles of incidence transmission will decrease.

Screens are characterized by their porosity, which influences their shading effect, and it is determined by the diameter and physical characteristics of the thread and the density of the screen (number of threads per centimetre), which in turns, affects its durability, global weight, strength, elasticity and insect exclusion (Castellano *et al.*, 2006).

Screens with black threads are the most common coverings used for shading structures, even though the use of non-coloured and coloured threads screens is increasing because of their capacity to modify solar radiation quality characteristics, their more favourable aesthetic impact upon the countryside (Castellano *et al.*, 2008b) and their ability to improve the microclimate providing the physical protection required (Shahak *et al.*, 2002). Various studies have demonstrated that when crops are grown beneath coloured screening, there is a notable increase in production, reduction of stresses and a general improvement in the vitality of the plant canopy, ripening, fruit colour, size and quality, together with a reduction in sunburnt. Studies on the transmittance properties of screens reveal that the colour of the screen influences the spectral distribution of the light that passes through it by absorbing its complementary colours (Castellano *et al.*, 2008a).

Crops such as fruit vegetables with a high plant canopy utilize diffuse radiation better than direct radiation, as diffuse radiation penetrates the middle layers of a high-grown crop and results in a better horizontal radiation distribution in the greenhouse (Hemmig *et al.*, 2008). For this reason, as stated by Cabrera *et al.* (2009), since the pioneering work of Deltour & Nisen (1970), laboratory studies aimed at characterizing the diffusive properties of greenhouse cladding materials have become of paramount interest (Pearson *et al.*, 1995; Wang & Deltour, 1999; Montero *et al.*, 2001; Pollet *et al.*, 2005). Diffuse radiation represents a significant fraction of the total solar radiation that enters a greenhouse (Baille & Tchamitchian, 1993), although few detailed studies have been undertaken to characterize diffuse radiation under greenhouse conditions (Basiaux *et al.*, 1973, Burek *et al.*, 1989) and little information is available about this component and its proportion compared to direct radiation (Baille & Tchamitchian, 1993; Hanan, 1998). Most of the studies carried out *in situ* have compared outdoors and under protection global radiation to estimate the material transmission to global radiation, but this offers little information about the diffusive properties of the covering material. Abdel-Ghany & Al-Helal (2010) measured diffuse radiation under different screens with incident angles of solar beam varying from 0° to 5°.

Diffuse radiation within a plastic greenhouse can be estimated on the basis of the radiation outside and two parameters: the coefficient of enrichment of diffuse radiation (D_i/D_o) and the factor ($\tau_{b\text{-}d}$) for converting direct radiation into diffuse radiation (Baille *et al.*, 2003). Knowledge of the coefficient of diffuse radiation enrichment (D_i/D_o) is of prime importance in crop models studies (Spitters, 1986), especially in those aimed at assessing the advantages and efficiency of a determined greenhouse cladding material. As mentioned by Cabrera *et al.* (2009), the principle challenge in the future will be that of maximizing the beneficial effects of diffuse radiation upon the homogeneity of interception of radiation by the plant canopy and upon factors affecting crop yield.

A common commercial parameter defining a screen is the shading factor, which describes the ability of a screen to absorb and reflect the visible range (380 ~ 760nm) of the sun radiation (Castellano *et al.*, 2008b). But the commercial shading degree does not take into account diffuse radiation, which is an important component of global sunlight (above all on cloudy days, when it is the major fraction).

There is no European norm to define the spectroradiometric properties that screens for agricultural use should comply with (Sica & Picuno, 2008) or certify the quality of the screen to guarantee its homogeneity (Soler *et al.*, 2007). Worldwide, there is just one national norm in Italy (10335/94; UNI, 1994) that regulates the methodology for assessing screens used for shading. Other national standards deal partly with agricultural films such as the French standard NF EN 13206 (NF, 2002) and the Italian standards UNI 9738 (UNI, 1990) and UNI 9298 (UNI, 1988) (Castellano *et al.*, 2008c).

The aim of this study was an experimental evaluation of the performance of nine of the most commonly used in Spain screening materials of different densities, colours, thread diameters and porosities to: ①document the global and PAR transmittance values, ②evaluate the screen transmission for diffuse solar radiation as well as the alterations of the radiation diffuse component when passing through the screens. These measurements were performed on eleven different angles of incidence (from 15° up to 65°) using a simple frame structure.

2 Material and Methods

A metal framework® (Soriano *et al.*, 2008) was used to measure the solar radiation transmission of different plastic screens in various conditions and to determine the solar radiation transmittance values of shading screens. To quantify solar radiation transmittance at various angles of incidence of direct sunlight upon the surface of the materials, the panel where the screens were located was movable.

The framework (Figure 1) was a black metal structure made of hollow, square, metal bars (4cm ×4cm) with a square base of 3m ×3m. Each corner of the structure supported a 3m high metal bar (4cm ×4cm). The arms supporting the material being tested were inserted into the top of two rear metal supports in such a way as to be able to swivel up and down at different known angles to the horizontal. The framework was also fitted with wheels to allow it to be directed towards the sun at all times according to the azimuth angle. The base of the framework had a horizontal platform to incorporate the measuring radiation sensors.

The study was conducted at the "IFAPA Camino de Purchil" Centre of Granada, located in the agricultural plain of Granada (37°10' 21" N 3°38' 10" W; 600m asl). Radiation measurements were made on fully sunny days during February, March and April 2009 between 8: 00 and 12: 00 a. m. solar time and on fully cloudy days in January to obtain τ_{b-d} (direct to diffuse radiation conversion factor, Baille *et al.*, 2003):

$$\tau_{b-d} = D^*/I_o = (D_i - \tau_d D_o)/I_o \qquad [1]$$

Where D^* = direct solar radiation converted into diffuse radiation under the cladding material ($W\ m^{-2}$); I_o = outside direct solar radiation ($W\ m^{-2}$); D_i = diffuse solar radiation under the cladding material ($W\ m^{-2}$); τ_d = transmission of diffuse solar radiation (equivalent to global radiation transmission on completely cloudy days); and D_o = outside diffuse solar radiation ($W\ m^{-2}$).

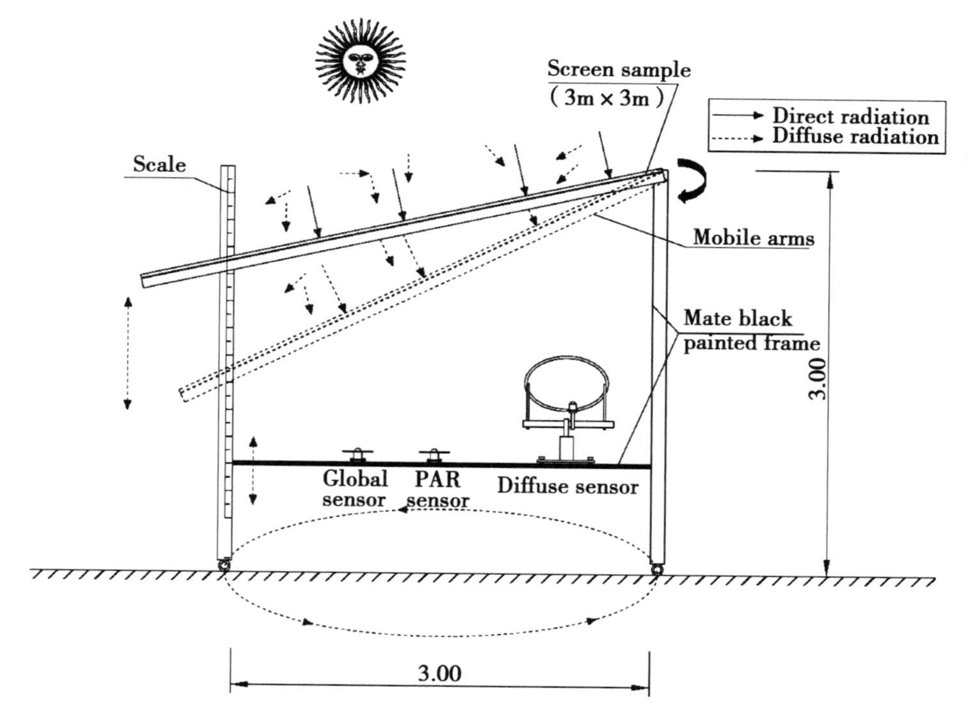

Figure 1 Framework to measure the transmission of solar radiation of plastic netting

To characterize the transmission of solar radiation in the screens, we used global radiation sensors (CM6B, Kipp & Zonen), PAR sensors (SKP215/S, Sky Instruments) and shadow rings (CM121B, Kipp & Zonen) for the diffuse radiation. Thus, three sensors were installed on the chassis of the framework beneath the screens and three more on a flat platform just outside the frame. All the sensors were previously calibrated. A correction factor (S) is given by the manufacturer of the ring of shade and was applied in the calculation of the diffuse radiation to compensate for the reduction in diffuse radiation in the resulting band of shade (Kipp & Zonen BV, Holland).

All sensors were connected to a CR10X datalogger (Campbell), programmed to simultaneously measure every 2min and record mean values every 10min (five repetitions). The transmission of the screens was calculated using the ratio of the radiation values measured by the sensors beneath the screens and outside the framework.

Nine different kinds of screens were tested. These screens are commonly found for shading in commercial screenhouses of inland areas in southern Spain. They were composed of a weave of high-density polyethylene threads of different diameter, colour and density: both black and non-coloured (white) screens with 20 × 10, 16 × 10, 9 × 6 y 6 × 6 threadscm^{-2}, and a green screen with 6 × 6 threadscm^{-2}, all of which were manufactured by Condepols. The geometric characterization was determined for the nine screens, counting the number of threads, measuring their diameter (Digital Caliper 0 ~ 150nm), and calculating the number of holes and the size of the hole and, finally, the porosity of the screen (Table 1). The recorded angles of incidence varied from 15° up to 65° in a 5° step.

Table 1 Main characteristics of the plastic screens studied

Density (threadscm^{-2})	Colour	Thread diameter (mm)	Weight (g m^{-2})	Hole size (mm^2)	Porosity (%)
20 × 10	Black	0.23	150	0.18	31.00
	Non-coloured	0.25		0.15	25.00
16 × 10	Black	0.23	122	0.30	40.20
	Non-coloured	0.25		0.26	35.00
9 × 6	Black	0.28	95	1.45	58.00
	Non-coloured				
6 × 6	Black	0.28	75	2.57	66.40
	Non-coloured				
	Green				

3 Results and Discussion

Transmittance values of the screens for global radiation and PAR

The global radiation and PAR transmittance values of the tested screens are shown in Figure 2. The global radiation and PAR behaviour of the various screens depending on the angle of incidence, was similar: an increase in the angle of incidence resulted in a reduction in radiation transmittance in all the samples. These results agree with those of Montero *et al.* (2001) for global radiation. At equal porosity, the non-coloured screens were the most transmissive to global radiation and PAR, followed by the green screen and finally the black screens. Similar results were also obtained for global radiation by Soriano *et al.* (2006). The threads of the black screens absorbed all the radiation wavelengths, contributing to a relevant reduction in transmittance as compared with the green or the non-coloured threads. Non-coloured and in a lower level, the green screen, were partially translucent contributing to increasing global and PAR transmittance relative to the black screens. Möller *et al.* (2010) also found that non-coloured and clear screens scattered radiation significantly.

The highest global and PAR transmittance values were reached with the 6 × 6 non-coloured screens, which attained values of 88% for global radiation transmittance and close to 86% for PAR transmittance at the lowest angle of incidence studied (15°). On the other hand, the highest values for black screens were close to 66% and 61% for global radiation and PAR respectively, the least dense weave of 6 × 6 being once more the most transmissive (Figure 2).

The green screen behaved more closely to the non-coloured types in terms of transmittance to global radiation although as far as PAR was concerned, their transmittance values were quite low which could be attributed to the fact that the green threads reflect a considerable amount of radiation in the green wavelength (500 to 600nm). Castellano *et al.* (2008a) measured transmittance under field conditions. They found out that PAR transmittance under the green screen behaved very closely to the non-coloured one within the range of 500 ~ 550nm. According to these authors, the

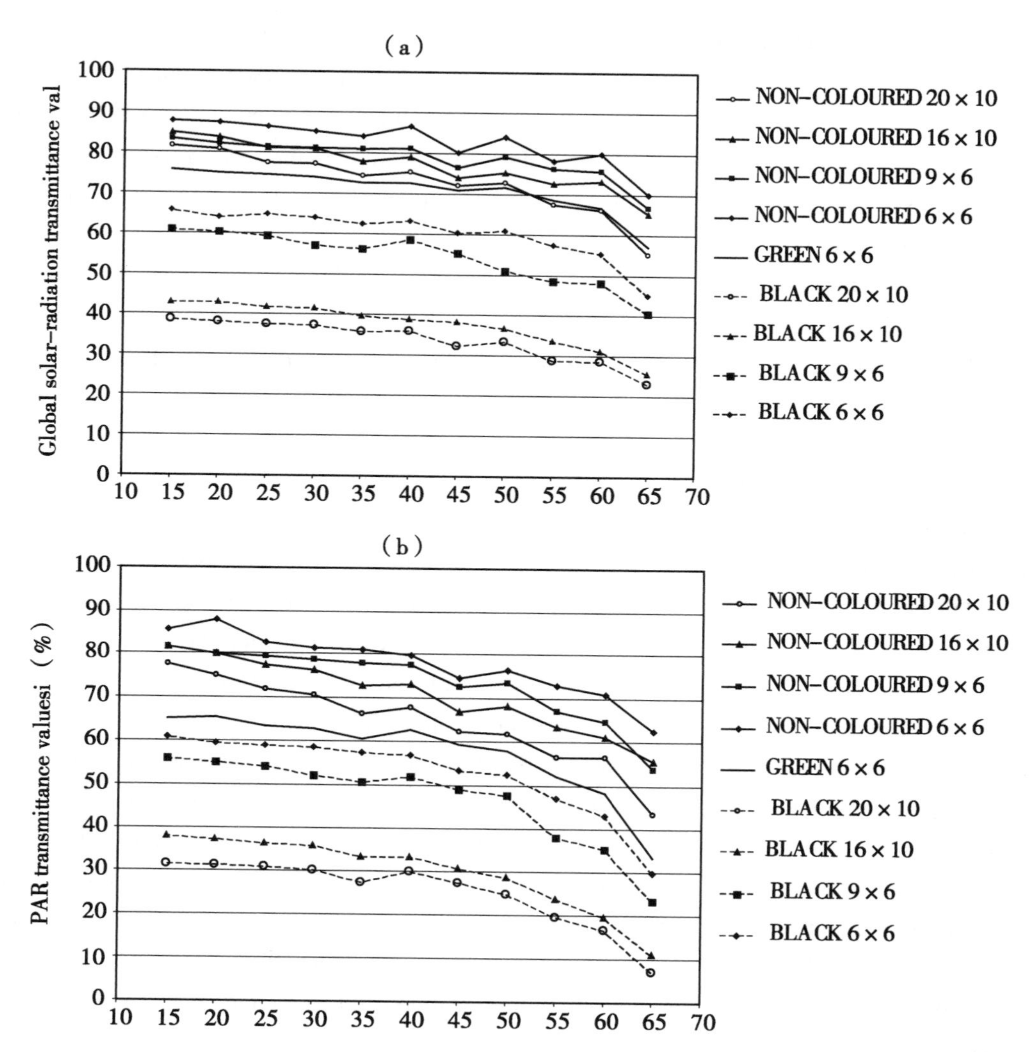

Figure 2 (a) Global radiation transmittance values (%) and (b) PAR transmittance values (%) *versus* angles of incidence (15°、20°、25°、30°、35°、40°、45°、50°、55°、60° and 65°) for both 20 × 10、16 × 10、9 × 6 and 6 × 6 threadcm^{-2} black and non-coloured screens, and a 6 × 6 threadscm^{-2} green screen

difference in PAR transmittance between these two screens was greater in the 550 ~ 600nm range, with a maximum difference of 6% at 600nm.

Figure 2 shows that the global and PAR radiation transmittance of the lower porosity non-coloured screen (20 × 10) was lower than that of the higher porosity green screen (6 × 6) at high angles of incidence. The maximum global and PAR transmittance values for the green screen (76% and 65% respectively) were close to those obtained with the densest non-coloured screen (20 × 10). Variation in transmittance according to the diameter and density of the threads was greatest in the black screens.

Diffuse component of solar radiation transmitted by screens (D_i/G_i)

Diffuse radiation on a sunny day is very low as compared with the direct radiation in our conditions. Daily average of 10.7% (JD 71) was registered for diffuse solar radiation, while the remaining 89.3% was direct global radiation. These values were similar to those recorded by the group of the Applied Physic Department at the University of Granada (12.6% diffuse-87.4% direct solar radiation; Tovar *et al.*, 2001).

Transmitted diffuse radiation with respect to transmitted global radiation (D_i/G_i in %) for the different screens tested and the various angles of incidence are displayed in Figure 3. As can be observed, the densest non-coloured screens greatly enriched diffuse radiation at all angles of incidence, followed by the more open weave non-coloured screens and for the green screen. In particular, at the highest angle of incidence, $D_i/G_i \approx 90\%$ for the 20 × 10 non-coloured screen, 62.4% for the 16 × 10 non-coloured screen followed by the more open weave non-coloured screens (50.7% for the 9 × 6 and 45.6% for the 6 × 6 at 65°) and for the green screen (≈ 28% at 65°). This effect is mainly due to the reflective and diffusive properties of the non-coloured and the green screen. Möller *et al.* (2010) found a further enrichment in diffuse radiation because of the higher scattering in the densest screens (largest for the 50mesh screens than for the 25mesh screens). Screening made of non-coloured threads maximized the transmission of diffuse radiation and the conversion of direct into diffuse radiation (Castellano *et al.*, 2008a) because these threads were translucent and thus allow part of the solar radiation through them in the same way as a light-diffusing plastic film. The higher the thread density, the higher proportion of diffuse radiation could be found in the transmitted radiation in these screens. The indicated values of the ratio (D_i/G_i) for non-coloured screens, especially in the lower porosity types (up to ≈ 90%) (Figure 3), were higher than those of light-diffusing polyethylene covering films reported by Cabrera *et al.* (2009). Möller *et al.* (2010) found that diffuse radiation below a non-black screen was much larger than that above the screens because of the contribution of scattered direct radiation in comparison to the diffuse radiation below the screens.

At high angles of incidence, we may expect that the translucent characteristics of the threads contribute to scatter the transmitted radiation, the whole screen acting similarly to a plastic film.

As far as black screens are concerned, however, transmitted diffuse radiation was considerably lower than the non-coloured screens, showing maximum values for D_i/G_i of 18.6% at the highest angles of incidence. Black threads are opaque and solar radiation is not modified by the screen (Castellano *et al.*, 2008b) because the threads absorb all wavelengths of light and no radiation scattering is produced. The lowest reductions in transmission with black screening were found in the 6 × 6 screen at all angles of incidence, with very similar values for the 20 × 10、16 × 10 and 6 × 6 screens at low angles.

The percentage of diffuse radiation transmitted by screens changed from 60.3% for the 20 × 10 non-coloured screen and 21.1% for the green screen to 10.4% for the 20 × 10 black screen, at an incident angle of 15°. Other authors found values from 170% for the dark-green screen to 17% for

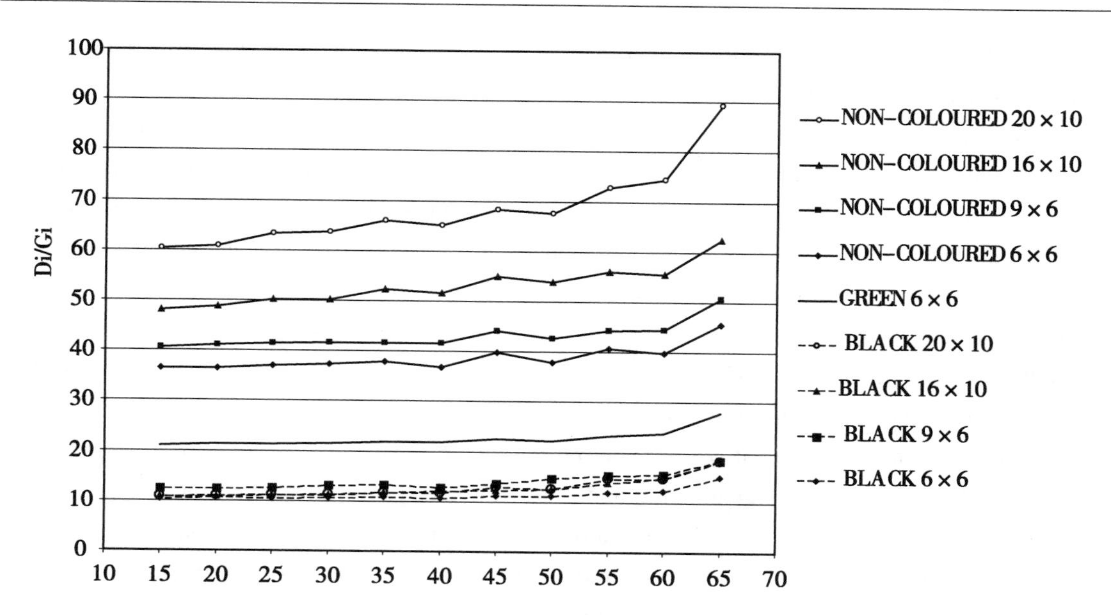

Figure 3 Diffuse component of solar radiation transmitted (D_i/G_i) *versus* the angles of incidence (15°、20°、25°、30°、35°、40°、45°、50°、55°、60° and 65°) for both 20 × 10、16 × 10、9 × 6 and 6 × 6 threadcm^{-2} black and non-coloured screens, and a 6 × 6 threadscm^{-2} green screen

the black screen with incident angles of solar beam varying from 0° to 5° (Abdel-Ghany & Al-Helal, 2010).

These results agree with those of other authors, who have determined that the diffusion of the covering material varies greatly according to differences in its structure (Deltour & Nisen, 1970; Basiaux *et al.*, 1973; Raviv & Allingham, 1983; Pearson *et al.*, 1995; Abdel-Ghany & Al-Helal, 2010).

Therefore, relative to diffuse radiation, three effects could be highlighted: ①the densest non-coloured screens enriched diffuse radiation more than the less dense types because of the light scattering effect caused by their higher number of threads; ②the black screens reduced diffuse transmitted radiation because their threads are opaque to solar radiation and the impinging diffuse radiation (coming from all sky directions) was therefore, only partially transmitted; ③the angle of incidence showed higher influence on diffuse radiation transmission for the high density non-coloured screens than for the lower density ones.

Transmission of diffuse solar radiation (τ_d), measured in completely cloudy conditions, was highest in the densest non-coloured screens, with values around 70%. In the lowest density (6 × 6) types, the non-coloured screen, τ_d was 64%, higher than the green (58%) and black (54%) types.

The higher value of direct to diffuse radiation conversion factor ($\tau_{b\text{-}d}$) calculated from equation 1, was reached for the 20 × 10 non-coloured screen (0.81). In the black screens, the proportion of direct solar radiation falling on the screen that is converted into diffuse radiation was much lower, with values between 0.01 and 0.05.

Within this context, when choosing screens it is also useful also to bear in mind criteria such as ventilation and visual and environmental impact, besides the insect exclusion criteria (when appropriate) because less dense screens reach a degree of shading similar to denser ones of clearer colours. For instance, a 6 × 6 green screen, as compared with a 20 × 10 non-coloured one, will presumably allow more ventilation (Teitel *et al.*, 2007; 2009) and because of its colour, might improve the aesthetic environmental impact upon the rural landscape generating a similar shading (Castellano *et al.*, 2008b), besides using less plastic material (for the threads) in its fabrication process and contributing to limit the global impact upon the environment.

The non-coloured translucent screens showed the highest levels of transmittance to global radiation and PAR, followed by the green types and finally the black screens for the same threads density and porosity. The green screen had similar transmittance values for global radiation as the non-coloured translucent screens, whereas for PAR radiation it had somewhat lower transmittance levels because of the effect of the green wavelength reflection.

The relative behaviour of the different screens regarding global radiation and PAR transmittance depending on the angle of incidence was similar: transmittance values fall concomitantly with an increase in the angle of incidence. Variation in transmittance according to the diameter and density of the threads was greatest in the black screens.

The amounts of transmitted diffuse radiation were higher in the non-coloured and the green screens than in the black ones. This effect was particularly notable in the densest non-coloured screens (20 × 10 and 16 × 10), where diffuse radiation forms a substantial fraction of the transmitted global radiation, especially at high angles of incidence.

Therefore, the non-coloured-thread screens may be the best option for maximising the transmission of diffuse radiation. This enrichment of diffuse radiation is very important for climbing plants because it enables the crop to catch more radiation, increasing the potential photosynthetic surface effectively and thus, the total production.

In this context, choosing the most appropriate screen will depend mainly on the crop growing season and the climatic conditions (e.g., frequency of cloudy days and/or wind). The green screen 6 × 6may be the best option for vegetables cultivation during the summer in inland areas with latitudes close to 36°N, where radiation is not a limiting factor, due to its lower environmental and visual impact. In contrast, the higher density and non-coloured screens pose an interesting option in regions with strong winds.

Greenhouse crops grown during autum-winter cycles have shown a positive response on the use of diffuse radiation by the plants, but this effect is unknown in summer conditions. Therefore, it would be interesting to investigate the response of these crops under different colored screens in future studies.

Acknowledgements

This research was funded by the "Instituto Nacional de Investigación y Tecnología Agraria y Alimentaria" (INIA) through the project RTA2006-00062, IFAPA and EU (FEDER).

References

[1] Abdel-Ghany AM, Al-Helal IM, 2010. Characterization of solar radiation transmission through plastic shading nets. Sol Energ Mat Sol Cells 94: 1 371 ~ 1 378

[2] Baille A, Tchamitchian M, 1993. Solar radiation in greenhouses. Workshop Crop structure and light microclimate. Characterization and Applications, 23 ~ 27 Sept 1991, INRA, Paris, pp: 93 ~ 105

[3] Baille A, Gonzalez-Real MM, López JC, Cabrera J, Pérez-Parra J, 2003. Characterization of the solar diffuse component under "parral" type Greenhouses. Acta Hort, 614: 341 ~ 346

[4] Basiaux P, Deltour J, Nisen A, 1973. Effect of diffusion properties of greenhouse covers on light balance in the shelters. Agric Meteorol, 11: 357 ~ 372

[5] Burek SAM, Norton B, Probert SD, 1989. Transmission and forward scattering of insolation through plastic greenhouse cladding materials. Acta Hort, 245: 498 ~ 504

[6] Cabrera FJ, Baille A, López JC, González-Real MM, Pérez-Parra J, 2009. Effects of cover diffusive properties on the components of greenhouse solar radiation. Biosyst Eng, 103: 344 ~ 356

[7] Castellano S, Russo G, Scarascia Mugnozza GS, 2006. The influence of construction parameters on radiometric performances of agricultural nets. Acta Hort, 718: 283 ~ 290

[8] Castellano S, Hemming S, Russo G, 2008a. The influence of colour on radiometric performances of agricultural nets. Acta Hort, 801: 227 ~ 235

[9] Castellano S, Candura A, Mugnozza GS, 2008b. Relationship between solidity ratio colour and shading effect of agricultural nets. Acta Hort, 801: 253 ~ 258

[10] Castellano S, Scarascia G, Russo G, Briassoulis D, Mistriotis A, Hemming S, Waaijenberg D, 2008c. Plastic nets in atriculture: A general review of types and applications. Appl Eng Agr, 24: 799 ~ 808

[11] Deltour J, Nissen A, 1970. Les verres diffusants en couverture de serres. Bulletin de la Recherche Agronomique, Gembloux, Belgium 1 ~ 2: 232 ~ 255

[12] Hanan JJ, 1998. Greenhouses. Advanced technology for protected horticulture. CRC Press, 684 pp

[13] Hemming S, Dueck T, Janse J, Van Noort F, 2008. The effect of diffuse light on crops. Acta Hort, 801: 1 293 ~ 1 300

[14] Möller M, Cohen S, Pirkner M, Israeli Y, Tanny J, 2010. Transmission of short-wave radiation by agricultural screens. Biosyst Eng, 107: 317 ~ 327

[15] Montero JI, Antón A, Hernández J, Castilla N, 2001. Direct and diffuse light transmission of insect proof screens and plastic films for cladding greenhouses. Acta Hort, 559: 203 ~ 209

[16] NF, 2002. Covering thermoplastic films for use in agriculture and horticulture. French Standard NF EN 13206. Norme Française, France

[17] Pearson S, Wheldon AE, Hadley P, 1995. Radiation transmission and fluorescence of nine greenhouse cladding materials. J Agr Eng Res, 62: 61 ~ 70

[18] Pollet IV, Pieters JG, Deltour J, Verschoore R, 2005. Diffusion of radiation transmitted through dry and condensate covered transmitting materials. Sol Energ Mat Sol Cells, 86: 177 ~ 196

[19] Raviv M, Allingham Y, 1983. Characteristics of modified polyethylene films. Plasticulture, 59: 3 ~ 12

[20] Shahak Y, Gussakovsky EE, Gal E, Ganelevin R, 2002. ColorNets: Crop protection and light ~ quality manipulation in technology. Acta Hort, 659: 143 ~ 151

[21] Sica C, Picuno P, 2008. Spectro-radiometrical characterization of plastics nets for protected cultivation. Acta Hort, 801: 245 ~ 252

[22] Soler A, Van Der Blom J, López JC, Gázquez JC, Cabello T, 2007. Eficacia de las mallas de 20 × 10 hilos ante el paso de *Bemisia tabaci.* XXXVII Seminario de Técnicos y Especialistas en Horticultura, pp: 1 005 ~ 1 009. [In Spanish]

[23] Soni P, Salokhe VM, Tantau HJ, 2005. Effect of screen mesh size on vertical temperature distribution in naturally ventilated tropical greenhouses. Biosyst Eng, 92 (4): 469 ~ 482

[24] Soriano T, Morales MI, Hita O, Romacho I, 2006. Cultivos estivales bajo mallas plásticas. Horticultura, 192: 14 ~ 18. [In Spanish]

[25] Soriano T, Hernández J, Morales MI, Escobar I, Castilla N, 2008. Bastidor para la medida de transmisividad a radiación solar de materiales de protección de cultivos. Boletín Oficial de la Propiedad Industrial, 16 de marzo de 2008, pag, 2490. [In Spanish]

[26] Spitters CJT, 1986. Separating the diffuse and direct components of global radiation and its implications for modelling canopy photosynthesis. Part II. Calculation of canopy photosynthesis. Agr Forest Meteorol, 38: 231 ~ 242

[27] Teitel M, Barak M, Tanny J, Cohen S, Ben-Yaakov E, Gatker J, 2007. Comparing greenhouse natural ventilation to fan and pad cooling. Acta Hort, 761: 33 ~ 39

[28] Teitel M, Dvorkin D, Haim Y, Tanny J, Seginer I, 2009. Comparison of measured and simulated flow through screens: Effects of screen inclination and porosity. Biosyst Eng, 104: 404 ~ 416

[29] Tovar J, Olmo FJ, Batlles FJ, Alados-Arboledas L, 2001. Dependence of one-minute global irradiance probability density distributions on hourly irradiation. Energy 26: 659 ~ 668

[30] UNI, 1988. Colourless transparent plastics film suitable for greenhouses and similar equipment for the forcing and semi-forcing of vegetable, fruit and flower growing cultures-requirements and test methods. UNI 9298, Unificazione Italiana, Italy

[31] UNI, 1990. Low density polyethylene flexible film for mulching of vegetables, flowers and fruits growing cultures-dimensions, requirements and test methods. UNI 9738, Unificazione Italiana, Italy

[32] UNI, 1994. Nets for agricultural applications-determination of the shading power of nets of polyethylene fibre. UNI 10335, Unificazione Italiana, Italy

[33] Wang S, Deltour J, 1999. Studies on thermal performance of a new greenhouse cladding material. Agronomie, 19: 467 ~ 475

气吸式穴盘育苗播种生产线研制开发

卓杰强[1*]，李建平[2]，周增产[1]，王会学[1]，商守海[1]，陈英伟[1]
（1. 北京京鹏环球科技股份有限公司，北京 100094；2. 中国农业大学工学院，北京 100083）

摘要：通过气吸式穴盘育苗播种生产线研制设计开发，实现工厂化育苗基质的自动搅拌，穴盘的自动装填、播种、覆土、淋水，解决现有育苗播种劳动生产率低，劳动强度大，产出效率低等问题，气吸式穴盘育苗播种生产线通过测试试验，平均播种合格率≥95%；重播率≤5%；空穴率≤5%；播种生产率达到200～400盘/h。气吸式穴盘育苗播种生产线因其省种、节本、高效，成为工厂化育苗的重要装备，通过该装备的使用实现工厂化育苗高效生产及种苗生产者的增产、增收、增效。

关键词：气吸式；穴盘；育苗；播种

Research and Development on the Air-suction Tray Seeding Production Line

Zhuo Jieqiang[1], Li Jianping[2], Zhou Zengchan[1], Wang Huixue[2],
Shang Shouhai[1], Chen Yingwei[1]
(1. *Beijing Kingpeng International Hi-Tech Corporation*, *Beijing* 100094, *China*;
2. *College of Engineering China Agricultural University*, *Beijing* 100083, *China*)

Abstract: Through the air suction tray seeding production line design development, realize automatic mixing matrix of factory seedling, tray self-loading, sowing, soil, water, solve the seeding low labor productivity of existing plants, high labor intensity, low output efficiency, gas suction tray seeding production line through the test, the average pass rate of seeding ≥95%; rebroadcast rate ≤5%; hole rate ≤5%; sow productivity reached 200～400disk/h. Air suction tray seeding production line because of its province, cost, efficiency, become the important equipment of factory production, achieve high efficiency factory production and seedling nursery producers, increase yield, efficiency through the use of the equipment.

Key words: Air-suction; Tray; Nursery; Seed

1 引言

工厂化育苗在国际上是一项成熟的农业先进技术，是现代工厂化农业的重要组成部分，随着农业结构调整的不断推进，工厂化育苗技术得到快速发展[1]。国外工厂化育苗机械研究起步较早，已经有40年以上的发展研制历程，技术比较成熟，功能完善，配套设施齐全，自动化程度较高。在工厂化育苗播种生产线的研究方面，荷兰、美国、韩国、日本都有相关的产品，其中，最有代表性的有荷兰Visser、美国Blackmore、Speedling、韩国大东机电育苗播种生产线等[2]。

---

* 作者简介：卓杰强（1979—），男，广西宾阳，高级工程师，从事设施农业装备研究，硕士，E-mail：zhuojq@163.com

通过气吸式穴盘育苗播种生产线研制设计开发，解决现有工厂化育苗播种劳动生产率低，劳动强度大，产出效率低等问题，实现工厂化育苗高效生产及种苗生产者的增产、增收、增效。气吸式穴盘育苗播种生产线可以实现基质的搅拌，穴盘的装填、播种、覆土、淋水，因其省种、节本、高效，成为工厂化育苗的重要装备。

2 气吸式穴盘育苗播种生产线工艺流程和总体技术要求

2.1 气吸式穴盘育苗播种生产线工艺流程（图1）

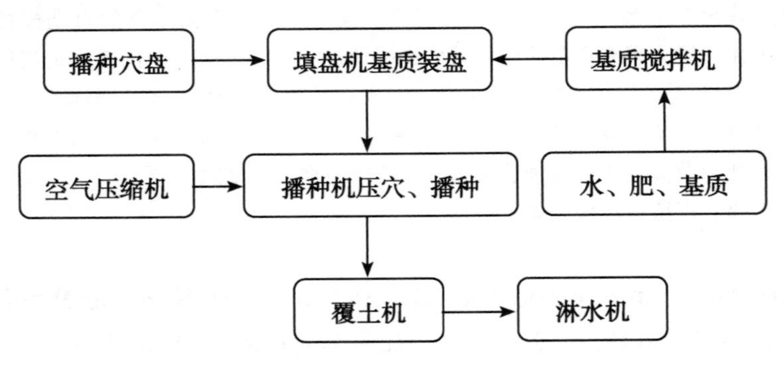

图1 生产线工艺流程图

气吸式穴盘育苗播种生产线工艺流程包括将草炭与蛭石等基质及水和肥在播种室由搅拌机搅拌均匀并装入填盘机，由填盘机对穴盘进行装填，育苗盘再经过打孔、播种、基质覆盖、淋水等工序完成播种。

2.2 气吸式穴盘育苗播种生产线总体技术要求（表1）[3]

表1 气吸式穴盘育苗播种生产线总体技术要求

生产效率	每穴播种数	重播率	空穴率	基质搅拌机容量	标准穴盘参数
200～450 盘/h	1～4	≤5%	≤6%	≥500L	标准 72 穴（54cm×28cm×4.5cm）穴孔间距 4.2cm，穴孔总体积 2 520cm^3

3 气吸式穴盘育苗播种生产线

3.1 基质搅拌机

基质搅拌机主要完成基质的搅拌工序，即将育苗土通过机械搅拌为均匀、细碎的基质营养土。该装置 475kg、功率 1.5kW、电压 380V、容量 650L。工作时，该机可以实现基质、肥料和水分的均匀混合，主要特点是混合均匀，能实现操作的安全性，在增加基质的过程中将上盖打开的时候，通过接触式开关实现搅拌电机的自动断电停止搅拌，防止操作过程中危险发生，同时，在出料口舱门采用气动提升方式，便于操作。输送搅龙叶片根据搅拌和出料合理布置叶片位置，实现有序操作，加快搅拌效率。

3.2 基质填盘机

基质填盘机主要完成穴盘基质的填充工序，即在穴盘的每一个穴孔里添加基质营养土。该装置425kg、功率2.5kW、电压380V。工作时，粉碎并混匀后的基质从搅拌机装入填盘机的料斗，在输送刮板的带动下，基质被提升至一定高度并在穴盘输送带的正上方落下，对穴盘进行填充。基质填盘机具有灵敏而准确的基质装填系统，可适应不同填料质地的疏密程度和需要。机器设有旋转刮平装置、旋转毛刷清扫装置，能够依次将穴盘上堆积的基质摊平、对穴盘上多余的基质进行粗刮、并对穴盘表面进行精细的清扫，去除穴盘表面多余的基质，保持盘面的清洁和基质的均匀性，多余的基质又被自动送回基质料斗，重复填装。

3.3 穴盘基质打孔装置

通过PLC控制器对全自动步进系统输送装置输送的装好基质的穴盘进行打孔，在每个孔穴的正中心完成一个压实点，保证均匀性和等深度，以便于下一步的精确播种操作；依据种子和穴盘孔大小的不同，打孔器模具头的大小可相应的调换；光电传感器能精确调节打孔的位置。

3.4 气吸式穴盘播种机

气吸式穴盘播种机通过PLC控制器对全自动步进系统输送装置输送的装好基质打孔好的穴盘进行播种，播种方式采用气吸针式播种方式。可确保种子在孔穴的中心位置和其播种深度，达到均匀一致的效果；每一个穴孔里，可选择设定播种1～4粒；净化气系统可以有效地防止吸嘴阻塞；可适用不同宽度和高度的穴盘；可记忆存储50种不同穴距规格的穴盘并可方便地调用。播种精度高，一次播种两行，播种速度可调。

工作时，穴盘随输送带前进，由光电传感器I通过检测穴盘第一排穴孔的前缘来确定穴盘到达播种位置，从而停止穴盘输送，等待这排穴孔播种；与此同时，播种机构摆动气缸通过固定在播种箱壳上的吸嘴摆臂齿轮，实现摆臂前吸种和摆臂后投种工作。摆臂前，吸嘴气路与真空发生器相连，气体振动器在播种工作周期振动带动种盘的振动，使种盘内种子处于“沸腾”状，便于吸嘴的单粒吸种；摆臂后，切断真空发生器气路，吸嘴负压消失，吸嘴吸附的单粒种子以自由落体方式落入穴盘孔内[4]。吸种摆臂的时间刚好和穴盘步进时间一致，因此，播完一排种子后穴盘移动的时间和到下一次吸种到播种时间一致，重复上述动作，可连续进行穴盘的精密播种。

3.4.1 输送带线速度和输送带滚筒转速（表2）[3]

播种机链轮速比12∶18，输送带线速度、输送滚筒转速可由下式计算得出：

$$V_s = \frac{C \times L}{60}$$

式中：V_s——输送带线速度 m/min；

C——理论生产效率；

L——穴盘步进长度 m（采用0.28m）；

$$n = \frac{V_s}{2 \times \pi \times R}$$

式中：n——输送带滚筒转速；

R——输送带传动端滚筒外径（含输送带厚度）m，本设计取数值为0.038m。

表 2　6 个理论生产效率的输送带线速度、输送滚筒转速

理论产量	输送带线速度 V_s（m/min）	输送带轮转速 n（rpm）
200 盘/h	0.93	3.91
250 盘/h	1.17	4.89
300 盘/h	1.40	5.87
350 盘/h	1.63	6.84
400 盘/h	1.87	7.82
450 盘/h	2.10	8.80

3.4.2　种盘气动振动器

振动气吸播种装置应用振动气吸的原理，设计了种子振动盘、吸嘴、气源及其他辅助部分组成的结构。种子经种盘振动后产生向上抛掷运动，呈“沸腾”状态，吸嘴在负压的作用下将被抛掷的种子吸在吸嘴上，并停在播种机输送带的穴盘上方适当位置，切断真空后，种子靠自重离开吸嘴落入穴盘。本设计采用台湾微型气动振动器，该振动器体积小，振动力大，振动频率高，耗气量小。

3.5　覆土装置

覆土装置完成覆土工序，即把每个苗穴的种子用营养土覆盖。工作时，接触式开关检测到被输送带输送到的穴盘后，刷土辊在转动状态下把营养土覆盖在播好种子的穴盘上。调节辊轮的间隙，就可以控制覆土层的厚度为 1 ~ 5mm；对播种盘的表面覆土有修饰、整理的功能；多余的覆土可以循环利用；传送带的速度具有可调节功能。

3.6　淋水装置

淋水装置完成淋水工序，接触式开关检测到被输送带输送到的穴盘后，电磁阀打开，淋水喷嘴与自来水接通，喷头均匀对穴盘淋水。

3.7　气吸式穴盘育苗播种生产线控制系统

气吸式穴盘育苗播种生产线控制系统由播种机系统硬件和软件、可编程控制器（PLC）、执行元器件、传感器及限位开关等组成，其组成框架和控制执行框图如图 2 和图 3 所示。

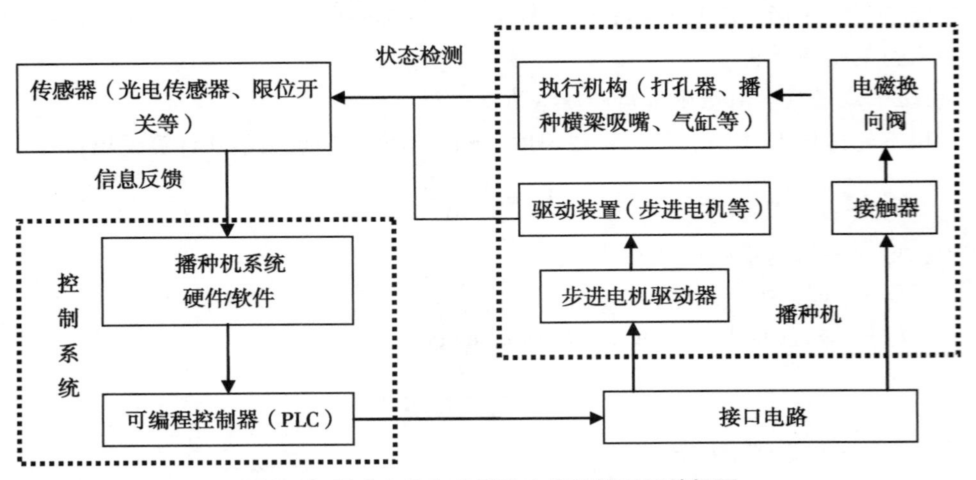

图 2　气吸式穴盘育苗播种生产线控制系统框图

在播种机播种过程中，由PLC控制系统统一协调各执行机构完成相应动作。主要是控制填盘机输送穴盘装填基质，打孔器气缸带着打孔头向下打孔和向上收回，停止气缸杆的伸出使穴盘得以定位，播种梁在摆动气缸的作用下，向上摆动到种盘进行吸种，向下摆动到穴盘上表面进行落种，以及控制覆土机电机驱动辊覆土和淋水机电磁阀淋水作业等[3]。

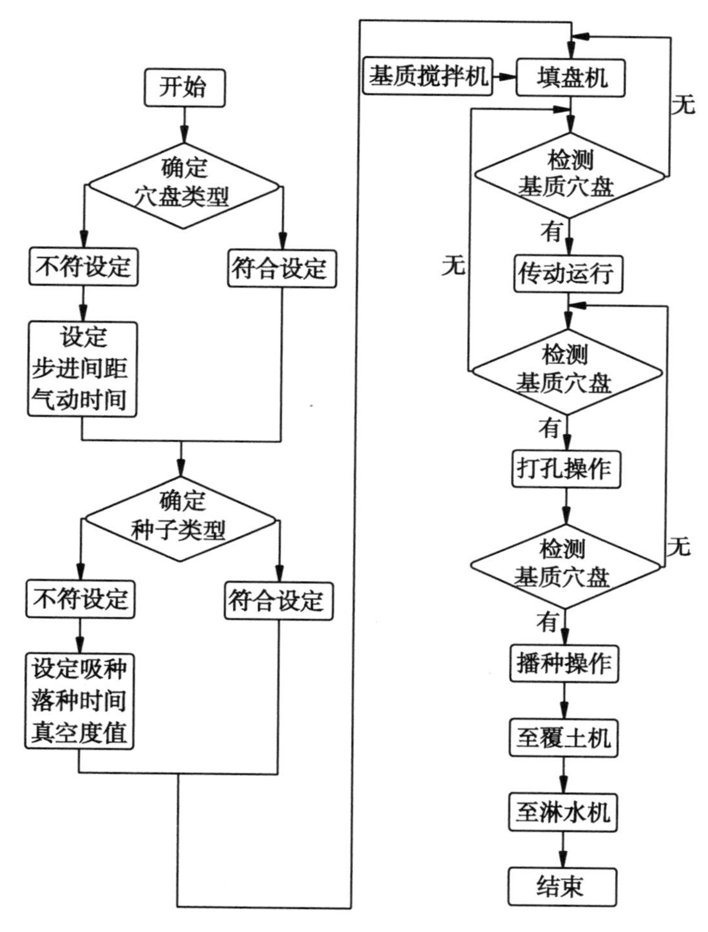

图3　气吸式穴盘育苗播种生产线控制执行框图

4　结论

气吸式穴盘育苗播种生产线通过测试试验，平均播种合格率≥95%；重播率≤5%；空穴率≤5%；播种生产率达到200～400盘/h。气吸式穴盘育苗播种生产线因其省种、节本、高效，成为工厂化育苗的重要装备，通过该装备的使用实现工厂化育苗高效生产及种苗生产者的增产、增收、增效。我国农业正处于一个从传统农业向现代化农业转变阶段，如何提高劳动生产率，提高土地产出率，提高资源利用率以及提高农产品商品质量和农业劳动者的效益，是我国农业是否健康、持续发展的关键。工厂化育苗在国际上是一项成熟的农业先进技术，是现代工厂化农业的重要组成部分。实现种苗的工厂化生产、商品化供应将成为传统农业走向现代农业的必然途径。因此，气吸式穴盘育苗播种生产线具有非常广阔的发展前景。

参考文献

[1] 漆向军，陈云辉，李首成．基于机器人技术的工厂化育苗生产线模型设计与制作．农业网络信息．2008（1）：32～42
[2] 程欢庆，张祖立，张为政．蔬菜穴盘播种装置的研究进展．农业科技与装备．2010（4）：31～33
[3] 沈美雄．2BSX300 型真空穴盘育苗播种生产线系统参数设计．机电技术．2010（3）：31～33
[4] 程卫东．苗盘气吸精密播种机的研制．农机化研究．2004（6）：141～143

面向空间生命保障的喇叭形蔬菜连续生产装置

刘　红[1*]，付玉明[1]，刘　慧[1]，邵玲智[1]，Yu. A. Berkovich[2]，A. N. Erokhin[2]

（1. 北京航空航天大学生物与医学工程学院环境生物学与生命保障技术实验室，100191　北京；
2. 俄罗斯联邦国家科学中心——生物医学问题研究所，123007　莫斯科）

摘要： 目的 设计开发一个新型高效的空间叶菜连续生产装置原理样机，并且评估其在长期运行过程中的生产蔬菜产量、品质、效率。方法基于新的栽培室、根部模块、光照子系统与供水子系统设计，研制新型的空间蔬菜高效连续培养装置。通过连续的生菜栽培实验确定装置的产量。以粗纤维、维生素、矿物质、硝酸盐为指标，以市售生菜为对比，评价长期运行下装置生产蔬菜的品质变化。以 Q 指数评价装置的生产效率。结果 开发的喇叭形空间蔬菜连续生产装置原理样机能够在模拟微重力效应的同时实现连续模块化栽培叶菜。该装置的栽培室体积为 $0.12m^3$，其内部呈喇叭形的 3 段式，能够满足植物生长过程中的空间扩展需求。该装置的种植表面为 $0.154m^2$，由内部填充栽培基质的六个环形根部模块组成。根部模块固定在中轴供水管上，可随着植物的生长不断向前移动。栽培室内部设有红色和白色 LED 组成的光板，提供光照面积为 $0.567m^2$。在 24h 连续光照、能耗 0.3kW 条件下，装置生菜的生产率为 254.3g/周。该装置长期运行过程中，生产的生菜的营养成分并没有发生较大的变化。装置基于 Q-指数的生产效率为 $7\times10^{-4}\ g^2\cdot m^{-3}\cdot J^{-1}$，高于先前的同类型设备。结论 喇叭形蔬菜连续生产装置在生产蔬菜方面具有高产高效、品质稳定的特点，对于空间蔬菜生产极具应用前景。

关键词： 地面原理样机；喇叭形；蔬菜连续生产；生命保障系统

A High-performance Ground-based Prototype of Horn-type Sequential Vegetable Production Facility for Life Support System in Space

Liu Hong[1], Fu Yuming[1], Liu Hui[1], Shao Lingzhi[1], Yu. A. Berkovich[2], A. N. Erokhin[2]

(1. *Laboratory of Environmental Biology and Life Support Technology*, *School of Biological Science and Medical Engineering*, *Beihang University*, *Beijing* 100191, *China*; 2. *State Scientific Center of the Russian Federation-Institute for Biomedical Problems*, *Moscow* 123007, *Russia*)

Abstract: Aims: The purpose of this work was to design a new prototype of vegetable production facility with enhanced efficiency, and to estimate its vegetable product quality under long-term operation. Methods: Development of a sequential vegetable production facility with high-efficiency for life support system in space was based on the new design of growth chamber, root modules, light subsystem, and water supply subsystem. Quality change of the vegetable under long-term operating conditions of the new facility was examined according to the crude fiber, vitamin, and mineral content of the harvested lettuce. As a comparison, the same lettuce cultivar purchased from a local market was also determined. Efficiency of the new facility was assessed using a Q-criterion. Results: A ground-based prototype of horn-type

* 作者简介，刘红，教授，俄罗斯自然科学院外籍院士，国际宇航科学院通讯院士，北京航空航天大学生物与医学工程学院 空间生命科学与生命保障技术中心主任，主要研究领域：生物再生生命保障系统与环境生物技术。E-mail：LH64@ buaa. edu. cn

sequential vegetable production facility named Horn-type Producer (HTP) developed, which was capable of simulating the microgravity effect and the continuous cultivation of leaf-vegetables on root modules. The growth chamber of the facility had a volume of 0.12m^3, characterized by a three-stage space expansion with plant growth. The planting surface of 0.154m^2 was comprised of six ring-shaped root modules with a fibrous ion-exchange resin substrate. Root modules were fastened to a central porous tube supplying water, and moved forward with plant growth. The total illuminated crop area of 0.567m^2 was provided by a combination of red and white light emitting diodes on the internal surfaces. In tests with a 24-h photoperiod, the productivity of the HTP at 0.3 kW for lettuce achieved 254.3 g eatable biomass per week. Long-term operation of the HTP did not alter vegetable nutrition composition to any great extent. Furthermore, the efficiency of the HTP, based on the Q-criterion, was 7×10^{-4} g^2 · m^{-3} · J^{-1}, which was better than similar facilities designed previously. Conclusion: HTP exhibited high productivity, stable quality, and good efficiency in the process of planting lettuce, indicative of an interesting design for space vegetable production.

Key words: Ground-based prototype; Horn-type; Vegetable sequential Production; Life support system

1 前言

蔬菜栽培在载人航天生命保障统中发挥着至关重要的作用。在应用于长期远距离载人航天任务（如月球前哨与火星基地）的生物再生生命保障系统（Bioregenerative Life Support System，BLSS）中，蔬菜作为系统自养生单元的一部分能够为航天员的生命维持去除 CO_2、再生 O_2、提供食物、净化废水[1]。在空间站和登火星飞船的环控生命保障系统内，蔬菜栽培主要用于增加航天员的饮食多样性，满足航天员对绿色维生素和粗膳食纤维的需求[2]。同时，食用富含抗氧化剂的蔬菜也能够减轻空间辐射对航天员健康的不利影响[3]。除显著的饮食意义外，蔬菜栽培也被视为改善航天员的舱内生活环境，提供心理慰藉的有效手段。

在轨栽培蔬菜，满足航天员的生理需求是 BLSS 真正应用于空间环境的第一步。在过去的几十年里，许多研究者都致力于空间站等载人航天器配备的蔬菜生产装置开发研究。当前，在国际空间站俄罗斯舱已配备了小型的蔬菜栽培机——Lada，它不但为航天员提供了偶尔的新鲜叶菜供给，也提供了一个心情放松场所[4]。然而，由于航天器内资源有限，能够连续供给并满足多位航天员日常需求的蔬菜生产装置，迄今尚未建立。如何提高单位消耗资源（能源、体积、航天员劳动等）的生产效率，仍然是空间蔬菜生产装置开发的核心问题。本项工作的目的是设计开发一个新型高效的空间叶菜连续生产装置原理样机，并且评估其在长期运行过程中的生产蔬菜产量、品质、效率。

北航研制的高效率空间蔬菜连续培养装置原理样机——喇叭型生产者（Horn-type Producer，HTP），其在栽培生菜的过程中表现出高产高效、品质稳定的特点。这一研究结果表明，类似 HTP 的构造设计在未来载人航天器内蔬菜栽培装置开发方面具有很大应用潜力。

2 HTP 的构造与操作

2.1 HTP 的构造设计

HTP 主要包括四个部分：①栽培室；②根部模块；③光照子系统；④供水子系统（图 1）。为了减少装置的复杂程度与能耗要求，HTP 只提供蔬菜栽培所需的光照、水分输送、营养供给功能，而温湿度和 CO_2 控制则依靠飞船内部环境。HTP 的主要技术指标如表 1 所示。

从体积和光能利用的角度而言，圆柱曲面种植的效率均要高于传统的平面种植效率[5]。因此，HTP 的栽培室和种植面被设计成两个同轴的圆筒结构。栽培室的直径 61cm，长度 42cm，体积 0.12m^3。圆柱形种植表面由六个相同的圆环形的根部模块组成，面积 0.154m^2。所有的环形根部模块固定到可以供水的中轴供水管上，并可随中轴供水管绕水平轴线旋转（表 1）。

表 1　HTP 的技术指标

参数	数值
栽培舱室直径 mm	610
栽培舱室长度 mm	410
种植面积 m^2	0.154
光照面积 m^2	0.567
舱室体积 m^3	0.12
根部模块数	6
光源类型	红 LED（660nm），白 LED（2700K）
LED 灯珠数量	红：872　白：694
风扇数量	8
用电功率 KW	0.3

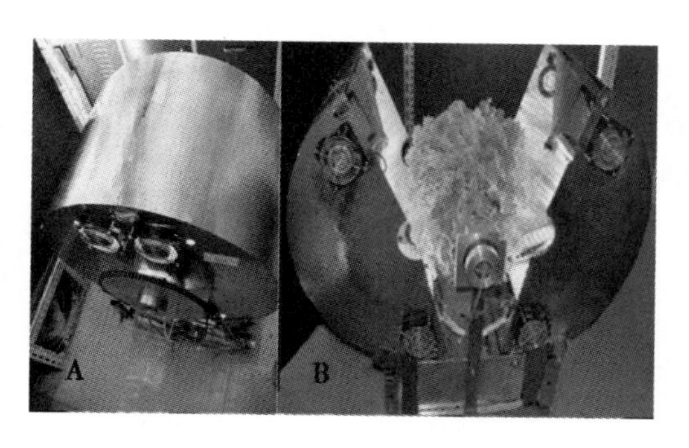

图 1　HTP 的总体外观：（A）关闭状态（B）打开状态

环形根部模块外径 15cm，由两个相同的不锈钢半圆部件组成的，每个部件各包含一个上盖和一个底部托架（图 2）。上盖宽留有宽度为 1.5cm 的缝隙用于播种种子。底部托架上具有 4 条 1cm 宽的间隙，用于确保水被填充于上盖和底部托架间的栽培基质吸收。上盖和底部托架之间为栽培基质层，由外至内分依次是保护膜、纤维离子交换树脂 BIONA-V3、微滤膜、吸水膨胀材料。这样的设计既保障了植物根部对水、营养、O_2 的需求，又保障了模块在种植周期中的顺利移动。

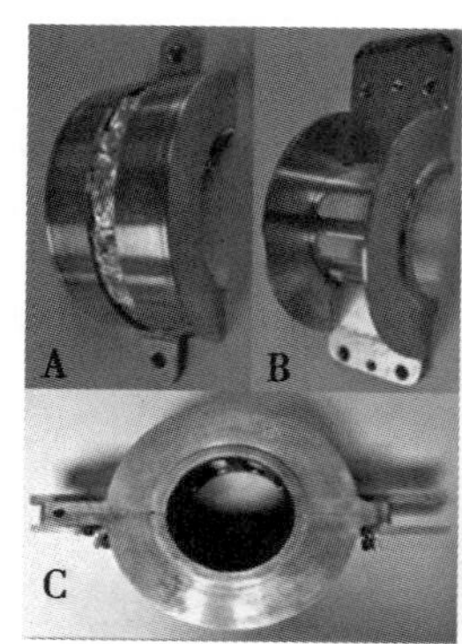

图 2　（A）上盖（B）底部托架（C）组装模块（D）收获模块

LED 光源具备空间蔬菜培养的光照系统所需的轻质、稳定、耐用特征。在受控环境中，多种作物而言，红光（600～700nm）和蓝光（400～500nm）的 LED 组合能够满足多数作物的生长需求。但是，绿色、远红外和紫外灯等其他波段光谱能够触发植物的生理反应，从而对植物生物量的积累具有正向作用[6]。而白光 LED 发出的是连续光谱，包括了植物生长需要的各种波段光谱。因此，HTP 以红光（660nm）和白光（2700 K）的组合 LED 作为其光照子系统的光源。依据作物生长的“S”曲线模型，为了尽量减少在光从光源发射到达植物冠层的能量损失，栽培室内表面的光照子系统被设计成“小圆柱-圆台-大圆柱”3 段式喇叭形结构（图 3A）。第一段由 10 块包含 200 颗红光、150 颗白光 LED 灯珠的矩形光源板拼接成，形成一个直径 32cm、高 13.5cm 的“小圆柱”。第三段是直径 60cm、高 7cm 的“大圆柱”，由 10 块包括 192 颗红光、144 颗白光 LED 灯珠的光源板拼接而成。而在第一段和第三之间高 20cm 的“圆台”则是由 10 块包括 480 颗红光、400 颗白光 LED 灯珠的光源板拼接成。HTP 光照子系统的总光照面积为 $0.567m^2$。从第一段到第三段光源板下 5 厘米处的平均光强分别是 $200\mu mol \cdot m^{-2} \cdot s^{-1}$，$350\mu mol \cdot m^{-2} \cdot s^{-1}$ and $500\mu mol \cdot m^{-2} \cdot s^{-1}$（图 3B）。这样的光强设计符合蔬菜在幼苗期、快速增长期和收获期的光强需求。

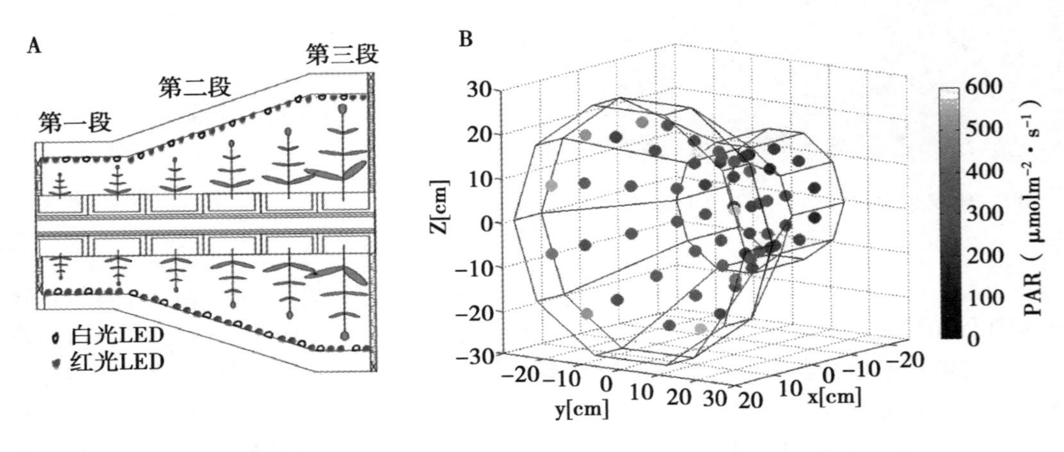

图 3　HTP 光照子系统（A）三段式喇叭形设计（B）光强 3D 图

HTP 的供水子系统采用多孔管负压供水方式。该方式已被证明是维持空间微重力条件下植物根区水分、营养输送的可靠方式[7]。多孔钛管外部包裹吸水膨胀材料后嵌入的带有孔隙不锈钢管内作为根部模块的固定中轴（图 4）。中轴不锈钢管与根部模块内的吸水膨胀材料吸水后通过孔隙接触充分，保障了水分和营养有效传递到栽培基质内。HTP 的水分和营养的供给在多孔管内的负压在 -0.5 ~ 0.2kPa 的条件下进行。

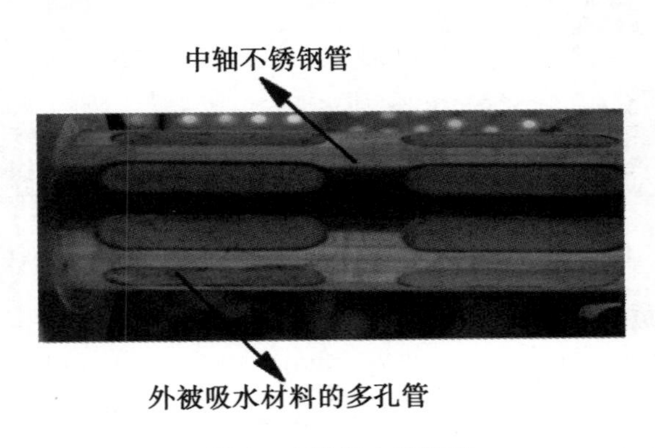

图 4　中轴供水管结构

HTP可利用电机通过铰链带动中轴不锈钢管外端的链轮旋转，从而实现模拟蔬菜培养过程中的空间微重力效应。栽培室两端的8个风扇能够有效地通风散热，保障光源产生的热量及时散出，不影响蔬菜生长。

2.2 HTP的操作

根据事先确定的蔬菜生长周期与模块数量，将生长周期均分成六个相等种植时间间隔，每经过一个种植间隔在中轴加上下一个植入种子模块，同时凭借吸水膨胀材料的润滑作用，在中轴不锈钢管上推动前期栽培模块向前移动。当第六个植入种子的模块固定于中轴上时，第一个模块的植物刚好移动到中央不锈钢管的末端（图3A）。再经过一个种植时间间隔，打开舱室收获第一个栽培模块的植物。然后将第一个根部模块重新填充新的栽培基质，植入种子，并重新固定到中轴不锈钢管的起始位置使用。如此往复循环，实现蔬菜的连续性生产。同时，为了模拟微重力效应并保障蔬菜在栽培室内的均匀生长，根部模块固定在中轴不锈钢管上以12rpm/h的速度旋转。

3 HTP的栽培试验

3.1 材料和方法

3.1.1 生菜种植、收获和预处理

HTP的种植实验以生菜（*Lactuca sativa* var. *dasusheng*）为栽培对象。每个模块栽培植物8株，每批种植时间间隔为7d，即当第一批生菜生长到42d时收获，而后每隔7d收获一批。栽培舱室的环境温度为23±2℃，湿度为30%～35%，CO_2浓度为常规大气浓度。光周期为24h连续光照。1/2浓度的Hoagland溶液作为营养液通过供水子系统供给。每次根部模块收获后，取各模块上部的可食生物量进行鲜重测量。而后部分生菜105℃杀青1h后，在60℃烘干至恒重，用于含水率和粗纤维的分析；部分样品立即冷冻干燥并-80℃贮存，用于维生素、矿物质与硝酸盐含量的检测。取自市售的相同品种生菜作为实验对照。

3.1.2 生菜品质评价

（1）粗纤维素含量测定。将冷冻干燥样品用研钵磨细并通过60目筛，称取样品1g放入250ml三角瓶中，采用经典方法用CTAB水解、湿润、乳化、分散样品中的蛋白质、多糖、核酸等组分，依照酸性洗涤剂法[8]进行测定。

（2）维生素和矿物质的含量测定。根据食品标准GB/T 5009.159—2003、GB/T5413.16—1997、GB/T5009.89—2003、GB/T5413.17—1997、GB/T5009.84—2003、GB/T5009.85—2003测定生菜的维生素C、叶酸、烟酸、泛酸、维生素B_1、维生素B_2、胡萝卜素。根据食品标准GB/T 5009.87—2003、GB/T 5009.91—2003、GB/T 5009.90—2003、GB/T 5009.14—2003、GB/T 5009.92—2003、GB/T 5009.13—2003分别测定生菜中钾、钠、铁、镁、锰、锌、钙与铜、磷的含量。

（3）硝酸盐含量测定。将冷冻干燥样品用研钵磨细并通过60目筛，称取样品1g放入150ml三角瓶中用热水从样品中提取样品中的硝酸根离子，采用离子色谱测定。

3.1.3 统计分析

使用SPSS 20.0 for Windows的软件对所有数据进行统计分析。

3.2 结果与分析

3.2.1 HTP内生菜生产

在HTP运行105d、连续收获10批生菜过程中，生菜生长状况良好（图5）。每个模块收获生菜的可食生物量鲜重范围为202.2~316.5g，平均为253.4g。单株生菜的平均质量为31.7g，含水率在92%~94%，这与市售的同品种生菜没有显著性差异（$P>0.05$，图6）这一结果表明，HTP的构造设计能够满足的生菜正常生长的需要。计算可得HTP在耗能0.3kW的情况下，生菜的平均与最大生产率分别是36.2 g和45.2 g。先前的研究表明，在环境密闭舱内或空间植物栽培舱“Vitacycle”连续培养蔬菜过程中，系统最大的生产率150~160g·d·kW[9]。显然，HTP的最大生菜生产率能够达到这一水平。

图5 HTP运行40d后生菜生长情况

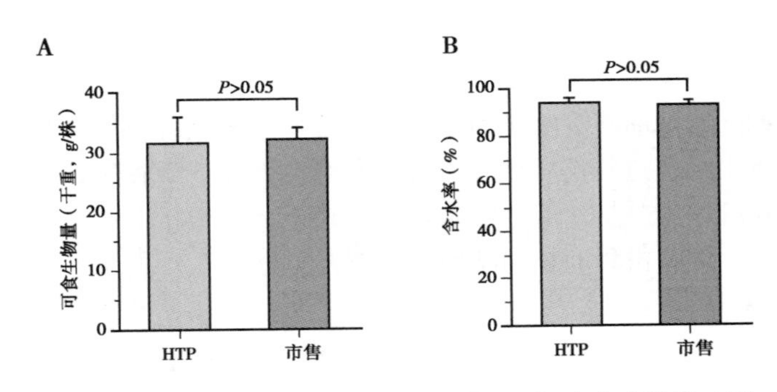

图6 HTP生菜与市售生菜可食生物量（A）含水率比较（B）

3.2.2 生菜的粗纤维、维生素与矿物质含量

绿叶蔬菜是人体所需的中粗纤维、维生素、矿物质的主要来源。这3类物质对保护航天员的健康具有重要作用影响（Berkovich *et al.*，2009）为了分析HTP长期运行下生产的生菜品质是否发生变化，我们以上述3个指标分析了第一批、第五批与第十批收获的生菜品质，并与市售的生菜品质进行了比较。图7A表明，HTP生产的3批生菜粗纤维的含量处于同一水平（125~153g·kg^{-1}），明显低于市售生菜的粗纤维含量（164~198g·kg^{-1}）。造成这种差异的原因可能与二者的种植条件、收获时间有关。粗纤维含量较高的蔬菜一般口感较差，难以发挥吸引味觉、诱人食欲的作用[10]。因此，与市售生菜相比，HTP内生产的生菜更加鲜嫩柔软，口感更佳。如表2和表3所示，HTP生产的不同批次生菜的维生素和矿物质含量没有明显的差异。而且，与市售生菜相比，HTP生产的生菜在各种维生素含量与大多数矿物质含量上表现出一致性，只有铜、铁微量元素含量差异较大。这些结果表明HTP长期连续种植下的生菜品质没有

发生大的变化，处于稳定状态。这进一步证实了 HTP 构造设计的合理性。

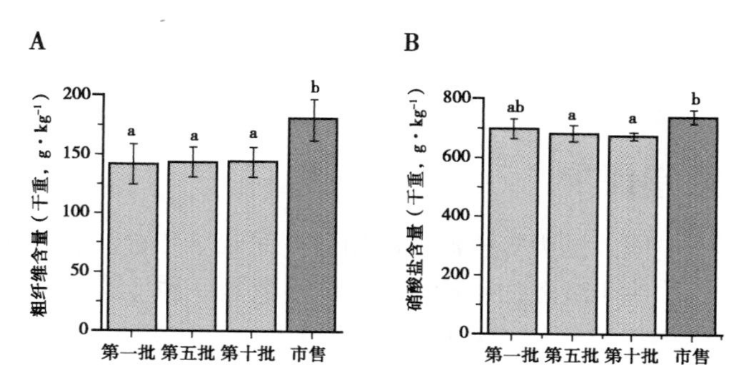

图 7　HTP 生菜与市售生菜（A）粗纤维（B）硝酸盐含量比较

表 2　HTP 生产的生菜的维生素含量

样品	β-胡萝卜素 mg/100 g(1)	维生素 C mg/100 g	维生素 B_1 μg/100 g	维生素 B_2 μg/100 g	维生素 B_3 mg/100 g	维生素 B_5 mg/100 g	维生素 B_9 μg/100 g
第一批	4.58 ± 0.48[a(2)]	16.47 ± 3.15[a]	30.27 ± 8.89[a]	57.88 ± 2.42[a]	0.38 ± 0.02[a]	0.14 ± 0.01[a]	33.76 ± 3.41[a]
第五批	4.27 ± 0.32[a]	18.93 ± 3.19[a]	44.00 ± 7.55[a]	71.13 ± 14.77[a]	0.37 ± 0.04[a]	0.14 ± 0.01[a]	35.16 ± 5.26[a]
第十批	4.87 ± 0.71[a]	16.87 ± 5.91[a]	33.33 ± 6.65[a]	70.13 ± 14.2[a]	0.35 ± 0.05[a]	0.15 ± 0.02[a]	33.73 ± 4.16[a]
市售	4.48 ± 0.55[a]	18.80 ± 2.33[a]	38.33 ± 11.06[a]	69.82 ± 5.46[a]	0.35 ± 0.02[a]	0.15 ± 0.01[a]	37.93. ± 1.87[a]

（1）维生素含量数据均是鲜重 ± 标准差

（2）相同字母表示不具有显著性差异（$P > 0.05$）

表 3　HTP 生产生菜与市售生菜的矿质元素比较

样品	大量元素（mg/100 g fw ± sd(1)）				
	K	Na	Ca	P	Mg
第一批	226.1 ± 19.5[a(2)]	13.0 ± 2.2[a]	88.5 ± 6.7[a]	23.8 ± 3.0[a]	32.0 ± 1.4[a]
第五批	227.6 ± 10.6[a]	15.4 ± 3.6[a]	88.2 ± 1.1[a]	26.1 ± 2.2[a]	32.7 ± 3.4[a]
第十批	210.5 ± 27.8[a]	14.3 ± 2.1[a]	80.2 ± 6.1[a]	23.5 ± 1.5[a]	31.7 ± 3.7[a]
市售	218.5 ± 23.0[a]	15.6 ± 2.7[a]	58.6 ± 2.7[b]	25.6 ± 2.4[a]	32.0 ± 5.4[a]
样品	微量元素（mg/100 g fw ± sd）				
	Fe	Mn	Cu	Zn	
第一批	1.54 ± 0.2[a]	0.32 ± 0.03[a]	0.031 ± 0.004[a]	0.28 ± 0.01[a]	
第五批	1.95 ± 0.3[ab]	0.34 ± 0.00[a]	0.030 ± 0.002[a]	0.27 ± 0.03[a]	
第十批	2.09 ± 0.6[ab]	0.31 ± 0.02[a]	0.032 ± 0.003[a]	0.28 ± 0.02[a]	
市售	5.28 ± 0.4[b]	0.34 ± 0.02[a]	0.046 ± 0.004[a]	0.28 ± 0.03[a]	

（1）fw = 鲜重，sd = 标准差

（2）不同字母表示具有显著性差异（$P < 0.05$）

3.2.3　生菜的硝酸盐含量

食品中的硝酸盐对人体健康有着重要的影响，高量的硝酸盐摄入会诱发癌症、高铁血红蛋白症、甲状腺增生等多种疾病。由于人体摄入的硝酸盐主要来源于蔬菜，因此，硝酸盐是蔬菜安全卫生的一个重要限制指标。如图7B所示，HTP生产的不同批次生菜硝酸盐含量仅为685mg·kg^{-1}鲜重，略低于市售生菜的硝酸盐含量747mg·kg^{-1}鲜重，远远低于生菜的硝酸盐含量上限设定为中国（3 000mg·kg^{-1}鲜重，GB18406.1/2001）和欧盟（mg·kg^{-1}鲜重，EC1881/2006）。这表明在HTP下长期连续生产的生菜具有食用安全性。

4 HTP的生产效率评价

由于载人航天器空空间、动力等资源的限制，空间栽培装置性能优劣不仅取决于其生产蔬菜的产量与品质，也取决于装置的生产效率。Q指数是一种计算空间栽培装置的生产效率的简便方法，采用单位体积与单位能耗所产生的最大可食生物量生产率表示[5]。计算可得HTP的Q指数为7×10^{-4} $g^2\cdot m^{-3}\cdot J^{-1}$，优于美国肯尼迪航天中心设计的“Salad Machine”空间蔬菜培养装置以及俄罗斯生物医学问题研究所研制的“Vitacycle”空间蔬菜培养装置（表4）。

表4 基于Q指数的不同栽培装置的生产效率比较

装置	参数			Q指数 ($g^2\cdot m^{-3}\cdot J^{-1}$)	文献来源
	体积（m^3）	能耗（W）	最大生产率（g/day）		
Vitacycle	0.75	1 000	150	3.5×10^{-4}	[1]
Salad machine	1.3	1 300	86	5.1×10^{-5}	[1]
HTP	0.12	300	45.2	7.0×10^{-4}	本文

5 结论

针对空间站、火星飞船生命保障系统研制的喇叭形蔬菜连续栽培装置的原理样机，能够模拟微重力效应的同时能够实现连续模块化栽培叶菜。在耗能0.3kW的情况下，样机每周可以生产253.4 g鲜重的生菜。HTP长期连续生产过中，生菜的营养品质并没有发生改变较大程度的改变。与市售生菜相比，HTP生产的生菜粗纤维和硝酸盐含量更低，口感品质更佳。HTP基于Q-指数的生产效能为$7\times10^{-4}g^2\cdot m^{-3}\cdot J^{-1}$，高于先前设计研发的同类型设备。HTP在栽培生菜的过程中表现出高产高效、品质稳定的特点，表明了类似HTP的构造设计在未来载人航天器内蔬菜栽培装置开发方面具有很大应用潜力。

致谢

本研究得到科学技术部国际科技合作项目（2012DFR30570）的资助。

参考文献

[1] Wheeler, R. M. Potato and Human Exploration of Space: Some Observations from NASA-Sponsored Controlled Environment Studies. Potato Res. 2006, 49: 67 ~ 90

[2] Berkovich, Yu. A., Smolyanina, S. O., Krivobok, N. M., Erokhin, A. N., Agureev, A. N., Shanturin, N. A. Vegetable production facility as a part of a closed life support system in a Russian Martian space flight scenario. Adv. Space Res. 2009, 44: 170 ~ 176

[3] Levine, L., Pare, P. W. Antioxidant capacity reduced in scallions grown under elevated CO_2 independent of assayed light in-

tensity. Adv. Space Res. 2009, 44: 887 ~ 894

[4] Sychev, V. N., Levinskikh, M. A., Gostimsky, S. A., Bingham, G. E., Podolsky, I. G. Spaceflight effects on consecutive generations of peas grown onboard the Russian segment of the International Space Station. Acta Astronaut. 2007, 60: 426 ~ 432

[5] Berkovich, Yu. A., Chetirkin, P. V., Wheeler, R. M., Sager, J. C. Evaluating and optimizing horticultural regimes in space plant growth facilities. Adv. Space Res. 2004, 34: 1 612 ~ 1 618

[6] Briggs, W. R., Beck, C. F., Cashmore, A. R., Christie J. M., Hughes J., Jarillo J. A., Kagawa T., Kanegae H., Liscum E., Nagatani A., Okada K., Salomon M., Rüdiger W., Sakai T., Takano M., Wada M., Watson J. C. The phototropin family of photoreceptors. Plant Cell. 2001, 13: 993 ~ 997

[7] Berkovich, Yu. A., Krivobok, N. M., Smolianina, S. O., Erokhin, A. N. Development and Operation of a Space-Oriented Salad Machine Phytoconveyer. SAE Technical Paper # 2005-01-2842, 2005

[8] Guevara, J. C., Yahia, E. M., Brito de la Fuente, E., Biserka, S. P. Effects of elevated concentrations of CO_2 in modified atmosphere packaging on the quality of prickly pear cactus stems (*Opuntia* spp.). Postharvest Biol. Technol. 2003, 29: 167 ~ 176

[9] Berkovich, Yu. A., Krivobok, N. M., Sinyak, Y. Y., Smolyanina, S. O., Grigoriev, Yu. I., Romanov, S. Yu., Guissenberg, A. S. Developing a vitamin greenhouse for the life support system of the international space station and for future interplanetary missions. Adv. Space Res. 2004, 34: 1 552 ~ 1 557

[10] Hounsome, N., Hounsome, B., Tomos, D., Edwards-Jones, G. Plant Metabolites and Nutritional Quality of Vegetables. J. Food Sci. 2008, 73: R48 ~ R65

构件集热式日光温室蓄热系统的研究*

陈 亮**，马承伟***，程杰宇，张建宇，孙国涛
（中国农业大学 农业部设施农业工程重点实验室，北京 100083）

摘要：为了提高日光温室内夜间气温以及增强墙体和地下的保温蓄热性，在吸收前人研究成果的基础上，从日光温室有效利用太阳能和减少投资的角度出发，设计并建造了构件集热式日光温室墙体与地中蓄热系统，即将管道埋设于墙体表面和墙体内部，构成矩形环路，利用墙面吸收太阳能，并以水作为热传递工质，将太阳辐射热能通过集热水管向内层墙体传递并蓄热；另外，温室管架与地中埋管连通构成循环管路，管路中的水在水泵作用下进行强制循环流动，利用管架吸收太阳辐射热量及室内富余热量，通过水流将热量传递并蓄积到地中土壤。试验结果表明，温室内日最低气温可提高 1.5℃，在阴雪天等不利天气条件下集热温室表现出一定优势；墙体不同深度处温度能提高 1～1.5℃，深度越深温度提高越多；地下不同深度处地温可提高 1℃左右。

关键词：日光温室，太阳能，构件集热，蓄热

Research on Component Heat Storage System of Solar Greenhouse

Chen Liang，Ma Chengwei*，Cheng Jieyu，Zhang Jianyu，Sun Guotao
（*key Laboratory of Agricultural Bio-Environmental Engineering*，*Ministry of Agriculture*，
China Agricultural University，*Beijing*，100083 *China*）

Abstract：In order to rise the night temperature and enhance the thermal insulation performance and heat storage performance of the walls and underground of the solar greenhouse，this paper aims at effectively using solar energy and reducing investment to design and build a solar greenhouse with component heat collector system on the basis of previous research achievements. The pipelines of this system were buried on the surface and in inward of the walls，forming a rectangular loop. This system used the wall to absorb solar energy and water as heat transfer vector，through the collector pipes passing the solar radiation to the inner walls. Moreover，the greenhouse's skeleton and underground pipes were connected to form a loop pipe system，with the water circulating in the pipes，so that the heat absorbed by the skeleton from solar radiation and excess heat indoor can be transferred and storage to the soil. The data showed that the daily minimum temperature in the greenhouse can be increased by 1.5℃. At the same time，this system showed certain advantages in bad weather. The temperature at different depths of walls can be increased 1～1.5℃，and the deeper the temperature was higher. The temperature at different depths of underground can be increased by about 1℃.

Key words：Solar greenhouse；Solar；Heat collector assistant structure；Heat storage

---

* 基金项目：现代农业产业技术体系建设专项资金（CARS－25－D－04），“十一五”国家科技支撑计划项目高效设施农业生产技术集成与示范课题（2009BADA4B04－01）资助

** 作者简介：陈 亮（1987—），男，湖北武汉人，在读硕士研究生。E－mail：luoye0687@163.com

*** 通讯作者：马承伟（1952—），男，重庆人，教授，博士生导师，中国农业工程学会高级会员，主要从事设施农业环境工程方面的研究。中国农业大学农业部设施农业工程重点实验室，北京 100083。E－mail：macwbs@cau.edu.cn

日光温室已成为中国最实用和广泛应用的主流园艺设施，目前的发展面积达到 $7\times10^5hm^2$，占我国园艺设施总面积的1/4。尤其在北方地区，大力推广高效节能型日光温室，冬季生产反季节蔬菜，基本满足了冬春对蔬菜的需求，取得了较好的经济效益与社会效益。但我国北方广大地区由于深秋、冬季和早春季节气候寒冷，且昼夜温差较大，日光温室的夜间温度在不加热条件下仍然过低，不能完全满足作物生长的要求。如果加温，传统的加温方式需要耗费大量的不可再生的化石燃料，同时产生的大量的 CO_2、CO、SO_2 和 NO_x 等有害气体，污染环境[1,2]。

太阳能是一种清洁能源，量大且分布广，应用于温室增温中，替代不可再生的化石燃料加热，是最有前景的能源之一[3]。太阳能加热系统不但能提高温室内的气温和地温，克服传统地上加热方式所存在的问题，使作物实现速生、高产、优质和高效[4,5]。

对于温室生产中，节能和保温蓄热性能是最关键的问题。国外针对大型温室，从材料选择、能源供应、环境控制措施等方面采用节能新技术[6]，而对于我国日光温室这种主要利用太阳能作为热源的温室来说，最大限度地获得太阳能同时尽可能减少内部的热量向室外扩散，也就是最大限度地提高太阳能的利用率，是日光温室节能的主要途径[7]。正确地设计，开发适用的复合墙体材料和合理的墙体构造，增强墙体的保温蓄热性能，一直是设施园艺工程界研究的重点[8,9]。

目前，国内外有关温室热环境调节的研究主要侧重太阳能集热管、地源热泵、地中热交换、地板加热等技术。陈青云等[10]利用太阳能集热管将白天收集的热量储存在放置于温室墙体内侧的水箱中，水箱在夜间释放热量以满足作物生长的需要。Onder Ozgner（2005）在土耳其将太阳能和地源热泵联合起来用于温室加温，系统供暖COP约为2.27[11]。2009年，国外的Hüseyin Benli和Aydıṇ Durmus评价了地源热泵系统和潜热蓄能装置[12]结合起来用于温室加温的加温效果[13]。马承伟[14]研究了单栋塑料大棚地中热交换的应用效果，并利用数学物理方程描述了地下热交换温室的地温场变化，为后期相关研究采用数学分析计算机模拟的方法提供了参考。对于日光温室保温蓄热性能的研究，在墙体结构设计上，王晓东，马彩雯，吴天乐等[15]研究针对塔城地区温室现状进行资料收集和数据测试，根据热工学原理，结合地域、土壤、经济等因素，对不同结构墙体进行总结和对比，引用结构相同、材料不同的温室环境温度测试结构，选择最优化的保温材料，提出最优化的温室墙体结构方案。在墙体材料方面，郭慧卿[16]、李振海[17]等人的研究结果表明，加气混凝土砌块、岩棉、粉煤灰砖作为墙体内表面蓄热材料，产生的昼夜温差小，热稳定性好，蓄热效果优于红砖，可作为取代红砖的备用品。崔秋娜[18]将相变材料镶嵌于温室后墙内侧形成相变墙板，实验结果显示：含有相变材料的温室在夏季白天可平均降低室内空气温度2~3℃，冬季夜间可平均提高室内空间温度1~2℃。

目前在温室中利用太阳热能的试验，因系统复杂和装置成本过高，多数尚未达到实用的程度。日光温室主要是通过墙体和地面的蓄热实现对太阳能的利用，但有关研究表明，墙体和地面在工作中，有效发挥蓄热和放热作用的材料层，主要仅限于从其内表面至深度20~30cm的部分，如果能增加蓄热的深度，使参与蓄热和放热的材料增多，将能增加蓄热量和放热量，使之在连阴天能持久放热，室温不致过低。本文作者在前人的研究基础上，针对上述问题，设计了一种构建集热式日光温室，既可增强墙体和地面蓄积太阳能的能力，同时，利用了温室自身的构件进行集热，成本也相对较低。

1　试验条件

1.1　试验温室概况

试验于2011年12月初至2012年3月初在北京市通州区中农富通园艺有限公司基地的两栋日光温室内进行，温室为东西向，坐北朝南，方位角为南偏西5°，长度为58m，跨度为7.8m，脊高4m，后墙高3m。东、西、北三面墙均采用混凝土空心砌块砌筑，室内墙面空心砌块以水泥砂浆填实，墙体中部以土填充，建有后坡，采光面（前屋面）采用半透明薄膜覆盖，夜间加盖保温被进行保温。

1.2　集热系统的构造及运行原理

构建集热式日光温室由墙体埋管集热系统和地中埋管集热系统组成，两部分相互独立。墙体埋管系统主要由墙中管道和补水箱组成；地中埋管系统由管架（包括拱架与拉杆）、屋脊主管、地中主管、立管、地中支管等部分组成（图1）。

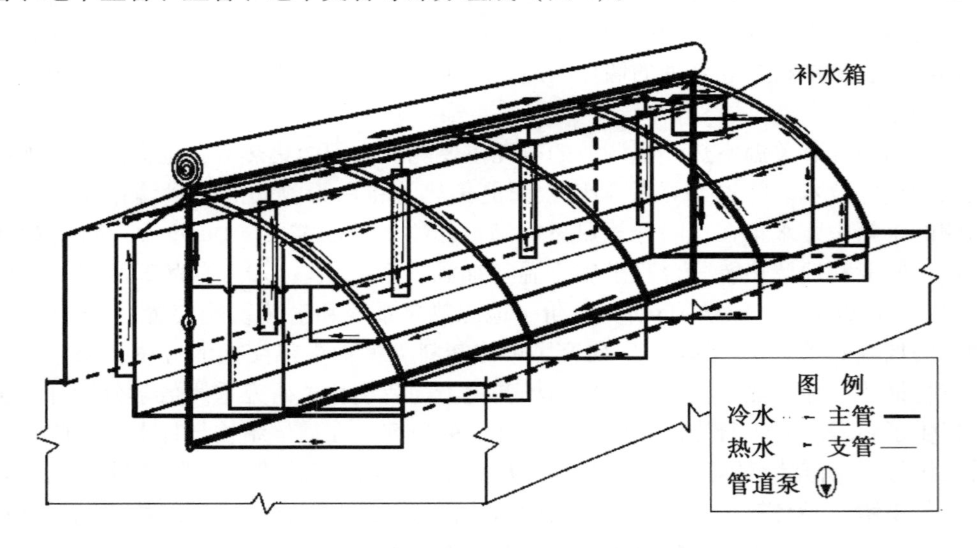

图1　构件集热式日光温室总体系统图

墙体中的埋管集热系统是以水作为热传递工质，充分利用墙面吸收的太阳辐射热能，并且将热量更多更快地向内层墙体传递和积蓄，增强墙体在夜间的持久放热能力，从而提高温室内夜间的空气温度。试验温室中采用自动补水箱进行补水，减少水压对集热效果的影响。供水管道进口与温室东面墙上的自动补水箱连接，补水箱的位置高于供水管道，利用水的重力压给供水管道补水，补水箱与自来水管连通，通过进水阀门控制其上水，为保证集热管道中的充满，进水阀门保持开启。埋管集热系统中水的流动是依靠墙面和墙内水管中的水具有不同温度而形成自然压差，使水在管道中循环流动，同时进行热量的传递，不消耗任何动力。墙体中集热管道间隔是0.5m，墙内和墙表管道形成回路，上部和下部各由一根管道连通，可补水与放水。温室墙体表面白天接收太阳辐射，温度升高很快，墙表的温度会高于墙表水管中水的温度，墙表会将热量传递给墙表水管，使墙表水管中水的温度高于墙内水管，由于墙表水管和墙内水管中的水存在温度差，会产生密度差，墙表水管中水的密度小于墙内水管，从而产生压差，使墙表水管中水向上流，墙内水管中水向下流，形成循环，同时会将墙表的热量通过水流与墙内水管管壁以对流换热的方式传递并积蓄到墙体内层。

地中埋管集热系统是通过传热介质水在水泵作用下的强制流动，将太阳辐射热量及温室内富余热量传递并蓄积到较深层土壤中，增强地面在夜间或多云、连阴天等情况下持久放热能力。整栋温室蓄热系统分为两个相同且相对于温室中间横断面对称的循环管路，两台水泵分别安装在温室中部立管处，为东、西两个循环管路中水流提供动力。图 1 所示为温室东部循环管路示意图，各管路相互连通，整个系统采用了同程式设计，保证了管路系统阻力平衡和水力工况稳定。根据本系统设计原理，支管埋于地下约 40cm 深的位置，以实现热量在土壤深处的蓄积。温室拱架设计间距为 1m，为增大地中管路与土壤换热面积，地中支管埋设间距定为 0.5m，每相邻两根地中支管通过温室南端立管与一根拱架（包括上下弦）连通。

天气晴好的白天，当温室内太阳辐射照度和温度升到较高时，水泵开启，循环管路中水沿图 1 中箭头所示方向流动。管架吸收太阳辐射热量，并将热量传递给水流，同时由于白天温室内气温较高，通过骨架外表面也可以吸收一部分空气热量并传递给水流。因此，水流经管架过程中吸热，温度升高，然后管架中水汇集到温室屋脊处的主管中。主管中的水流向温室中部水泵所在的立管，在水泵强制力作用下，高温水进入地中埋管。由于温室内深层土壤温度一般较低，高温水流经地中管路过程中，将热量传递给地中土壤，实现了地中蓄热。下午，温室内温度降低到一定程度时，水泵停止工作，集热过程停止。夜晚，室内气温不断下降，蓄积了更多热量的土壤通过地表放热也相对较多，从而使室内气温不致降至过低。

2 日光温室热环境测试试验方案

测点布置与实验仪器：

温室东西向选取三个截面，分别为温室中部、中部偏东 12m、中部偏西 12m，在墙体每个截面三个高度上布置热电偶，高度分别为离地面 0.7m、1.5m、2.3m 处，每个截面上布置热电偶测点的埋设深度分别为距离内侧墙体表面 5cm、15cm、30cm、50cm、75cm、110cm、155cm，共计 7 个深度，试验温室每处再设置一个测定集热水管表面的温度测点。在地面对应墙体的三个截面上，每个截面南北向布置四处热电偶，由北向南距离墙体分别为 1.5m、3.4m、5.3m、7.2m 处，每处设 5 个不同深度测点，距地表深度分别为 0cm、15cm、30cm、40cm、50cm。温室东西方向上的跨中位置放置四个温湿度记录仪，分别为温室中部偏东 6m、中部偏东 18m、中部偏西 6m，中部偏西 18m，南北向中间位置，距离地面 1.5m 处。太阳辐射传感器布置于温室偏东 6m 左右的跨中位置，距离地面 1.5m 处。

地下集热系统中水泵由 XMT626 智能控制仪外接热电偶输入温度信号进行控制，热电偶测试端悬挂于温室东侧山墙偏西 12m 处拱架中部下方 0.3m 位置处，测试空气温度，未加防辐射罩。当测试温度大于设定上限值时，水泵开启，当小于设定下限值时，水泵关闭。因未经防辐射处理，经过观察，发现在白天同一时刻 XMT626 智能控制仪显示温度值比室内温湿度记录仪高 1～2℃，因此设定 XMT626 控制上限温度为 20℃，即控制水泵开启温度，为避免继电器频繁释放与吸合，下限温度设定为 18℃，即控制水泵关闭温度。根据室内气温的变化情况可以确定天气晴好时集热系统运行时间大概为 11：00～16：00。

试验中用到的仪器有：CR1000 数据采集仪 2 台、CMP3 辐射传感器 2 个、HOBO U14 温湿度记录仪 8 个、TAFF 型 2mm×0. 3mm T 型铜-铜镍热电偶若干、HOBO H21 小型自动气象站 1 套、XMT626 温度智能控制仪 1 个。

3　试验结果分析

3. 1　室内气温比较

日光温室保温蓄热效果主要体现在室内空气温度环境，室内空气温度是日光温室内植物生长好坏的决定性因素，因而它是重点研究指标。两温室结构和建造上相同，只是试验温室增加了集热管道，因而可大体看做两温室条件相同（图 2）。为方便起见，规定 1#温室为试验温室，即集热温室，2#为对照温室，即非集热温室。

选取 1 月 27～29 日 1#和 2#温室气温进行对比分析，其中，28 号为小雪转阴，27、29 号均为晴转阴，27 号之前为连续晴天。27、29 号温度升高后，试验温室水泵开始运行，温度降低后停止运行，28 号温度未达到水泵运行要求，水泵未运行。

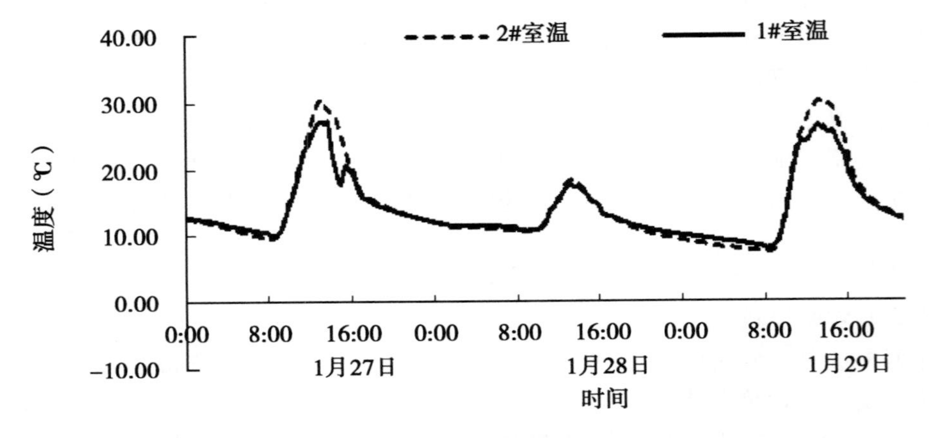

图 2　1#与 2#温室室内气温对比

由图 2 可以看出，1 月份天气连阴情况下，1#和 2#温室室内最高气温可以达到 25℃，最低气温在 7℃左右；每天最高温度出现在对照温室，最低温度也出现在对照温室；每天最高温度出现在 13：00 左右，最低温度出现在 8：00 左右；试验温室夜晚温度均高于对照温室，白天温度均低于对照温室，夜晚差值不大，但白天有时差值比较显著；晴阴雪天对应温度反应比较明显；图中有些变化异常点系温室管理中操作引起。

温室气温变化规律：早上揭开保温被之前，室内气温一直处于缓慢降低过程，7：30 揭开保温被时，因室内外气温相差较大，室内向室外快速传导热量，室内温度有一个急剧下降的过程，随着太阳的辐射作用，温度马上回升，在这一过程中，1#温室温度一直略高于 2#温室。之后一段时间温室持续升温，两温室气温基本相同。当室内气温上升到 20℃时（大概 11：00），1#温室水泵启动，管架中水开始循环，流经地下支管，向地下土壤中蓄积热量，之后 1#温室相比 2#温室温度上升缓慢，且最高温也较低，在水泵停止运行前，1#温室温度一直低于 2#温室，正是由于管道内水的循环转移了部分热量，才产生以上结果，也证明该系统确实可起到蓄积热量的作用，同时降低了温室内正午时刻的高温。再之后太阳辐射减弱，室内气温降低，当温度低于18℃时，水泵停止运行，16：30前后盖上保温被，在室外气温影响下，室内气温缓慢下降，1#温室温度下降速率小于2#温室。前半夜2#温室温度高于1#温室，后半夜1#温室温度高于2#温室，在后半夜墙体和地下土壤反过来向温室内散发热量，减弱温室气温降低趋势，这充分体现集热温室中集热系统的效果。

1月27日天气较晴朗，加上之前连续晴天，因而当天温度较高。1#温室下午13：00后温度急剧下降后又急剧上升，期间1#、2#温室最大温差达到8℃，明显不正常，这是由于在温室管理中，工人工作时打开保温膜放风，导致温度急剧下降，当关上风口时，温度又马上回升，这样做一定程度上损失了部分热量，蓄热量减少。28号小雪天气，阳光不充足，且前一天蓄热不充分，导致当天1#温室内气温偏低，特别是28号0：00至7：00，两温室温差不明显，平均温差仅0.33℃。但经过一个阴雪天气后，29号0：00至7：00，1#温室温度比2#温室高0.90℃，这正体现出集热温室的优势来，蓄积晴天时的多余热量，在阴天时保持夜间较高温度。29号白天很好的体现了集热系统运行后温室内气温变化情况，水泵运行期间，集热温室最多比非集热温室低4℃，很好的缓解了温室中温度过高情况。

3.2 墙体温度比较

墙体埋管集热系统中水管位于墙体内离墙表50cm深度处，与同深度处填土的温度测点距离为10cm，通过测量墙体50cm处土壤温度和埋管管壁温度，观察热量蓄积与释放过程。

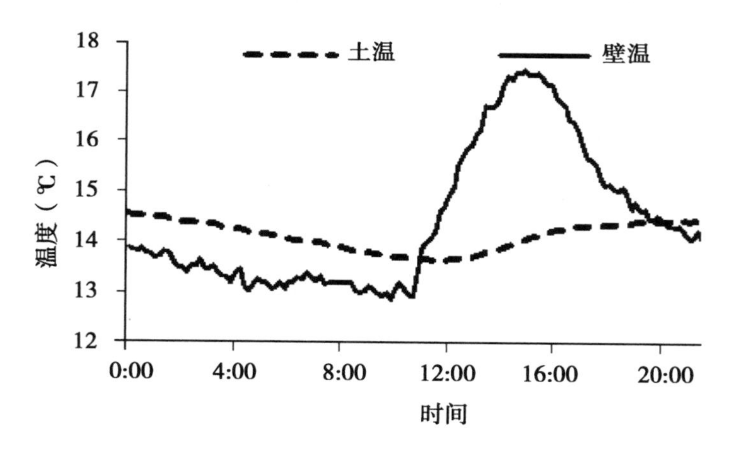

图3　1#温室墙内50cm处土温与壁温对比（1月2日）

图3为1月2日1#温室墙体中50cm处土壤温度和埋管管壁温度的变化曲线。由图可知，土温一直处于很稳定状态，管壁温度在中午到下午这段时间有明显的波动状态，具体来说，0：00至11：00土温一直高于管壁温度，且两温度有相同的变化趋势，之后管温迅速上升，超过土温，在15：00左右达到最大值，再然后慢慢降低，20：00之后又低于土温。在早晨和上午阳光不是很充足时，土温高于管温，但温差不大，这是由于管道中水比土壤散热快的缘故。当太阳辐射增加时，墙体表面水管中水温度升高，在重力作用下墙体管道内水开始循环，故50cm处管温持续上升，存在一定温差时，管中水的热量会向土壤传导，使墙体深层蓄积热量。太阳辐射减弱后，水管中温度下降，直至略低于土温。这一过程正好反映

了墙体管道中水温随太阳辐射变化而变化的结果，显示中午和下午系统确实可将水温提升，通过循环向深层墙体蓄积，证明了系统运行的合理性。

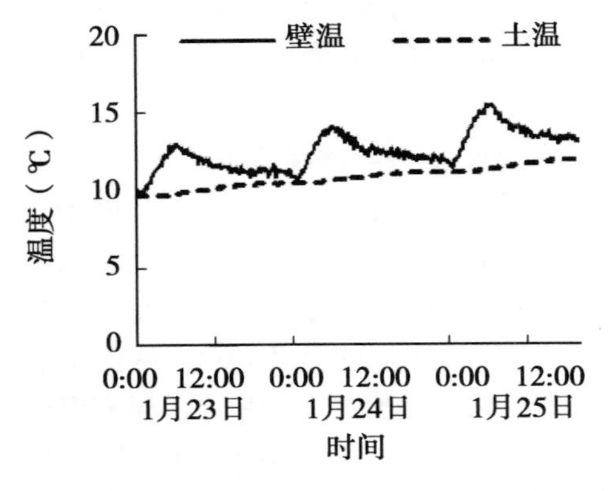

图4　连续晴天墙内50cm处土温与管温对比

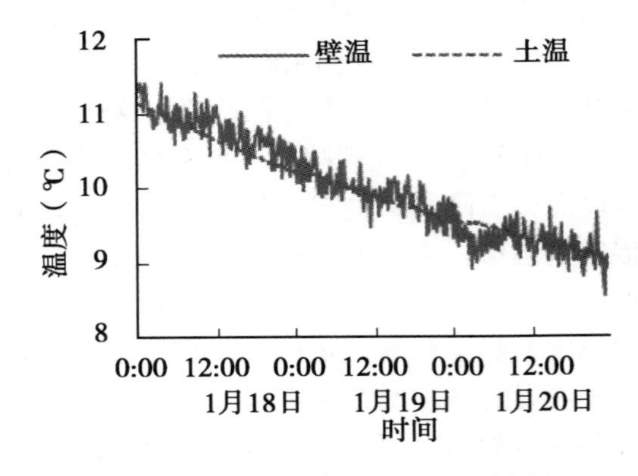

图5　连续阴天墙内50cm处土温与管温对比

1月23～25日为连续晴天（图4），1月18～20日为连续阴天（图5），连续晴天时土温有稍微的上升趋势，而壁温呈现周期性变化总体而言也是上升趋势，并且壁温一直都比土温高；连续阴天时土壤温度和管壁温度都呈下降趋势，土温变化比较平滑，而壁温较为曲折，壁温变化曲折是由于管道中水比热大，吸热放热较快导致。对这两个图分析得出埋管集热系统墙内水管与同深度土壤的日逐时温度和不同天气条件下的日平均温有相同的变化趋势，温度的差值也会随天气变化而有所变化。晴天时系统运行，管道内水循环，使同深度土壤温度升高；阴天时，墙内管道中水和墙表管道中水不存在温差，不能形成循环，不能蓄积热量，故墙体一直处于放热过程，温度持续降低。

3.3　地下土壤温度比较

3.3.1　不同深度处地温比较

通过观察图6至图9可以发现，集热温室不同深度地温均高于非集热温室，且二者地温变化趋势相同。1#温室不同深度间比较可知，50cm深度地温最低，40cm深度地温略高于30cm深度，可见地中蓄热对相同深度地温提高有一定效果；通过对比，没有看出集热系统运行对地温变化产生明显的波动影响，可能原因有两个：第一，地温测点位于两支管中间的竖直平面上，距离支管较远，与埋管同深度（40cm）测点距离埋管也有20cm以上，经过热量传递其日温度变化影响已不明显；第二，地中埋管换热在短时间内对深层地温变化影响不明显。

3.3.2　进出口水温比较

1#温室地下管道北侧为进水口，南侧为出水口，晴天时，在水泵的作用下，水流经管架，接受太阳辐射后汇集到屋脊主管，在进入地下前的主管上测量其温度，即为进水口温度，当水流过地下支管后，在流出地面处再测量其温度，即为出水口温度，这两处温度变化即能体现地下集热系统运行规律。

2月11日，天气晴朗，温室内气温较高，集热系统正常运行，选取这一天地下进出口水温来探讨。由图10可知，水温最高可达到25℃以上，北管最低温在15℃左右，南管最低

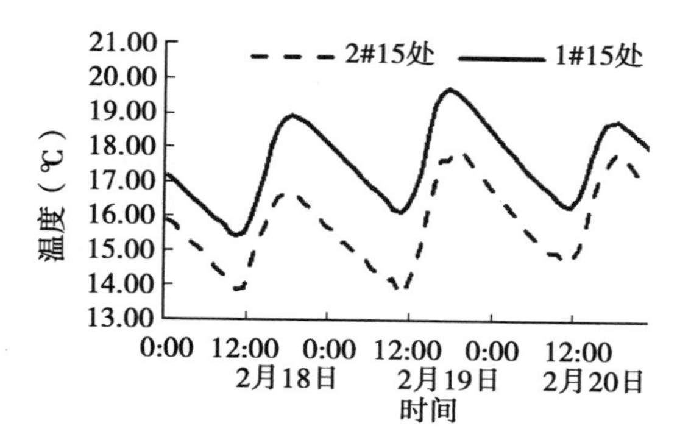

图 6　1#与 2#温室地下 15cm 处土温对比

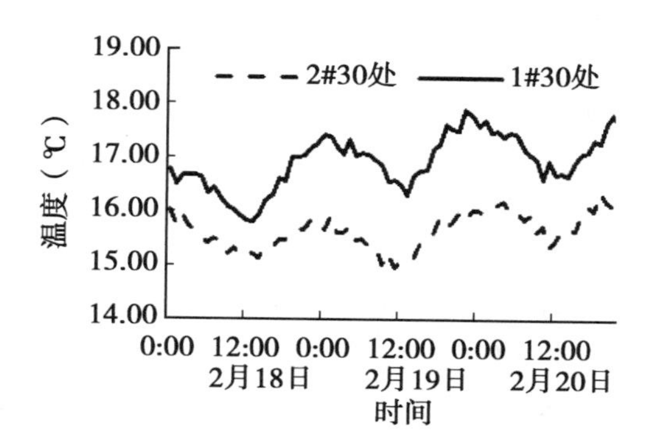

图 7　1#与 2#温室地下 30cm 处土温对比

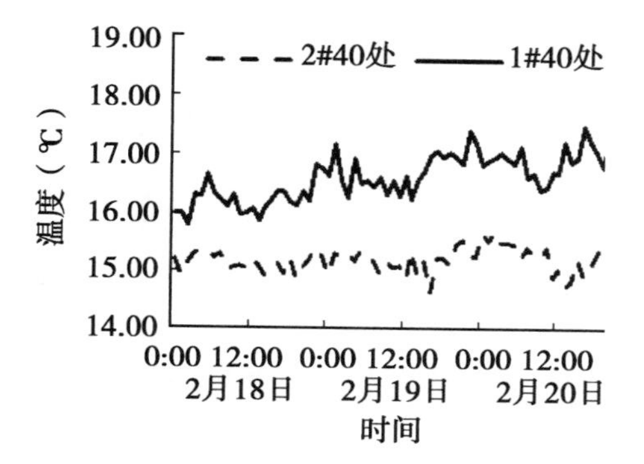

图 8　1#与 2#温室地下 40cm 处土温对比

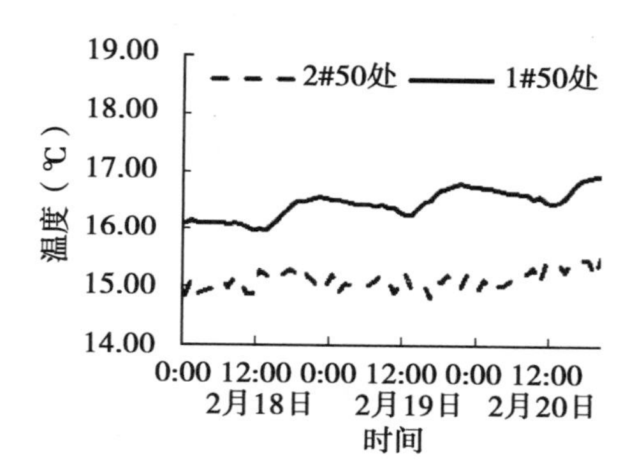

图 9　1#与 2#温室地下 50cm 处土温对比

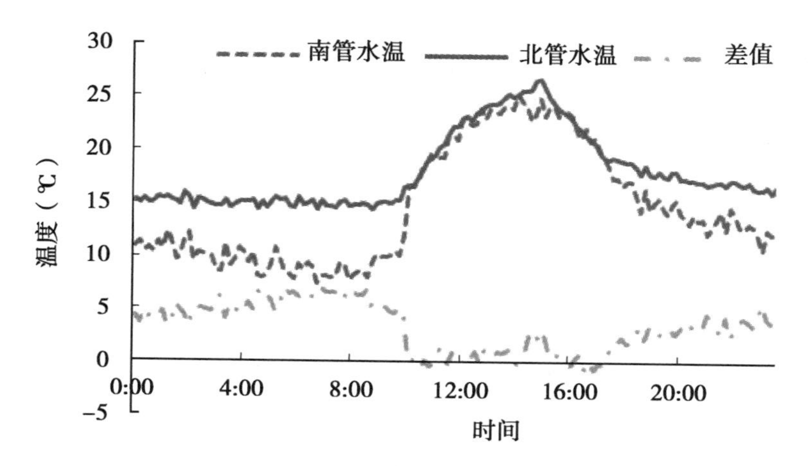

图 10　1#温室南北管水温对比（2 月 11 日）

温在 8℃左右，且北管水温一直高于南管水温。上午 10：00 之前，北管水温比南管高 5℃左右，之后南管水温急剧上升，直至和北管水温相差不大，在 10：00～17：00，两处水温同时上升又下降，但温差在 1℃左右，17：00 后，南管水温下降速率大于北管水温，两处温差增大直至5℃左右。

经分析数据可知，图中开始阶段和最后阶段南北管水温存在5℃的温差是因为南管在温室南端，靠近室外，北管在温室屋脊正下方，离后墙有一定距离，且又有后墙的保温蓄热效应，故集热系统运行前南管水温高于北管水温。当气温上升到启动水泵时，南管水温快速回升，10：00时两处水温大致相等，在17：00之前水温一直处于同一变化趋势，说明此段时间管道中水处于动态循环过程，表明地下集热系统运行正常，在水泵的作用下，水流经地下支管，向深层地下传递热量。地下40cm处土壤温度为16～17℃，水管中水温20℃以上有6h，存在5℃左右温差，热量由水通过管壁传递给土壤，实现了土壤蓄热；虽然经过了热交换，但水温变化很小，水流经地中管路前后差值平均不到1℃。

4 结论

本文作者对所研究开发的构件集热式日光温室进行了试验，对日光温室内气温、墙体温度、土壤温度变化情况进行了测试分析，主要得出以下几点结论。

第一，构件集热式日光温室在系统运行一段时间后，温室内日最低气温会提高1.5℃，在阴雪天等不利天气条件下集热温室表现出一定优势。

第二，白天达到最高气温前后的一段时间中，非集热温室室内气温高于集热温室，平均温差为2～3℃，而夜间集热温室室内气温要高于非集热温室，平均温差为0.6～1.5℃，此变化与集热系统运行效果相一致。

第三，构件集热式日光温室在系统正常运行一段时间后，墙体不同深度处温度能提高1～1.5℃，深度越深温度提高越多。

第四，构件集热式日光温室不同深度地温变化幅度很小，集热系统地中埋管换热对于地温变化在短时间内没有明显影响，但从长期变化来看，集热温室蓄热系统对提高地温有一定效果。

总体来说，构件集热日光温室在改进温室热环境方面有一定效果，且利用了自身构件集热，成本较低，但效果还不够理想，主要原因：墙体表面集热水管较少，集热量少；骨架集热不足，地下水管散热面积有限；管理操作不善，运行不稳定。进一步可在骨架上涂聚光漆和增加地中埋管方面改进。

参考文献

[1] 王彦华．农业生产机构调整与设施园艺发展趋势．北方园艺，2001（1）：2～3

[2] 李萍萍，毛罕平．我国温室生产现状与亟待研究的技术问题探讨．农业机械学报，1999，27（3）：135～139

[3] 张述英，主向东．蔬菜保护地高产高效栽培技术．北京：中国农业出版社，1997

[4] 李 宾．太阳能地中热交换塑料大棚的生产效应．中国蔬菜，1997（4）：39～41

[5] 张海莲，熊培桂，赵利敏等．温室地下蓄集太阳热能效果研究．西北农业学报，1997，6（1）：54～57

[6] BOT G P A. Developments in indoor sustainable plant production with emphasis on energy saving. Computers and Electronics in Agriculture，2001，30：151～165

[7] 佟国红，罗新兰，刘文合等．半地下式日光温室太阳能利用分析．沈阳农业大学学报，2010，41（3）：31～32

[8] 张立芸．新材料墙体日光温室的试验研究［硕士论文］．北京：中国农业大学，2006

[9] 杨仁全，马承伟，刘水丽．日光温室墙体保温蓄热性能模拟分析．上海交通大学学报，2008，26（5）：449～453

[10] 陈青云．日光温室发展的科技提升战略．设施农业与集约化养殖工程科技中长期发展战略研讨会，北京：2010

[11] Onder Ozgener, Arif Hepbasli. Experimental performance analysis of a solar assisted ground-source heat pump greenhouse heating system. Energy and Buildings, 2005, 37: 101 ~ 110
[12] 王宏丽，邹志荣，陈红武等. 温室中应用相变储热技术研究进展. 农业工程学报. 2008, 24 (6): 6
[13] Hüseyin Benli, Ayd ,n Durmus. Evaluation of ground-source heat pump combined latent heat storage system performance in greenhouse heating. Energy and Buildings, 2009, 41: 220 ~ 228
[14] 马承伟. 塑料大棚地下热交换系统的研究. 农业工程学报，1985, 1 (1): 54 ~ 65
[15] 王晓东，马彩雯，吴天乐等. 日光温室墙体特性及性能优化研究. 新疆农业科学，2009, 46 (5): 1 016 ~ 1 021
[16] 郭慧卿，李振海，崔引安等. 日光温室北墙构造与室内温度环境的关系. 沈阳农业大学学报，1995, 26 (2): 193 ~ 199
[17] 李振海，郭慧卿. 日光温室几何参数与室内温度环境的关系. 沈阳农业大学学报，1995, 26 (1): 53 ~ 58
[18] 崔秋娜. 相变储能墙板在温室中的应用研究 [硕士学位论文], 北京: 中国农业大学, 2006

光伏温室建筑一体化设计与应用

董 微*，周增产，卓杰强，兰立波，李迎忠，程 龙
（北京京鹏环球科技股份有限公司，北京 100094 *China*）

摘要：随着世界范围内的能源短缺和公众环保意识的增强，太阳能被公认为是最重要的新能源，光伏建筑物一体化已成为研究的热点。阐述了太阳能光伏发电原理及光伏建筑一体化优点，探讨了太阳能光伏发电如何与设施农业相结合，并以北京市农业机械研究所与北京京鹏环球科技股份有限公司计划共建的低碳物联网温室光伏发电项目为例，详细论述了光伏发电温室的结构设计。光伏温室建筑一体化将是现代农业未来的发展方向，其发展前景十分广阔。
关键词：光伏建筑一体化；光伏发电；光伏温室

Design and application of building integrated photovoltaic in the greenhouse

Dong Wei, *Zhou Zeng-chan*, *Zhuo Jie-qiang*, *Lan Li-bo*, *Li Ying-zhong*, *Cheng Long*
（*Beijing Jingpeng International Hi-Tech Corporation*, *Beijing* 100094, *China*）

Abstract: With the enhancement of the worldwide energy shortage and the public awareness of environmental protection, solar energy is recognized as the most important new energy, PV buildings integration has become a hot research. It describes a solar photovoltaic principle and photovoltaic building integration advantages, and explore how the solar photovoltaic facilities agriculture combined, Beijing Agricultural Machinery Institute and Beijing Jingpeng International Hi-Tech Corporation plans to build low-carbon material networking greenhouse photovoltaic power generation projects, for example, discusses in detail the structural design of the photovoltaic power generation greenhouse. Building integration of photovoltaic greenhouse will be the future direction of development of modern agriculture and its development prospects are very broad.
Key words: BIPV; Photovoltaic power generation; Photovoltaic greenhouse

在全球能源日益短缺的今天，开发可再生能源已成为全世界面临的重大课题。光伏建筑一体化（BIPV）发电系统，就是利用安装在建筑物或与建筑物结合在一起的太阳能电池的光电效应，直接把太阳能这种可再生能源的辐射能转变成电能的一种发电方式，它所生成的电能经过与其相配套的逆变控制器的转换，直接满足该建筑的用电需求[1]。

太阳能电池方阵采用与建筑结合结构安装，既解决发电装置用地要求，又不影响屋顶原有使用功能。从建筑、技术和经济角度来看，光伏建筑有以下诸多优点。

第一，可以有效地利用建筑物外表面，同时遮阳形式增强建筑屋顶功能，无需占用宝贵的土地资源，这对于土地昂贵的城市建筑尤其重要。

---

* 第一作者简介：董 微，1980年1月24日出生，女，北京京鹏环球科技股份有限公司，农艺师，现主要从事设施农业方面研究工作

第二，可原地发电、原地用电，在一定距离范围内可以节省电站送电网的投资。对于联网用户系统，光伏阵列所发电力既可供给本建筑物负载使用，也可送入电网。

第三，能有效地减少建筑能耗，实现建筑节能。光伏并网发电系统在白天阳光照射时发电，该时段也是电网用电高峰期，从而舒缓高峰电力需求。

第四，光伏组件安装在建筑的屋顶上直接吸收太阳能，因此建筑集成光伏发电系统不仅提供了电力，而且还降低了建筑物的温升。

第五，并网光伏发电系统没有噪音、没有污染物排放、不消耗任何燃料，具有绿色环保概念，可增加建筑物综合品质。

本文中以北京市农业机械研究所与北京京鹏环球科技股份有限公司计划共建的低碳物联网温室光伏发电项目作为实例，说明光伏温室建筑一体化系统的发电原理与设计方案。

1　太阳能光伏发电基本原理

太阳能光伏发电是利用光伏电池板直接将太阳辐射能转化为电能的发电方式，通常由太阳能光伏电池组件、功率控制器、逆变器、蓄电池以及负载等部件组成（图 1）。太阳能光伏模组接受阳光，产生直流电流；功率控制器防止光伏电池组件对蓄电池过充电以及对负载过放电；逆变器将光伏电池产生的直流电转变为相应的交流电，使光伏系统发出的电可以直接并入电网。因此，当前太阳能光伏发电的过程，通常是直流电转化为交流电的运作过程[2]。

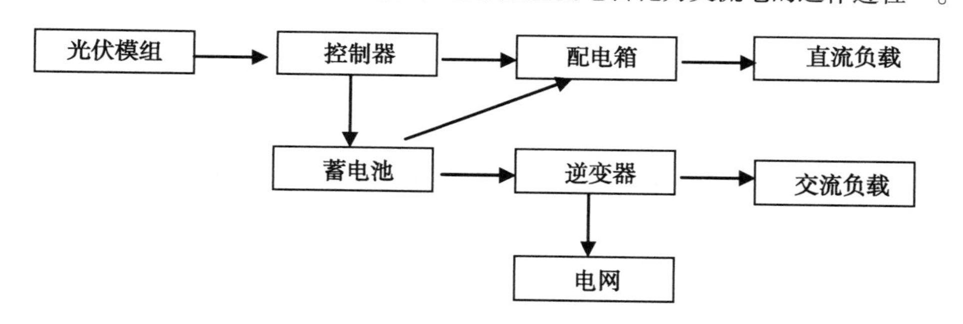

图 1　太阳能光伏发电系统组成

2　太阳能光伏电池组件的安装结构设计

2.1　安装结构分类

目前推广应用的太阳能光伏发电工程项目中，太阳能电池板的安装方式有两种，一种是地面安装式光伏发电系统，即在地面实施土建安装基础，然后将太阳能电池板的安装支架结构在地面基础上安装。另一种是太阳能光电建筑，即将光伏发电与建筑物相结合，在建筑物的外围结构表面上布设光伏器件产生电力，从而使“建筑物产生绿色能源”[3]。

2.2　光伏与建筑的结构设计

太阳能光电建筑光伏与建筑的结合有如下两种方式。

2.2.1　一种是建筑与光伏系统相结合

把封装好的的光伏组件安装在居民住宅或建筑物的屋顶上（BAPV），组成光伏发电系统。

2.2.2 另外一种是建筑与光伏器件相结合

将光伏器件与建筑材料集成化，用光伏器件直接代替建筑材料，即光伏建筑一体化（BIPV），如将太阳光伏电池制作成光伏玻璃幕墙、太阳能电池瓦等，这样不仅可开发和应用新能源，还可与装饰美化合为一体，达到节能环保效果，是今后的发展光伏建筑一体化的趋势[4~5]。

3 项目实施地气候资源情况

北京市通州区位于北纬39°36′~40°02′，东经116°32′~116°56′，太阳辐射量全年平均为112~136kcal/cm。通州大部分地区日照时数在2 600h左右。全年日照时数以春季最多，月日照在230~290h；夏季正当雨季，日照时数减少，月日照在230h左右；秋季日照时数虽没有春季多，但比夏季要多，月日照230~245h；冬季是一年中日照时数最少季节，月日照不足200h，一般在170~190h。

4 光伏温室建筑一体化设计方案

4.1 设计依据

GB37 / T729—2007 光伏电站技术条件。

GB50009—2001 建筑结构荷载规范。

10J908-5 建筑太阳能光伏系统设计与安装。

IEC61730.1 光伏组件的安全性构造要求。

IEC61730.2 光伏组件的安全性测试要求。

4.2 总体设计方案

北京京鹏环球科技股份有限公司与北京市农业机械研究所计划共建的低碳物联网温室屋顶以采光顶方式安装透光型薄膜太阳能电池组件，达到总装机容量为18.74kWp；利用并网逆变器将太阳能电池组件产生的直流电逆变为交流电，并入到办公楼接入的市电网低压配电侧中使用；同时系统具备储能系统，市电停电时系统能继续供给重要负载用电，并且继续使用太阳能产生的电能（图2和图3）。

图2 低碳物联网温室效果图

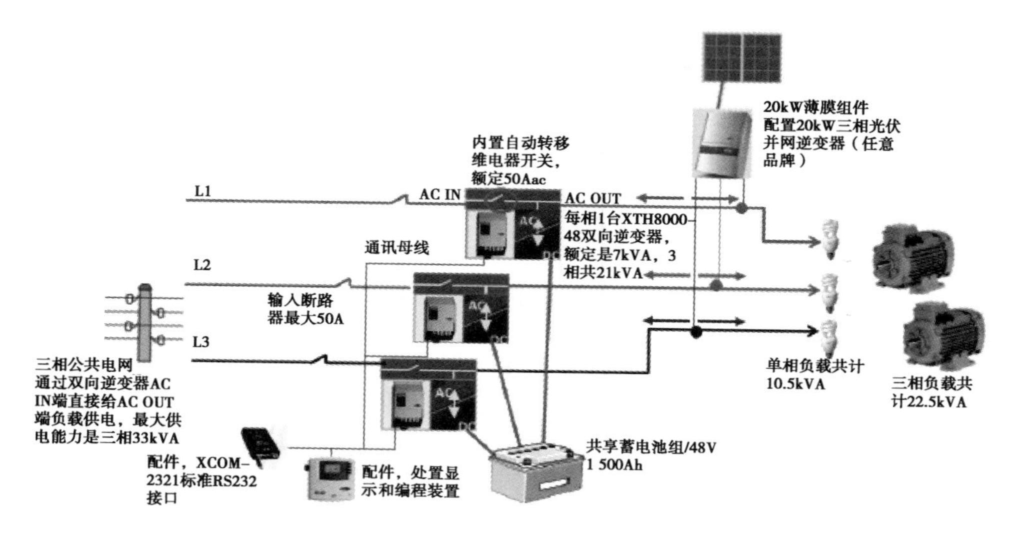

图3　光伏发电系统原理图

4.3　安装结构设计

（1）低碳物联网温室太阳能光伏发电系统由薄膜太阳电池组件、屋面安装钢架结构、并网逆变器、交（直）流配电设备、数据采集监控系统、双向逆变充电一体机、蓄电池组、线缆及电缆桥架等组成。

（2）整个光伏电站安装薄膜太阳能光伏电池组件208块，选用每块光伏电池组件的峰值功率为90Wp，玻璃尺寸长1 300mm，宽1 100mm，光伏电站总装机容量为18.72kWp。薄膜太阳电池组件以采光顶形式安装屋顶钢架上，朝向南侧，与地面间倾角23°，在不改变原有建筑风格和外观的前提下，设计安装太阳能光伏阵列的结构和布局，增强屋顶使用功能（图4和图5）。

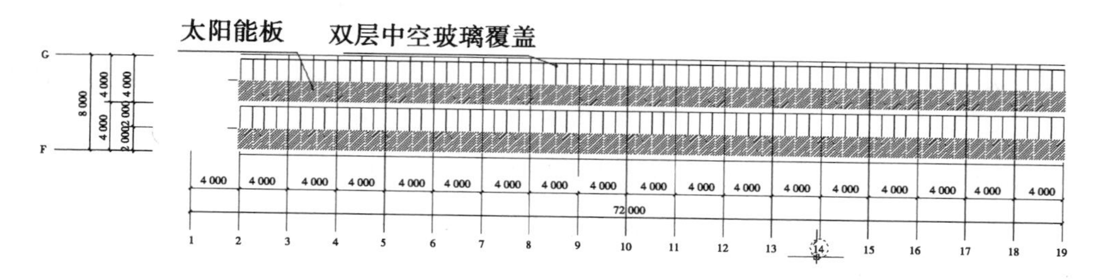

图4　屋面光伏系统太阳电池板排布示意图

4.4　防雷、接地设计

4.4.1　防雷设计

防雷击设计包括防直击雷、防雷电感应，主要措施有设置避雷装置和防雷接地。本项目采用如下措施，以保护设备免受直击雷和雷电侵入波的危害。

（1）电气设备直击雷保护。直击雷保护包括光伏电池组件和交流和直流配电系统的直击雷保护。光伏组件安装支架、交流和直流配电设备金属外壳均可靠地与接地网相连接。

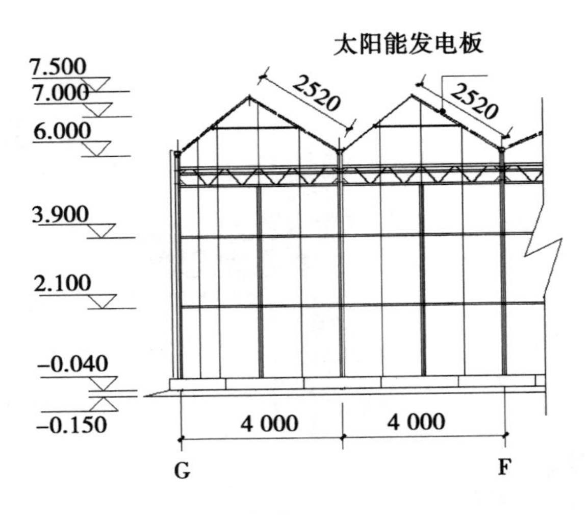

图5　安装立面示意图

(2) 站内光伏电池组件防直击雷措施，光伏电池组件边框为金属，将光伏电池组件边框与支架可靠连接，然后与建筑接地网连接，光伏电池组件可防止半径为30m的滚雷，为增加雷电流散流效果，可将站内所有光伏电池组件支架可靠连接。

(3) 为防止感应雷、浪涌等情况造成过电压而损坏配电室内的并网设备，其防雷措施主要采用防雷器来保护。太阳能光伏电池串列经电缆接入直流防雷配电单元，配电箱内配置防雷器。

4.4.2　接地设计

充分利用每个太阳能光伏电池组件支架的钢筋作为自然接地体，根据现场实际情况及土壤电阻率敷设不同的人工接地网，以满足接地电阻的要求，重点区域加强均匀布置以满足接触电势和跨步电压的要求。保护接地的范围：

根据《交流电气装置的接地》(DL/T621—1997) 规定，对所有要求接地或接零的设备均应可靠地接地或接零。

所有电气设备外壳、开关装置和开关柜接地母线、架构、电缆支架和其他可能事故带电的金属物都应可靠接地。

本系统中，支架、太阳能板副框以及连接件均是金属制品，每个子方阵自然形成等电位体，所有子方阵之间都要进行等电位连接并通过引下线与建筑接地网就近可靠连接，接地体之间的焊接点应进行防腐处理。

5　前景展望

温室用电是温室能耗的重要内容，温室中的各类环境调控设备、生产设施与装备、灌溉设备等，都需要电力驱动。尤其是一些采用新型技术与装备的温室，如地源热泵加温的温室，在保证栽培温度需求的前提下，冬季用电量开支达0.5元/(m^2·d)以上，成为温室生产者主要的成本负担。太阳能光伏发电为设施农业提供了低价的能源，降低了农业生产成本。

太阳能光伏温室建筑一体化既在温室设计上增加了创新意识，体现了清洁、绿色的理念，又为温室运行所需的电力供应提供了选择和保障，无论从美学还是实用角度来看，都具有很大的应用价值。太阳能光伏温室建筑一体化技术必将具有广阔的发展前景。

参考文献

[1] 姜志勇．光伏建筑一体化（BIPV）的应用．建筑电气，2008（6）：7~10
[2] 赵争鸣，刘建政，孙晓英等．太阳能光伏发电及其应用．北京：科学出版社，2005
[3] 王其恒．太阳能与建筑一体化的探索与应用．安徽建筑，2009（3）：45~48
[4] 韩利，艾芊．光伏技术在节能建筑中的应用．建筑节能，2009（2）：4~12
[5] 肖潇，李德英．太阳能光伏建筑一体化应用现状及发展趋势．节能，2010（2）：12~18

后墙立体栽培对日光温室内温度场的影响

栗亚飞[1*]，何华名[1]，郑　亮[1]，邢文鑫[1]，宋卫堂[1,2**]
（1. 中国农业大学水利与土木工程学院，北京　100083；
2. 农业部设施农业工程重点（综合）实验室，北京　100083）

摘要：在日光温室的后墙上，采用管道无土栽培方式进行蔬菜或草莓生产，可以提高温室空间利用率和作物种植量，但可能会出现因为管道和植物的挡光而减少后墙蓄热、降低冬季温室内部温度的问题。为此，通过连续31d的温度监测，在3种典型气候（晴天、阴天、雪天）条件下，对比分析了有后墙立体栽培的日光温室（ESG）和无后墙立体栽培的日光温室（NSG）温度场的变化。结果表明，ESG的月平均气温较NSG高0.84℃，其中，最大日温差为2.22℃，最小日温差为0.14℃。晴天条件下，ESG的日平均冠层温度和空气温度分别是12.72℃和13.04℃，NSG的日平均温度是10.68℃和11.04℃，ESG冠层温度最低值是4.68℃，NSG最低值是4.10℃，因此，ESG较NSG的气温要高一些；阴天和雪天条件下，两种温室内的温度场无明显差别。因此，日光温室后墙的立体栽培，没有降低反而提高了冬季温室内部的温度，是一种值得推广应用的温室高效栽培技术。
关键字：日光温室；后墙立体栽培；温度场

Effect of the Back Wall Soilless Cultivation on the Temperature Field in Solar Greenhouse

Li Yafei[1], He Huaming[1], Zheng Liang[1], Xing Wenxin[1], Song Weitang[1,2]
(1. *College of Water Resources and Civil Engineering*, *China Agricultural University*, *Beijing* 100083, *China*;
2. *Key Laboratory of Agricultural Engineering in Structure and Environment*,
Ministry of Agriculture, *Beijing* 100083, *China*)

Abstract: In solar greenhouse production, the use of pipeline soilless cultivation of vegetables or strawberry on the back wall in greenhouse can increase the greenhouse space utilization, however, shading of pipes and plants may reduce the back wall heat storage and the internal temperature of the greenhouse in winter. Therefore, in this project, we continuously monitored the temperature condition of the two different kinds of greenhouses for 31 days, and analyzed the temperature field of the equipped solar greenhouse (ESG) and non equipped greenhouse (NSG) of three typical Climatic conditions (sunny, cloudy and snowy). The result showed that the monthly mean temperature of ESG was 0.84℃ higher than NSG, and the maximum of daily temperature difference was 2.22℃, minimum was 0.14℃. In sunny day, ESG's daily mean canopy temperature and air temperature was 10.68℃ and 11.04℃, while the NSG's was 10.68℃ and 11.04℃, the lowest canopy temperature of ESG was 4.68℃ and NSG was 4.10℃. So, ESG was better than NSG, while? there was not significant difference in the cloudy and snowy days. Therefore, the

* 作者简介：栗亚飞（1988—），女，中国农业大学农业生物环境与能源工程专业2012级在读硕士
** 通讯作者：宋卫堂（1968—），男，博士，教授，主要从事设施园艺栽培技术与设备研究。中国农业大学水利与土木工程学院，100083。E-mail：songchali@cau.edu.cn

back wall soilless cultivation of the greenhouse did not reduce but enhance the internal temperature of solar greenhouse in winter. It is an efficient cultivation technology worthy of popularization and application in solar greenhouse.
Key words: Solar greenhouse; The back wall cultivation; Temperature field

日光温室是我国北方地区独有的一种温室类型[1]，冬季室内不加热，即使在最寒冷的季节，也只依靠太阳能来维持室内一定的温度水平，以满足蔬菜作物的生长需要[2]。日光温室具有造价低、节能，结构优化、性能良好等特点。采光和保温性能的优劣是日光温室进行实际生产的基础[3]。墙体作为温室的围护结构之一，对温室内的热环境有直接的影响，在北方冬季温室生产中，墙体的蓄热、保温性尤其重要。

立体栽培又称垂直栽培，是在尽量不影响地面栽培的前提下，通过竖立起来的栽培柱、栽培管道等作为植物生长的载体，充分利用温室空间和太阳能的一种栽培方式。主要种植一些矮秧类作物如叶菜、草莓等，可以提高土地利用率 3 ~ 5 倍，提高单位面积产量 2 ~ 3 倍。20 世纪 60 年代，立体无土栽培在发达国家首先发展起来，美国、日本、西班牙、意大利等国家研究开发了不同形式的立体无土栽培技术[4]。

后墙立体栽培是利用特定的栽培设备附着在建筑物的墙体表面，有效地利用了空间，节约了土地，实现了单位面积上更大的产出比。蔬菜栽培在墙上，菜农可以站立进行管理和收获，劳动强度大大降低，特别适合发展采摘和观光农业。在日光温室后墙上固定一定间隔距离的通长栽培管道，根据后墙的高度可设置 3 ~4 排[5]。后墙管道的采光条件好，可充分利用太阳光，有利于作物生长和果实品质的提高。但温室后墙管道栽培在增大空间利用率的同时，又对后墙产生了部分遮挡，减少了后墙对太阳光的吸收面积，有可能会影响到温室后墙的蓄热能力，降低冬季温室内部的温度。

为此，在冬季三种典型气候（晴天、阴天、雪天）条件下，通过连续 31d 的监测，对比分析有后墙立体栽培的日光温室（ESG）和无后墙立体栽培的日光温室（NSG）温度场的变化，研究后墙立体栽培对日光温室内的温度场有何影响，并评估后墙立体栽培应用于日光温室栽培的可行性。

1　试验条件与方法

1.1　试验对象

供试日光温室位于北京市昌平区小汤山天润园草莓专业合作社。选取其他条件一致的有后墙立体栽培的温室（ESG）和无后墙立体栽培的温室（NSG）各一座。温室长 50m，跨度 8m，北墙高 2.3m，脊高 3.5m；东、西山墙和北墙均为 500mm 厚砖墙；屋面覆盖材料为 0.12mm 厚的 EVA 无滴膜；夜间保温覆盖物为 30mm 厚的保温被。保温被每天上午 9：00 开启，下午 16：30关闭。其中，ESG 的后墙管道栽培设备为内径为 D＝160mm 的建筑输水 PVC 管，采用 4 排水平管道栽培，管道边缘距后墙为 6cm，各排管道之间的垂直距离为 40cm；管道内采用基质栽培，栽培基质深 10cm，灌溉方式为滴灌。其他条件 ESG 和 NSG 均一致。

1.2　试验方法

本试验主要测定两种温室的温度和湿度变化状况，测试内容为温室内的空气温度和相对湿度。为使试验数据更加可靠，每个温室布点 5 个。各测点高度分别为草莓生长冠层和距离地面 1.5m 处，各测点均采取防辐射处理。测点布置如图 1 所示。

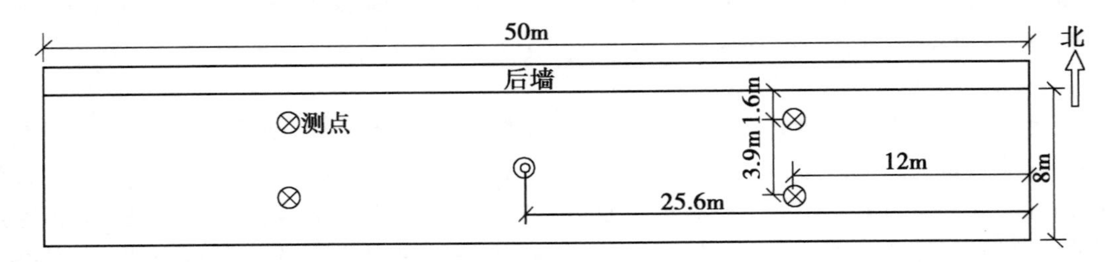

图 1　温室内测点布置平面图

Figure 1　Measuring points in greenhouses

实验材料为草莓红颜品种，采用高畦栽培，株距为 15cm。

测试仪器采用日本 Esupekkumikku 有限公司的 RS-13 温湿度记录仪，RT-13 温度记录仪，自动采集记录温度和湿度数据，数据采集记录间隔 10min，测量精度为 ±0.3℃，±5%。

测试时间为 2013 年 1 月 20 日至 2013 年 2 月 20 日。

2　试验结果与分析

2.1　室外气温

图 2 所示是测试期间（2013 年 1 月 20 日至 2013 年 2 月 20 日）北京市昌平区室外最高气温、最低气温、日平均气温的变化曲线。由图 2 可知，测试期间，室外空气温度范围是 -13℃到 7℃，平均气温为 2.89℃。

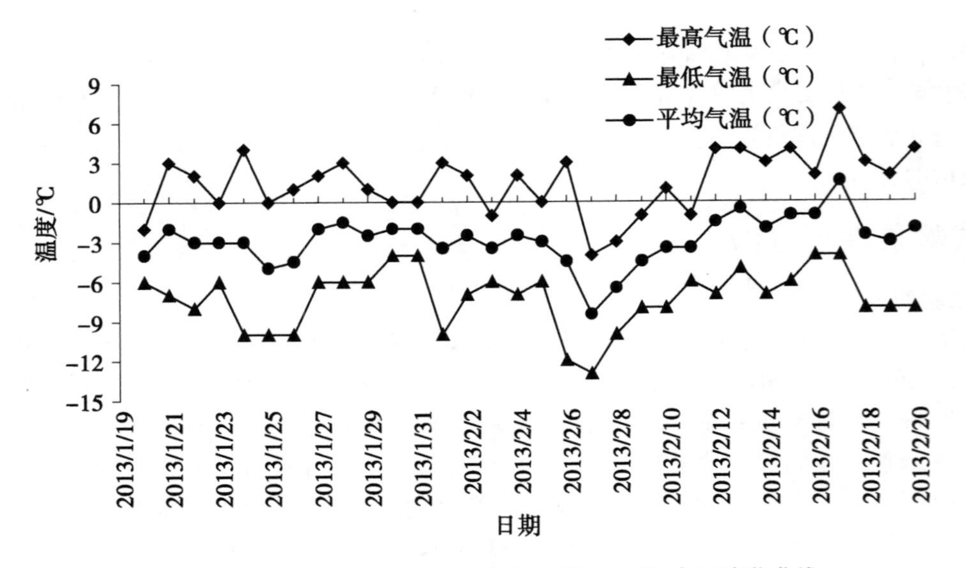

图 2　2013 年 1 月 20 日至 2 月 20 日北京市昌平区气温变化曲线

Figure 2　Temperature curves from January 20, 2013 to February 20, 2013in the Changping district, Beijing

2.2 温室内气象条件

2.2.1 温室内的气温状况

温室的环境温度，受温室周围墙体和土壤表面辐射温度及空气温度共同作用[6]。通过对各测点的温度记录仪 RT-13 同一时刻的数据进行算术平均，得出自 2013 年 1 月 20 日至 2013 年 2 月 20 日的两个温室的日平均气温，如图 3 所示。

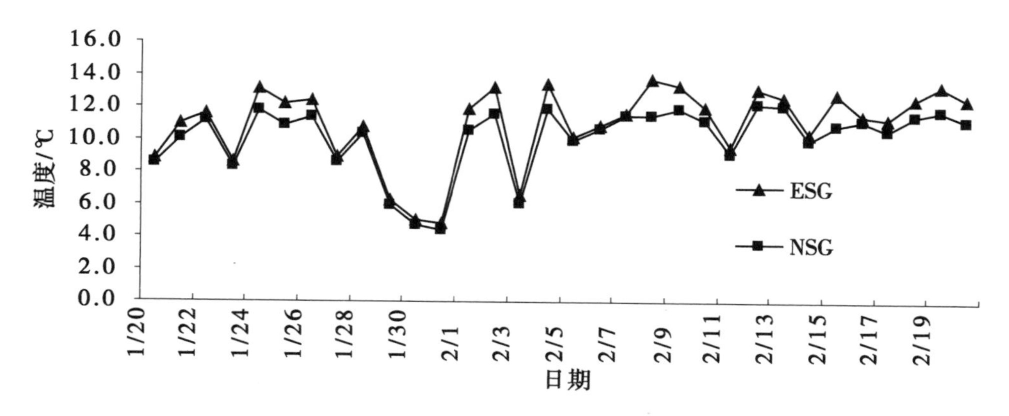

图 3　温室内日平均气温变化曲线

Figure 3　Daily mean temperature curves in greenhouses

从图 3 可以看出，ESG 和 NSG 的日平均气温变化趋势基本一致，均随外界气温的变化而变化。测定期间，ESG 的气温波动范围是 4.78～13.72℃，NSG 的气温波动范围是 4.45～12.15℃；ESG 和 NSG 的月平均气温分别是 10.94℃和 10.10℃，ESG 的月平均气温较 NSG 高 0.84℃，其中最大日温差出现在 2 月 8 日，为 2.22℃；最小日温差出现在 2 月 7 日，为 0.14℃。ESG 的月最高气温出现在 2 月 8 日，为 13.72℃；NSG 的月最高气温出现在 2 月 12 日，为 12.15℃；由于受到外界雨雪天气的影响，ESG 和 NSG 的月最低气温均出现在 1 月 31 日，分别为 4.78℃和 4.45℃。因此可以看出，有后墙管道栽培的温室的气温要略高于无后墙管道栽培的温室。

2.2.2 温室内的湿度状况

温室内的湿度同时受到温室的温度和栽培环境的影响。

通过对温湿度记录仪 RS-13 同一时刻各测点的相对湿度进行算术平均，得出自 2013 年 1 月 20 日至 2013 年 2 月 20 日的两个温室的日平均相对湿度，如图 4 所示。

如图 4 所示，ESG 和 NSG 的日平均相对湿度的变化趋势基本一致，同样受到外界气候条件的影响而波动。ESG 的相对湿度范围是 81.90%～99.00%，NSG 为 71.21%～99.00%，月平均相对湿度分别为 89.47% 和 85.89%。而日光温室内草莓生长的适宜相对湿度在 70%～90%[7]，在该湿度范围内不会影响草莓的正常生长。由图 2 可知，ESG 较 NSG 的日平均气温要高，而 ESG 较 NSG 的相对湿度也高，在其他条件一致的情况下，出现这种现象的原因有可能是后墙管道栽培灌溉过程中的水分蒸发，以及基质中的水分散失，增加了 ESG 内的相对湿度。

2.3 典型气候条件下的温度状况

为了进行针对性分析，特选取典型晴天、阴天以及阴雨雪天气条件下的测试结果进行对

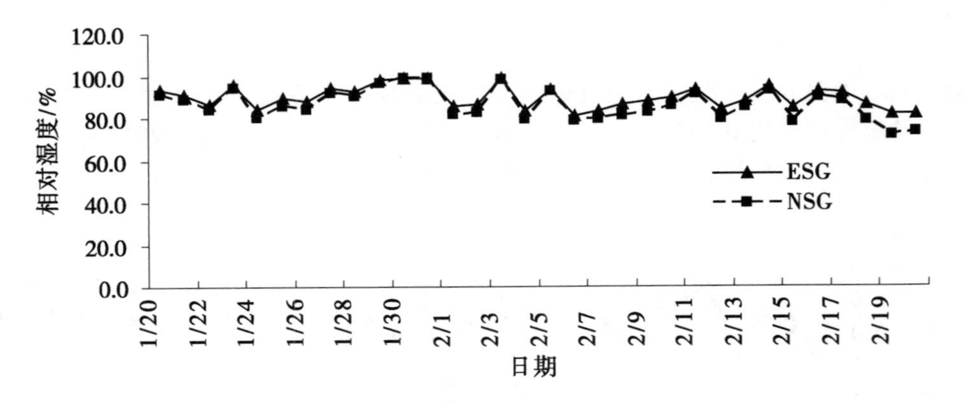

图4 温室内日平均相对湿度变化曲线

Figure 4 Daily mean humidity curves in greenhouses

比分析。根据试验期间的气象资料记载，晴天选取2013年2月15日，阴天选取2013年1月31日，雪天选取2013年2月11日作为典型日进行分析。植物冠层是植物叶片进行光合作用和蒸腾作用的重要空间[8]。冠层温度是快速、非破坏性检测整株植物是否受到水分胁迫的有效参数，反映出植物体内的水分状况[9]。因此，测点高度选在草莓生长冠层和距地面1.5m处。

2.3.1 典型晴天条件的温度状况 图5（a）为晴天条件（2013年2月15日）下温室内草莓生长冠层处0：00～24：00的温度变化曲线。图5（b）为温室内距地面1.5m处的温度变化曲线。

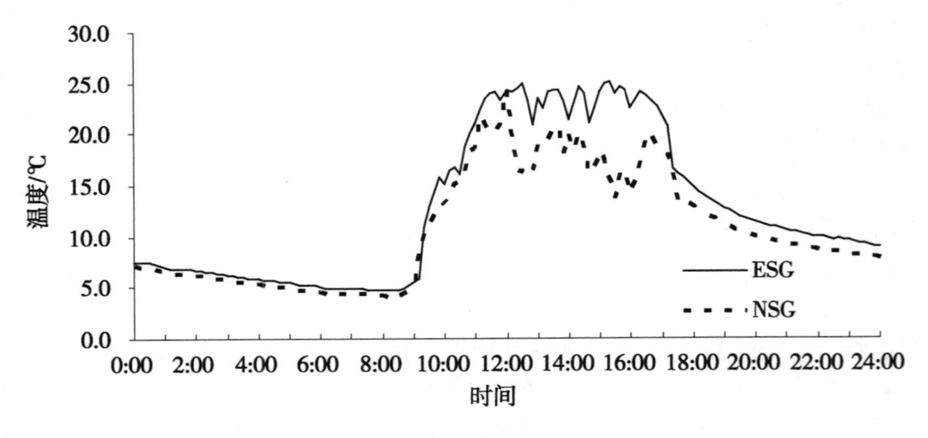

图5（a） 晴天草莓生长冠层处日温度变化曲线

Figure 5（a） Daily mean temperature curves of strawberry canopy in sunny day

由图5（a）可知，晴天条件下草莓生长冠层ESG和NSG的气温日变化趋势基本一致，均昼高夜低。ESG的日平均冠层温度是12.72℃，NSG的日平均温度是10.68℃。ESG冠层温度在8：00达到最低值4.68℃，NSG在8：10达到最低值4.10℃，这是由于温室揭开保温被后，室内气温与外界气温之间的总热阻值减少，室内向室外散热量增加，而短时间内太阳辐射还较弱，从而导致室内气温下降。ESG 冠层温度在 14：20 达到最高值 24.70℃，NSG 在 13：50 达到最大值 20.70℃。草莓生长最适温度是 18～25℃，ESG 冠层温度在 18～25℃ 范围内的时间为 6h，NSG 为 3.5h。因此，晴天条件下，ESG 较 NSG 更适宜草莓生长。

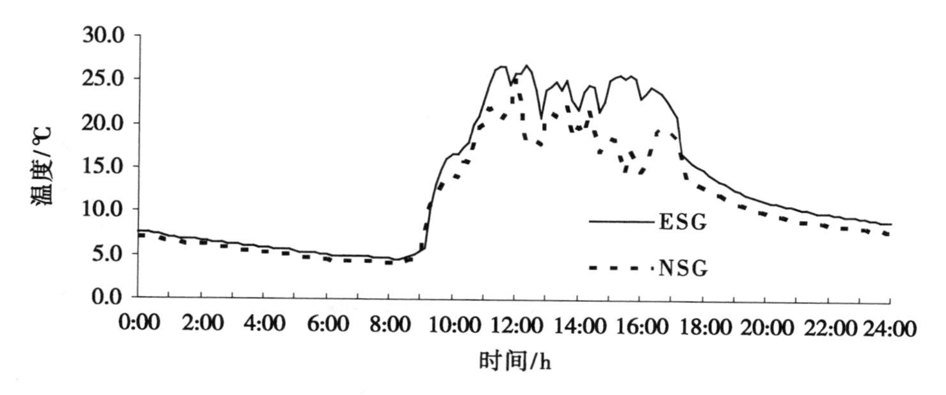

图 5（b） 晴天温室内距地面 1.5m 日温度变化曲线

Figure 5（b） Daily mean temperature curves 1.5m above ground of the sunny day in greenhouses

由图 5（b）可知，ESG 内的日平均气温为 13.04℃，NSG 的日平均气温是 11.04℃。ESG 在 8：10 达到最低气温为 4.65℃，NSG 在 8：20 达到最低值为 4.13℃；ESG 的最高气温为 26.95℃，NSG 的最高气温为 22.73℃；ESG 的室内气温在 18～25℃ 范围内的时间为 6.5h，ESG 为 2h。因此，晴天条件下，ESG 较 NSG 维持草莓适宜生长温度的时间较长。

2.3.2 典型阴天条件的温度状况 图 6（a）为阴天条件（2013 年 1 月 31 日）下温室内草莓生长冠层处 0：00～24：00 的温度变化曲线。图 6（b）为温室内距地面 1.5m 处的温度变化曲线。其中，室外最高气温为 0℃，最低气温是 -4℃，温差较小。

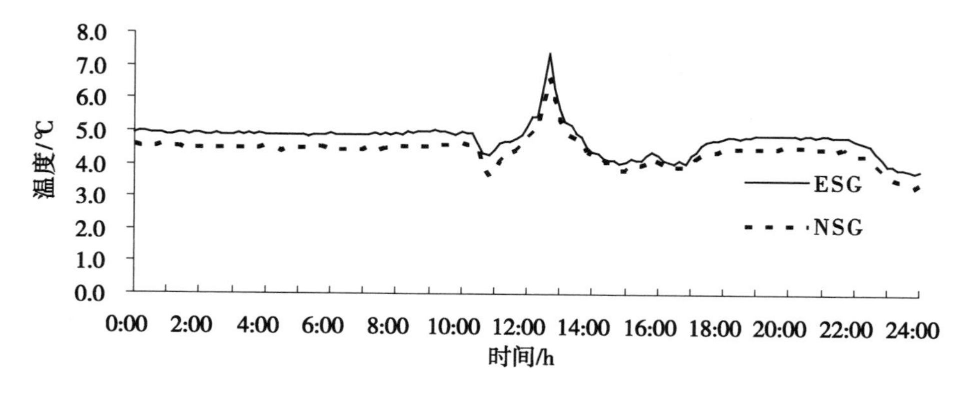

图 6（a） 阴天草莓生长冠层处日温度变化曲线

Figure 6（a） Daily mean temperature curves of strawberry canopy in cloudy day

阴天条件下，草莓生长冠层的温度变化趋势与晴天条件下基本一致，如图 5（a）所示，同样是昼高夜低。ESG 的日平均冠层温度 4.77℃，NSG 的日平均温度是 4.40℃，温差为 0.3℃；ESG 和 NSG 冠层温度均在 10：50 达到最低值为 4.28℃ 和 3.60℃；同时，ESG 和

NSG 冠层温度均在 12：40 达到最高值为 7.40℃和 6.60℃。与晴天相比，阴天室内温度变化幅度较小。在此温度范围内，ESG 和 NSG 的温度均不利于草莓的生长。

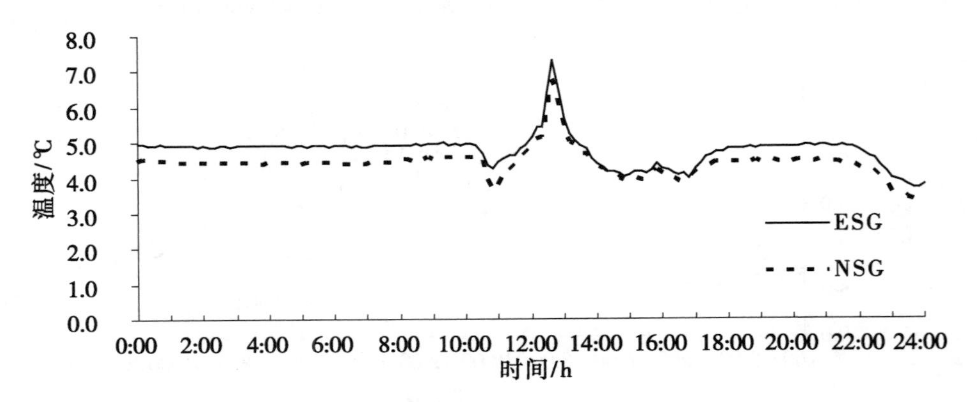

图 6（b） 阴天温室内距地面 1.5m 日温度变化曲线

Figure 6（b） Daily mean temperature curves 1.5m above ground of the cloudy day in greenhouses

由图 6（b）可知，ESG 的日平均气温为 4.76℃，NSG 的日平均气温是 4.40℃，与草莓冠层温度基本一致，且两温室的室内气温相差较小；ESG 和 NSG 室内气温均在 10：50 达到最低值为 4.28℃和 3.60℃；同时，ESG 和 NSG 室内气温均在 12：40 达到最高值为 7.28℃和 6.73℃。室内气温变化幅度较小，这与陈端生的研究相符[10]，即光环境是影响温室气候环境的第一因素。

2.3.3 典型雪天条件的温度状况

在温室的使用过程中，低温冻害经常在连续阴雨雪天气时发生，因此，对温室内温度在雨雪天气时的变化进行分析极其重要。

图 7（a）为雪天条件（2013 年 2 月 11 日）下温室内草莓生长冠层处 0：00～24：00 的温度变化曲线。图 7（b）为温室内距地面 1.5m 处的温度变化曲线。其中，室外最高气温是 -1℃，最低气温是 -6℃。

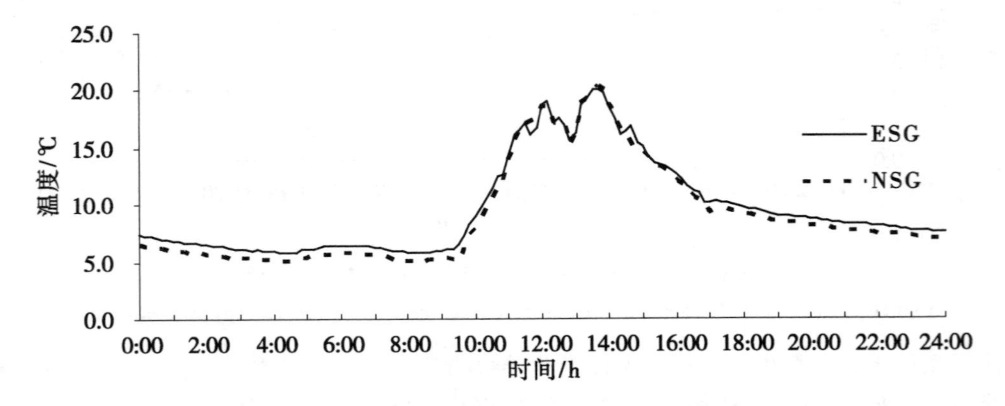

图 7（a） 雪天草莓生长冠层处日温度变化曲线

Figure 7（a） Daily mean temperature curves of strawberry canopy in snowy day

由图7（a）可知，雪天气候条件下，温室内草莓生长冠层温度与晴天一致。ESG 的日平均冠层温度 9.58℃，NSG 的日平均冠层温度是 9.01℃；ESG 冠层温度在 8：30 达到最低值为 5.75℃，NSG 在 8：20 达到最低值为 5.10℃；同时，ESG 和 NSG 冠层温度均在 13：30 达到最高值为 21.6℃和 21.2℃，两个温室自 10：00～16：00 冠层温度和变化趋势基本一致。雪天条件下，白天两温室的冠层温度基本一致，而夜间，NSG 的温度略低于 ESG。

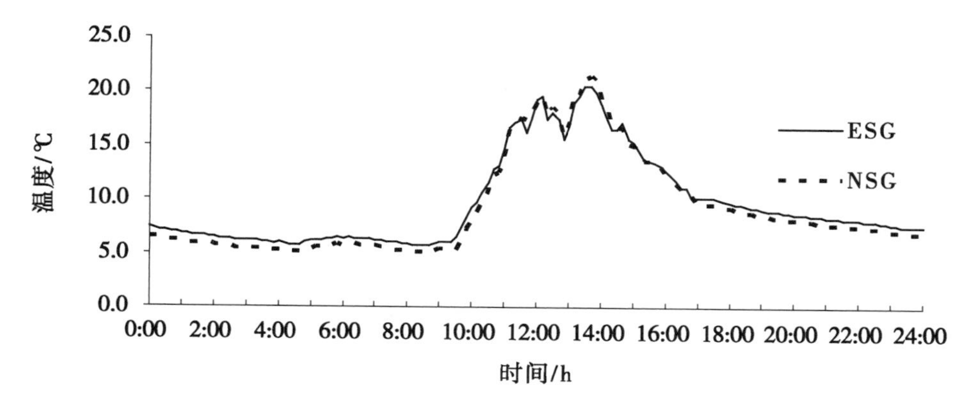

图 7（b） 雪天温室内距地面 1.5m 日温度变化曲线

Figure 7（b） Daily mean temperature curves 1.5m above ground of the snowy day in greenhouses

由图7（b）可知，ESG 和 NSG 的日平均气温分别是 9.64℃和 9.17℃；ESG 内气温在 8：30达到最低值 5.70℃，NSG 在 8：20 达到最低值 5.10℃，与冠层温度基本一致；ESG 气温在 13：30 达到最高值 20.60℃，NSG 在 13：40 达到最高值 21.43℃。两个温室自 11：00 至 16：00 的气温和变化趋势基本一致。雪天条件下，两温室的温度变化无明显差异。

3 结论和讨论

作为一种新型的日光温室栽培技术，在日光温室的后墙上采用管道无土栽培的方式进行蔬菜和草莓的生产，是一种高效的、可以提高温室空间利用率、进而提高温室内作物产量的栽培方式[5]。试验温室长度为 50m，跨度为 8m，温室内草莓采取高畦栽培，共 50 垄，每畦两排，栽培畦长 7m，因而普通温室实际栽培长度为 50×2×7＝700m；后墙立体栽培管道长 49m，共 4 排，因而后墙栽培长度是 49×4＝196m。通过日光温室后墙的立体栽培，栽培长度增加了 28%，草莓的栽培株距为 15cm，可增加草莓的栽培株数 1 300多株，大大提高了草莓的种植面积。

本研究是就后墙立体栽培对后墙挡光从而可能影响温室蓄热、降低温室内温度的问题开展研究。持续监测冬季温室内的温度变化的结果表明，利用后墙空间，进行立体栽培，并没有降低冬季温室内温度；晴天时，有后墙栽培的温室，其内部的冠层温度和气温均略高于无后墙管道的温室；阴天和雪天时，两种类型的温室内的温度变化无明显差别，温度都较低，均低于草莓生长的适宜温度范围[7]，在此期间应注意适时加温，避免冻害出现，影响草莓的生长，产量和品质。

在温室后墙上通过安装管道，在管道内进行无土栽培草莓，试验结果显示温室内的温度并没有下降，反而略有升高，分析原因：①可能是太阳光照射到管道表面，管道传热给内部基质，基质有一定的蓄热能力，热量蓄积在基质中；管道不与后墙墙体接触，夜间热量不会通过管道、墙体的热传导向室外散失，基质中白天蓄积的热量主要向温室内部散热，从而使温室内的温度提高；②可能是因为基质的比热较墙体大，蓄热能力强，从而能够在白天蓄积更多的热量；③可能是因为基质中蓄积的水分的比热较墙体大，水的蓄热能力强，从而能够在白天蓄积更多的热量。有关产生这种现象的原因，有待继续进行研究。

总之，初步测试的结果表明，后墙立体栽培不但提高了作物的种植面积和温室空间的利用率，并且没有降低温室内部的温度，是一种可以推广使用的栽培技术。

参考文献

[1] 陈青云．日光温室的实践与理论．上海交通大学学报（农业科学版）．2008（5）：343～350
[2] 刘志杰，郑文刚，胡清华等．中国日光温室结构优化研究现状及发展趋势．中国农学通报．2007（2）：449～453
[3] 何雨，须晖，李天来等．日光温室后墙内侧温度变化规律及温度预测模型．北方园艺．2012（7）：34～39
[4] 姜新法．立体无土栽培技术浅述．农机服务．2007，24（12）：102
[5] 邢文鑫，赵永志，曲明山等．草莓立体栽培概况．河北农业科学．2011（7）：4～7
[6] 管勇，陈超，李琢等．相变蓄热墙体对日光温室热环境的改善．农业工程学报．2012（10）：194～201
[7] 赵玉科．日光温室草莓栽培技术．北方园艺．2009（12）：150～151
[8] 何芬，马承伟．华北地区冬季温室植物冠层温度建模．农业机械学报．2009（5）：169～172
[9] Klocke F，Eisenblatter G. Dry cutting. Annals of the CIRP. 1997，46（2）：519～526
[10] 陈端生．日光温室的温度环境．农村实用工程技术．2003（5）：26～28

基于 LabVIEW 雾培根际水肥环境控制系统设计*

刘义飞，程瑞锋，杨其长

（1. 中国农业科学院农业环境与可持续发展研究所，北京　100081；
2. 农业部设施农业节能与废气物处理重点实验室，北京　100081）

摘要：精准间歇喷雾调控技术是雾培系统根际水肥环境控制的关键技术。以图形化编程软件 LabVIEW 为核心，结合数据采集卡，设计了雾培根际水肥环境控制系统。该系统选取根际湿度作为调控依据，以实现对于植物生长过程比较精确控制。能够实现根际湿度、根际温度、营养雾滴 pH、EC 等根际参数的实时显示与存储、调控。便于对雾培生产系统性能进行分析以及对控制方式进行优化改进。

关键词：雾培；根际水肥环境控制；LabVIEW

The Design of Aeroponics Root Environment Monitoring and Control System Based on LabVIEW

Liu Yifei, Cheng Ruifeng, Yang Qichang

(1. *Institute of Environment and Sustainable in Agriculture*, *Chinese Academy of Agricultural Science*, *Beijing* 100081, *China*; *Key Lab. for Energy Saving and Waste Disposal of Protected Agriculture*, *Ministry of Agriculture*, *Beijing* 100081, *China*)

Abstract: Precise regulation of intermittent spray is key technology in aeroponics root environment monitoring and control system. This paper uses the graphical programming software LabVIEW as the core and combines the data acquisition card to design the system. The regulation and control is based on rhizosphere humidity. The root environment performance real-time display and storage such as rhizosphere humidity, rhizosphere temperature, nutrition droplet pH, and EC can be realized, in order analyze the aeroponics system performance and improve the control algorithm

Key words: Aeroponics; Root environment monitoring and control; LabVIEW

雾培（Aeroponics）是指植物的根系悬挂生长在封闭、不透光的容器内，营养液经特殊设备处理后形成雾状，间歇性喷到植物根系上，以提供植物生长所需的水分和养分的一种无土栽培技术[1~3]。雾培具有节水节肥，根系处于最佳的水、气、肥环境中，从而使作物发挥出更大的生长潜能，是速生高产、环境友好型的植物生产模式[4]。

精准间歇喷雾调控技术是雾培系统根际水肥环境控制的关键技术之一[5]。早期的雾培控制系统由美国宇航局开发，它通过在栽培槽内加装一对光学的发射和接收器件作为传感器，这种方法适合于与超声波雾培装置配合使用，因为超声波产生的雾颗粒较小，直径在 1～10μm，采用这种控制方法，苗床内的相对湿度总是维持在 95% 以上，只适用于对水分

* 本研究由中央级公益性科研院所基本科研业务费课题（BSRF201305）、国家自然科学基金项目（31000689）、国家科技支撑计划（2011BAE01B10）、国家“863”计划（2013AA103004）等课题资助

需求特别大的植物[6]。国内雾培技术起步较晚，控制方式上有些采用较粗放的定时喷雾的方式，这种方法依赖于操作人员的客观经验，无法进行精确控制。也有部分采用喷雾量间歇控制方法，就是在计算机程序中设定了与外界环境温度相关的时间模块，外界温度越高，喷雾越频繁。这种方法是以采集的环境参数来参照喷雾时间和间歇时间，并不考虑植物本身的需水要求，科学性较差。

以植物生理传感器采集数据位基础，以作物生长模型为依据的雾培喷雾先进调控技术正在被发达国家采用。这种方法是以茎胀速与昼夜变量为参数进行调控，采用微米级的微位移传感器夹置于植株的茎干或叶片上，利用水分变化胀落对茎干直径产生影响的机理来运算确定喷雾时间的控制。美国的 AgriHouse 公司依据这个原理研制出专用于气雾培的叶片传感器，它的技术关键在于不同植物生长模式或专家系统的科学建立。这种技术在日本植物工厂中也有了较大程度的应用，其科学性和精确性明显高于以上方式，但过高的传感器价格和开发成本限制了这种技术的推广。

本系统设计选取与植物生长直接相关的参量根际内湿度作为调控依据，选择图形化编程软件 LabVIEW 作为系统构建核心，以实现对于植物生长过程比较精确控制。

1 控制系统的组成

控制系统由 PC 机、数据采集卡以及雾培系统的执行机构组成。主要的设计任务是上位机的测控显示系统软件。在工作系统过程中，传感器对雾培根际环境因子和营养液的各种参数进行监测，数据采集卡可完成检测信号的数据采集和信号预处理，并将信号传给上位机 PC 机，并能够接受上位 PC 机的数字信号，对继电器或电磁阀等驱动设备进行开关操作，控制高压水泵、营养液罐酸 pH 值调节、冷却系统等执行机构。

该系统选取与植物生长直接相关的参量根际内湿度作为调控依据，以实现对于植物生长过程比较精确控制。能够实现根际湿度、根际温度、营养雾滴 pH 值、EC 等性能参数的实时显示、存储与调控。便于对雾培生产系统性能进行分析以及对控制方式进行优化改进。

上位机以虚拟仪器 LabVIEW 为软件平台，系统使用模块化的编程方式，开发环境测控显示系统，实现对环境参数的采集、存储、显示、打印等功能，设置环境参数的上下限，并通过控制策略调节环境。整套系统不间断循环使用，实现对雾培根际水肥环境的自动实时监测、显示和控制。

测试系统硬件系统主要由信号检测部分、信号控制、集线器、串口、执行机构以及 PC 机组成，整个系统设计结构框图如图 1 所示。

2 系统控制模块的构建和硬件选择

2.1 数据采集卡

数据采集卡是将模拟信号转换为数字信号，形成计算机能够处理的数据，并可以直接进行数字信号的输入和输出。本系统选择美国国家仪器（NI）有限公司生产的 NI USB6009 多功能数据采集卡。该采集卡是用于 USB 的 14 位，48 kS/s 的多功能数据采集卡，有 8 路模拟输入通道，14 位分辨率，12 条数字 I/O 线，2 路模拟输出，1 个计数器。作为一个便携式总线供电型设计，可用于 Windows，Mac OS X 和 Linux OS 的驱动软件。还具有体积小，即插

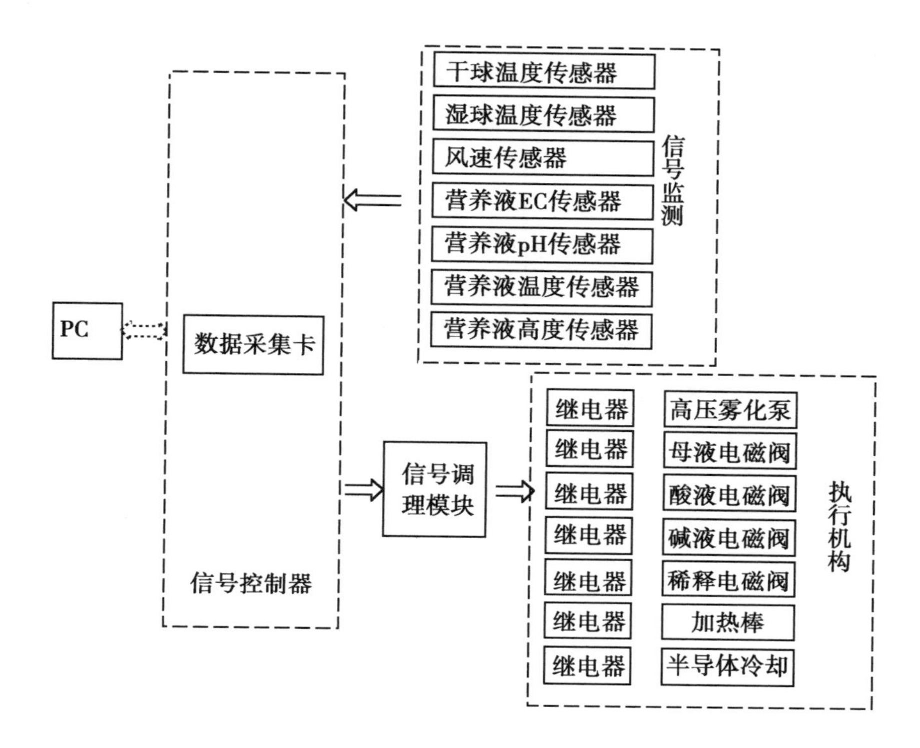

图 1　雾培根际水肥环境测控系统框图

Figure 1　The frame of aeroponics root environment monitoring and control system

即用等特点，特别适合本测试系统的车载电脑的输入输出接口。而其功能强大，足以用于复杂的测量控制的应用。使用 LabVIEW 用于 Windows 的 NI-DAQmx 为 USB-6009 编程（图 2）。

图 2　USB 6009 多功能数据采集卡

Figure 2　data acquisition card USB 6009

2.2　根际湿度控制模块设计方案与硬件选择

本系统选取雾培栽培床内根际湿度作为调控依据，以实现对于植物生长过程比较精确控制。因此，对于根际湿度的准确测量是系统正常运转的关键。而雾培系统也对传感器的选择提出了特殊的要求。根际湿度是与植物生长直接相关的参量，为满足植物生长的需要，雾培栽培床内的根际相对湿度普遍要维持在 80% ~100%，当相对湿度低于初步的设定值 80%，高压雾化水泵开始工作，雾化喷头开始喷雾，栽培床内的相对湿度会立刻达到 100%，并且伴随着很严重的结露现象。这种环境超出了许多湿敏元件湿度传感器的量程，结露形成的水滴滴在湿度传感器表面，会使传感器短时期内失去功效，即使传感器表面的水分自然蒸发以后，也需要很长一段时间进行校准[7]。

为实现该环境下相对湿度的实时准确的测量，并避免结露导致的传感器失效等问题，选择更为准确的气象和工业中应用干湿球测湿法。干湿球测湿原理是：用一只温度传感器检测空气温度（干球温度），用另一只相同的传感器检测被蒸馏水浸湿的面纱套内的温度（湿球温度），通过计算得出相对湿度。在实际使用中，干湿球测湿法只需定期给湿球加水及更换湿球纱布即可；并且干湿球测湿法不会产生老化、精度下降等问题。所以，干湿球测湿方法更适合于在高温及恶劣环境的场合使用[8]。

下面通过对干湿球测湿原理进行分析，以阐述该控制模块的设计思路。干湿球法测湿的关键是获得准确的干、湿球温度，然后借助干湿球方程换算出湿度值。干、湿球温度的测量使用北京九纯健科技公司生产的铂电阻温度传感器 PT100，测量范围为 0 ~60℃，准确度为 0.5%。

相对湿度 U 的计算公式：

$$U = (E/Etw) \times 100\% \qquad (1)$$

U 为相对湿度，E 为水气压（hPa），Ew 为干球温度 t（℃）所对应的纯水平液面（或冰面）饱和水气压（hPa）。

水气压 E 的计算公式：

$$E = Etw - APh\ (t - tw) \qquad (2)$$

（2）式中 Ph 为气压（hPa）；A 为干湿表系数，湿球未结冰时的通风干湿表系数（通风速度≥2.5m/s）是 0.662×10^{-3}℃，湿球结冰时的通风干湿表系数（通风速度≥2.5m/s）是 0.584×10^{-3}℃。

基于通风速度的要求，需要根际湿度的测量装置加上轴流风机以强制通风。同时要通过风速传感器的实时监测反馈调整通风速度在 2.5m/s。强制通风风机选择宁波九龙电讯电机有限公司生产的轴流风机 100FZY2-S，功率为 18W，额定电压为 220V。风速传感器选择北京市检测仪器厂生产的智能热球式风速仪，范围：0.05 ~30m/s，误差 ±3%。

（2）式中 Etw（hPa）为湿球温度 tw（℃）所对应的纯水平液面的饱和水气压，湿球结冰且湿球温度低于 0℃时，为纯水平冰面的饱和水气压；饱和是一种动态平衡态，在该状态下，气相中的水汽浓度或密度保持恒定。在整个湿度的换算过程中，对于饱和水蒸气压公式的选取显得尤为重要。

下面介绍气象学常用的 Goff-Grattch 饱和水汽压公式。从 1947 年开始，世界气象组织就推荐使用 Goff-Grattch 的水汽压方程。该方程是以后多年世界公认的最准确的公式。它包括两个公式，一个用于液 - 汽平衡，另一个用于固 - 汽平衡。

对于水平面上的饱和水汽压：

$$\lg e_w = 10.79586\ (1 - T_0/T) - 5.02808\ \lg\ (T/T_0) + 1.50475 \times 10^{-4}\ [1 - 10^{-8.2969(T/T_0 - 1)}] + 0.42873 \times 10^{-3}\ [10^{4.76955(1 - T_0/T)}] + 0.78614$$

式中，T_0 为水三项点温度 273.16K

对于冰面上的饱和水汽压：

$$\lg e_i = 9.096936\ (1 - T_0/T)\ - 3.56654\ \ \lg\ (T/T_0)\ + 0.87682\ (1 - T/T_0)\ + 0.78614$$

以上两式为 1966 年世界气象组织发布的国际气象用表所采用。

上述的饱和水汽压公式均比较繁杂，为了适应大多数工程实践需要，特别是利于计算机、微处理器编程需要，工程师总结了一组简化饱和水汽压公式。

对于水面饱和水汽压：

$$\ln e_w = (10.286T - 2\,148.4909)\ /\ (T - 35.85)$$

对于冰面饱和水汽压：

$$\ln e_i = 12.5633 - 2\,670.59/T$$

上式与 Goff-Gratch 公式的最大相对偏差小于 0.2% 。因此，该根际湿度控制模块编程中选择了该简化公式。

关于根际湿度测量的编程代码流程图如图 3 所示。

雾培根际湿度测量传感器初步构建如图 4 所示。

2.3 营养液 pH 值、EC 控制模块设计方案

营养液配制系统设四只母液罐，分别为 A 液罐、B 液罐、酸液罐和碱液罐。其中，A 液罐、B 液罐为含有植物生长所必需的矿物离子化合物的母液，用于调整营养液中的 EC 值；酸、碱液则用于调整营养液中的 pH 值。营养液调配采用 PWM（Pulse Width Modulation——脉冲宽度调制技术）控制方式，由计算机带动执行机构完成。当营养液混液罐中的营养液 EC 值低于设定下限时，母液罐双腔计量泵启动，将 A 液、B 液同时等量注入储液池中。当 EC 值高于设定上限时，补水电磁阀打开，向储液池中补入清水。酸碱液则按 pH 值设定要求，分别通过酸碱液电磁阀控制，利用液位高度差的重力作用实现自流注入。

2.4 营养液温度的控制模块设计方案

营养液温度控制采用比例控制方式，它的优点是调节时间较短。控制设备主要由电加热器、制冷机组、冷却水蒸发器和温度传感器等组成。温度传感器 1 固定在营养液混液罐中，负责监控营养液混液罐内营养液温度。温度传感器 2 为可移动式，负责检测雾培栽培床内温度。当夏季营养液温度超过设定值时，制冷机和冷却水蒸发器启动工作；当冬季营养液温度低于设定值时，加热棒启动工作，把营养液温度控制在设定温度范围内。

2.5 根际环境温度的控制模块设计方案

利用营养液混液罐的温度调节来控制根际环境的温度。当夏季根际环境温度超过设定值时，制冷机和冷却水蒸发器启动工作，雾化营养液温度降低，通过雾化从而降低根际环境的温度。当冬季根际环境温度低于设定值时，加热棒启动工作，把雾化营养液温度升高，通过喷雾从而把根际温度控制在设定温度范围内。

3 应用前景与分析

农业微气候环境控制大致经历了人工手动控制、机械设备控制、电子环境控制、计算机综合控制这几个发展阶段。目前，我国微气候环境的测控系统多依靠经验或半经验的方式进行管理，以人工操作或半自动化操作，也有单片机、PC 或其他手段实现自动控制的。虚拟

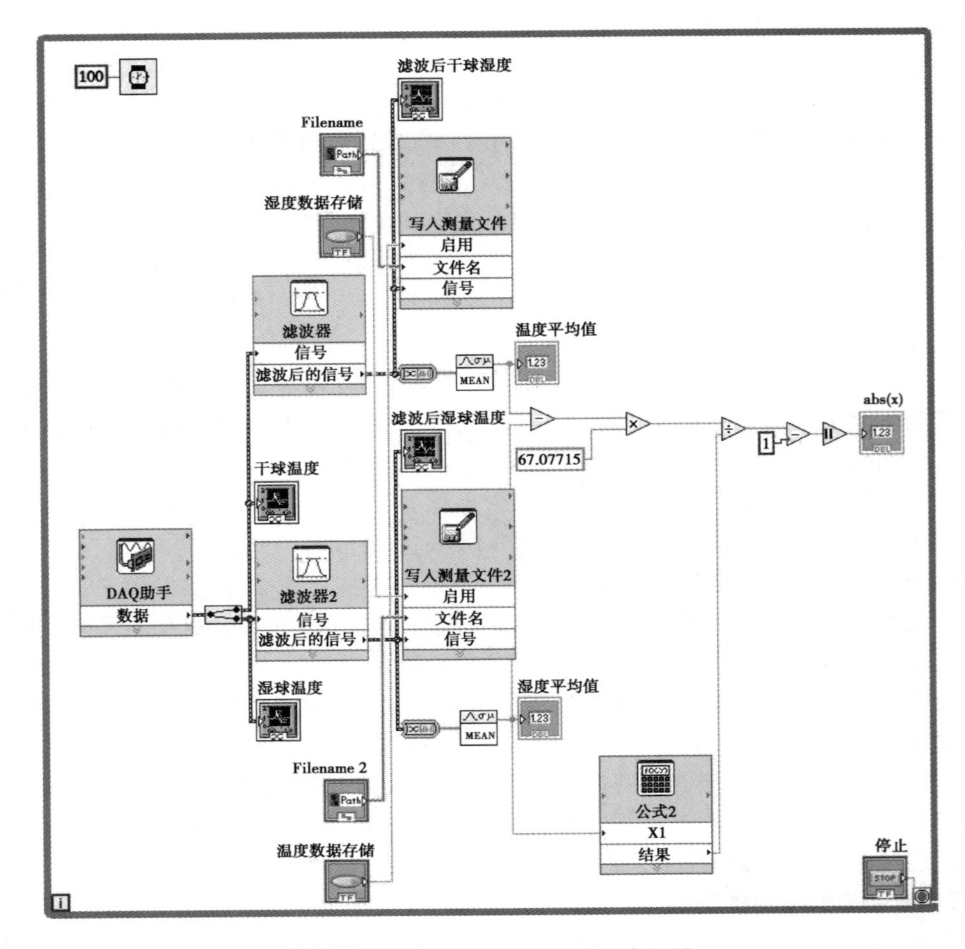

图 3　根际湿度测量的编程代码流程图

Figure 3　The flowchart of rhizosphere humidity data acquisition

仪器技术的出现为测控系统的研制开辟了一条新的途径。应用虚拟仪器有开发时间短、人机界面好、操作方便等优点（图 3 和图 4）。LabVIEW 在工业方面已经有着广泛的应用，而在农业领域，特别是设施园艺环境测控与显示方面，应用才刚刚起步。LabVIEW 交互式图形化的用户界面和数据流编程方法为测控系统管理软件的实现带来便利。

选择 LabVIEW 构建雾化栽培的控制系统具有很大优势。该系统不仅测试自动化程度高，测量范围宽，且能根据测试对象不同的需求随时改变测量和控制方案。基于 LabVIEW 编程平台，使系统具有很强的数据处理功能和与其他测试仪器的通信能力[9]。初步的实验证明，该系统具有很好的测试、控制、显示和存储性能。下一步将结合雾培植物的反应，优化控制系统的设计，并进一步探讨以根际湿度为调控依据的合理性。

参考文献

[1] 郭世荣. 无土栽培学. 北京：中国农业出版社，2003

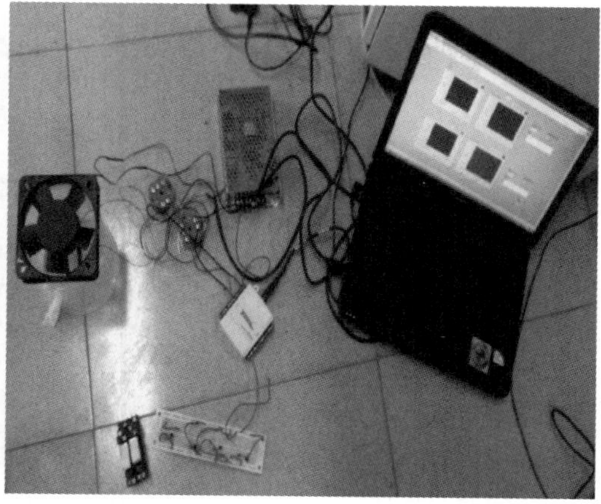

图 4　雾培根际湿度测量传感器

Figure 4　The rhizosphere humidity measurement sensor

[2] 徐伟忠，王利炳，詹喜法等．一种新型栽培模式：气雾培的研究．广东农业科学．2006 (7)：30～34

[3] 王玲，刘广晶，孙周平．雾培对黄瓜植株生长、产量和品质的影响．江苏农业科学，2009 (1)：174～177

[4] 闻婧，程瑞锋，杨其长等．超声波雾化栽培装置的研制和应用效果．2012，24 (1)：23～25

[5] 高建民，任宇，顾峰等．低频超声雾化喷头优化设计及试验．江苏大学学报（自然科学版）．2009，30 (1)：1～4

[6] Pecision agrotechnology in vegetable farming through aeroponics—the Sime Darby's experience. Gan L. T.；Tek J. C. Y.；Lee C. Y. Planter. 2000，5 (76)：890

[7] 洪南，程凯．基于 AVR 单片机温室气雾培控制系统的设计与实现．中国海洋大学学报，2008

[8] 黄晓因，张连根．干湿球法测量相对湿度算法研究及单片机实现．云南民族大学学报（自然科学版）．2003，12 (3)：155～157

[9] 刘君华，贾惠芹等．虚拟仪器图形化编程语言 LabVIEW 教程．西安：西安电子科技大学出版社，2001

基于二维传热过程的日光温室热模型及其验证*

王　楠[1**]，马承伟[1]，曹晏飞[1]，赵淑梅[1***]，蒋程瑶[1]，魏家鹏[2]

（1. 中国农业大学农业部设施农业工程重点实验室，北京　100083；

2. 新世纪种苗有限公司，寿光　262700）

摘要：运用工程热物理和温室环境工程等理论，构建了日光温室墙体和土壤二维传热的热环境动态数值模型。为检验模型的准确性，在山东寿光日光温室中对温室内空气温度、土壤温度、墙面温度等进行了测试。采用实测数据对模型进行验证的结果显示，连续5d模拟得到的室内空气、墙体表面以及土壤表面温度与实测数据呈现相同的变化规律，相关系数均在0.9以上，且模拟值与实测值之间的平均绝对误差在1.5～2.5℃范围内，表明所构建二维传热模型能够比较准确模拟预测地面下沉型日光温室的空气温度以及墙面和土壤表面温度的变化规律。

关键词：日光温室；热环境；二维传热；数值模型；验证

Thermal Environment Model of Solar Greenhouse Based on the Two-Dimensional Heat Transfer Process and Model Validation

Wang Nan[1], Ma Chengwei[1], Cao Yanfei[1], Zhao Shumei[1], Jiang Chengyao[1], Wei Jiapeng[2]

(1. *Key Lab of Agricultural Bio-environmental Engineering*, *China Agricultural University*, *Beijing* 100083, *China*;

2. *New Century Seeding Co.*, *Ltd*, *Shouguang* 262700, *China*)

Abstract: Based on the theories of engineering thermophysics and greenhouse environmental engineering, the author built a dynamic model for the thermal environment in solar greenhouse. We tested the air temperature, soil temperature and wall temperature in solar greenhouses in Shouguang. The measured temperatures of 5 days had same variation with the simulated ones, and the average relative errors between them were about 1.5～2.5℃. It shows that the model can accurately predict the temperature in greenhouse environment from experimental results.

Key words: solar greenhouse; thermal environment; two-dimensional heat transfer; numerical model; verification

1　引言

日光温室是我国特有的温室形式，具有节能、高效、低成本等特点，近年来在全国各地发展迅速。但目前大多日光温室的设计和建造主要依据有限的经验，普遍存在设计水平不高、环境性能较差等问题。要改变这一现状，首先有必要阐明日光温室热环境的形成机理，然后通过科学严密的理论，建立准确的物理和数学模型，并且在模拟的基础上，结合一定的

---

* 基金项目：适合西北非耕地园艺作物栽培的温室结构和建造技术研究与产业化示范（201203002）；现代农业产业技术体系建设专项资金（CARS－25）；高效设施农业标准化工程技术集成示范（201130104）

** 第一作者：王楠（1991—），男，在读硕士，研究方向为设施园艺。E-mail：wangn1991@126.com

*** 通讯作者：赵淑梅（1967—），女，博士，副教授，现主要从事设施农业生物环境工程研究工作。E-mail：zhaoshum@cau.edu.cn

经验和分析计算方法，预测和评价不同气象条件及不同构造方案下的日光温室的环境性能。这样就可以在设计过程中优化日光温室建造方案，从而提高我国日光温室的设计建造水平。

我国研究人员在日光温室热环境理论方面开展了较多研究工作，李元哲、郭辉卿、陈青云、李小芳、吴春艳、佟国红、孟力力、杨其长等都先后研究和提出了一些日光温室的热环境模型[1~7]。但其中有关日光温室墙体传热的分析方法大多建立在墙体一维传热假定的基础上，这对于目前大多数高度小于5倍厚度的日光温室墙体，并不适用；另外，大部分研究是将土壤看作半无限大域的一维传热问题处理，这对于日光温室这种跨度相对较小、土壤温度很容易受边界影响的计算对象，也会产生较大的误差，不能比较真实地反映实际情况。

论文在前人模型研究的基础上，采用二维模型描述墙体、后屋面和土壤中的传热过程，建立了温室热环境动态数值模型[8]，并通过实测数据对模型的准确性进行了验证。

2　二维热环境模型

日光温室热环境模型包括对室内热环境产生影响的墙体、土壤、覆盖层、通风和蒸发蒸腾五大传热分析模块，以及太阳辐射和室外气象条件等模块。采用以下微分方程反映各模块以及与室内热环境状况（室内气温等）的总体联系。

$$\rho_a C_p V \frac{dt_i}{d\tau} = Q_w - Q_s - Q_g - Q_v - Q_e \tag{1}$$

（1）式中：t_i—室内气温，℃；τ—时间，s；V—室内容积，m^3；ρ_a—空气密度，kg/m^3；c_p—空气比热容，J/（kg·℃）；Q_w—墙体与后屋面向室内的散热量，W；Q_s—地面向室内的散热量，W；Q_g—前屋面损失热量，W；Q_v—空气渗透损失热量，W；Q_e—植物与地面腾发耗热量，W。

墙体、后屋面与土壤区域采用二维非稳态传热模型，其微分形式的控制方程如公式(2)。二维模型可以更好地反映目前常用的梯形断面墙体、复合材料砌筑墙体，以及下沉式日光温室土壤区域中的传热过程。

$$\rho c \frac{\partial t}{\partial x} = \lambda \frac{\partial}{\partial x}\left[\lambda \frac{\partial t}{\partial x}\right] + \frac{\partial}{\partial y}\left[\lambda \frac{\partial t}{\partial y}\right] + S \tag{2}$$

有关模型的详细描述及其解析过程、初始条件的确定等，在文献［8］有详细介绍，这里不再赘述。

3　模型验证

任何理论模型在应用于实践之前，都有必要经过实测数据的检验。因此为了验证上述模型的准确性，于2011年1月在山东省寿光进行了试验。

3.1　实测温室概况

试验温室位于寿光市韩家庄（北纬36°52’，东经118°44’），东西走向，方位角为0°。温室采用钢结构骨架，EVA无滴膜覆盖，外覆盖保温材料为草帘，采用机械卷帘，墙体为土墙。温室详细构造参数如表1所示。其中，温室于2010年12月10日定植黄瓜。

表1 试验温室的构造参数

Table 1 Variables of experiments greenhouse

跨度	长度	脊高	后墙高度	墙体厚度	下沉深度	种植作物
12.5m	69.0m	3.9m	3.0m	上2.2m，下6.0m	1.6m	黄瓜

3.2 测试内容及方法

测定的室内环境因素有空气温度、空气相对湿度、墙体厚度方向温度分布、温室跨度和深度方向的土壤温度分布、土壤热流量、太阳辐射照度；测定的室外环境因素有空气温度、空气相对湿度、太阳辐射照度、风速、风向。

墙体、土壤及室内空气温度由T型热电偶测量，测量精度为±0.15℃，空气温度均采用辐射罩遮挡；室内相对湿度采用日本Especmic公司的RS-12温湿度计测量并自动采集数据，测量精度为±0.3℃及±5%；太阳辐射采用美国Onset公司的S-LIB-M003辐射传感器测量和H21-002数据采集器，太阳总辐射精度为±10 W/m^2；墙体、土壤热流量传感器为荷兰Hukseflux公司生产的HFP01热流量传感器，测量精度为±5%；室内热电偶等测试数据利用美国安捷伦科技有限公司的34970A进行自动采集；室外气象数据由美国Onset公司的HOBO-H21-002室外气象站测量，温度精度为±0.7℃，湿度精度为±4%，风速精度为±1.1m/s，太阳总辐射精度为±10 W/m^2。自动数据采集的时间间隔均为10min。

3.3 模型检验

在有关温室热环境模型验证评价指标方面，有文献采用了平均相对误差和平均绝对误差两个评价指标[6]。温度数据相对于其他数据比较特殊，对于两个不同水平的温度值（如2℃和20℃），同样一个温度波动（如2℃）下的相对误差，若以摄氏零度为基准，则两个相对误差可能会有很大差异（如100%和10%），若以绝对温度零点为基准，则两个相对误差（0.73%和0.68%）几乎没有差别，因此，很难准确反映出实际的波动程度。鉴于此，以下模型验证采用平均绝对误差和最大绝对误差来评价模拟值与实测值之间的相差程度，以相关系数考核模拟值和实测值间的线性相关性。平均绝对误差、最大绝对误差、相关系数，各自定义如公式（3）~（5）所示：

$$\overline{Vx} = \frac{1}{n}\sum_{i=1}^{n} | X_{pi} - X_{mi} | \quad (3)$$

$$Vx_{max} = max | X_{pi} - X_{mi} | \quad (4)$$

$$r = \frac{1}{cov(X_{pi}, X_{pi}) \sqrt{cov(X_{mi}, X_{mi})}} \quad (5)$$

式中：Xpi-模拟值；Xmi-测量值。$|r|$越接近1，线性关系越密切[9]。

小寒（2011年1月5日）和大寒（2011年1月20日）是一年相对较冷的一段时间。测试结果显示该段时间室外最低温度出现在1月15日，达-17.4℃，因此，选取1月13~18日这段时间的数据进行分析。且这段时间的天气状况包括了多云、多云、晴、晴、多云转晴等天气，比较典型。

以实测的室外空气温度、太阳辐射等数据作为室外气象条件，利用模型模拟计算了室内空气温度、温室后墙内表面温度以及室内土壤表面温度，模拟结果与实测数据的对比如图

1a～1c 所示。统计的模拟值与实测值之间的平均绝对误差、最大绝对误差以及相关系数，如表 2 所示。

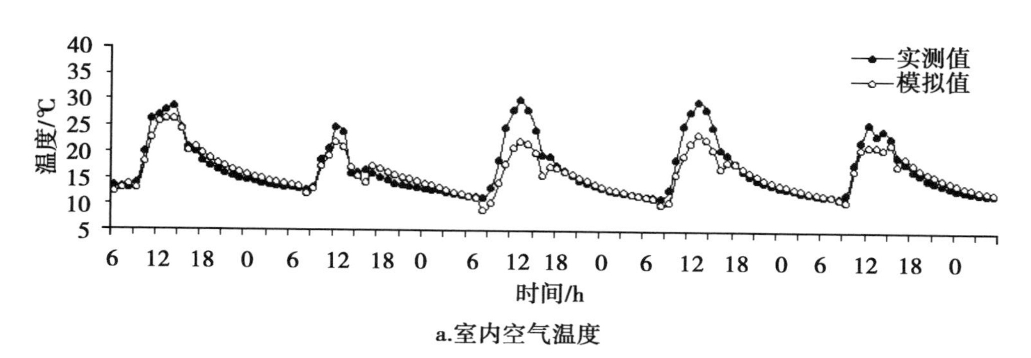

a.室内空气温度

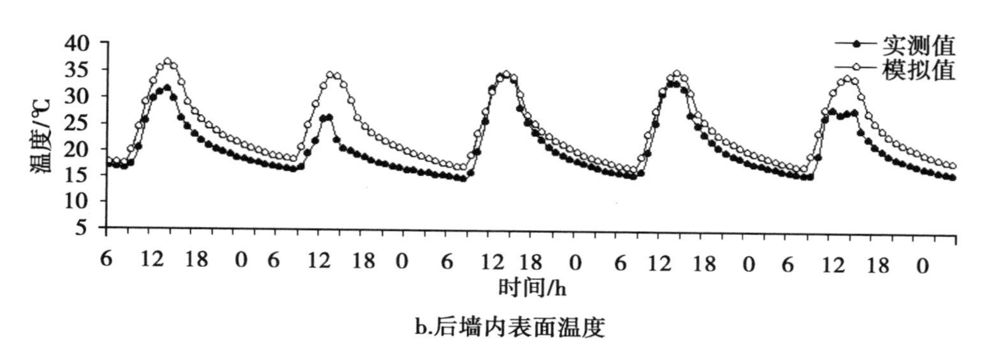

b.后墙内表面温度

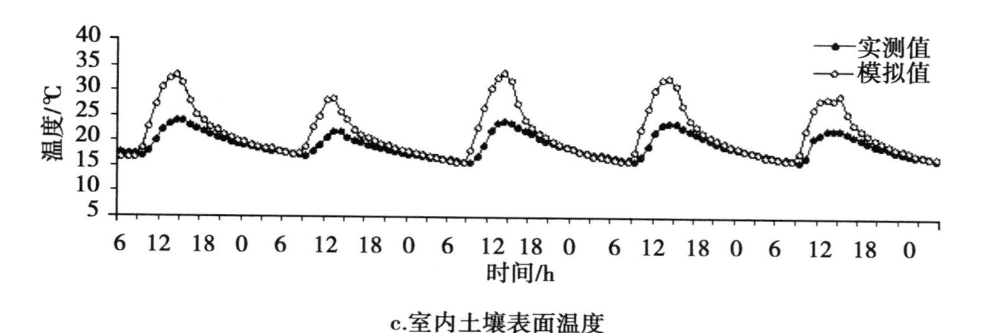

c.室内土壤表面温度

图 1　室内温度模拟值及实测值（2011 年 1 月 13～18 日）

Figure 1　Simulated and measured temperature distributions inside the greenhouse

表 2　温度模拟值与实测值平均绝对误差及相关系数（2011 年 1 月 13～18 日）

Table 2　Average relative error and correlation coefficient of the simulated and measured temperatures

位置	平均绝对误差/℃	最大绝对误差/℃	相关系数
室内空气温度	1.51	8.05	0.928
后墙内表面温度（距地面 1.5m 高）	3.17	12.04	0.938
温室土壤表面温度	2.46	9.76	0.906

从图1和表2可以看出，连续5d的模拟数据与相应的实测数据之间具有良好的一致性，基本能反映出温室的实际温度环境状况及变化规律。其中，室内空气温度在多云天气和夜间，二者非常吻合，但在晴天的白天，实测值略高于模拟值，分析原因，可能是传感器遮阳不够理想，受到了太阳辐射的影响；在墙体内表面温度方面，晴天情况下的拟合效果良好，但是多云天气下的模拟数据略高于实测，分析原因可能是温室保温被的实际管理模式与模型内部设定的管理模式之间存在差异导致的；另外，在土壤表面温度方面，二者日变化的规律相同，只是所有白天的模拟温度都高于实测温度，而导致这一差异的原因，应该是土壤表面覆有地膜、且地膜内表面严重结露，影响了土壤表面实际接受到的太阳辐射，但模型中对于地膜的影响考虑还不是很细致。总体看来，模型模拟值与实测值吻合程度较好，可以比较准确得模拟日光温室热环境的变化。

4 结论与讨论

运用工程热物理和温室环境工程等理论，基于墙体和土壤二维传热过程，构建了日光温室热环境的动态数值模型，并通过实测数据对模型进行了验证。通过使用连续5d的实测数据进行检验的结果显示，该模型能够比较准确预测地面下沉型日光温室室内空气温度；能够比较准确预测室内后墙表面及土壤表面温度的变化规律，在温室环境预测及评价方面具有较好的推广应用前景。

由于模型目前在植物与土壤的蒸腾蒸发、温室换气以及前屋面传热（包括保温被管理模式）等方面的描述还不是很细致，墙体材料以及土壤的特性参数主要参考相关资料，部分模拟结果与实测有一定的出入，因此在今后的研究中将针对这些问题做进一步完善。

参考文献

[1] 李元哲，吴德让，于竹．日光温室微气候的模拟与实验研究．农业工程学报．1994，10（1）：130～136
[2] 郭慧卿，李振海，张振武等．日光温室温度环境动态模拟．沈阳农业大学学报．1994，25（4）：438～443
[3] 陈青云，汪政富．节能型日光温室热环境的动态模拟．中国农业大学学报．1996（1）：67～72
[4] 李小芳．日光温室的热环境数学模拟及其结构优化（博士论文）．北京：中国农业大学，2005
[5] 吴春艳，赵新平，郭文利．日光温室作物热环境模拟及分析．农业工程学报．2007，23（4）：190～195
[6] 佟国红，李保明，Christopher David M. 等．用CFD方法模拟日光温室温度环境初探．农业工程学报，2007，23（7）：178～185
[7] 孟力力，杨其长，Bot Gerard. P. A. 等．日光温室热环境模拟模型的构建．农业工程学报．2009，25（1）：164～170
[8] 马承伟，韩静静，李睿．日光温室热环境模拟预测软件研究开发．北方园艺．2010（15）：69～75
[9] 宇传华，颜杰．Excel与数据分析．北京：电子工业出版社，2002
[10] 曹晏飞，张建宇，赵淑梅等．山东寿光日光温室冬季热环境测试．国农业科学院．设施园艺创新与进展——2011第二届中国·寿光国际设施园艺高层学术论坛论文集．2011

密闭式育苗设施热湿环境测试与分析*

金文卿**，胡　彬，马承伟***，付彦彦，阳　萍，陈　亮
（中国农业大学水利与土木工程学院，北京　100083）

摘要：密闭式育苗设施是高度科学化、生产精细化农业发展出的新型植物生产设施。作者团队在引进和吸收日本植物工厂技术的基础上，设计开发了中农型密闭式育苗设施，使用荧光灯作为人工光源，采用多层高密度育苗架以及气流分配腔等气流调整结构。作者对该设施内的温度、相对湿度进行了调控试验和测定。结果表明，整体环境控制达到了较高水平，针对冬季暗期环境相对湿度过高的问题，采用了引进室外干燥空气的方法，可以有效降低相对湿度到设定目标。因此，该育苗设施能够达到环境调控目标，可以满足各种不同育苗环境的需求，适用于高品质与高附加值作物的生产。

关键词：植物工厂；密闭设施；育苗；温度；湿度；测试

Tests and Analysis of Temperature and Humility Environmental Factors of Closed Nursery Facility

Jin Wenqing, Hu Bin, Ma Chengwei, Fu Yanyan, Yang Ping, Chen Liang
(*College of Water Resources and Civil Engineering*, *China Agricultural University*, *Beijing* 100083, *China*)

Abstract: Closed nursery facility is a new type of plant production facilities with highly scientific technology and meticulous production in development of agriculture. China Agricultural University designed a closed nursery facility named CAU-closed nursery facility on the basis of the technology of Kozai in closed plant production. It runs well with fluorescent lamp as artificial light source, multilayer and highly density shelf for seedling and structures to adjust the airflow. The tests adjust and control the indoor temperature (T) and relative humidity (RH) during the process of the production. The result shows the control of the integrated environment is steady and efficient. The plant facility is brought in the outside dry air to solve the problem of over high relative humidity (RH) during the dark time in winter. The measure is effective to diminish the relative humidity (RH) to appropriate range. The CAU-closed nursery facility can achieve the target of controlling the environment, meeting the demands of seedling environment for multiple plants and being suitable for the production of plants with high quality and high added value.

Key words: Plant Factory; Closed Facility; Nursery; Temperature; Humility; Test

---

* 基金项目：现代农业产业技术体系建设专项资金（CARS—25），公益性行业（农业）科研专项（201203002），“十一五”国家科技支撑计划项目（2009BADA4B04—01）资助

** 作者简介：金文卿（1987—），男，朝鲜族，吉林长春人，硕士研究生，现从事设施农业工程工作。E-mail：moonkeng.kim@gmail.com

*** 通讯作者：马承伟（1952—），男，重庆人，教授，博士生导师，中国农业工程学会高级会员，主要从事设施农业环境工程方面的研究。中国农业大学农业部设施农业工程重点实验室，北京 100083。E-mail：macwbs@cau.edu.cn

1 引言

植物工厂是指在利用现代科学技术实现的高度环境控制的封闭或半封闭生产设施空间内，进行植物周年生产的一种植物生产系统[1]。其生产对象包括蔬菜、花卉、水果、药材、食用菌以及一部分粮食作物等。而密闭型植物工厂的概念由日本学者古在丰树提出[2]：即使用不透光的绝热壁板做围护结构形成封闭的空间，其中的光照、温度、湿度、CO_2 浓度、气流速度等全部环境因素均可进行精细调控，为植物提供理想化的生长环境。经日本千叶大学研究组对其开发的研究型密闭式植物工厂的试验研究，得出密闭式植物工厂内 CO_2 与水的利用率达到90%以上，荧光灯光能利用率是自然光的7倍以上，耗电成本仅占总成本3%~5%的结论[2~7]。可以预见的是，密闭式植物工厂因其具有不受外界环境影响、全部环境条件可进行人工精细调节的特点，伴随环境控制技术、设施设备性能的不断提高，其作为未来培育优质种苗或特殊物种的应用前景不可限量。

近年来，针对密闭式植物工厂的科学研究已在各国开展，中国在种苗工厂化领域已进行大量的基础研究，但一直未能走向商业化[8]。中国农业大学于2011年引入了密闭式植物工厂的技术，研究开发了中农型密闭式育苗设施。本文以该中农型密闭式育苗设施为例介绍密闭式植物工厂内湿热环境的实际测试与数据分析。

2 试验设施与测试方法

2.1 密闭式育苗设施结构设计

中农型密闭式育苗设施建立在中国农业大学水利与土木工程学院楼西侧，占地39.54m²，以发泡聚氨酯彩钢板为外围护结构。主体部分（育苗间）高为3.9m。其中，育苗间为20.25m²（4 500mm × 4 500mm），催芽室及环境控制室为6.75m²（4 500mm × 1 500mm），营养液灌溉与操作准备间为12.54m²（3 800mm × 3 300mm）（图1）。100mm厚发泡聚氨酯彩钢板的外围护结构主要起到遮光与保温、隔热的作用，各部分之间也由100mm厚发泡聚氨酯彩钢板作为隔断。

育苗设施中的催芽室及环境调控室中，催芽室所占面积为1.8m²（1 200mm × 1 500mm），内含22个穴盘架，每层高为120mm，其主要功能是用于种子催芽。在催芽期，即播种后到出芽的过程中对穴盘的温湿度进行管理，可容苗数为15 000株左右。当同期进行不同作物育苗时，如出芽时间或催芽温度等要求不同，可分别管理。

育苗设施中的育苗间主要由8个7层的立体育苗架组成，每个育苗架单层尺寸为1 300mm × 700mm × 400mm。整个育苗间中共分为南北两个育苗区和3个过道。在南北两育苗区，每个育苗区的南北两侧均使用了锡箔纸与薄棉布遮挡，对育苗区起到保温和减少光照散射的作用。每个育苗区7层苗架的顶部为控制该育苗区域部分的空调、风机等设施。其中 CO_2 浓度的控制点在南育苗区顶端。育苗间中还设计了气流分配腔，即育苗间西侧有一个用板材隔出的4 500mm × 400mm × 3 900mm空间，来自空调和风机的气流进入该空间后再从板材上分布的直径为16mm、18mm、20mm的圆孔中流出到育苗间内，以达到各高度风速均匀一致的效果。在整个育苗间的东部，育苗架端部距离东墙500mm，设计了汇集从育苗架中回流的气流的汇集腔。育苗间中的气流路线如图1所示。

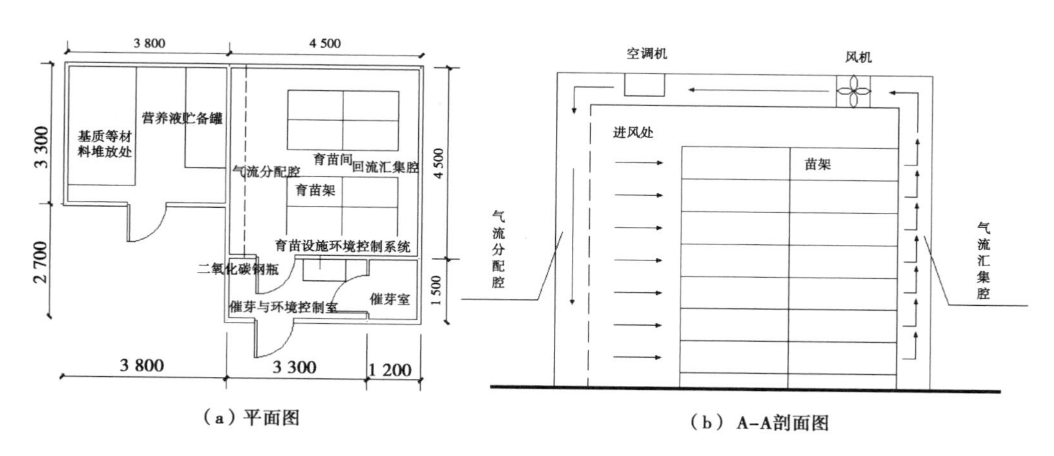

（a）平面图　　（b）A-A剖面图

图 1　中农型密闭式育苗设施平面图及育苗间剖面图

Figure 1　The plan of CAU-closed nursery facilities and the profile of nursing area

此外，育苗系统的营养液灌溉与操作准备间，主要用来放置储备营养液的营养液罐，并留有较大的空间可以放置育苗的基质、材料等物件，且在育苗播种、操作等过程中可以作为操作间。

2.2　湿热环境测试内容及方法

试验设定催芽期采取连续暗期，高温高湿（28℃，90% 以上相对湿度）的环境直至种苗探头发芽；育苗期采取 12h 明期、12h 暗期进行管理：其中，明期为 8：00 至 20：00，温度控制为 26.0 ± 1.0℃；暗期为 20：00 至第二日 8：00，温度控制为 19.0 ± 1.0℃。湿度的控制目标是明期相对湿度在 65 ± 10%，暗期相对湿度在 80 ± 10%。

温、湿度采用温湿度传感器（RS-12、RS-13，Espec Mic 公司，东京，日本）进行测试，可同时记录布置点的温度与湿度。

温湿度的测试从总体上分为两部分测试内容。

①测定整个育苗设施内不同空间点空气温度与湿度随时间的变化情况，需要对整个育苗设施进行连续的监控。

②测定整个育苗设施内在育苗期某时刻的空气温度与湿度的空间分布情况。主要针对育苗区，包括对育苗区的所有进风口温、湿度差异；所有出风口的温、湿度差异；南北各苗架的进出风口的温、湿度差异；育苗区的 2 层、4 层、6 层各层平面内部的温、湿度分布。

3　测试结果与数据分析

3.1　设施内空气温度与湿度随时间变化的连续监测结果与分析

从连续监测的数据中，可以看到整个育苗设施在实际运行过程中的温、湿度控制情况，如图 2 和图 3 所示，选取了育苗设施中 2011 年 12 月 31 日 19：00 至 2012 年 1 月 2 日 8：00 的温湿度数据，设置了 1 个室外测点、室内分布于 4 层苗盘进风口、4 层苗盘出风口的 21 个测点，将 21 个室内点进行综合求平均得出育苗设施内的平均温度和湿度。

测试结果与分析。

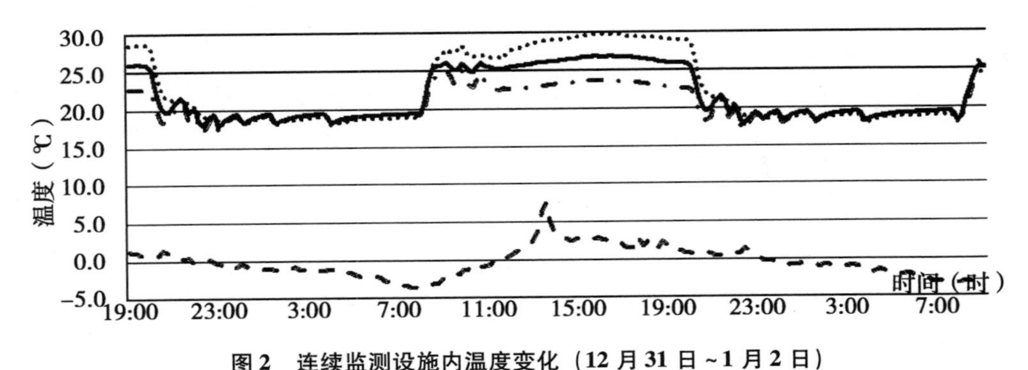

图 2　连续监测设施内温度变化（12 月 31 日 ~1 月 2 日）

Figure 2　Temperature changes in continuous tests（Feb. 31 ~ Jan. 2）

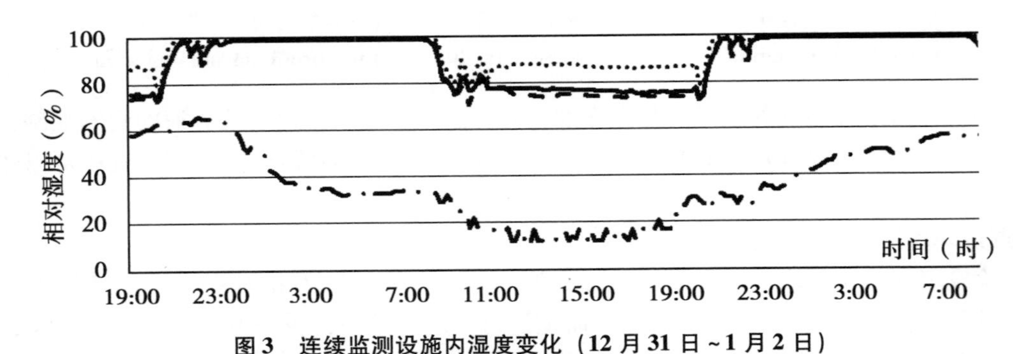

图 3　连续监测设施内湿度变化（12 月 31 日 ~1 月 2 日）

Figure 3　Humility changes in continuous tests（Feb. 31 ~ Jan. 2）

3.1.1　明期时

设施内的平均温度整体为 26.2℃ ±0.5℃，在调控目标范围内；苗盘进风口温度 23.4℃ ±0.6℃，低于苗盘出风口温度 28.7℃ ±1.0℃；室外温度为 −2.3 ~7.6℃；暗期时，设施内的平均温度为 19.1℃ ±0.4℃，在调控目标范围内；苗盘进风口温度为 19.0℃ ±0.5℃，与苗盘出风口温度 18.8℃ ±0.3℃相近；测试期室外温度为 −3.6 ~0.3℃。

3.1.2　从暗期到明期

温度逐渐从 19℃升到 26℃，历时 0.5 ~1h；从明期到暗期，温度从 26℃降到 19℃，历时 1 ~1.5h。

3.1.3　明期时

设施内的平均相对湿度整体在 77% ±2% 之内，略高于调控目标：苗盘进风口相对湿度 87% ±2%，出风口相对湿度 75% ±2%，进风口一侧高于出风口一侧；明期室外相对湿度为 11% ~32%；暗期时，设施内的平均湿度达到 99%，明显高于调控目标，苗盘进风口湿度与苗盘出风口湿度几乎一致，差别不大。暗期室外湿度为 32% ~65%。

3.1.4　从暗期到明期

相对湿度逐渐从 99% 降到 80%，历时 1 ~1.5h；从明期到暗期，相对湿度从 75% 升到 95%，历时 1 ~1.5h。

3.1.5 室内温

湿度完全不受室外环境条件变化的影响，变化趋势也不存在任何关系。而设施内温度与湿度存在着负相关的关系。

3.1.6 从整体环境来看

温度与明期的湿度调控效果较为理想，同时也适合种苗的生长，但是暗期的湿度却没有得到控制，高湿度易造成种苗的徒长和病菌感染。

3.2 设施内空气温度与湿度的空间分布测试结果与分析

根据测试结果，育苗设施内在暗期中的温度与湿度，因无光照产生的热源，在整个区域空间内分布较为均匀，以下主要需要研究明期时各区域各分区的温湿度分布情况。

3.2.1 育苗区进风口平面内与出风口平面内温、湿度分布的测试

2011 年 12 月 28 日 11：00～12：00 及 15：00～16：00 分别对育苗区明期同一时刻进风口、出风口进行了温湿度测试。进风口处与出风口处测点各 24 个，位于育苗区的各层，分别距进风口 5cm 与出风口 5cm 处（图 4 和图 5）。

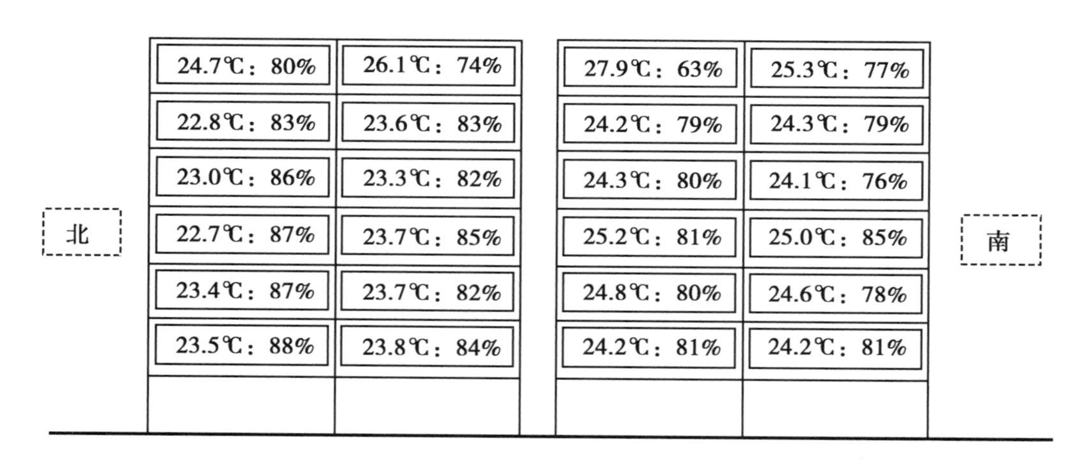

图 4 明期育苗区进风口温、湿度（12 月 28 日 11：00～12：00）

Figure 4 The temperature and humility of ventilates inputs in lighting period（Dec. 28）

测试结果与分析。

（1）进风口处 北侧苗架平均温度为 23.7℃，最高为 26.1℃，最低为 22.8℃，相差 3.3℃；南侧苗架平均温度为 24.8℃，最高为 27.9℃，最低为 24.1℃，相差 3.8℃。进风口处南侧苗架比北侧苗架平均温度略高，各苗架内部上下层相差 3～4℃。2～6 层的温度与层数没有明显的关系，主要是由于采取了“混后入冠”的气流组织方式。气流进入冠层前的温度和湿度，主要与混合的程度有关系。进风口处相对湿度较高，主要是与进风口处的气流温度较低的情况相关。

（2）出风口处 北边育苗架的平均温度为 28.4℃，最高为 30.4℃，最低为 26.4℃，相差 4.0℃；南边育苗架的平均温度为 29.8℃，最高为 31.9℃，最低为 26.8℃，相差 5.1℃。出风口各点温度差大于进风口，主要是因为各种灯具对湿空气的加热与蒸发蒸腾作用存在差异。

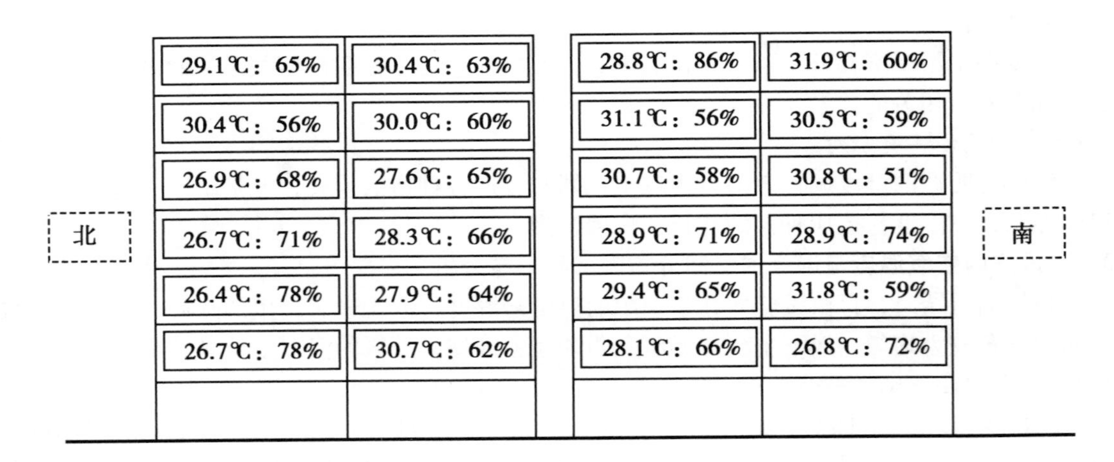

图5　明期育苗区出风口温、湿度（2011年12月28日15：00~16：00）

Figure 5　The temperature and humility of ventilates inputs in lighting period（Dec. 28）

3.2.2　育苗区在高度方向（不同层间）温、湿度分布的测试

如表1、表2所示，在不同时刻分别测得在高度方向（不同层间）南北两育苗架进出风口温、湿度情况。测点为育苗区进、出风口处。北侧苗架的测试时间为2011年12月30日9：00~10：00，南侧苗架的测试时间为2011年12月29日11：30~12：30。

表1　北侧苗架进出风口温湿度

Table 1　The temperature and humility of ventilated inputs and outputs in north side of the nursery shelves

层数	进风口温度（℃）	与第7层差值（℃）	出风口温度（℃）	进出风口温差（℃）	进风口湿度（%）	出风口湿度（%）
7	24.5 ±0.5	0.0	28.2 ±0.4	3.7	87 ±4	70 ±3
6	24.0 ±0.5	−0.5	29.0 ±0.4	5.0	86 ±3	68 ±2
5	24.0 ±0.4	−0.5	27.2 ±0.5	3.2	85 ±4	71 ±2
4	24.2 ±0.5	−0.3	28.1 ±0.6	3.9	86 ±4	77 ±2
3	24.4 ±0.5	−0.1	27.3 ±0.5	3.0	86 ±3	75 ±4
2	24.5 ±0.5	−0.1	28.1 ±0.4	3.7	88 ±4	64 ±3
1	24.3 ±0.5	−0.2	26.1 ±0.4	1.8	88 ±3	82 ±2

测试结果与分析。

（1）出风口的温度高于进风口温度　进出风口的温度相差随着层数、植物蒸发蒸腾作用、所使用的灯具的不同而有所不同，育苗区进出风口温差为1.8~5.8℃。

（2）湿空气从育苗区的进风口到出风口　温度上升，相对湿度下降。但其绝对湿度增加，主要原因为：湿空气流经苗盘时，吸收了苗盘中种苗各层多余的水分。因此，对整个设施内来说，尽管苗盘区的出风口相对湿度较低，但湿空气中含湿量相对较高。

表 2 南侧苗架进出风口温湿度

Table 2 The temperature and humility of ventilated inputs and outputs in south side of the nursery shelves

层数	进风口温度（℃）	与第 7 层差值（℃）	出风口温度（℃）	温差（℃）	进风口湿度（%）	出风口湿度（%）
7	24.0 ±0.2	0.0	28.8 ±0.2	4.8	84 ±1	79 ±0
6	23.3 ±0.1	-0.7	29.1 ±0.3	5.8	83 ±0	64 ±1
5	24.1 ±0.2	0.1	28.9 ±0.3	4.8	79 ±1	61 ±1
4	24.9 ±0.2	0.9	27.3 ±0.4	2.5	79 ±0	78 ±1
3	25.9 ±0.4	1.9	28.8 ±0.1	2.9	75 ±0	68 ±1
2	24.3 ±0.2	0.3	26.3 ±0.1	2.0	83 ±0	74 ±1
1	23.2 ±0.2	-0.8	25.6 ±0.1	2.4	85 ±0	77 ±1

3.2.3 育苗区水平面内温、湿度均匀性测试

图 6 所示为育苗区中第六层平面育苗区内部的温、湿度测试平均值，测试时间为 2011 年 12 月 27 日 15：30 ~ 16：00。

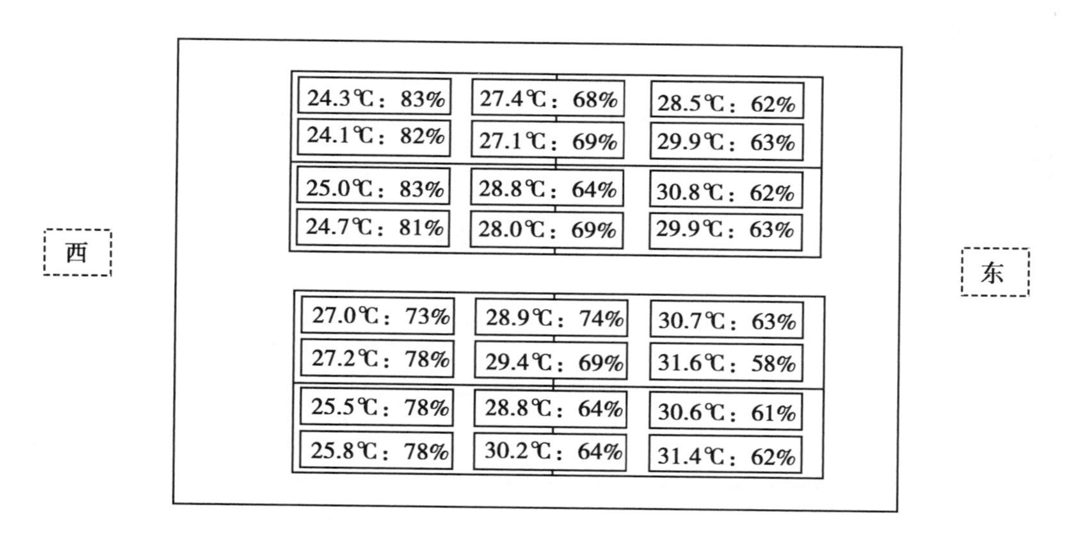

图 6 育苗区第六层平面内温、湿度测试结果

Figure 6 The test results of temperature and humility in 6th floor of nursery area

测试结果与分析。

（1）对整体而言 进风口到出风口湿空气温度逐渐升高、湿度逐渐降低；苗盘内靠近苗盘内部的温度高于靠近过道点的温度，湿度则相反。

（2）平面内平均温度 为 28.1℃，最大值为 31.6℃，最小值为 24.1℃，相差 7.5℃，平均温度比控制目标高 2℃。但由于平面的测点均在光照区内，虽然对传感器进行了防辐射处理，但并不能完全避免由光照产生的误差，可能是导致平面测试温度较高的原因，同时由于各区域灯具本身特性的不同以及种苗品种的不同，而导致平面内均匀性较差。设施平面均

匀性的测试还需要在规模化统一育苗时进行进一步的测试。

3.2.4 暗期湿度调整措施试验验证

前期试验测试结果表明，冬季暗期设施内湿度过高：暗期大部分时间各点相对湿度为99%。

设施内湿度过大的主要原因为暗期时灯具热负荷不存在，夜间外部相对温度较低，设施内的热量通过墙壁向外缓慢散热，使得室内的空调处于少量加热或者不运行的状态，导致整个设施内没有冷凝除湿的过程。湿空气经过育苗区时通过植物的蒸发蒸腾作用，使含湿量不断增加，并且在循环过程中几乎没有冷凝除湿的过程，最终导致湿空气相对湿度达到接近100%。

为解决这一问题，根据本设施结构的特点，采取了与室外干燥空气交换的方式降低暗期室内湿度，即使用新风换气机实现这一过程。根据理论计算，室外暗期平均湿度为41%，温度为 -1.5℃，其含湿量根据计算可得 $d_1 = 0.00136$kg/kg，而同期室内平均温度为19℃，湿度为100%，其含湿量根据计算可得 $d_2 = 0.013777$kg/kg，由此可见室外空气含湿量远远低于设施内，通过空气交换和引进室外干燥空气，排出设施内的水汽具有可行性。为了在排出室内水汽的同时，减少设施内热量的损失，采用热交换装置回收部分热量。改造后暗期气流组织图如图7所示。

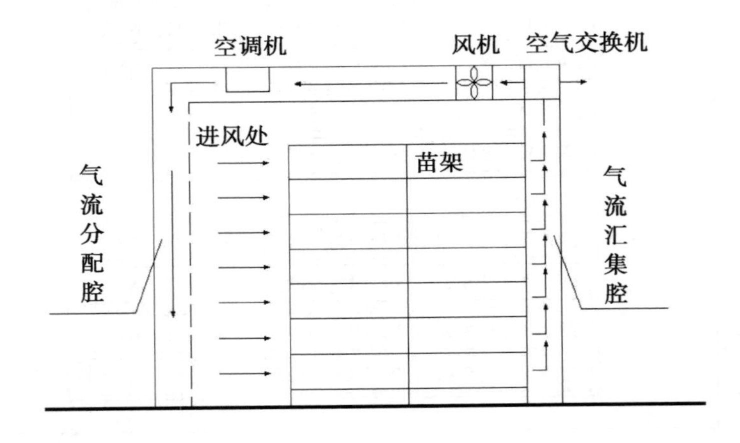

图7 改造后暗期气流组织图

Figure 7 The schematic illustration of airflow during the dark time after reconstruction

改造后对设施进行了空运行测试，即设施内并没有植物，只是苗盘中相对方式少量的水分，以模拟其设施内的环境。空测时间段为2012年3月1~4日，经测定，风机进风口处室外温、湿度0.9℃，53.5%；风机出风口处室外温湿度10.1℃，82.8%；风机出风口处室内温湿度20.6℃，54.4%。因此，在理论上暗期湿度过高的问题可以在引入空气交换机的前提下解决，并且避免了设施内热量损失。前期试验表明，引入室外干燥空气有效降低了设施内的相对湿度。但仍高于理想值，进一步试验分析有待进行。

4 结论

通过一段时间的实际运行表明：中农型密闭式育苗设施总体上明期的温、湿度、暗期的温度调控结果符合要求，总体满足育苗条件。但由于气流组织方式存在不合理之处，且苗盘由进风口至出风口距离过大等原因导致暗期的湿度过大，严重影响了植物的正常生长。现阶段为解决这一问题，通过更改设施内通风组织形式，增加与室外空气交换的通风装置，在设施内相对湿度过高时引入室外干燥空气的方法。经前期试验测定，新风换气机较好的降低了设施内部相对湿度。因此，密闭型育苗设施可以提供适合种苗生长的环境，人工调控可以达到精细水平，从而实现利用最小物质和能源投入取得最大生产效果的目标。

参考文献

[1] 马承伟，苗香雯．农业生物环境工程．北京：中国农业出版社，2005

[2] Chun C, Kozai T. A closed-type transplant production system. Molecular Breeding of Woody Plant, 2001, 18: 375 ~ 384

[3] Ohyama K, Kozai T, Yoshinaga K. Electric energy, water and carbon dioxide utilization efficiencies of a closed-type transplant production system Transplant production in the 21st century. Netherlands: Kluwer Academic Publishers, 2000: 28 ~ 32

[4] Ohyama K, Fujiwara M, Kozai T. Water consumption and utilization efficiency of a closed-type transplant production system. An ASAE meeting presentation. 2000, No. 004085

[5] Ohyama K, Yoshinaga K, Kozai T. Energy and mass balance of a closed-type transplant production system (Part 2) -water balance. SHITA, 2000, 12 (4): 217 ~ 224

[6] Yoshinaga K, Ohyama K, Kozai T. Energy and mass balance of a closed-type transplant production system (Part 3) -Carbon dioxide balance. SHITA, 2000, 12 (4): 225 ~ 231

[7] Ohyama K, Kozai T, Yoshinaga K. Electric energy, water and carbon dioxide utilization efficiencies of a closed-type transplant production system. Transplant production in the 21st century. Netherlands: Kluwer Academic Publisher, 2000: 28 ~ 32

[8] 贺东仙，朱本海，杨珀等．人工光型密闭式植物工厂的设计与环境控制．农业工程学报，2007，23（3）：151 ~ 157

热风炉热效率测试系统的设计*

刘　娜**，王国强，张　丽，王　彦，齐新洲，刘　涛

（新疆农业科学院农业机械化研究所 新疆设施农业工程与装备工程技术研究中心，
新疆林果棉及设施农业装备技术实验站，乌鲁木齐　830091）

摘要：本文针对日光温室冬季补温设备（热风炉）热效率评价问题，设计并研发了一套可实时监测热风炉热效率的测试系统：自动采集热风炉的烟道温度、热风道出口温度、热风道风压、冷风进口温度等信号，基于工控组态软件编制的热风炉热效率测试软件可以显示并记录上述各参数，结合热风炉热效率算法，自动得到热风炉的热效率值。本文探明了热风炉风速与风压的关系及热风炉热效率的影响因素，热风炉的热效率与热风炉的热风道风速、热风道出口温度、热风道管径及进风口温度等因素有关，推导出热风炉的热效率算法，开发了热风炉热效率测试系统软件。

关键词：热效率；热风炉；组态；日光温室

Design on the Thermal Efficiency Test System for Hot-blast Stove

Liu Na，Wang Guoqiang，Zhang Li，Wang Yan，Qi Xinzhou，Liu Tao

(Institute of Agricultural Machinery Xinjiang Academy of Agricultural science Agricultural Engineering and Equipment Engineering Research Center of Agricultural Facilities Xinjiang, Equipment and Technology Experiment Station of both Forestry Fruit Cotton and Agricultural Facilities Xinjiang, Urumqi 830091, *China*)

Abstract: Aimed at the evaluation problem of thermal efficiency for the equipment (the hot-blast stove) that complements temperature in winter, the test system is researched and developed in the thesis, which can monitor the thermal efficiency of the stove in real time. The system is able to automatically collect the signals, including the flue temperature of the hot-blast stove, the outlet temperature of the hot air flue, the velocity pressure of the hot air flue, the inlet temperature of the cold air flue, etc. Additionally, the system software, which is based on industrial control software, can also show and record the above-mentioned parameters. Finally, the thermal efficiency values will automatically be figured out, combining with the thermal efficiency arithmetic of the hot-blast stove. The relationship between wind velocity and wind pressure of the hot-blast stove, together with the influencing factor of the thermal efficiency, are summarized through literature reviews. The thermal efficiency of hot-blast stove is influenced by the wind velocity of the hot air flue, the outlet temperature of the hot air flue, the pipe diameter of the hot-blast stove, the hot-blast temperature and the inlet temperature of the air flue, etc. The thermal efficiency test system is developed, which utilizes the industrial control configuration software to convert the analog signals to the host computer via the data acquisition module. The analog signals is mainly comprised of the flue temperature of the hot-blast stove, he outlet

---

* 基金项目：新疆维吾尔自治区科技计划 PT1003，201130104

** 作者简介：刘　娜（1983—），女，江苏泰兴人，硕士研究生，助理研究员，E-mail：liuna8316@163.com

temperature of the hot air flue, the velocity pressure of the hot air flue, the inlet temperature of the cold air flue, etc.

Key words: Thermal efficiency; Hot-blast stove; Configuration; Greenhouse

1 引言

煤炭是我国主要的能源资源，我国的煤炭产量和消费量在世界上居于首位，占世界总量的30%[1]。2010年，我国的原煤需求量达到25.00亿t，到2020年全国煤炭需求量预计为29.00亿t。我国煤炭的使用占全部能源使用的将近70%，而我国的煤炭利用效率是很低的。新疆煤炭预测储量占全国的40%[2,3]（约为2.1万亿t），是全国最主要的煤炭产区，但是煤炭企业以中小型煤企为主，煤炭资源回采率很低，平均不到30%[4,5]。今天，我国的能源人均占有量很低，而我国的农业产业又是产出比较低的产业，这就要求在农业上的能耗要更低。热风炉是设施农业冬季生产必配的能耗设备之一，如何提高热风炉热效率，降低能耗是亟待解决的重要问题之一。

新疆维吾尔自治区（以下简称新疆）地区多年来大力发展设施农业，2011年底发展设施农业103.82万亩，建造温室约40多万座，温室大棚的数量每年还在不断增加。在新疆发展设施农业冬季就需要使用热风炉。现在市场上热风炉种类很多，热风炉生产厂家为了提高热风炉的换热系数，减少换热面积，减少生产投资，还在加大换热侧空气的流速，这使得热风炉的运行成本增加很多[6]。热风炉在使用中又由于多种原因热效率不高，耗能大，生产投入高。本文主要以列管式燃煤热风炉为研究对象，利用现代的传感技术、计算机技术等建立热风炉热效率测试系统，给热风炉生产及使用提供现场测试数据，为优化热风炉的结构设计、风机型号和参数的选择、热风炉供热面积等提供理论分析依据。

2 热风炉热效率算法

2.1 热风炉简介

热风炉主要应用于冬季温室的生产，它放置于温室中间位置，热风通过送风软管向温室两面供热，供热距离为左右各30~40m。在新疆北部地区冬季室外极限温度很低（如塔城极限温度低于-40℃），在这些极限温度的温室内使用热风炉是非常有必要的；在新疆南部地区冬季会出现连续10天到1个月的阴雨天气，在这种天气里日光温室白天没有吸收太阳辐射的热量以至于夜间温度过低，热风炉的使用也是非常有必要的。列管式热风炉是间接加热炉中的一种常见的结构形式，换热管按一定次序排列组成换热器，烟道内的烟气与进口冷空气在管外和管内流动换热。新疆农业科学院农业机械化研究所研制了一种农用列管式燃煤热风炉，专利号为ZL201020195063.4。目前，列管式燃煤热风炉已在南北疆推广超过6 000台，使用效果良好。本文主要以列管式燃煤热风炉为研究对象。

2.2 热风炉热效率算法

通过查阅文献[7]整理归纳热风炉风速与风压的关系及热风炉热效率的计算公式，热风炉的风压与风速的关系；热风炉的热效率与热风炉的热风道热风流速、热风道管径、热风温度及进风口温度等因素有关。

2.2.1 热风道风速计算

热风出口流速尽量快些，可以将热量更快的输送出去，正常的空气流速控制在

6～8m/s，但是热风炉热风管的风速应当较快。

（1）输出热风密度：

$$\rho = 2.176 \times 10^{-3}\left(\frac{H}{273+t}\right)\left(\frac{1+x}{0.62+x}\right) \tag{2-1}$$

式中，ρ-输出热风密度，kg/m^3；H-气体绝对压力，Pa；t-输出热风平均温度,℃（取测定周期内全部所测输出热风温度的算术平均值）；x-进风湿含量按附录B（标准的附录）求出，kg/kg。

（2）输出热风平均风压：

$$Z_P = \frac{1}{n}\left\{\sum_{j=1}^{n}\left(\frac{\sum_{i=1}^{m}\sqrt{z_i}}{m}\right)^2\right\} \tag{2-2}$$

式中，Z_i-i点微压计动压读数，Pa；M-测点数；n-试验期间测定的次数；Z_P-测定截面上m点输出热风平均动压，Pa。

（3）输出热风平均流速：

$$Vp = \sqrt{2}Kd\sqrt{K}\frac{\sqrt{Zp}}{\sqrt{P}} \tag{2-3}$$

式中，V_P—输出热风平均流速，m/s；K-风速测量常数因子；k_d-冷凝管系数，k_d取值0.8～0.85。

根据公式（2－1）、（2－2）和（2－3）可以得出风压和风速的关系：

$$Vp = \sqrt{1.8ZP} \tag{2-4}$$

2.2.2 热风炉热效率算法

（1）输出热风热量：

$$Q_{yx} = qv\rho(C_{pm0}^{t}t - C_{pm0}^{t_0}t_0) \tag{2-5}$$

式中，Q_{yx}-输出热风换热量，kJ/h；t_0-进风平均温度,℃；C_{pm0}^{t}－温度t时输出热风平均定压质量比热容，kJ/（kg·K）；$Q_{GG} = Q_{DW}^{Y}B$－温度t_0时输出热风平均定压质量比热容，kJ/（kg·K）。

（2）输人热量：

$$Q_{GG} = Q_{DW}^{Y}B \tag{2-6}$$

式中，Q_{GG}-输入热量，kJ/h；Q_{DW}^{Y}－燃煤应用基低位发热值，kJ/kg；B-每小时平均燃煤量，kg/h。

（3）热风炉的热效率：

$$\eta = \frac{Q_{YX}}{Q_{GG}} \times 100\% \tag{2-7}$$

式中，η—换热效率,％。

（4）根据公式（2－5）、（2－6）和（2－7）总结出热风炉热效率算法：

$$\eta = \frac{3593 \times (t1 - t0) \times Vp \times \left(\frac{D^2}{4}\right)}{[(273 + t1) \times Q_{GG} \times 1000]} \times 100\% \tag{2-8}$$

式中，t_1 为热风道出口温度,℃；t_0 为冷风进口温度,℃；D 为热风道的直径，mm；V_P-输出热风平均流速，m/s；Q_{GG}-输入热量，KJ/h。

分析热风炉热效率算法（公式 2－8）可以得出：热风炉热效率与热风炉热风道管径、热风道温度及空气流速、冷风进口温度、加入煤的质量及热值等参数相关。

3 热效率测试系统设计

3.1 软件需求分析

设计热风炉热效率测试系统的主要目的是通过传感获取参数信息，结合热风炉热效率算法，得到热风炉的热效率值。首先，软件能准确显示、实时记录并形成报表用以储存各参数信号[8,9]；其次，软件可计算各参数信号的平均值并带入热风炉热效率的计算公式。热风炉热效率测试软件，需要分 4 个界面显示，界面之间可以来回切换，每个界面上可以按照需求设立多个子面板。

3.2 人机界面

人机界面是指计算机上组态软件操作界面。组态软件可通过电脑系统运行，开始测试后，计算机就不断发送和接收信号，与数据采集模块进行实时通信。组态的优势在于能应用灵活多样方式进行系统集成，缩短系统集成的时间，提高集成效率。热风炉热效率测试系统，包括主界面、实时报表界面、温度监控界面、退出界面 4 个界面。

在组态软件界面上，点击“农科院热风炉 529”文件，点击菜单栏的运行，即可看到热风炉实时监控界面，主界面上热风道平均温度、烟道温度、冷风进口温度、热风道风压等数据均为实时测量的数据，且可形成波形图。风速的数值通过热风道风压实时计算获得，并且随热风道风压的数值实时变化。标准煤的单位热量、测试用标准煤、检测时间、热风管管径、烟管管径的数值可利用人工手动输入获得，按上下箭头可修改也可直接在白框处填入数值。热风道温度、进口温度、热风道风速是检测时间到达后对应记录数据的平均值。热风炉热效率是通过热风炉热效率算法将热风道温度、进口温度、热风道风速、热风管管径、检测时间等数据带入并计算得到的最终结果。

3.3 平台搭建

热风炉测试系统由 9 组热风道温度传感器、1 组冷风温度传感器、1 组风压传感器和 1 组烟道温度传感器组成，如图所示。12 路模拟量输入通过数据采集模块转换为标准的 4～20mA 的电信号通过 485 通讯方式将数据传输给计算机，运行热风炉热效率测试软件实现测试数据的实时显示、结果计算和生成报表等。

4 结论

本论文从我国人均能源占有率较低入手，结合我国热风炉的结构和使用中存在的问题，设计并验证了一种检测热风炉热效率的测试系统原型，该系统可应用于相关热风炉热效率测试研究和性能评定领域。本论文主要工作及成果如下。

第一，总结出热风炉的风压与风速的关系：热风炉的热效率与热风炉的热风道风速、热风道出口温度、热风道管径及进风口温度等因素有关，推导出热风炉热效率算法。

第二，搭建了热风炉热效率测试系统，传感器阵列采集热风炉的烟道温度、热风道出口

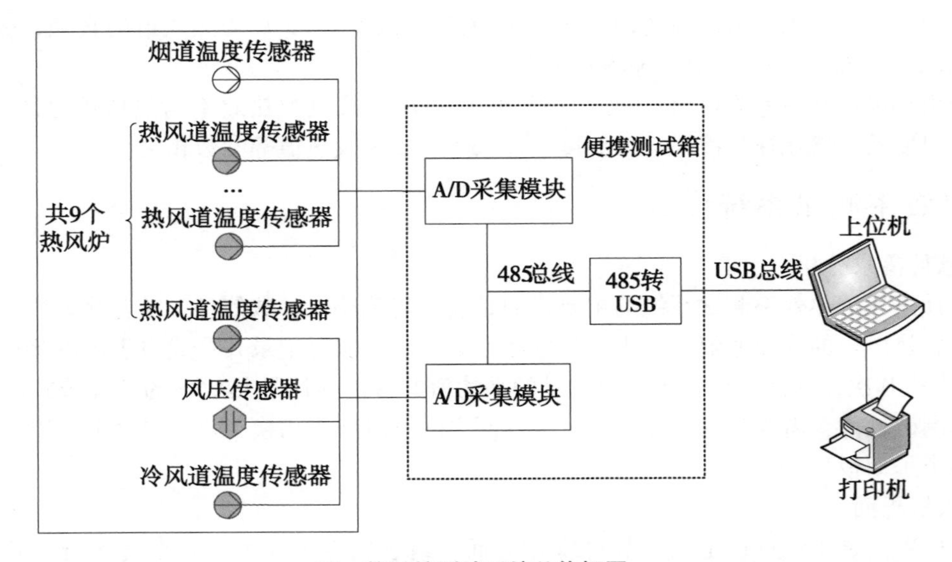

图　热风炉测试系统总体框图

Figure　Diagram of hot-blast stove test system

温度、热风道风压、冷风进口温度等信号，并将采集的模拟信号传输到数据采集模块。数据采集模块将模拟信号转化为电信号传输给计算机，通过热风炉热效率算法将测试结果显示在热风炉测试系统软件的人机界面上。

第三，热风炉热效率测试系统为新疆地区提供了一种热风炉热效率评定的现场测试方法和测试系统，解决了市场上热风炉种类繁多不易认证的问题。

参考文献

[1] 史斗，郑军卫．我国能源发展战略研究．地球科学发展，2000，15（4）：406～414

[2] 周凤起，周大地．2020年中国能源战略分析．中国高校科技与产业化，2007（11）：52～55

[3] 张晓燕．能源利用与能源开发．能源工程，2001（5）：30～34

[4] 王先斌．开发能源资源的思考与选择．科学通报，1999，44（5）：550～560

[5] 王莉，王玲．能源利用与环境保护．化工进展，1998，17（2）：36～45

[6] 孙锋，邱立春，王秀珍．燃煤热风炉在温室生产中的应用．农机化研究，2006（6）：179

[7] 林金天，崔远勃，刘德望等．燃煤热风炉试验方法．中华人民共和国机械行业标准：JB/T 6672.2—2001

[8] 隋洪岗．电脑开发与应用，2011（4）：64～65

[9] 胡汉辉．三维组态软件的应用．湖南工业职业技术学院学报，2003，3（4）：5～7

日光温室热环境分析专用气象数据集的创建*

徐　凡**，马承伟***，刘　洋，胡　彬，王双瑜
（中国农业大学水利与土木工程学院/农业部设施农业工程重点实验室，北京　100083）

摘要：日光温室环境的动态模拟离不开室外气象要素，为了提供更有代表性的室外气象资料，研究根据已获得的华北五省区9个站点的气象资料，阐述了典型数据的选取方法，建立了日光温室主要使用月份（11月至次年3月）的标准气象年、最高温气象年、最低温气象年、辐射最高气象年、辐射最低气象年气象数据集。通过该数据集，可以查询标准年或特征年任意一天的详细室外气象资料，为不同气候条件下日光温室热环境的精确模拟提供基础数据。

关键词：日光温室；热环境分析；标准年；典型年；气象数据集

The Establishment of Meteorological Data Set Dedicated for Solar Greenhouse Thermal Environment Analysis

Xu Fan，Ma Chengwei*，Liu Yang，Hu Bin，Wang Shuangyu
(*College of Water Resources & Civil Engineering*，*China Agricultural university/Key laboratory of Agricultural Engineering in Structure and Environment*，*Ministry of Agriculture*，*Beijing* 100083，*China*)

Abstract：The dynamic simulation of solar greenhouse environment based on outside meteorological factors，in order to provide more typical outside meteorological data，this paper use the data of 9 weather stations in North China，stated the selection method of typical data，established meteorological data set in solar greenhouse using months (from November to March next year)，including standard year，the highest temperature year，the lowest temperature year，the highest radiation year，and the lowest radiation year. The data set can view the detailed outside meteorological data in any day in standard or characteristic years，providing the basic data for solar greenhouse thermal environment exactly simulation under different climate situation.

Key words：Solar greenhouse；Thermal environment analysis；Standard year；Characteristic year；Meteorological data set

日光温室是我国北方地区冬季主要的园艺生产设施，可为植物提供适宜的生长环境，截至2010年，我国日光温室面积超过38万 hm^2[1]。随着日光温室研究的深入，关于日光温室

* 基金项目：现代农业产业技术体系建设专项资金（CARS－25），中国农业大学研究生科研创新专项（KYCX2010099），“十一五”国家科技支撑计划项目（2009BADA4B04），高效设施农业标准化工程技术集成示范（201130104），公益性行业（农业）科研专项（201203002），资助

** 作者简介：徐　凡（1982—），女，辽宁人，博士生，主要从事设施园艺环境工程研究。E-mail：luckyfan@126.com

*** 通讯作者：马承伟（1952—），男，重庆人，教授，博士生导师，主要从事设施园艺环境工程研究。E-mail：macwbs@cau.edu.cn

热环境性能的分析已经从试验研究更多的转向动态理论模型的构建[2-8]，这就需要更为详尽的气象及温室参数资料作为基础。室外气象数据是日光温室热环境模拟中必不可少的部分，数据的质量直接影响模拟的精度。模拟中具有代表性的室外气象数据的选择是研究的关键，合理的选择数据既可以减少模拟的工作量，又可以在模拟结果应用于指导实践的过程中减少材料等的浪费。

目前，在建筑领域，张晴原等根据建筑内空调及供暖的特点，创建了《中国建筑用标准气象数据库》[9]，但由于日光温室有其自身的结构和使用特征，因此，建筑领域的标准气象数据库并不能直接应用于日光温室动态模拟中。本研究拟根据华北地区的气象资料，探讨日光温室热环境分析中典型气象数据的选取方法，并初步建立日光温室热环境分析专用标准年及特征年的气象数据集。

1　资料的来源

1.1　气象资料来源

通过气象局气象资料室、气象数据共享网，获得华北地区日光温室主要建设地北京、天津、河北、山东、山西 5 省市具有辐射站点的 9 个气象站近 31 年（其中，个别站点为 16 ~ 20 年）的日值气象数据（1980 ~ 2010 年）。数据包括温度、相对湿度、气压、太阳辐射、云量、日照时数等参数，受云量资料的限制，研究中主要取 1980 ~ 2005 年数据进行统计学分析，各站点资料年份如表 1 所示。

表 1　各站点资料年份统计

台站号	站名	省份	资料年份
53487	大同	山西	1980 ~ 2005
53772	太原	山西	1980 ~ 2005
53963	侯马	山西	1991 ~ 2005
54511	北京	北京	1980 ~ 2005
54527	天津	天津	1980 ~ 2005
54539	乐亭	河北	1992 ~ 2005
54764	福山	山东	1996 ~ 2005
54823	济南	山东	1980 ~ 2005
54936	莒县	山东	1990 ~ 2005

1.2　统计指标

计算各气象站每年在日光温室主要使用季节（11 月至翌年 3 月）各月的日平均气温、日最高气温、日最低气温、日总辐射、日平均相对湿度、日最小相对湿度六项指标的月平均值。

将上述统计好的数据按月份归类整理，即得到各站点每月的逐年气象资料。再以此计算各月份上述的历年平均值、最高值、最低值。将依此构建标准月、标准年气象数据集。

2　气象数据集的构建

在日光温室热环境模拟中，通常考虑平均状况及一些特殊情况作为代表性室外气候。因此，研究中界定了气象标准年和特征年的概念，来构建日光温室分析用气象数据集。日光温室的热环境受室外太阳辐射影响最大，其次为室外气温，因此根据太阳辐射和气温特点，特征年分为辐射最高年，辐射最低年，气温最高年、气温最低年。下面将就标准年和特征年数据的选取方法及构成进行详细的阐述。

2.1 标准年气象数据

标准年为最接近历年平均值的情况，代表了该地区的平均室外气象条件，由标准月数据构成。

以上述6项指标作为评价标准，日光温室主要使用月份中，主要考虑平均气温和总辐射，并参照考虑最高气温、最低气温及相对湿度，每个站点逐月选取当月各指标接近该站点历年平均值的月份作为该月份的标准月。例如，北京，11月份，计算结果如表2所示。可以看出，1984年各项指标的月平均值最接近历年平均，因此，北京11月份的标准月即定为1984年11月。

表2 北京气象站点11月份标准月选取

年份	平均气温（℃）	最高气温（℃）	最低气温（℃）	平均相对湿度（%）	最小相对湿度（%）	总辐射（w/m^2）
历年平均	4.9	10.2	0.3	63	39	808
1982	5.1	10.9	0.2	66	43	671
1984	4.8	10.3	0.2	65	42	783
1989	4.6	10.1	0	54	32	787
1991	4.6	9.8	-0.2	60	40	781
1996	4.2	8.8	0	60	35	789
2005	7.5	13.4	2.6	60	30	846

其他站点及月份均按此方法选取气象标准月。标准月数据相连接构成标准年气象数据。标准月之间取6天以3日滑动平均法进行平滑处理。标准月数据的选取结果如表3所示。

表3 标准月数据的选取

台站号	省份及站名	11月	12月	1月	2月	3月
53487	山西大同	1991	1990	1992	1995	1983
53772	山西太原	1982	1988	1990	1991	1989
53963	山西侯马	2004	1992	1997	1997	1993
54511	北京北京	1984	1981	2003	1987	1996
54527	天津天津	1985	1986	1986	1987	1983
54539	河北乐亭	1998	1999	2005	1993	1996
54764	山东福山	2001	1999	2004	1995	1995
54823	山东济南	2001	1993	1998	2003	1989
54936	山东莒县	1999	2003	2003	1994	1995

2.2 特征年气象数据

日光温室使用季节，1月份最寒冷，因此也最受关注，所以日光温室特征年的选取以1

月份数据为基础。

气温最高年气象数据：取 1 月平均气温最大值所对应的年份为最高温气象年，由上年 11～12 月份及该年 1～3 月份数据构成气温最高年气象数据集。

气温最低年气象数据：取 1 月平均气温最小值所对应的年份为最低温气象年，由上年 11～12 月份及该年 1～3 月份数据构成气温最低年气象数据集。

辐射最高年气象数据：取 1 月总辐射最高值所对应的年份为辐射最高气象年，由上年 11～12 月份及该年 1～3 月份数据构成辐射最高气象年数据集。

辐射最低年气象数据：取 1 月总辐射最低值所对应的年份为辐射最低气象年，由上年 11～12 月份及该年 1～3 月份数据构成辐射最低气象年数据集。

特征年数据的选取结果如表 4 所示。

表 4　特征年数据的选取

台站号	省份及站名	气温最高年	气温最低年	辐射最高年	辐射最低年
53487	山西大同	2002	1993	1985	1996
53772	山西太原	2002	1993	1986	2004
53963	山西侯马	2002	2000	1993	2004
54511	北京北京	2002	2000	1984	2001
54527	天津天津	2002	2000	2002	1998
54539	河北乐亭	2002	2000	1995	2001
54764	山东福山	2002	1997	2002	2000
54823	山东济南	2002	2000	1999	2001
54936	山东莒县	2002	1993	2005	2001

2.3　建立气象数据集

将上述选好的标准年和特征年气象数据，导入 ACCESS 数据库，建立日光温室热环境分析专用气象数据集（Access 文件），对于其中缺失数据采用插值法补齐。包含已获得的华北地区 9 个气象站的日光温室主要使用月份（11 月至次年 3 月）的标准气象年、最高温气象年、最低温气象年、辐射最高气象年、辐射最低气象年气象数据集。

利用该气象数据集，可以查询已有站点标准年或特征年任意一天的详细气象数据。

例如，查询北京（54511），1 月 8 日，在标准年及特征年中，与日光温室热环境模拟相关的气象要素值，结果如下。

标准气象年（图 1）：可用于通常情况下日光温室热环境的模拟。

温度特征年（图 2）：可用于室外温度处于极端条件时，温室内环境的模拟预测。

辐射特征年（图 3）：可用于室外辐射处于极端条件时，温室内环境的模拟预测。

3　结论与讨论

通过对已获得的 9 个气象站气象资料的统计分析，研究了日光温室代表性气象数据的选取方法，并初步建立了的华北 5 省日光温室模拟分析专用气象数据集，通过该数据集可查询标准年及特征年任意一天的详细室外气象资料。可为不同气候年份下，日光温室内热环境的

标准气象年 | 最高温气象年 | 最低温气象年 | 辐射最高气象年 | 辐射最低气象年

标准气象年

ID:	522	总辐射:	672
台站:	54511	平均相对湿度:	76
月:	1	最小相对湿度:	53
日:	8	平均风速:	13
平均本站气压:	10233	最大风速(10分钟平均风速):	25
日最高本站气压:	10278	最大风速的风向:	8
日最低本站气压:	10195	日平均总云量:	0
平均气温:	-67	日平均低云量:	0
日最高气温:	-5	日照时数:	62
日最低气温:	-128		

日光温室热环境分析专用气象数据集说明文档

1. 台站说明

台站号	站名	省份	经度	纬度	拔海高度
53487	大同	山西	11320	4006	10672
53772	太原	山西	11233	3747	7783
53963	侯马	山西	11122	3539	4338
54511	北京	北京	11628	3948	313
54527	天津	天津	11704	3905	25
54539	乐亭	河北	11853	3926	105
54764	福山	山东	12115	3730	326
54823	济南	山东	11703	3636	1703
54936	莒县	山东	11850	3535	1074

说明：

1. 经度：由5位数组成，前3位为度，后2位为分；
2. 纬度：由4位数组成，前两位为度，后2位为分；
3. 拔海高度：观测场拔海高度，单位为0.1m。

2. 数据说明

观测项目	单位
平均本站气压	0.1hPa
日最高本站气压	0.1hPa
日最低本站气压	0.1hPa
日平均气温	0.1℃
日最高气温	0.1℃
日最低气温	0.1℃
总辐射	0.01MJ/m²
日平均相对湿度	1%
日最小相对湿度	1%
平均风速	0.1m/s
最大风速（10分钟平均风速）	0.1m/s
最大风速的风向	16个方位
日平均总云量	0.1成
日平均低云量	0.1成
日照时数	0.1小时

图1　北京1月8日标准年气象数据查询结果及数据说明文档

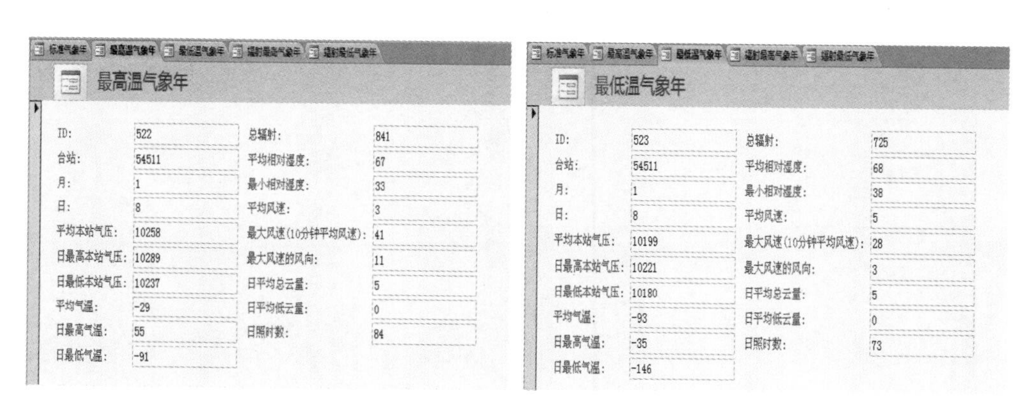

图2　北京1月8日温度特征年气象数据查询结果

标准气象年 | 最高温气象年 | 最低温气象年 | 辐射最高气象年 | 辐射最低气象年

辐射最高气象年

ID:	522	总辐射:	879
台站:	54511	平均相对湿度:	45
月:	1	最小相对湿度:	21
日:	8	平均风速:	18
平均本站气压:	10144	最大风速(10分钟平均风速):	70
日最高本站气压:	10187	最大风速的风向:	14
日最低本站气压:	10106	日平均总云量:	3
平均气温:	-7	日平均低云量:	0
日最高气温:	77	日照时数:	78
日最低气温:	-80		

辐射最低气象年

ID:	525	总辐射:	156
台站:	54511	平均相对湿度:	78
月:	1	最小相对湿度:	66
日:	8	平均风速:	30
平均本站气压:	10229	最大风速(10分钟平均风速):	59
日最高本站气压:	10248	最大风速的风向:	1
日最低本站气压:	10206	日平均总云量:	100
平均气温:	-5	日平均低云量:	50
日最高气温:	10	日照时数:	0
日最低气温:	-13		

图3　北京1月8日辐射特征年气象数据查询结果

精确动态模拟提供详尽的室外气象数据。

本研究侧重于在诸多室外气象要素中找出日光温室热性能的主要影响因素，初步建立应用于日光温室的气象数据集，但数据中标准月数据间的平滑连接问题、数据集的不断完善和丰富，以及各气象要素对温室热环境的具体贡献、如何应用在日光温室环境模拟预测中，将在今后的研究中继续深入探讨。

参考文献

[1] 全国设施农业发展“十二五”规划（2011~2015年）. 温室园艺，2011（11）：36~42

[2] 李元哲，吴德让，于竹. 日光温室微气候的模拟与实验研究. 农业工程学报，1994（3）：130~136

[3] 王捷，邱仲华，康永劼等. 甘肃省高效节能日光温室光热环境分析. 西北农业学报，1996，5（1）：76~81

[4] 田军仓，韩丙芳，李建设等. 宁夏三类温室与其小气候关系的试验研究. 宁夏农学院学报，2002，23（4）：46~49

[5] 陈青云，汪政富. 节能型日光温室热环境的动态模拟. 农业工程学报，1996，12（1）：67~71

[6] 辛本胜，乔晓军，滕光辉. 日光温室环境预测模型构建. 农机化研究，2006（4）：96~100

[7] 孟力力，杨其长，Gerard. P. A 等. 日光温室热环境模拟模型的构建. 农业工程学报，2009，25（1）：164~170

[8] 马承伟，韩静静，李睿. 日光温室热环境模拟预测软件研究开发. 北方园艺，2010（15）：69~75

[9] 张晴原，Joe Huang. 中国建筑用标准气象数据库. 北京：机械工业出版社，2004

日光温室蓄放热装置增温试验*

方　慧[1,2]**，杨其长[1,2]***，张　义[1,2]，孙维拓[1,2]

（1. 中国农业科学院农业环境与可持续发展研究所，北京　100081；

2. 农业部设施农业节能与废弃物处理重点实验室，北京　100081）

摘要：日光温室冬季夜晚低温现象时有发生，严重影响作物的产量和品质。针对这一问题，设计了一种温室蓄放热增温装置，白天利用该装置吸收太阳辐射热，并通过介质水将热量储存起来；夜间当室内空气温度较低时，再通过水的循环将蓄积的热量释放到温室中，以提高夜间室内空气温度。试验结果表明：晴天时该蓄放热装置能提高温室夜间温度4.6℃，阴天时能提高温室夜间温度4.5℃；室外空气温度为 -12.5℃时，对照温室温度仅为5.4℃，而试验温室温度为10.1℃；该装置在阴天平均集热效率为44.2%，在晴天时平均能将55.5%的太阳能蓄积起来；与电加热方式相比该装置的节能率达到69.8%以上。

关键词：日光温室；节能；蓄放热

Research on Warming Effect of Heat Storage-release Curtain in Chinese Solar Greenhouse

Fang Hui[1,2], Yang Qichang[1,2*], Zhang Yi[1,2], Sun Weituo[1,2]

(1. *Institute of Environment and Sustainable Development in Agriculture*, *Chinese Academy of Agricultural Sciences*, *Beijing* 100081, *China*; 2. *Key Lab of Energy Conservation and Waste Treatment of Agricultural Structures*, *Ministry of Agriculture*, *Beijing* 100081, *China*)

Abstract: In a Chinese Solar Greenhouse (GSG), during cold winter night, air temperature can be very low and this considerably decreases crop production. To increase this low nighttime air temperature, a heat collection and storage-release system was studied. During daytime the absorbed solar energy from the system was transferred to the water by the circulation pump. At night, when the greenhouse air temperature dropped below 10℃ the system was used to transfer the low temperature heat from the water to the greenhouse air. The results showed that the average nighttime air temperature in the CSG was 4.6and 4.5℃ higher than that in the reference CSG on cloudy and sunny day respectively. When the outside air temperature was -12.5℃, the air temperature inside the normal greenhouse was just only 5.4℃, while that in the experiment greenhouse was 10.1℃. The average efficiency of heat storage-release system can reached 44.2% and 55.5% on cloudy and sunny day respectively. Compared with electricity heating equipment, over 69.8% energy had been saved.

---

* 基金项目：863计划（2013AA102407）；国家自然科学基金资助项目（31071833）；国家科技支撑计划（2011BAE01B10）

** 作者简介：方　慧（1983—），女，硕士，助理研究员，主要从事设施农业环境工程方面的研究。北京中国农业科学院农业环境与可持续发展研究所，100081。E-mail：fh2002124@163.com

*** 通信作者：杨其长（1963—），男，博士，研究员，博士生导师，主要从事设施园艺环境工程研究。北京中国农业科学院农业环境与可持续发展研究所，100081。E-mail：yangq@ieda.org.cn

Key Words：Chinese solar greenhouses；Energy saving；Heat release and storage

1　引言

日光温室是我国特有的一种温室形式，单面为透光覆盖材料，由于其良好的保温蓄热和透光性能，在我国北方地区得到了大面积的推广。日光温室的显著特征之一是其具有“北墙”结构，即是重要的承重体，又是维持室内温度的保温蓄热材料。白天，墙体表面通过接收透过前屋面照射进来的太阳辐射进行热量蓄积，晚上通过墙体表面与室内空气形成的温度差不断向室内释放热量以提高空气温度[1~12]。然而，受墙体材料热物理特性的限制，温室的蓄热能力有限，到后半夜时温室内的温度往往较低，低温冷害现象时有发生，影响作物的高效生产。针对上述问题，多年来，众多学者在有关日光温室墙体蓄热保温性能提升、工程集热等方面进行了不懈的探索。王宏丽等将相变材料与建筑材料混合制成相变蓄热砌块，并以其为墙体建造相变蓄热温室，通过测试相变温室室内气温波动幅度比对照温室小4.1℃，最低气温比对照高1.7℃，而最高气温则比对照低2.4℃[13~15]。王顺生等在冬季晴天条件下，利用内置式集热调温装置白天蓄积的热量，夜间用来提高温室气温时，加温部分可比不加温部分平均提高1.7℃到2.6℃[16]。白义奎等人建立了日光温室燃池-地中热交换系统，测试结果表明，采用燃池-地中热交换系统一侧的土壤平均温度比对比侧的土壤平均温度高约2.0℃，昼夜平均气温高2.6℃，夜间平均气温高4.2℃[17]；戴巧利等人试验研究了一套太阳能温室增温系统，白天利用太阳能空气集热器加热空气，由风机把热空气抽入地下，通过地下管道与土壤的热交换，将热量传给土壤储存，夜间热量缓慢上升至地表，实现增温。该系统的蓄热量可达228.9~319.1MJ，可使夜间温室的气温平均升高3.8℃，地温平均升高2.3℃[18]。H. ALl-HUSSAINI [19]和GRMIADELLIS[20]等在温室内建造了蓄热水池，水池深5~20cm，水池内表面涂黑，水池中装满一定浓度的盐水，外面覆上透明塑料膜，利用蓄热水池蓄集的热量加热温室，但试验表明其蓄热能力有限。虽然目前的研究已经提出了一些增加日光温室蓄热量的方法，但由于成本高或蓄热量有限，在日光温室中的推广应用受到一定的限制。

为此，笔者提出一种采用蓄放热装置增温的新型蓄放热方式，白天通过蓄放热装置尽可能多的蓄积太阳能并将这部分热量转移储存起来，夜间当室内温度较低时再通过蓄放热装置将这部分热量释放出来，以提高温室夜晚温度。

2　材料与方法

2.1　温室概况

试验日光温室位于北京市昌平区小汤山现代农业科技示范园西区，温室为东西走向，长49m，跨度8m，后墙高2.5m，脊高3.7m，后坡长1.5m，采用钢骨架结构，前坡覆盖材料为单层0.08mmPVC塑料薄膜，温室外覆盖材料为自防水保温被，后墙内测为12cm厚红砖，外侧为24cm厚红砖，中间为10cm厚聚苯板，后坡内侧为10cm厚预制板，外侧为10cm厚聚苯板。对照温室结构、材料和建造时间均与试验温室相同。两温室南北方向间距10m。试验温室主要用于育苗，对照温室种植草莓。

2.2 试验装置及原理

蓄放热装置包括集放热板、循环管路、水泵和蓄热水池。集放热板为 PE 双黑膜，温室安装 29 块集放热板，板的宽为 1.35m，高为 2.0m，固定于后墙内表面距地面高度为 0.4m。循环水泵 2 台，流量均为 $7m^3/h$，扬程 10m，功率为 750W。

蓄放热装置增温装置工作原理如图 1 所示。晴天，随着室内太阳辐射增强和气温升高，温室后墙温度升高，集热板截获太阳辐射并将其转化为热量，通过蓄热介质水的循环将热量收集并储存起来；在夜晚，当温室内温度较低时，通过水的循环将蓄热介质水蓄积的热量释放出来加热温室，调节室内温度。

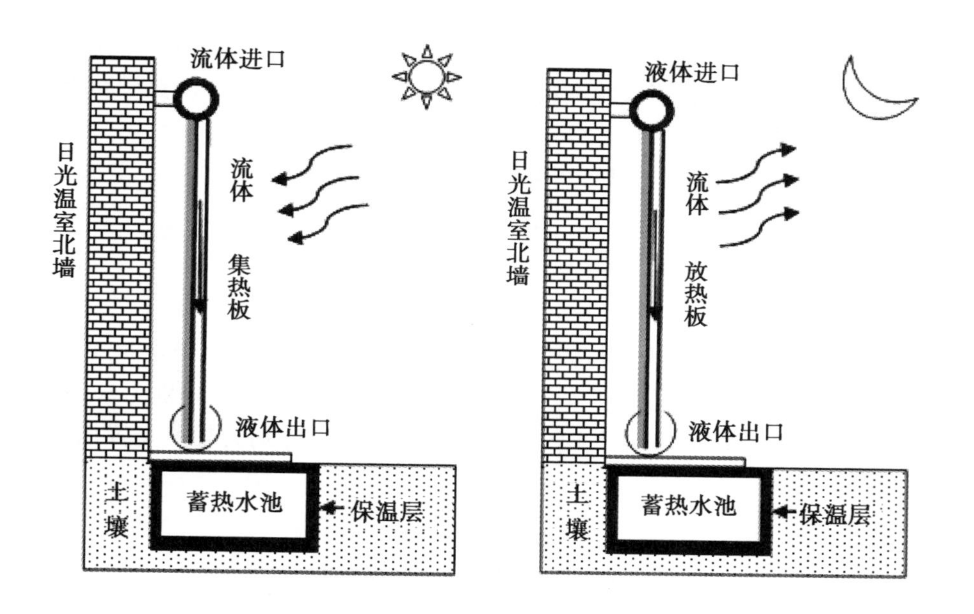

图 1　蓄放热装置运行原理图

Figure 1　Schematic diagram of thermal release-storage equipment

2.3 试验仪器

单点温度测试选用德图公司生产的 testo174T 型温度自动记录仪，精度为 ±0.2℃，测量范围为 -30 ~ 70℃。

太阳辐射测试选用美国坎贝尔公司生产的太阳辐射传感器，准确度为 0.5%，测量范围为 0 ~ 2 000W/m^2。

多点温度测试选用中国计量院生产的热电偶，测量精度为 ±0.2℃。

数据采集仪选用美国坎贝尔公司生产的 CR1000，用于自动记录热电偶采集的温度值和辐射传感器采集的太阳辐射值。

2.4 试验方法

2.4.1　试验设置

试验监测时间为 2012 年 11 月 15 日至 2013 年 1 月 20 日，选取 1 月 11、12、13 日的 3d 数据进行分析。温室揭保温被时间为 8：30，盖保温被时间为 15：30，主动蓄热装置蓄热时间为 8：30～15：30，放热时间根据温度控制，当室内温度低于 10℃时开始加热。蓄水池实测蓄水量为 7.35m^3。每间隔 10min 记录一次各测点的温度。

试验温室与对照温室空气温度测点布置：在试验温室与对照温室分别距北墙 4m，距东墙 12m、24m、36m 处共设置 6 个测点，在距东墙 24m，距北墙 2、6m 处共设置 4 个测点，测点均距地面 1.5m 高。

蓄放热板温度测点布置：在蓄放热板的外表面和中间水流层设置 2 个测点，同时在后墙内表面距东墙 24m 距地面 1.5m 高处设置太阳辐射测点。

蓄水池温度测点设置：在水池的进水口、出水口和水池几何中心设置 3 个温度测点。

2.4.2　蓄放热装置性能测试

测试期间统计蓄放热装置的蓄热时间段、放热时间段、水泵消耗的电功率，如表 1 所示。蓄放热装置性能分析包括指系统在运行中消耗的能量，白天收集的热量和夜间放出的热量以及系统运行中的集热功率。计算公式为：

$$Q_s = \rho_w V_w C_w (T_o - T_i) \tag{1}$$

$$E_s = \sum Q_s \tag{2}$$

$$Q_r = \rho_w V_w C_w (T_i - T_o) \tag{3}$$

$$E_r = \sum Q_r \tag{4}$$

式中，E_s、E_r 分别为装置吸收与释放的热量，kJ；Q_s、Q_r分别为单位时间装置吸收与释放的热量，kW；V_w 为集热装置循环水的体积流量，m^3/s；ρ_w，C_w 分别为水的密度与比热容，分别取为 1.0×10^3kg/m^3 和 4.2 kJ/（kg·K）；T_o、T_i 分别为集热装置循环管路上送、回水的温度,℃。

表 1　试验参数

Table 1　Experiment parameter

试验时间	蓄热时间	放热时间	耗电量/kW·h	天气
2013-01-11	8：30～15：30	23：00～08：30	24.8	阴
2013-01-12	8：30～15：30	1：00～08：30	21.8	晴

集热板的集热效率可通过水获得的热量与照射在集热器上的辐射量的比值求得。

$$\eta_c = \frac{Q_s}{A_c I_c} = \frac{\rho_w V_w C_w (T_o - T_{\backslash mathop\{\backslash rm\ i\}\backslash nolimits\}})}{A_c I_c} \times 1\,000 \tag{5}$$

式中，η_c 为集热效率；A_c 为有效集热面积，m^2；I_c 为投射至集热器采光面积上的太阳辐射照度，W/m^2。

2.4.3　节能效果分析

根据蓄放热装置的运行时间与 2 个水泵的运行功率即可得出消耗的电量 Q_{wp}，根据公式（3）、（4）可推算出有效的放热量 Q_r，若 Q_r 为电加热产生的热量，则可推算出节能率：

$$Q_{wp} = 2 \times W_{wp} \times h \times 3\,600 \tag{6}$$

$$\eta = \frac{E_r - E_{wp}}{E_r} \tag{7}$$

式中，η 为温室采用蓄放热板加热后的节能率；W_{wp}为水泵的功率，kW；h 为水泵运行时间，h；Q_{wp}为水泵消耗的电能，kJ。

3 结果与分析

3.1 室内气温

试验温室与对照温室温度变化如图 2 所示。在 2013 年 1 月 11 到 13 日连续 3d 的测试中，由于温室白天开通风口进行通风，对温室内温度有一定的影响。测试期间试验温室内的温度一直高于对照温室。盖上保温被后到系统放热前，两温室温差较小，分别为 1.6℃、1.9℃，在蓄放热装置开启后，试验温室内的温度明显升高，在加热时间段两温室的平均温差分别为 4.6℃、4.5℃。1 月 12 日室外温度较低，最低温 -12.5℃，此时对照温室温度为 5.4℃，而试验温室温度为 10.1℃。

3.2 室内气温南北分布

从图 3 可以看出，试验温室在南北方向温度分布比较均匀，1 月 11 日蓄热阶段温室南北方向平均温度分别为 19.1℃、19.1℃、19.3℃，在盖上保温被到开启蓄放热装置前南北方向平均温度分别为 13.8℃、14.0℃、14.1℃，在开启蓄放热装置到揭开保温被前温室北面的温度略高于南面温度，此阶段从南到北平均温度分别为 11.0℃、11.4℃、11.6℃。1 月 12 日蓄热阶段温室南北方向平均温度均为 25.0℃，在盖上保温被到开启蓄放热装置前南北方向平均温度分别为 13.4℃、13.5℃、13.6℃，在开启蓄放热装置到揭开保温被前南北方向平均温度分别为 13.0℃、13.1℃、13.6℃。试验测试数据表明，白天温室空气温度主要受太阳辐射的影响，温度分布比较均匀；夜间，在开启蓄放热装置后，温室内的热量是通过蓄放热装置传递到空气中，所以，温室北面的温度略高于南面。

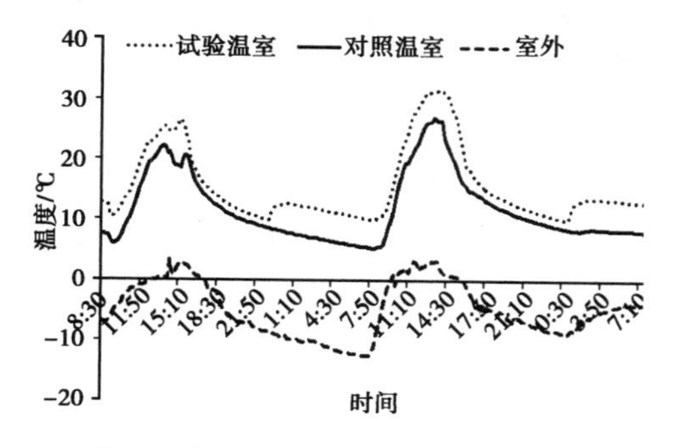

图 2 试验温室与对照温室温度变化

Figure 2 Temperature curves in experiment and normal greenhouses

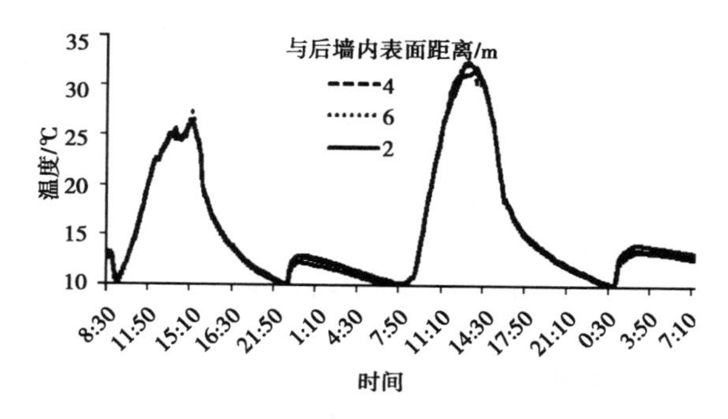

图 3 温室南北方向温度变化曲线

Figure 3 Temperature curves in south-north direction

3.3 进出水温差

1 月 11 日为阴天，太阳平均辐射量为 168.6W/m^2，最大为 245.5W/m^2，此时进出水温差也最大为 0.6℃。水温最高为 28.4℃，出现在 15：30，此时太阳辐射量为 222.0W/m^2，室温为 26.0℃，之后水温开始下降。1 月 12 日为晴天，太阳平均辐射量为 264.0W/m^2，最大为 393.4W/m^2，此时进出水温差最大为 1.3℃。水池水温最高为 31.3℃，出现在 15：30，此时太阳辐射量为 120.0W/m²，室温为 24.7℃，之后水温开始下降。通过测试数据可以看出，白天水温升高取决于太阳辐射量，受制于室内空气温度，水池进出水温差主要受太阳辐射的影响，进出水温差也高，说明蓄热量也大。在后半夜系统放热时，供回水温差绝对值逐渐减小，放热越来越慢（图 4 和图 5）。

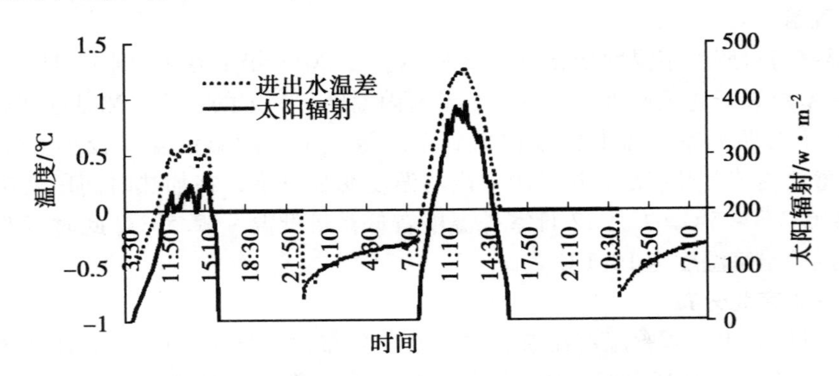

图 4　试验温室内进出水温差与太阳辐射变化

Figure 4　Inlet/outlet water temperature difference and the solar radiation

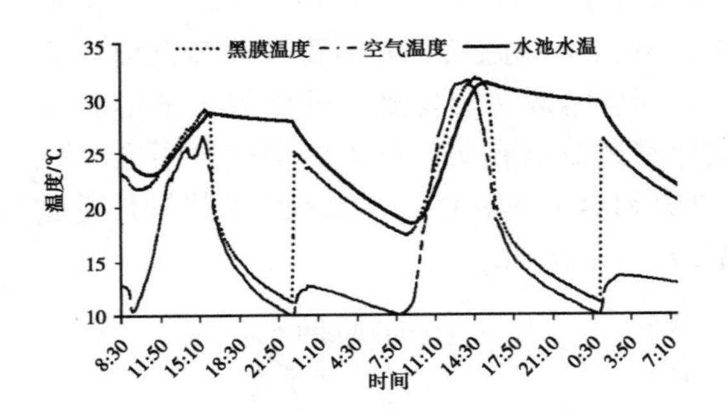

图 5　试验温室内水池水温度、空气温度与黑膜温度变化

Figure 5　Temperature curves of water, air and black membrane

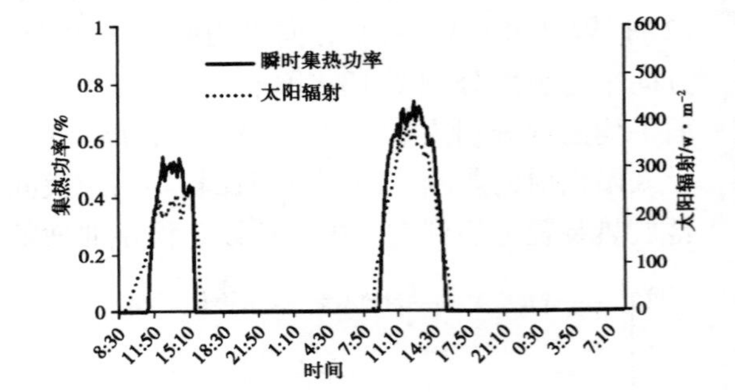

图 6　蓄放热装置瞬时集热功率

Figure 6　Equipment instantaneous efficiency

3.4　黑膜温度

白天揭开保温被后温室开始蓄热，此时温室内的水池水温高于气温。蓄热阶段，黑膜吸收太阳辐射使温度升高，通过水的循环将黑膜吸收的热量收集并储存起来，此时黑膜温度高于蓄水池水温，1 月 11 日与 12 日黑膜与水的温差平均分别为 0.4℃、1.0℃；放热阶段，通过水的循环将热量传递到黑膜，再通过黑膜与室内空气的对流换热将热量释放到空气中，此时水温高于黑膜温度，1 月 11 日与 12 日黑膜与水的温差均为 -1.7℃。

当供回水温差为 0℃，即水温达到最大值，此时存在一个动态平衡，循环水吸收的太阳辐射能与向空气传递的热量相同。以黑膜为中介，水温升高阶段，黑膜吸收太阳辐射，温度高于水温和室温，黑膜分别向循环水和空气传递热量；到达此热平衡时，黑膜吸收的辐射量等于向空气散失的热量，黑膜温度等于水温（图 6）。

3.5 蓄放热装置性能与节能效果

通过公式（1）～（7）计算蓄放热装置增温装置白天的蓄热量、夜间的放热量、集热功率与节能率如表2所示。

表2 蓄放热装置蓄放热量与节能率

Table 2 Heat got and released from this device and the energy-saving rate

试验时间	白天蓄热量/MJ	夜间放热量/MJ	单位面积平均即热功率/W·m^{-2}	节能率
2013-01-11	176.0	295.7	89.2	69.8%
2013-01-12	395.1	270.7	200.3	71.0%

蓄放热装置的瞬时集热效率如图6所示。温室阴天蓄积的热量较少为176.0 MJ，但温室夜间的放热量为295.7MJ，大于白天的蓄热量，主要原因是白天开始蓄热时蓄水池水温较高为23.0℃，蓄热结束时水温为28.7℃；夜间放热时由于温室内的温度较低，放热量较大，经过放热后蓄水池水温只有18.6℃。温室晴天时蓄积的热量较多为395.1MJ，远大于温室夜间的放热量270.7MJ。通过计算，与电加热方式相比，温室晴天与阴天的平均集热功率分别为44.2%和55.5%，比之前的集热功率有明显提升[21]，节能率分别为69.8%和71.0%，节能效果明显。

4 结论与讨论

温室白天通过蓄放热装置吸收太阳辐射能，在夜间进行控制释放，能有效实现热量的收集、存储与释放。

第一，应用该系统可将温室内夜间的温度提高4.5℃以上，当室外最低气温为－12.5℃时，对照温室温度仅为5.4℃，而试验温室温度为10.1℃。

第二，夜间对温室进行加温时，虽然热量是通过蓄放热装置传递到空气中，越靠近装置的地方温度也略高，但南北方向的温度梯度并不大，温室空气温度分布较均匀。

第三，通过试验蓄放热装置在阴天能将后墙上44.2%的太阳辐射热量收集并储存，晴天能将55.5%的太阳辐射热量收集并储存。通过与电加热方式相比，在晴天与阴天的节能率分别到达了69.8%和71.0%，节能效果明显。

在阴天，蓄放热装置蓄积的热量较少，除去天气原因外也与装置本身蓄热性能有关，由于集热板直接与室内空气接触，阴天温室内空气温度较低，集热板蓄积的热量有一部分传递到了温室中，因此在下一步的研究中应考虑集热板的热损失。

参考文献

[1] 佟国红，David M Christopher. 墙体材料对日光温室温度环境影响的CFD模拟. 农业工程学报，2009，25（3）：153～157

[2] 张立芸，徐刚毅，马成伟等. 日光温室新型墙体结构性能分析. 沈阳农业大学学报，2006，37（3）：459～462

[3] 李天来，韩亚东，刘雪峰等. 日光温室内传热筒与空气间的换热量及对气温的影响. 农业工程学报，2011，27（2）：237～242

[4] 李建设，白青，张亚红. 日光温室墙体与地面吸放热量测定分析. 农业工程学报，2010，26（4）：231～236

[5] 马承伟，卜云龙，籍秀红等．日光温室墙体夜间放热量计算与保温蓄热性评价方法的研究．上海交通大学学报（农业科学版），2008，26（5）：411～415

[6] 高艳明．宁夏不同类型日光温室温光性能观测与评价．北京：中国农业大学农学与生物技术学院，2006

[7] 房树田，商跃春．EPS外保温复合墙体的保温层厚度设计研究．黑龙江工程学院学报（自然科学版），2004，18（1）：57～59

[8] 蒋志刚，龙剑．复合外墙内外保温的传热特性研究．制冷，2006，25（2）：44～47

[9] 王晓冬，马彩雯，吴乐天等．日光温室墙体特性及性能优化研究．新疆农业科学，2009，46（5）：1 016～1 021

[10] 王思倩，张志录，侯伟娜等．下沉式日光温室南侧边际区域土壤温度变化特征．农业工程学报，2012，28（8）：235～240

[11] 张峰，张林华．下沉式日光温室土质墙体的保温蓄热性能．可再生能源，2009，27（3）：18～20

[12] 杨建军，邹志荣，张智等．西北地区日光温室土墙厚度及其保温性的优化．农业工程学报，2009，25（8）：180～185

[13] 王宏丽，李晓野，邹志荣．相变蓄热砌块墙体在日光温室中的应用效果．农业工程学报，2011，27（5）：253～257

[14] 王宏丽，邹志荣，陈红武等．温室中应用相变储热技术研究进展．农业工程学报，2008，24（6）：304～307

[15] 王宏丽，李晓野，邹志荣．相变蓄热砌块墙体在日光温室中的应用效果．农业工程学报，2011，27（5）：253～257

[16] 王顺生．日光温室内置式太阳能集热调温装置的研究．北京：中国农业大学（博士论文），2006

[17] 白义奎，迟道才，王铁良等．日光温室燃池：地中热交换系统加热效果的初步研究．农业工程学报，2006，22（10）：178～181

[18] 戴巧利，左然，李平等．主动式太阳能集热/土壤蓄热塑料大棚增温系统及效果．农业工程学报，2009，25（7）：164～168

[19] AL-Hussaini H，Suen K O. Using shallow solar ponds as a heating source for greenhouse in cold climates. Energy Convers，1998，9（13）：369～1376

[20] Ì. GRMIADELLIS，E. TRAKA-MAVRONA. Heating greenhouses with solar energy-new trends and developments．CIHEAM-Options Mediterraneennes vol. 31：119～134

[21] 方慧，杨其长，梁浩等．日光温室浅层土壤水媒蓄放热增温效果．农业工程学报，2011，27（5）：258～263

日光温室主动蓄放热方法的实践应用*

张　义[1,2]**，杨其长[1,2]***，方　慧[1,2]，李　文[1,2]，孙维拓[1,2]
（1. 中国农业科学院农业环境与可持续发展研究所，北京　100081；
2. 农业部设施农业节能与废弃物处理重点实验室，北京　100081）

摘要：日光温室的显著特征是利用北墙白天蓄热与夜晚放热，但这种依靠墙体自然蓄积的被动式蓄放热方式，由于材料传热特性的限制，有效蓄放热量有限，温室低温及冷害现象时有发生。针对这一问题，作者研究团队以最大化收集并储存到达墙体的热能，提高日光温室热能利用效率为目标，提出一种以水为媒介的日光温室主动式蓄放热方法，白天利用水媒循环不断将到达墙体表面的太阳辐射能吸收并蓄积起来，夜晚再通过水循环释放热量，变被动为主动，使热能蓄放效率成倍提升。试验结果表明：日光温室主动蓄放热方式能有效增加日光温室蓄热量，显著提高日光温室夜间温度；应用该原理研发的水幕帘主动蓄放热试验系统可将温室内夜间温度提高5.4℃以上，可将作物根际温度提高1.6℃以上；该系统夜间通过水幕帘的放热量达到4.9～5.6MJ/m²。日光温室主动蓄放热方法的研究成果对日光温室结构的改进、温度调控有较大的科学意义。

关键词：日光温室；液态介质；主动蓄热；蓄能；增温

Practical Application of active heat storage-release system in Chinese Solar Greenhouse

Zhang Yi，Yang Qichang，Fang Hui，Li Wen，Sun Weituo
（1. *Institute of Environment and Sustainable Development in Agriculture*，*Chinese Academy of Agricultural Sciences*，*Beijing* 100081，*China*；2. *Key Lab. of Energy Conservation and Waste Treatment of Agricultural Structures*，*Ministry of Agriculture*，*Beijing* 100081，*China*）

The heat for keeping temperature needed for plant growth in Chinese Solar Greenhouse at winter night is supplied by the north wall using passive heat-storage style. But it is a slow transfer process of the heat into the north wall by the day and from it at night，for the heat transfer characteristic of the materials for the north wall. The limited heat stored in the north wall couldn't meet the needs of the crops in winter time. Confronted with this weakness of passive heat-storage style of the north wall，my research will focus on a solar energy utilization system which using active heat storage-release style to get more heat in Chinese Solar Greenhouse. This new method will increase the amount of heat supplied in winter night and realize the heat release process controllable. The experiments did in winter show that：the air

* 基金项目：863 计划资助项目（2013AA102407）；国家自然科学基金资助项目（31071833）；公益性行业科研专项资助项目（201203002）

** 作者简介：张义（1981—），女，吉林，博士，主要从事设施园艺环境工程研究。中国农业科学院农业环境与可持续发展研究所，100081。E-mail：xingfu_ 536@163.com

*** 通讯作者：杨其长（1963—），男，安徽，博士，研究员，博士生导师，主要从事设施园艺环境工程研究。中国农业科学院农业环境与可持续发展研究所，100081。E-mail：yangq@ieda.org.cn

temperature was increased by over 5.4℃ and the soil temperature at crop rhizosphere was increased by over 1.6℃; the heat released from plastic water curtain at night was 4.9 ~ 5.6MJ/m^2; the extra heat collected by water curtain system has put the cherry tomatoes ahead by 20 days. The active heat storage-release method is useful for structure improvement and temperature control in Chinese Solar Greenhouse.

1 引言

中国式日光温室（CSG）是中国独有的一种温室结构型式，它是以日光为主要能量来源。白天，太阳光进入温室后，即以热量的形式存储在墙体和土壤中；夜晚，当室内气温下降时，墙体和土壤中蓄积的热量又源源不断地向温室供应，从而实现在中国北方地区冬季不用加温也能进行果菜类作物的生产[1~23]。

但在极端低温情况下，日光温室蓄积与释放的热量往往难以满足作物生产需求，低温冷害现象时有发生，多年来，相关学者在提高日光温室蓄放热量和夜间气温方面进行了大量探索。在提高日光温室北墙蓄热保温性能方面，佟国红等对不同材料组成的600mm厚墙体的传热特性进行了分析，结果表明：在相同室外气象条件下，复合异质墙体夜间温室空气温度比夯实土墙平均提高3.0℃[2]；马承伟等人建立了日光温室墙体传热过程模拟与墙体放热量的计算方法，通过计算分析了日光温室常用墙体材料的蓄放热性能，在北京地区（北纬40°，东经120°）室外气象条件下，日光温室墙体冬季夜间累计自然散热量为0.35 ~2.5MJ/m^2[4]。针对日光温室北墙蓄热能力不足的问题，众多学者在提高日光温室蓄热能力方面进行了多年探索。张海莲等利用太阳能集热器实现了温室气温的提升，但由于没有辅助加热设备，连阴天难以使用[5]；陈威等研究了温室蓄热层的传热与流动特征，并对温室土壤或岩床吸收与贮存太阳能的性能进行了分析[6]；毛罕平等设计了温室太阳能加热系统，系统由集热、蓄热、供热和辅助热源等四部分组成，试验表明太阳能加热温室与热水锅炉系统相比更加经济可行[7]；王奉钦等设计了一套太阳能集热器辅助加热系统，在室外平均温度为 -1℃的情况下，系统能提高根区土壤温度2.2℃[8]；王顺生等在日光温室后墙内侧安装太阳辐射集热调温装置，并利用蓄积热量加热土壤，提高地温3.2 ~3.8℃[9]；李炳海等采用太阳能地热加温系统，使日光温室15cm深土温在晴天时平均比不加温的对照区提高2.9℃，阴天提高2.6℃，最低土温由11.0℃提高到13.9℃[10]。

大量的前期试验研究表明，我国北方地区到达日光温室北墙的太阳辐射能最大可达400 ~500W/m^2，但仅有20% ~30%的能量能被墙体有效蓄积与释放，这种墙体被动式蓄热方式常常不能满足温室的增温需求。为此，作者所在研究团队以最大化收集并储存到达墙体的热能，提高热能利用效率为目标，以太阳能光热转换原理为基础，提出了一种以流体为媒介的日光温室主动式蓄放热思想，白天利用流体循环不断将到达墙体表面的太阳辐射能吸收并蓄积起来，夜晚再通过流体的循环释放热量，变被动为主动，使热能蓄积释放效率成倍提升，可显著提高日光温室冬季夜晚温度。

2 日光温室主动蓄放热机理

日光温室主动蓄放热，是以太阳能为能量来源、以流动的液体为热量吸收与释放介质的一种热量蓄积与释放方式。白天，利用流动的液体介质不断将吸热层得到的太阳能带入储热液体容器中，使储热容器中的液体温度不断升高，用于热能的储存；夜晚，当温室温度较低时，通过液体介质的流动不断将储热容器中的热能释放出来，用于温室低温时段的加温，实现热量在空间、时间上的转移。日光温室主动蓄放热机理示意图如图 1 所示。

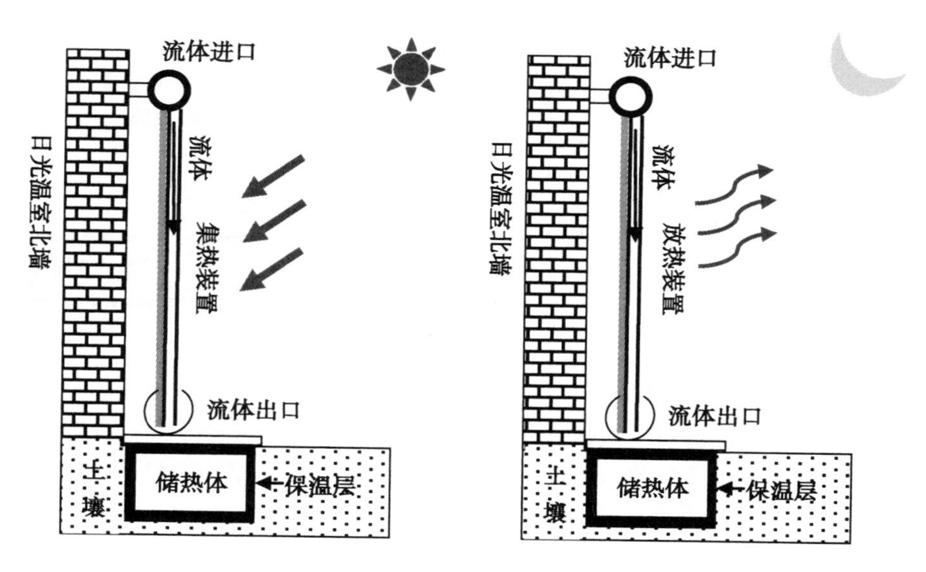

图 1　日光温室主动蓄放热机理示意图

3　实验验证

3.1　主动蓄放热实验系统构建

根据日光温室主动蓄热的机理，笔者设计了一套水幕帘主动蓄放热试验系统（图 2）。该系统的集热部分为水幕帘，其由 3 层薄膜组成的密闭空间，外层及内层为透明塑料薄膜，中间层为黑色薄膜，水沿着黑色薄膜流下，形成水流均匀的水幕，白天通过水流循环吸收水幕帘表面的太阳辐射热量；同时，该水幕帘也是夜晚热量释放的装置，夜晚当温室内气温降低到一定程度时，开启循环水泵，通过水幕帘将水池中的热量释放到温室中。该装置的蓄热部分为带有保温层的水箱，水箱白天储存热量，夜晚释放热量。

实验温室位于中国农业科学院农业环境与可持续发展研究所通州设施农业示范基地内，温室东西走向，长度 80m，跨度 8m，脊高 3. 8m，后墙高 2. 8m，采用全钢装配式骨架结构，温室北墙、后坡、山墙为拼接式聚苯乙烯泡沫板，表面喷涂抗裂砂浆。将温室从东向西方向等距划分为 4 个区域，中间用双层塑料薄膜（PE 膜）进行分隔，从东向西依次称为一、二、三、四区，选择一区安装水幕帘蓄放热系统，作为试验区；三区作为试验对照区。

3.2　结果分析

2011 年 1 月，作者在北京市郊区对水幕帘主动蓄放热试验系统进行了试验研究，试验结果与分析如下。

3.2.1　温室内气温变化

温室试验区与对照区的气温变化如图 3 所示。1 月 24 ~ 25 日测试数据表明，白天试验

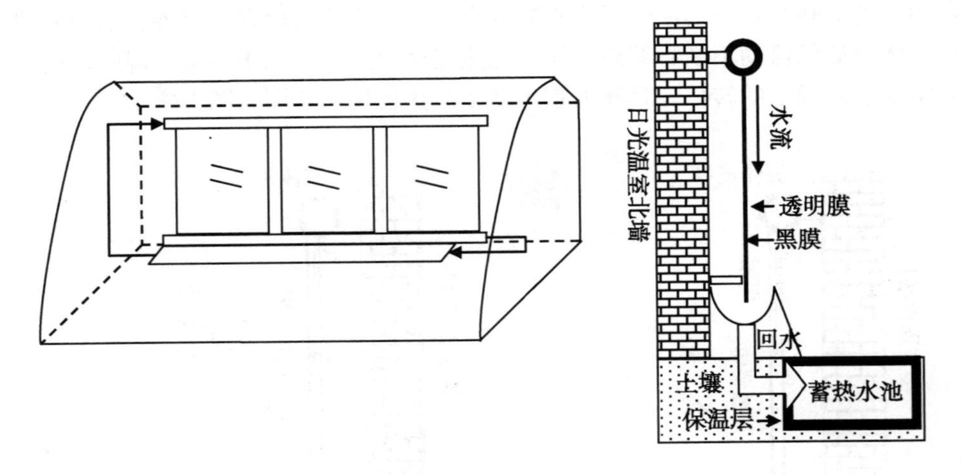

图2　日光温室水幕帘主动蓄放热试验系统

区的温度低于对照区，但夜间气温明显高于对照区，两区的平均气温差为3.5℃；当室外气温在凌晨6：00达到－10.3℃的低温时，试验区的气温仍维持在16.0℃以上，室内外温差达到26.3℃，主动蓄放热系统对温室的增温效果明显。

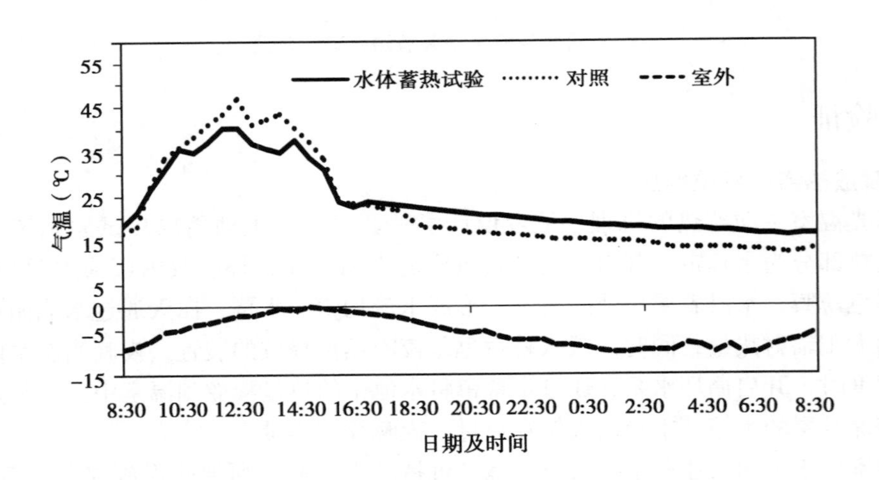

图3　试验区及对照区气温变化（2011年1月24～25日）

3.2.2　温室内地温分布

温室试验区的土壤温度测试表明（图5），土壤表面温度随太阳辐射强度的变化明显，越深层的土壤温度波动越小，在20cm深度以下，土壤温度几乎没有波动，相当于恒温层。试验中的恒温层深度要比普通日光温室的恒温层（60cm）要浅，这主要是因为试验区夜间仍有较高的气温，使得土壤的恒温层上移。深度40cm处，土壤温度可维持在19.1℃以上，深度50cm处，土壤温度可维持在18.1℃以上，较高的作物根际温度有利于作物的生长。

3.2.3　主动蓄放热系统的热量分析

试验中测试了主动蓄放热系统各部分的温度参数，测试结果如图5所示，在2011年1

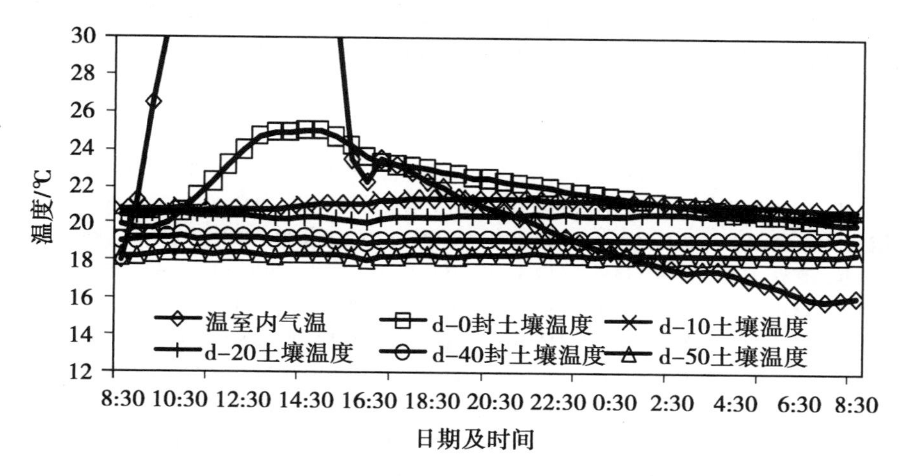

图4　试验区土壤温度纵向分布（2011年1月24~25日）

月24日的测试中，储热水箱在系统开启时的水温为24.6℃，结束时的水温为35.2℃，集热时间为6.5h。通过计算，该系统白天的蓄热量为168.2MJ，夜晚的放热量为178.4MJ，水幕帘的日平均集热功率为210.2W/m²。在夜间，水幕帘蓄放热装置通过水幕帘的放热量为4.46 MJ/m²，而一般日光温室夜间北墙自然散热量为0.35~2.5MJ/m²，可见该装置的散热量远远大于普通日光温室北墙的散热量。

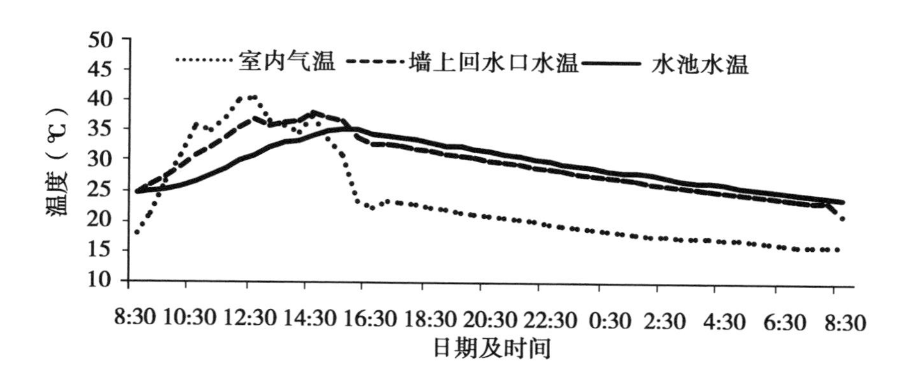

图5　水幕帘蓄放热系统中水循环系统及水池的温度参数变化

4　结论与建议

在日光温室中采用以太阳能为能量来源，以液体为介质进行热量的蓄积与释放的主动蓄放热方法是可行的。白天利用介质循环集热装置吸收太阳能，同时将能量储存在一定容积的液体中，夜晚通过介质循环放热装置释放热量，实现了热量在空间、时间上的转移。

日光温室主动蓄放热系统白天的蓄热量为168.2MJ，夜晚的放热量为178.4 MJ，日平均集热功率为210.2W/m²；夜间通过水幕帘的放热量为4.46 MJ/m²，大大高于普通日光温室夜间北墙的散热量0.35~2.5MJ/m²。

日光温室主动蓄放热方式大大增加了日光温室的蓄热量，为日光温室北墙结构及功能的改进提供了一种新的思路，即北墙仅具有较好保温性能即可，蓄放热功能由主动蓄放热系统完成。但由于主动蓄放热方法刚刚提出，其机理还有待进一步完善。

参考文献

[1] 陈端生. 中国节能型日光温室建筑与环境研究进展. 农业工程学报，1994，10（1）：123～129
[2] 佟国红，王铁良，白义奎等. 日光温室墙体传热特性的研究. 农业工程学报，2003，19（3）：186～189
[3] 王静，崔庆法，林茂兹. 不同结构日光温室光环境及补光研究. 农业工程学报，2002，18（4）：86～89
[4] 马承伟，韩静静，李睿. 日光温室热环境模拟预测软件研究开发. 北方园艺，2010（15）：69～75
[5] 张海莲，熊培桂，赵利敏等. 温室地下蓄集太阳热能的效果研究. 西北农业学报，1997，6（1）：54～57
[6] 陈威，刘伟，黄素逸. 温室及其蓄热层中传热与流动的研究. 工程热物理学报，2003，24（3）：508～510
[7] 毛罕平，王晓宁，王多辉. 温室太阳能加热系统的设计与试验研究. 太阳能学报，2004，25（3）：305～309
[8] 王奉钦. 太阳能集热器辅助提高日光温室地温的应用研究. 北京：中国农业大学，2004
[9] 王顺生，马承伟，柴立龙等. 日光温室内置式太阳能集热调温装置试验研究. 农机化研究，2007（3）：130～133
[10] 李炳海，须晖，李天来. 日光温室太阳能地热加温系统应用效果研究. 沈阳农业大学学报，2009，40（2）：152～155

三种主动蓄放热系统集热功率的比较分析*

卢　威[1,2]**，杨其长[1,2]***，张　义[1,2]，方　慧[1,2]，李　文[1,2]，孙维拓[1,2]，辛　敏[1,2]

（1. 中国农业科学院农业环境与可持续发展研究所，北京　100081；
2. 农业部设施农业节能与废弃物处理重点实验室，北京　100081）

摘要：针对日光温室冬季夜晚温度低、作物易发生冷害等问题，本研究团队基于日光温室主动蓄放热思想，结合日光温室特殊的光温环境特点，设计了 3 种以水为储热媒介的主动蓄放热系统。系统由集热装置、储热装置和控制装置等 3 部分组成。在试验室条件下测试了 3 种装置的集热功率，结果表明，透光双黑膜主动蓄放热系统与双黑膜主动蓄放热系统集热功率差异不大，分别为 247.47W/m^2 和 246.75W/m^2，透光水幕帘主动蓄放热系统集热功率较小，为 201.58W/m^2。

关键词：主动蓄放热；太阳能；蓄热量；集热效率

Study on the Heat Collecting Power of three Active Heat Storage and Release Systems

Lu Wei[1,2], Yang Qichang[1,2], Zhang Yi[1,2], Fang Hui[1,2], Li Wen[1,2], Sun Weituo[1,2], Xin Min[1,2]

(1. *Institute of Environment and Sustainable Development in Agriculture*, *Chinese Academy of Agricultural Sciences*, *Beijing* 100081, *China*; 2. *Key Lab of Energy Conservation and Waste Treatment of Agricultural Structures*, *Ministry of Agriculture*, *Beijing* 100081, *China*)

Abstract: In order to raise soil and air temperature in Chinese solar greenhouse, according to the features of strong solar radiation and high temperature in Chinese solar greenhouse in daytime, based on the theory of active heat storage and release, three automatic active heat storage and release systems have been designed, accomplishing the design selection of solar-thermal system, heat storage system and controlling system. Furthermore, three systems have been tested in the laboratory at the same working conditions. The results showed that the heat collecting power of the systems have not significant difference between the translucent double black plastic films system and the double black plastic films system, they are 247.47W/m^2 and 246.75W/m^2, followed by the heat collecting power of the translucent water curtain system, which can reach 201.58W/m^2.

Key words: Active heat storage and release; Solar energy; Heat storage; Heat collecting power

日光温室是中国特有的一种温室形式，其能量的主要来源是太阳辐射能。日光温室白天利用其后墙、山墙和土壤等自身结构储存热量，夜间当室内温度下降时，又通过温室蓄热结

* 基金项目：863 计划课题经费资助（2013AA102407）；国家自然科学基金资助项目（31071833）；公益性行业（农业）科研专项经费资助（201203002）

** 作者简介：卢　威（1989—），男，硕士研究生，主要从事设施园艺环境工程研究。中国农业科学院农业环境与可持续发展研究所，100081

*** 通讯作者：杨其长（1963—），男，博士，研究员，博士生导师，主要从事设施园艺环境工程研究。中国农业科学院农业环境与可持续发展研究所，100081。E-mail：yangq@ieda.org.cn

构源源不断地自然释放，提高室内空气温度[1~4]。

由于墙体等结构的蓄放热能力有限，夜间容易出现低温现象，难以满足作物高效生产的温度需求，所以部分地区的温室盲目增加后墙厚度，造成土地资源严重浪费。针对此问题，本研究团队提出了日光温室主动蓄放热思想，根据这一研究思路，本课题组张义等以水为蓄热介质进行热量的蓄积与释放，设计了一种水幕帘系统，可将温室内夜间空气温度提高5.4℃以上[5]。方慧等以水为介质，以太阳能为热源，以温室浅层土壤为蓄热体，白天通过水的循环将热量收集并储存到温室浅层土壤中，夜间通过土壤的自然放热将热量释放到温室中，既提高了空气温度，也提高了作物根部土壤温度[6]。随后方慧等又在日光温室内试验了一种基于热泵的浅层土壤水媒蓄放热装置，试验结果表明，该装置在阴天系统系数（Coefficient of performance，COP）能达到3以上，与燃煤热水锅炉相比节能33%，可将温室空气温度和土壤温度提高3.2℃和3.3℃，开启热泵机组后，可将空气温度和土壤温度提高5.7℃和2.9℃[7]。

本团队研发了多套主动蓄放热系统，但鉴于日光温室现场试验的条件限制，无法同时开展对比试验，本文作者在前人研究基础上，在试验室中对3种以水为储热媒介的主动蓄放热系统进行了性能测试。

1 材料与方法

1.1 试验室概况

该试验于中国农业科学院农业环境与可持续发展研究所设施农业工程中心试验室内进行，试验室长8m，跨度6.4m，脊高3.8m，檐高2.7m。试验测试时间为2013年3月10~27日，选取3月18日典型晴天的数据进行分析。

1.2 主动蓄放热系统结构及工作原理

主动蓄放热系统结构由集热装置、储热装置、控制装置等3个部分组成。

三种系统的区别在集热装置，其储热装置和控制装置完全相同。三种系统的集热装置分别为：双黑膜主动蓄放热系统的集热装置由2层黑色塑料薄膜组成，水沿黑色薄膜流下，形成水流均匀的水幕；透光水幕帘主动蓄放热系统的集热装置由3层薄膜组成，内、外层均为透明塑料薄膜，中间层为黑色薄膜，水流沿中间层黑色薄膜流下；透光双黑膜主动蓄放热系统的集热装置也由3层薄膜组成，外层为透明塑料薄膜，内层和中间层为黑色薄膜，水沿内层黑色薄膜流下，形成水流均匀的水幕。储热装置为蓄热水箱，由聚酯硬质板焊接而成，其四周外表面设置聚苯乙烯泡沫板保温层（厚度10mm），水箱体积0.27m^3；控制装置由水泵、进回水管路和控制器组成，水泵选用单相潜水电泵，功率0.37kW，循环水量0.05m^3/s。主动蓄放热系统的主要参数如表1所示。

1.3 试验测试方案

试验室内水平安装3种主动蓄放热系统，每种系统放置2个集热单元。室内空气温度测点布置在试验室几何中心点，距地面高度1m，室外空气温度测点距地面高度也为1m。水温测点分别布置在3个水池的几何中心。室内太阳辐射照度测点布置在集热装置外表面，距地面高度2.8m，室外太阳辐射照度测点水平布置，距地面高度也为2.8m。采用铜-康铜热电偶作为温度传感器，传感器探头做防辐射处理，水中传感器探头做防锈处理，使用数据采集仪（CR1000）进行数据采集和存储，数据采集时间间隔为30min。试验期间，8∶30开启水泵，16：30 关闭水泵。

表 1　主动蓄放热系统的主要参数

Table 1　Key parameters of Active Heat Storage and release system

参数	单位	数值
集热板长	单元/m	2.4
集热板高	单元/m	1.8
保温隔热板厚	m	0.03
给水管直径	m	0.05
带孔管道直径	m	0.032
回水管直径	m	0.16
带孔管道小孔孔径	m	0.0032
保温板导热系数	w/m. k	0.030
透光层光透过率	%	87
水流速	m^3/s	0.05
蓄热水池容积	m^3	0.27

2　结果与分析

2.1　测试期间室内外环境

2.1.1　室内外气温的变化

测试期间室内气温与室外气温变化如图 1 所示。在 2013 年 3 月 18 日连续 8h 的测试期间，室内外气温均呈先升高后降低的趋势。室内最高气温为 40.91℃，最低气温为 17.01℃；室外最高气温为 17.02℃，最低气温为 6.26℃。

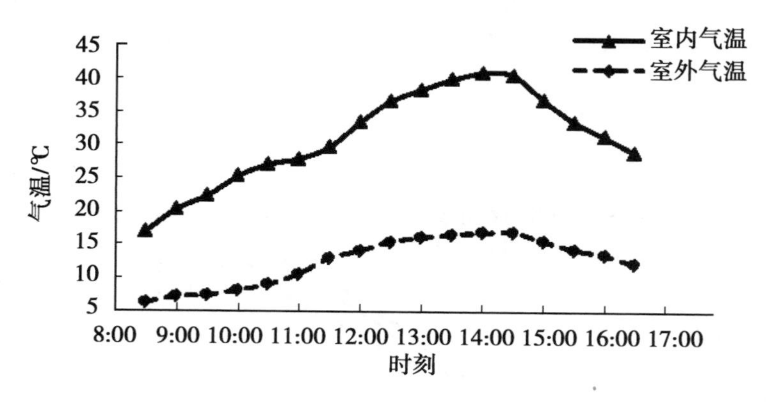

图 1　试验温室内外的气温变化

Figure 1　Air temperature curves inside/outside test greenhouse

2.1.2　室内外太阳辐射照度的变化

测试期间室内太阳辐射照度与室外太阳辐射照度变化如图 2 所示。在 2013 年 3 月 18 日连续 8h 的测试期间，室内外太阳辐射照度均呈先升高后降低的趋势。室内最大太阳辐射照度为 284.2W · m^{-2}，最小太阳辐射照度为 69.42W · m^{-2}；室外最大太阳辐射照度为 952W · m^{-2}，最小太阳辐射照度为 332W · m^{-2}。

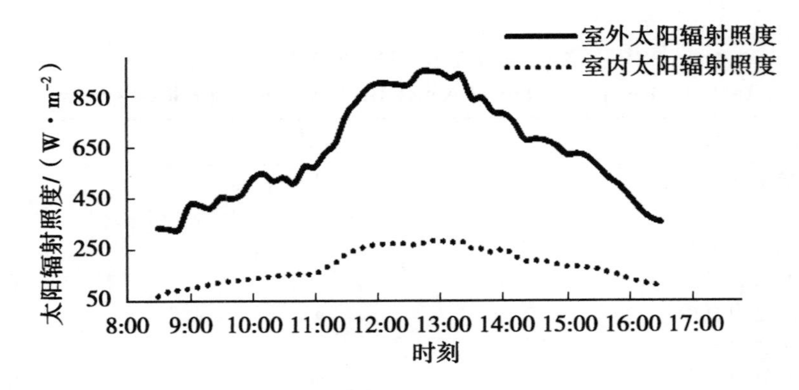

图 2　试验温室内外太阳辐射照度的变化

Figure 2　Solar radiation curves inside/outside test greenhouse

2.2　主动蓄放热装置性能分析

2.2.1　蓄水池水温的变化

3 种主动蓄放热系统的蓄水池水温变化如图 3 所示。测试期间，水池水温呈先上升后下降的趋势，在 15∶30 达到最大值，随后开始下降，说明在 15∶30 之前主动蓄放热装置为吸热量大于放热量，在 15∶30 之后由于室内空气温度较低，太阳辐射照度较小，主动蓄放热装置放热量大于吸热量。双黑膜主动蓄放热系统、透光水幕帘主动蓄放热系统和透光双黑膜主动蓄放热系统的水温的最大值分别为 44.18℃、40.02℃和 43.45℃。相同时间段内，透光双黑膜主动蓄放热系统的蓄水池温升最大，为 27.15℃，双黑膜主动蓄放热系统的蓄水池温升次之，为 27.07℃，透光水幕帘主动蓄放热系统的蓄水池温升最小，为 22.12℃。试验中关闭水泵的时间为 16∶30，实际 15∶30 后水温不再上升，水泵继续运行，系统放热量大于吸热量，造成能量的浪费。

2.2.2　蓄热量与集热功率计算与分析

试验测试了 3 个水池的水温，系统蓄热时间 8h，蓄热面积为 $4.32m^2$，通过公式（1）

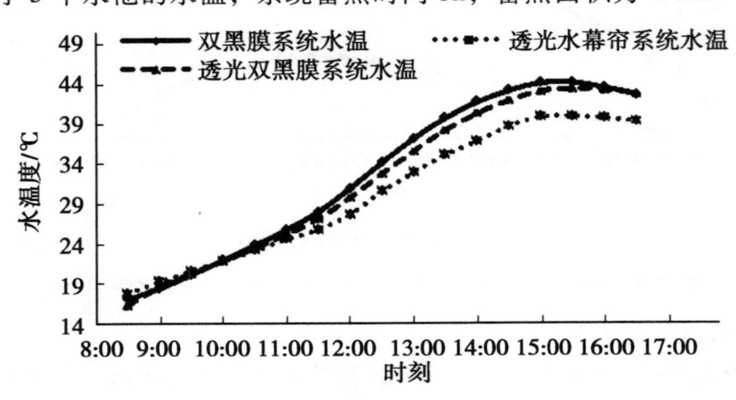

图 3　水池内水温随时间的变化

Figure 3　Water temperature curves in different pools with time

可以分别计算 3 种系统的蓄热量和集热功率，结果如表 2 所示。

$Q_w = C_w \rho_w V_w \Delta t$ （1）

式中，Q_w 为系统蓄热量，即水池储热量，J；C_w 为水的比热容，4.2×10^3J/（kg·℃）；ρ_w 为水的密度，1 000kg/m^3；V_w 为水池容积，m^3；Δt 为水池温升，℃。

表 2 主动蓄放热系统蓄热量及集热功率的比较

Table. 2 Comparison on the heat storage and the average heat collecting power of three systems

主动蓄放热系统	蓄热量/MJ	蓄热面积/m^2	蓄热时间/h	系统平均集热功率/W·m^{-2}
双黑膜	30.70	4.32	8	246.75
透光水幕帘	25.08	4.32	8	201.58
透光双黑膜	30.79	4.32	8	247.47

结果表明，透光双黑膜主动蓄放热系统与双黑膜主动蓄放热系统的蓄热量和平均集热功率差异不大，其蓄热量分别为 30.79MJ、30.70MJ，其平均集热功率分别为 247.47W/m^2、246.75W/m^2，透光水幕帘主动蓄放热系统的蓄热量和平均集热功率较小，分别为 25.08MJ、201.58W/m^2。

3 结论与讨论

通过本试验测试，可以得出以下结论。

第一，透光双黑膜主动蓄放热系统与双黑膜主动蓄放热系统的蓄热量和平均集热功率差异不大，其蓄热量分别为 30.79MJ、30.70MJ，其平均集热功率分别为 247.47W/m^2、246.75W/m^2，透光水幕帘主动蓄放热系统的蓄热量和平均集热功率较小，分别为 25.08MJ、201.58W/m^2。

第二，本试验测试期间，系统水温在 15：30 达到最大值，但系统水泵关闭时间为 16：30，在此时间段，系统放热量大于蓄热量，造成电能和热能的浪费，因此在实际应用中，应根据水温变化自动关闭水泵，这样相比人工设置节省劳动力，同时能提高系统热能利用效率。

通过优化系统结构参数，改进工程装配工艺，进一步提高三种主动蓄放热系统的性能，结合日光温室环境特点，制定更加科学的控制管理方案，预计其在日光温室生产中有广阔的应用前景。

参考文献

[1] 陈端生．中国节能型日光温室建筑与环境研究进展．农业工程学报，1994，10（1）：123～128

[2] Bot G P A. A validated dynamical model of greenhouse climate. Acta Hort，1980，106：149～158

[3] Horiguch I. The variation of heating load coefficient for the greenhouse. Acta Hort，1978，87：95～101

[4] Willits L H，Chandra P，Peet M M. Modeling solar energy storage system for greenhouses. J Agri Engng Res，1985，32（1）：73～93

[5] 张义，杨其长，方慧．日光温室水幕帘蓄放热系统增温效应试验研究．农业工程学报，2012，28（4）：188～193

[6] 方慧，杨其长，梁浩等．日光温室浅层土壤水媒蓄放热增温效果．农业工程学报，2010，27（5）：258～263

[7] 方慧，杨其长，张义．基于热泵的日光温室浅层土壤水媒蓄放热装置试验．农业工程学报，2012，28（20）：210～216

苏北日光温室分类及应用效果分析*

王军伟**，王　健，孙　锦，郭世荣***

（南京农业大学园艺学院，南京　210095；农业部南方蔬菜遗传改良重点开放实验室，南京　210095；南京农业大学（宿迁）设施园艺研究院，江苏宿迁　223800）

摘要：近几年苏北地区日光温室发展迅速，为了提出适合苏北地区地理位置和气候条件的日光温室类型和种植模式，因此对当地现有日光温室类型和应用现状进行调查具有重要意义。本文抽样调查了苏北地区 15 个乡镇共 98 栋日光温室，调查内容包括日光温室的的结构和材料参数、使用现状、土地利用率以及温室用途、主栽作物、茬口安排和年经济效益等。对日光温室的结构和材料参数按前屋面形状、后墙材料、前室立柱以及前屋面拱架材料进行分类，并从每一类型的日光温室中选择 1 栋作为使用现状和经济效益的调查对象。结果显示，苏北日光温室可分为三大类 19 种不同的构型。此外，苏北日光温室建设规模较大、特点突出、经济效益较好，但同时存在温室结构不合理、温室配套设备落后、经营理念滞后、作物栽培种类和茬口安排缺乏区域内的统筹布置等问题。在今后的发展中，需通过优化设施构型、改善经营思路、形成区域调控机制以及挖掘当地特有资源以提高经营者的经济效益。

关键词：日光温室；结构类型；经济效益

The Classification and Application effect of Solar Greenhouse in Northern of Jiangsu Province

Wang Junwei，Wang Jian，Sun Jin，Guo Shirong

(*College of Horticulture*, *Nanjing Agricultural University*, *Nanjing* 210095, *China*; *Key Laboratory of Southern Vegetable Crop Genetic Improvement*, *Ministry of Agriculture*, *Nanjing* 210095, *China*; *Nanjing Agricultural University* (*Suqian*) *Academy of Protected Horticulture*, *Jiangsu Suqian* 223800, *China*)

Abstract: Solar-greenhouse has developed rapidly nearly years in northern of Jiangsu province. In order to put forward suitable solar-greenhouse types and planting patterns for geographical location and climate conditions, it is significant to investigate the types and application of the existing greenhouse. This paper took 15 towns for sample to survey which had 98 solar-greenhouses in total. The survey content includes structures and material parameters, the application, land utilization rate of solar-greenhouse and function, crops, rotation and annual economic benefit, etc. According to the shapes of front roof, materials of rear wall, post of front room and truss material of front roof, these greenhouses were classified. For each type, one solar-greenhouse was chosen as the object for application and economic benefit investigation. The result showed that solar greenhouse could be divided into three main categories, in-

* 基金项目：现代农业产业技术体系建设专项资金资助〔CARS－25－C－03〕，江苏省农业三项工程项目〔SXGC〔2012〕391〕

** 作者简介：王军伟，博士研究生，主要从事设施园艺工程研究，E-mail：p1100p@126.com

*** 通讯作者（Corresponding author）：郭世荣，教授，博士生导师，主要从事设施园艺与无土栽培研究，E-mail：srguo@njau.edu.cn

cluding 19 different configurations. The solar-greenhouse had a huge scope, outstanding feature, better economic benefit in northern of Jiangsu province. However, some problems were also being, such as the ill-adapted of greenhouse types to the the local climate, the laggard equipment and management idea. Especially, the regional macro-control to crop cultivation and rotation arrangement was lacked. In order to improve the economic benefits, measures should be took to optimize facilities configuration, improve management idea, form the regional regulatory mechanism and explore local resources.

Key words: Solar greenhouse; Structure type; Economic benefits

20 世纪 80 年代以来，日光温室在我国北方地区迅速发展，有效地解决了我国北方地区冬季蔬菜供应问题，大幅度增加了农民的收入，带动了农业结构的快速调整。截至 2010 年，我国日光温室面积达 78.34 万 hm^2，占园艺设施总面积的 22.75% 左右[1]。日光温室是我国特有的园艺保护设施，其建造和运行成本低，合乎我国国情，适合我国经济发展和低碳、节能的需要，而且伴随着能源的短缺，日光温室将成为今后我国大面积温室园艺产业发展的必然选择[2]。

近几年苏北地区日光温室发展迅速，已成为促进农业增效、农民增收新的增长点[3]。此外，苏北地区地处我国北方地区南沿，冬季光热资源丰富，风荷载、雪荷载偏小，更有利于发展日光温室进行园艺作物的秋延后、越冬和春提早栽培。鉴于苏北地区在日光温室发展方面的成就，本研究拟对苏北日光温室建设较为集中、且整体水平相对较高的地区展开全面调查，总体上把握苏北地区日光温室的基本类型、使用现状和经济效益，并从中总结成功的经验和存在的问题，以期为苏北地区日光温室未来的发展方向和模式提供参考。

1 调研区域概况

苏北地区在行政区划上包括徐州市、连云港市、宿迁市、淮安市和盐城市，下辖 5 市 19 县 15 区，总面积 5.42 万 km^2，总人口 3 212 万。该地区纬度跨度为 N32.72° ~ N34.97°，属暖温带半湿润季风气候，四季分明；常年气温约 14℃，1 月为最冷月，7 月为最热月；年日照时数为 2 100 ~ 2 500h，太阳年辐射总量为 459.8 ~ 501.6kJ · cm^{-2}；年均降水量 900 ~ 1 300mm，年均无霜期 200 ~ 220d。气候资源较为优越，有利于作物生长。

2 调研内容与方法

2.1 苏北日光温室基本类型及分类

基本构型调查主要对苏北现有日光温室进行实体测量，测量项目主要包括日光温室的“五度”、“四比” 和 “三材” 等结构和材料参数，测量方法是对同一类型的日光温室随机选取 3 栋进行测量，之后取测量平均值作为该类型日光温室的结构和材料参数。在此基础上，对日光温室的结构和材料参数按前屋面形状、后墙材料、前室立柱以及前屋面拱架材料进行分类。

2.2 苏北日光温室使用现状

在对苏北日光温室进行分类之后，从每一类日光温室中选取有代表性的 1 栋作为使用现状的调查对象，调查内容包括温室的建造年份、建造地点、单位造价、卷帘设备、环境监测设备、灌溉方式、栽培方式、经营方式等，并通过实地测量其栽培面积和占地面积计算出该温室的土地利用率。

2.3 苏北日光温室茬口安排和经济效益

在对苏北日光温室进行分类之后，从每一类日光温室中选取有代表性的1栋作为茬口安排和经济效益的调查对象，调查内容包括日光温室的经营用途、茬口安排、主栽作物种类、固定资产年折旧费用、主栽作物的直接投入、作物产量、作物产值以及附加收入，最后计算出日光温室的年收益。其中，日光温室固定资产包括日光温室框架结构、覆盖材料和保温材料等；主栽作物的直接投入包括种子种苗、农药化肥、劳动力及燃料动力等；附加收入主要包括园区观光、产品采摘等。

3 调研结果

3.1 苏北日光温室基本类型及分类

先后调查了15个乡镇共98栋日光温室，按前屋面形状、后墙材料、前室立柱以及前屋面材料的不同对日光温室进行分类，共分为三大类19种不同的日光温室构型，如表1所示。在这些基本类型中，部分日光温室具有特殊结构，例如在传统日光温室的北侧，增加了一个同长度但采光面朝北的单屋面温室，两者共用一堵后墙，称为“阴阳型”日光温室；日光温室室内地面下挖，称为下沉式日光温室。

表1 苏北日光温室基本类型

Table 1 Structure types of solar greenhouse in Northern of Jiangsu province

编号	主要类型	后墙结构及材料	前室立柱及材料	前屋面结构及材料	后坡结构及材料	其他特殊结构
SR-1	拱圆型日	红砖墙及空	无立柱	镀锌钢管桁架	草苫、保温被、水泥板+	—
SR-2	光温室	心砌体墙	无立柱	镀锌钢管单管	草苫、木板、薄膜、水泥	—
SR-3			无立柱	镀锌钢管桁架和镀锌钢管单管	板+聚苯板	—
SR-4			水泥立柱	钢竹混合拱架	无后坡	“阴阳型”
SR-5			水泥立柱	钢竹混合拱架	薄膜、草苫、水泥板、无	—
SR-6			水泥立柱	水泥拱架	后坡	—
SR-7			水泥立柱	泡沫混凝土拱架		—
SR-8			水泥立柱	单层竹竿拱架		—
SR-9		机打土墙	无立柱	镀锌钢管桁架和镀锌钢管单管	草苫	—
SR-10			无立柱	钢竹混合拱架	水泥板+草苫	下沉75cm
SR-11			水泥立柱	钢竹混合拱架	草苫、无后坡	—
SR-12			水泥立柱	单层竹竿拱架	薄膜、水泥板、保温被	下沉100cm
SR-13	一立一坡式日光温室	空心砌块墙	水泥立柱	单层竹竿拱架	草苫	—

（续表）

编号	主要类型	后墙结构及材料	前室立柱及材料	前屋面结构及材料	后坡结构及材料	其他特殊结构
SR-14	二折式日光温室	秸秆后墙	水泥立柱	单层竹竿拱架	秸秆	—
SR-15		无后墙	水泥立柱	单层竹竿拱架	无	—
SR-16		空心砌块墙	无立柱	镀锌钢管桁架	薄膜	—
SR-17		机打土墙	水泥立柱	钢竹混合拱架	水泥板＋草苫	—
SR-18			无立柱	镀锌钢管桁架和镀锌钢管单管	水泥板＋草苫	下沉50cm
SR-19			水泥立柱	单层竹竿拱架	草苫	—

3.2 苏北日光温室使用状况

通过对苏北日光温室使用状况调查发现，该地区日光温室的建造年份集中于2004年之后，由于建筑材料、配套设备以及建造年份的差异，温室的单位造价差异较大，如用于园区观光及示范推广的日光温室造价偏高，这主要是该类温室对结构强度和配套设备要求较高，而农户建造的日光温室结构简单、选材便宜、基本不具有配套设备，而造价往往较低，如表

表2　苏北日光温室使用状况

Table 2　Application of solar greenhouse in Northern of Jiangsu province

温室编号	建造年份	建造地点	单位造价（元/m^2）	卷帘设备	环境监测设备	灌溉方式	栽培方式	经营方式	土地利用率（%）
SR-1	2009	铜山	125.6	机械卷帘	小型监测仪	膜下滴灌	基质槽培	公司经营	43.76
SR-2	2008	沭阳	93.3	机械卷帘	温湿度计	膜下滴灌	土壤栽培	个体经营	49.15
SR-3	2008	淮阴	102.3	机械卷帘	温湿度计	膜下滴灌	土壤栽培	个体经营	45.65
SR-4	2010	贾汪	51.8	机械卷帘	温湿度计	肥水漫灌	土壤栽培	个体经营	61.81
SR-5	2008	铜山	50.8	机械卷帘	无	肥水漫灌	土壤栽培	个体经营	57.20
SR-6	2010	贾汪	63.7	机械卷帘	无	膜下滴灌	土壤栽培	个体经营	49.38
SR-7	2010	铜山	64.4	机械卷帘	温湿度计	膜下滴灌	土壤栽培	个体经营	48.90
SR-8	2008	睢宁	72.1	机械卷帘	温湿度计	膜下滴灌	基质槽培	公司经营	53.13
SR-9	2005	东海	45.0	机械卷帘	小型监测仪	肥水漫灌	土壤栽培	个体经营	48.13
SR-10	2009	铜山	59.3	机械卷帘	膜下滴灌	膜下滴灌	土壤栽培	公司经营	40.66
SR-11	2005	丰县	41.5	机械卷帘	温湿度计	肥水漫灌	土壤栽培	个体经营	38.79
SR-12	2006	铜山	38.5	机械卷帘	温湿度计	肥水漫灌	土壤栽培	个体经营	39.73
SR-13	2004	沭城	43.8	机械卷帘	温湿度计	肥水漫灌	土壤栽培	个体经营	46.35
SR-14	2010	铜山	24.8	人工卷帘	无	肥水漫灌	土壤栽培	个体经营	49.72
SR-15	2010	铜山	18.5	人工卷帘	无	肥水漫灌	土壤栽培	个体经营	48.54
SR-16	2009	铜山	63.8	机械卷帘	温湿度计	膜下滴灌	土壤栽培	个体经营	49.38
SR-17	2009	灌云	78.8	机械卷帘	温湿度计	膜下滴灌	土壤栽培	个体经营	44.71
SR-18	2004	赣榆	49.9	机械卷帘	温湿度计	肥水漫灌	土壤栽培	个体经营	37.35
SR-19	2005	射阳	33.8	机械卷帘	温湿度计	肥水漫灌	土壤栽培	个体经营	42.97

注：土地利用率是指温室的栽培面积占温室总占地面积的百分比

2 所示。两栋日光温室之间存在较大的温室栋间距，造成了日光温室的土地利用率不足 50%，尤其是厚土墙日光温室，而 SR-4 型日光温室充分利用温室栋间区域，土地利用率较高（61.81%）。机械卷帘设备在日光温室上的使用，大大降低了劳动者的劳动强度，同时增加了温室的采光时间，该设备已在苏北温室生产中普及应用，但是 SR-14 和 SR-15 两种日光温室由于前屋面结构强度较差而不能使用机械卷帘设备。苏北日光温室常见的灌溉方式为膜下滴灌和肥水漫灌，作物的栽培方式和产品的经营方式主要取决于日光温室的功能定位，一般用于观光、示范推广的日光温室栽培方式常采用基质栽培，经营方式除产品批发出售外还可通过采摘、观光等方式获得附加收入。

3.3 苏北日光温室经济效益分析

苏北日光温室的茬口安排和经济效益如表 3 所示。在设施园艺发展的早期阶段，日光温室主要用于园艺作物的反季节栽培，而目前日光温室除进行作物生产之外，还可以进行功能拓展，如温室科技示范、产品采摘观光、园区休闲观光等，这些社会福利性设施园艺可提供附加收入以提高经营者的经济效益。在作物茬口安排方面，种植者主要采用一年两茬、三茬或长季节栽培。长季节栽培一般是葡萄和草莓的周年生产或避夏生产；两茬栽培主要是春提早栽培和秋延后-越冬栽培，而夏季高温季节常采用闷棚休整；三茬栽培一般为春提早、秋延后和越冬栽培，温室周年不间断使用。此外，部分经营者采用套种方式以增加温室内土地利用率。从栽培作物来看，苏北日光温室常见的栽培作物有黄瓜、番茄、辣椒、花椰菜、草莓、葡萄等，常见的栽培方式是 2～3 种作物轮作或单种作物连作；从经济效益来看，苏北日光温室的年收益一般在 15 000～25 000元/667m²，示范型日光温室可获得观光收入，“阴阳型”日光温室通过养殖家禽增加收益，采摘型日光温室可通过产品采摘获得额外附加值，如表 3 所示。

表 3 苏北日光温室经济效益表

Table 3 Economic benefits of solar greenhouse in Northern of Jiangsu province

温室类型	温室用途	温室作物栽培面积（m²）	茬口安排	主栽作物种类	固定资产年折旧费用元/（年·室）	主栽作物直接投入（元/室）	产量（kg/室）	产值（元/室）	设施及作物附加收入元/（年·室）	年收益（元/667m²）
SR-1	蔬菜生产及观赏	294.0	3～6 月	黄瓜	10 048	3 100	3 000	7 500	10 000	47 988
			7～9 月	长茄		3 900	4 000	20 000		
			10～2 月	甘蓝		4 200	3 000	4 900		
SR-2	蔬菜生产	654.5	4～10 月	黄瓜	8 036	1 320	5 800	16 000	0	22 261
			10～3 月	苦瓜		1 500	5 000	16 700	0	
SR-3	蔬菜生产	448.5	8～2 月	西红柿	6 232	1 350	4 000	12 000	0	21 145
			3～6 月	黄瓜		1 200	4 000	11 000	0	
SR-4	蔬菜生产家禽养殖	771.4	7～12 月	番茄	11 019	3 600	6 500	18 000	10 000	23 762
			12～6 月	黄瓜		3 900	6 500	18 000	0	
SR-5	水果生产	826.0	周年生产	葡萄	8 475	8 000	4 000	36 000	0	15 766
SR-6	水果生产	539.0	周年生产	葡萄	9 486	6 000	3 500	30 000		17 961
SR-7	水果生产	477.4	周年生产	草莓	9 604	5 000	3 000	25 000	5 000	21 511

（续表）

温室类型	温室用途	温室作物栽培面积（m^2）	茬口安排	主栽作物种类	固定资产年折旧费用元/（年·室）	主栽作物直接投入（元/室）	产量（kg/室）	产值（元/室）	设施及作物附加收入元/（年·室）	年收益（元/667m^2）
SR-8	蔬菜生产及观光	754.4	10～5月	辣椒	9 402	5 200	60 00	20 000	8 000	24 029
			6～10月	黄瓜		4 220	6 500	18 000	0	
SR-9	蔬菜生产	1600.2	9～6月	水果黄瓜	10 702	3 200	13 000	48 000	0	14 213
SR-10	蔬菜生产及观光	846.0	12～5月	西葫芦	8 176	4 700	4 500	9 500	4 000	16 260
			7～12月	辣椒		5 000	7 500	25 000		
SR-11	蔬菜生产	891.8	3～8月	辣椒	10 332	2 650	8 000	25 000		22 750
			10～4月	黄瓜		3 600	8 500	22 000		
SR-12	蔬菜生产	776.0	4～10月	黄瓜	7 573	1 320	6 500	17 000	0	20 033
			10～3月	苦瓜		1 500	5 000	16 700	0	
SR-13	蔬菜生产	554.8	9～11，12～6月	黄瓜	7 555	3 570	9 500	27 000		19 084
SR-14	水果生产	623.2	8～6月	草莓	4 813	4 000	2 800	24 000		16 254
SR-15	铜山	448.0	8～6月	草莓	2 934	3 000	2 000	18 000		17 964
SR-16	蔬菜生产及采摘	539.0	6～10月	耐热花椰菜	6 382	19 60	2 000	3 600	5 000	23 633
			10～6月	草莓		4 160	2 500	28 000		
SR-17	蔬菜生产	476.0	3～7，9～2月	番茄	7 045	3 800	7 000	22 000	0	24 319
SR-18	蔬菜生产	585.0	9～2月	豇豆(套种)		1 300	1 500	7 500	0	
			2～6月	早春番茄	8 953	3 350	5 000	15 000	0	15 366
			6～10月	耐热花椰菜		1 950	1 900	2 880	0	
SR-19	蔬菜生产	890.0	10～2月	秋延后辣椒		3 150	4 500	13 000	0	
			2～6月	早春番茄	8 213	3 650	8 000	21 000	0	19 273
			6～10月	耐热花椰菜		1 950	1 900	2 880	0	
			10～2月	秋延后辣椒		2 350	65 000	18 000	0	

注：表中的年收益是指日光温室内单位栽培面积的净收益

4 讨论

4.1 兼顾日光温室的安全性和经济性

在日光温室的设计建造和结构优化过程中，一定要注重考虑日光温室的安全性和经济性。苏北地区较我国东北、西北以及华北北部地区冬季风荷载和雪荷载偏小，较小的荷载减轻了日光温室结构用材，降低了温室的建造成本。苏北日光温室类型较多，建筑材料的选取多样，如前屋面骨架材料有镀锌钢管、毛竹等，后墙材料有粘土红砖、空心砌块、夯实粘土等。日光温室建造材料的不同直接影响了其结构安全性能和建造成本，如 SR-1 型日光温室前屋面骨架材料为热浸镀锌钢管桁架结构，建造成本较高，适用于现代农业科技示范园区内建造使用；SR-14 型日光温室前屋面骨架材料为毛竹，建造成本较低，但结构安全性较差，在大风大雪天气易发生结构变形甚至倒塌。因此，在对苏北日光温室进行结构优化时应双重考虑温室的安全性能和建造成本，通过合理结构以寻求二者之间的的最优组合，即在保证结构强度的情况下尽可能的降低温室造价。

4.2 苏北地区日光温室的配套设备和管理方式

总体而言，苏北日光温室配套设备还比较落后，除少部分日光温室内有小型的环境监测设备，其它温室只具有温度计、湿度计等。近年新建的温室中，所谓配套设备也仅仅是卷帘机和膜下滴灌系统，此外，诸如 CO_2 增施设备、补光设备、通风设备等均处于空白。这说明苏北日光温室整体上环境调控能力不足，优质、高产能力较差。此外，苏北日光温室基本上采用土壤栽培，病虫害和土壤盐渍化问题严重；经营方式多为小农户个体经营，缺乏大型公司经营和企业 + 农户合作经营，这制约了苏北日光温室的长期发展。解决上述问题，对苏北地区设施蔬菜的栽培、管理和经营水平的提升具有重要意义，因此这应成为苏北日光温室今后进一步研究的重点所在。

4.3 如何提高日光温室的土地利用率

为了保证种植作物冬季在日光温室内能正常生产，在光照时间最短的冬季 12 至翌年 1 月必须保证每栋温室至少有 4 ~ 6h 的全地面光照时间[4]。因此，两栋温室之间必须留有足够的采光空间，造成了日光温室土地利用率不足 50%，为增强保温性能，宽厚的后土墙也降低了土地利用率。本调查发现 SR-4 型日光温室（“阴阳型”日光温室）在传统日光温室的北侧，增加了一个同长度但采光面朝北的单屋面温室，该温室可进行耐荫作物栽培或家禽家畜养殖，与传统日光温室相比土地利用率较高（61.81%）。另外，西北农林科技大学设计了西北型双连跨日光温室，其总跨度为 16m，前、后跨度各位 8m，该结构可以充分利用土地资源[5]。此外，合理安排作物茬口，减少日光温室室内土地的露荒时间，也是提高日光温室土地利用率的途径。

4.4 栽培作物种类和茬口安排对日光温室经济效益的影响

近几年随着设施园艺面积的不断增加，设施园艺产品常因过于集中上市而出现价格大幅度下跌，价格的较大波动不仅直接影响经营者的经济效益，同时还降低广大农户发展设施农业的热情。解决该问题可通过对区域内栽培作物种类和茬口安排进行统筹布置，提倡“一村一品”或反季节栽培，即保证作物栽培面积不出现较大波动，又可确保产品上市时间不过于集中。可在一定区域范围内，以村为基本单位，按照国内外市场需求，充分发挥本地资源优势、传统优势和区位优势，挖掘出一个或几个市场潜力大、区域特色明显、附加值高的主导产品，从而大幅度提升农村经济整体实力和综合竞争力，又可以避免因盲目种植。此外，在园艺作物反季节栽培的基础上，提倡反季节栽培，合理安排作物茬口，错开蔬菜集中上市期以取得相对的价格优势。

4.5 经营方式对日光温室经济效益的影响

随着人们生活水平的提高，对蔬菜、水果、花卉等园艺作物的要求日趋多样化，设施园艺在原有的设施作物种植的基础上，又拓展了园区观光、采摘、餐饮等诸多作用。正如本调研中部分日光温室在经营过程中加入了上述要素，进一步提高了经营者的经济效益。此外，现代日光温室科技含量高，经营理念先进，可以起到先进科技的示范推广作用。在日光温室的发展过程中可以结合当地的自然、气候、土壤、水质等地方农业生物资源，进行研究、提升和开发；挖掘本地特色自然风光、人文景观、民俗文化、农家风情等特有资源，进行提炼、整合和打造，培育出具有原生优势和现代科技、文化、经营理念相结合的产业模式。

总之，苏北日光温室建设规模较大、特点突出、经济效益较好，可为其他地区日光温室的建设与发展提供很好的借鉴，同时，也存在温室类型与当地气候不相适应、温室配套设备落后、经营理念滞后、作物栽培种类和茬口安排缺乏区域内的宏观调控等问题。因此，在苏北地区日光温室今后的发展中可通过设施构型优化形成适应当地气候条件的日光温室类型，改善产品的经营思路以提高经营者的经济效益，形成区域内的调控机制以应对价格波动风险，挖掘当地特有资源以提高经营者的附加收益。

参考文献

[1] 吴凤芝．园艺设施工程学．北京：科学出版社，2012
[2] 刘建，周长吉．日光温室结构优化的研究进展与发展方向．内蒙古农业大学学报．2007，28（3）：264～268
[3] 巫健华，王宝海，杨意成等．江苏省设施园艺发展现状与对策．农业工程技术（温室园艺）．2009（9）：36～38
[4] 周长吉．周博士考察拾零（二）——阴阳型日光温室．农业工程技术（温室园艺）．2011（4）：48～52
[5] 李式军，郭世荣．设施园艺学（第二版）．北京：中国农业出版社，2011

太阳能热泵技术在日光温室中的应用*

孙维拓**，杨其长***，张 义，方 慧，李 文

（1. 中国农业科学院农业环境与可持续发展研究所，北京 100081；
2. 农业部设施农业节能与废弃物处理重点实验室，北京 100081）

摘要： 针对太阳能用于日光温室加温的间歇性和不稳定性，本研究结合主动蓄放热装置，设计了一套适用于日光温室的太阳能热泵系统。在晴天及多云天气，白天利用循环水通过主动蓄放热装置吸收太阳能，并将热量储存到蓄水池中，适时开启热泵将低品位热能进行高效提升，降低循环水温，进而提高主动蓄放热装置集热效率，夜间通过主动蓄放热装置释放热量；在阴、雪天气，热泵机组可用作应急电加热系统；在太阳辐射较强的晴天，也可单独采用主动蓄放热装置为温室加温。经过连续4d 的加温试验，结果表明：在不同天气条件下，太阳能热泵系统可将温室夜间平均温度提高3.5～4.1℃，维持作物根际平均温度在15℃以上，加温效果显著，加温性能稳定。

关键词： 日光温室；太阳能；热泵；加温；主动蓄放热

Application of Solar Heat Pump Technology in Solar Greenhouse

Sun Weituo, Yang Qichang, Zhang Yi, Fang Hui, Li Wen

(1. *Institute of Environment and Sustainable Development in Agriculture*, *Chinese Academy of Agricultural Sciences*, *Beijing* 100081, *China*; 2. *Key Lab of Energy Conservation and Waste Treatment of Agricultural Structures*, *Ministry of Agriculture*, *Beijing* 100081, *China*)

Abstract: Aiming at limitations of solar energy heating system, based on active heat storage-release device, a solar heat pump system (SHPS) applicable to solar greenhouse heating was designed in the present study. When sunny and cloudy weather, the solar energy obtained from the north wall was absorbed by the circulating water and stored in impounding reservoir during the daytime, meanwhile heat pump unit was used to reduce the circulating water temperature and improve heat collecting efficiency, the heat energy was released through the active heat storage-release device when indoor air temperature dropped to lower limit at night. When cloudy and snowy weather for several days running, the heat pump unit could serve for electric heating to meet an emergency. When solar radiation was strong, active heat storage-release device also could be used for heating alone. After continual running of SHPS for four days, it showed that the average air temperature at night in the experimental greenhouse was increased by 3.5～4.1℃ and the average soil temperature at crop rhizosphere was more than 15℃. The experimental results prove that SHPS has obvious heating effect, stable heating performance and a good prospect for applications.

---

* 基金项目：863 计划资助课题（2013AA102407）；国家自然科学基金资助项目（31071833）；国家科技支撑计划（2011BAE01B00）；公益性行业（农业）科研专项（201203002）

** 作者简介：孙维拓（1989—），男，山东邹城人，主要从事设施农业环境工程方面的研究。北京 中国农业科学院农业环境与可持续发展研究所，100081。E-mail：swt0226@163.com

*** 通信作者：杨其长（1963—），男，安徽无为人，博士，研究员，博士生导师，主要从事设施园艺环境工程研究。北京 中国农业科学院农业环境与可持续发展研究所，100081。E-mail：yangq@ieda.org.cn

Key words: Solar greenhouse; Solar energy; Heat pump; Heating; Active heat storage-release

1 引言

日光温室是我国独创的一种高效节能型温室型式，近年来发展迅速，已成为广大农民脱贫致富的重要生产设施。在我国北方特别是东北、西北地区冬季寒冷，日光温室仅靠自有结构吸收太阳辐射能并不能完全满足喜温果菜类蔬菜、花卉等作物生长发育对温度的需要。日光温室保温蓄热能力有限、温度环境调控能力低，冬季夜间低温高湿，严重影响产量和品质。一些地区为提升日光温室夜间温度，盲目增加北墙厚度，最厚处可达6～8m，严重浪费土地资源的同时其增温效果也极为有限。鉴于日光温室特有的结构形式和光热环境，多年来，国内众学者围绕如何提高日光温室夜间温度开展了大量研究。王顺生等设计了一小型日光温室内置式太阳能集热调温装置，白天使水温升高20℃以上，夜间可提高气温1.7℃[1]；于威等采用全玻璃真空管集热器、地热管道、水箱、循环系统等组成太阳能土壤加温系统，可提升地温4～5℃[2]；刘伯聪等利用全真空玻璃管集热器、蓄热水池、地下管网等组成太阳能蓄热系统，通过蓄热水池与集热器、蓄热水池与地下管网两个换热循环，可有效提高日光温室内气温和地温[3]；李炳海等采用自主研发的太阳能地热加温系统，在16：00～20：00对辽沈Ⅳ型日光温室土壤进行加温，结果表明，使用地热加温系统后，日光温室内15cm深处土温在晴天时提高2.94℃，阴天提高2.56℃[4]；刘圣勇等采用太阳能真空管集热器、保温蓄热水箱、循环水泵和地下散热器等部件构成了太阳能地下加热系统用以提高地温，可将10cm深度处的平均地温提高3.4℃，但系统对日光温室气温的提升效果不明显[5]；张晓慧等采用地源热泵空调系统为日光温室加温，明显提高了气温和地温，而且还能有效调节日光温室内的湿度，系统制热性能系数（COP）为4.16[6]；方慧等利用地源热泵加热日光温室，测试结果表明与传统燃煤锅炉相比，平均节能为29.6%[7]；徐刚毅等采用新型电锅炉供暖方式对土壤进行加温，并配套设计了土壤温度自动检测与控制系统，结果表明该系统是经济可行的[8]；白义奎等利用燃池提高温室地温和气温，室内外平均温差可达30℃以上[9]；马丹等对SG-1型高效热风炉在辽沈Ⅰ型日光温室内的加温效果进行了测试，结果表明，植株群体内部及植株冠层的温度绝大部分达到13℃以上[10]；王铁良等采用热水供热系统加温日光温室，加温效果明显，最低温度13℃[11]；周长吉博士通过走访考察，介绍了一种使用太阳能平板集热器提高日光温室地温的方法以及一种日光温室临时加温系统，可有效缓解极端天气特别是连阴天对作物造成的低温冷害[12～13]。

由于太阳能是一种清洁可再生能源，取之不尽、廉价、安全、无需运输，因此针对日光温室夜间低温问题，不难看出众学者把研究的重点放在了太阳能的开发利用上。利用太阳能为日光温室加温是未来发展趋势，也符合我国节能减排战略[14]，但太阳能加温系统也存在一些不足，使众学者的研究成果很难得到推广和应用，主要表现在：由于太阳能受季节和天气影响较大、热流密度低，采用各种形式的太阳能集热器为温室加温具有间歇性和不稳定性；随着集热器运行温度的升高集热效率下降，这将减少有效集热时间，降低太阳能利用率；并且传统的太阳能集热器初投资大，如采用廉价的集热装置加温效果更加有限。

为克服太阳能加温系统的局限性，高效节能的热泵技术受到了重视。太阳能集热器在低温时集热效率较高，而热泵机组在其蒸发温度较高时制热性能系数较大，那么可以考虑采用太阳能加温系统作为热泵机组热源。太阳能热泵将太阳能热利用与热泵节能技术有机结合起来，即可通过太阳辐射作用提高热泵机组的蒸发温度和COP，又可以弥补太阳能加温系统中存在的低密度、间歇性和不稳定性等缺点，提升集热器集热效率和整个系统COP，从而降低运行费用。根据太阳能集热器和热泵蒸发器的组合形式，太阳能热泵系统可分为直膨式和非直膨式。在直膨式系统中，集热器与蒸发器合二为一，即制冷工质直接在集热器中吸收太阳辐射能而得到蒸发，然后通过热泵循环将冷凝热释放给被加热物体。在非直膨式系统中，太阳能集热器与热泵蒸发器分离，通过集热介质（一般采用水、空气、防冻溶液）在集热器中吸收太阳能，并在蒸发器中将热量传递给制冷剂，或者直接通过换热器将热量传递给需要预热的空气和水[15~17]。

与传统的太阳能加温系统相比，太阳能热泵的最大优点是可以采用结构简易的集热器，集热成本非常低。在非直膨式系统中，太阳能集热环路作为蒸发器的低温热源，运行温度通常为20~30℃，集热器散热损失小，集热效率高。有研究表明，在非寒冷地区即使采用结构简单、廉价的普通平板集热器，集热器效率也高达60%~80%。本研究团队在增加日光温室蓄热量、提升日光温室夜间温度、实现温室温度可控方面做了大量研究，提出了主动蓄放热思想。主动蓄放热思想是在日光温室中，白天利用流体介质的循环不断将到达墙体表面的太阳辐射能吸收并蓄积起来，夜间再通过流体的循环释放热量，变被动蓄放热方式为主动蓄放热方式，实现热量在空间、时间上的转移，从而提高太阳能利用率，提升温室夜间温度。方慧等研究了日光温室浅层土壤水媒蓄放热增温效果，可提高室温4.0℃，30cm深处土壤温度提高3℃，60cm深处土壤温度提高5℃[18]；张义等研究了日光温室水幕帘蓄放热系统增温效应试验，可将温室内夜间温度提高5.4℃以上，作物根际温度提高1.6℃以上[19]。主动蓄放热系统加温效果明显，初投资和运行费用均较少，应用前景广阔，但系统的集热性能、加温稳定性以及应对极端天气的能力仍有待提高。因此，本研究利用非直膨式太阳能热泵原理，结合主动蓄放热装置设计建造了一套太阳能热泵系统，并进行了初步的试验与研究。

2 材料与方法

2.1 试验温室

试验日光温室位于北京市昌平区小汤山现代农业科技示范园西区，温室东西走向，长49m，跨度8m，后墙高2.5m，脊高3.7m，后坡长1.5m，采用钢骨架结构，前坡覆盖材料为单层0.08mmPVC塑料薄膜，北墙内侧为12cm红砖，外侧为24cm红砖，中间为10cm聚苯板，后坡内侧为10cm预制板，外侧为10cm聚苯板。试验期间，温室用于番茄育苗。

2.2 系统组成

太阳能热泵系统主要由主动蓄放热装置、热泵机组、蓄水池、循环水泵、循环管道、控制系统等组成。主动蓄放热装置安装于北墙内侧距地面0.4m高处，集热材料为双层黑色PE膜，循环水在双层膜间流动，白天用于集热，晚上放热，装置采用单元式结构，单元高2m，宽1.35m，共29个单元。循环水泵2台，额定流量分别为10m^3/h和7m^3/h，扬程10m。循环管道由不同口径的PVC管连接而成，管外覆盖保温套。热泵机组型号为DISMY DDR-192GSPA1-PA，额定制热量21kW，额定制热输入功率5.12kW，机组循环泵为格兰富CH4-20，蒸发器侧水流量3.3m^3/h，冷凝器侧水流量1.8~3.9m^3/h。蓄水池包括蓄水池Ⅰ

和Ⅱ，两者中间由Φ80的截止阀控制连通，蓄水池Ⅰ为热泵机组热源，实际蓄水量1.86m^3，蓄水池Ⅱ为热泵机组热汇，实际蓄水量6.00m^3，蓄水池主体材料为12cm厚粘土砖墙，外表面紧贴10cm高密度聚苯板，内表面涂抹0.3cm防渗水泥砂浆。热泵机组和蓄水池位于温室中部，图1为太阳能热泵系统布置图。

图1　太阳能热泵系统布置图

Figure 1　Layout diagram of solar heat pump system

2.3　系统工作原理

图2为太阳能热泵系统工作原理示意图。根据不同天气状况，系统运行可分3种情况：在晴天、多云和阴至多云天气，白天通过主动蓄放热装置将北墙的热量转移并储存到蓄水池中，根据天气情况，当蓄水池水温上升到一定温度时适时开启热泵机组，蓄水池Ⅰ和Ⅱ断开连通，蓄水池Ⅱ作为热汇水温逐渐升高，蓄水池Ⅰ为热泵提供热源，其水温逐渐下降，并作为循环水来源不断吸收太阳辐射能，夜间室温下降到设定的下限温度时循环水泵自动开启，通过主动蓄放热装置将蓄水池中热量释放到温室中；在连阴天和雪天，连续运行系统，热泵机组可用作应急电加热系统；为进一步节约耗能，在晴天可单独采用主动蓄放热装置为温室加温。

2.4　试验方法

单点温度测试选用德图公司生产的testo174T型温度自动记录仪，精度为±0.2℃，测量范围为-30～70℃；太阳辐射测试选用美国坎贝尔公司生产的太阳辐射传感器，准确度为0.5%，测量范围为0～2 000W/m^2；多点温度测试选用中国计量院生产的热电偶，测量精度为±0.2℃；数据采集仪选用美国坎贝尔公司生产的CR1000，用于自动记录热电偶采集的温度值和辐射传感器采集的太阳辐射值。

空气温度测点布置：在温室距北墙4m，距东墙12m、24m、36m处设置3个测点，在距东墙24m，距北墙2m、6m处设置2个测点，测点距地面1.5m高处；土壤温度测点布置：在温室中部距地面1cm、15cm、30cm深处设置3个测点。太阳辐射测点位于北墙内表面距地面1.5m高处。

3　试验结果与分析

2012～2013年，冬天对太阳能热泵系统进行了试验测试（图2）。选取2013年1月

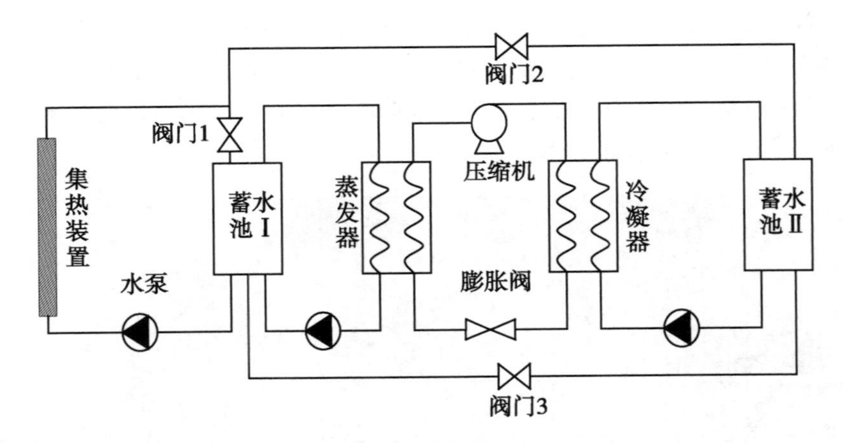

图 2　太阳能热泵系统工作原理示意图

Figure 2　Schematic diagram of SHPS operating principle

17～21 日连续 4d 的试验数据，分析不同天气条件下系统的加温效果。连续 4d 系统具体运行调控参数及天气情况如表所示。

表　系统运行调控参数

Table　Control parameters of the system operation

时间（年-月-日）	天气	揭保温被时间	放保温被时间	白天系统集热时间	热泵机组运行时间	夜间室温设定值		太阳辐射量均值/（W·m^{-2}）
						下限/℃	上限/℃	
2012-1-17～18	晴	8：30	16：00	8：30～16：00	12：30～15：30	13	15	322.7
2012-1-18～19	多云	8：30	16：00	8：30～16：00	12：30～15：10	12	15	212.2
2012-1-19～20	阴转多云	8：30	16：00	8：30～15：30	12：00～14：40	11	15	101.7
2012-1-20～21	雪	/	/	/	10：00～次日 8：30	11	15	/

3.1　室内温度变化

图 3 所示为连续 4d 试验温室与对照温室室温变化曲线。在晴天和多云天气的白天，两温室温度无明显差异。晴天夜间，系统 21：00 自动运行供热，21：00 至次日 8：30 试验温室平均室温 12.0℃，比对照温室提高 3.8℃，最低室温提高 4.0℃；多云天气夜间，系统 20：30 供热，20：30 至次日 8：30 试验温室平均室温 12.3℃，比对照温室提高 4.1℃，最低室温提高 4.1℃。在阴转多云天气，白天室温较低，系统在吸收太阳辐射的同时也在向温室放热，在 8：30 至 16：00 时间范围内两温室平均温差为 1.9℃，夜间系统 1：10 供热，到次日 8：30 试验温室平均室温 11.9℃，比对照温室提高 3.6℃，最低室温提高 3.6℃。在雪天无有效太阳辐射可以利用，系统全天运行，利用蓄水池中剩余热量及热泵电加热应对温室低温，1 月 20 日 8：30 至 21 日 8：30 试验温室平均室温 10.8℃，比对照温室提高 3.5℃，最低室温提高 4.1℃。因此，与对照温室相比，在不同的天气条件下，太阳能热泵系统对日光温室的增温效果均显著。

3.2　土壤温度变化

如图 4 所示，试验温室土壤表面温度受室温影响较大，越深层的土壤温度波动越小。黄

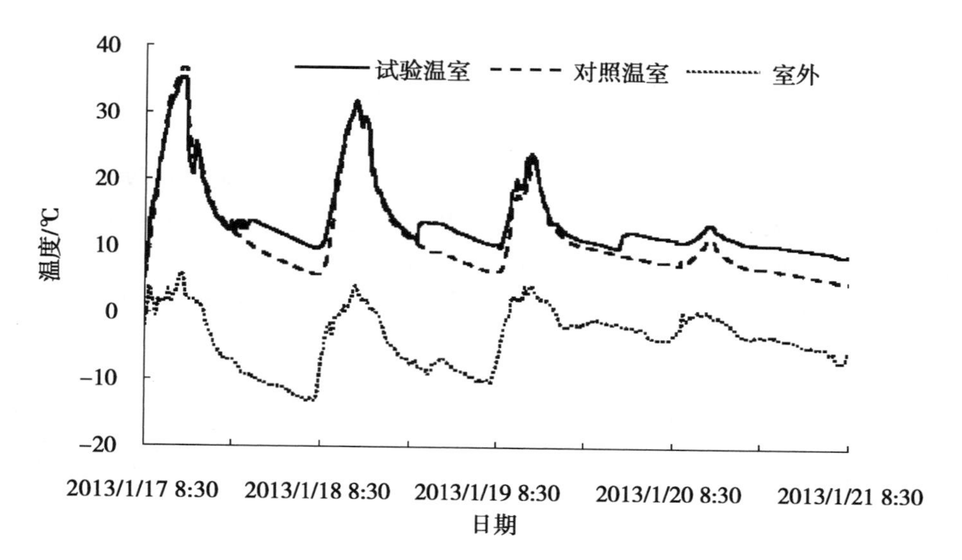

图 3　试验温室与对照温室室温变化曲线（1 月 17 日 8：30 至 1 月 21 日 8：30）

Figure 3　Air temperature curves in experimental and comparative greenhouse（8：30，Jan. 17～8：30，Jan. 21）

瓜、番茄等果菜类作物，根系大都在距离地面 20cm 左右的范围，试验温室 30cm 深处土壤温度基本无波动，平均温度在 15℃以上，较高的根际温度有利于提高作物产量和品质。

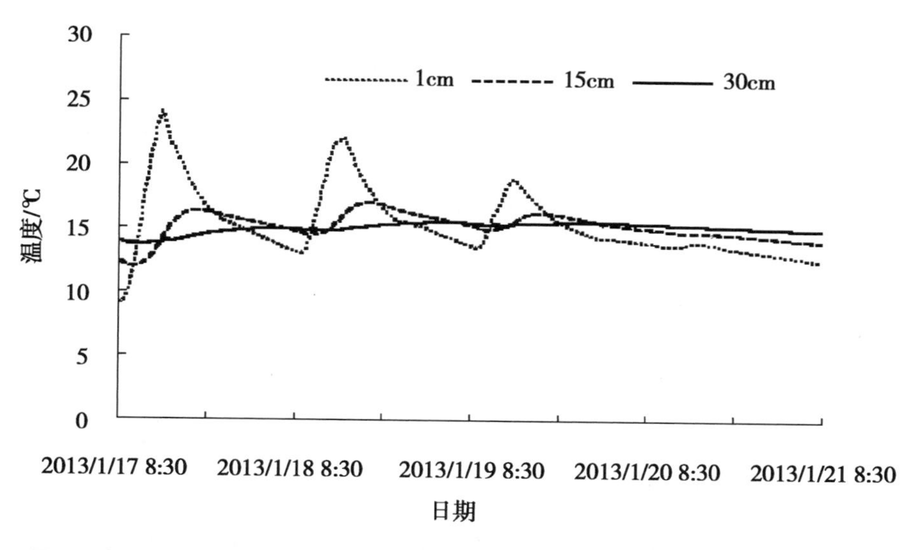

图 4　试验温室不同深度土壤温度变化曲线（1 月 17 日 8：30 至 1 月 21 日 8：30）

Figure 3　Soil temperature curves with different depths in experimental greenhouse（8：30，Jan. 17～8：30，Jan. 21）

4 结论与讨论

应用太阳能热泵系统提高日光温室夜间温度是可行的，通过本试验测试，得出以下结论。

第一，本试验中系统采用廉价的主动蓄放热装置用于白天集热和晚间放热，无需另外安装地埋散热管和风机盘管等散热设备；热泵机组以太阳能为低温热源，在提升主动蓄放热装置集热效率的同时还能应对极端天气，无需配置其他辅助加温设备。

第二，在晴、多云、阴转多云、雪 4 种不同天气条件下，系统可将夜间平均室温提高 3.5～4.1℃，作物根际平均温度在 15℃以上，加温效果显著，加温性能稳定。

第三，由于日光温室为中国特有，太阳能热泵技术用于日光温室的研究未见报道，本试验中针对日光温室设计的太阳能热泵系统及运行方案还处于研究的初级阶段，系统各组件参数配置还有待优化，系统性能、运行机制、能耗及其经济性都将会在后续的试验中加以分析和验证。

参考文献

[1] 王顺生，马承伟，柴立龙等．日光温室内置式太阳能集热调温装置试验研究．农机化研究，2007（3）：130～133

[2] 于威，王铁良，刘文合等．太阳能土壤加温系统在日光温室土壤加温中的应用效果研究．沈阳农业大学学报，2010，41（2）：190～194

[3] 刘伯聪，曲梅，苗妍秀等．太阳能蓄热系统在日光温室中的应用效果．北方园艺，2012（10）：48～53

[4] 李炳海，须晖，李天来．日光温室太阳能地热加温系统应用效果研究．沈阳农业大学学报，2009，40（2）：152～155

[5] 刘圣勇，张杰，张百良等．太阳能蓄热系统提高温室地温的试验研究．太阳能学报，2003，24（4）：461～465

[6] 张晓慧，陈青云，曲梅等．地源热泵空调系统在日光温室中的加温效果．上海交通大学学报（农业科学版），2008，26（5）：436～439

[7] 方慧，杨其长，孙骥．地源热泵在日光温室中的应用．西北农业学报，2010，19（4）：196～200

[8] 徐刚毅，刘明池，李武等．电锅炉供暖日光温室土壤加温系统．中国农学通报，2011，27（14）：171～174

[9] 白义奎，王铁良，刘文合等．燃池在日光温室应用的试验研究．可再生能源，2005（3）：11～13

[10] 马丹，须晖，韩亚东等．日光温室专用燃煤热风炉加温效果分析．农业工程技术·温室园艺，2010（5）：13～14

[11] 王铁良，白义奎，杨丽萍等．日光温室热水供热系统．可再生能源，2002（6）：25～28

[12] 周长吉．一种使用太阳能平板集热器提高日光温室地温的方法．农业工程技术·温室园艺，2012，(8)：18～22

[13] 周长吉．周博士考察拾零（一）一种日光温室临时加温系统．农业工程技术·温室园艺，2011，(2)：25

[14] 李锐，张建国，俞坚等．太阳能热泵系统．可再生能源，2004（4）：30～32

[15] 王振辉，崔海亭，郭彦书等．太阳能热泵供暖技术综述．化工进展，2007，26（2）：185～189

[16] 杨灵艳，姚杨，姜益强等．太阳能热泵蓄能技术研究进展．流体机械，2008，36（12）：65～69

[17] 张昌．热泵技术与应用．北京：机械工业出版社，2008

[18] 方慧，杨其长，梁浩等．日光温室浅层土壤水媒蓄放热增温效果．农业工程学报，2010，27（5）：258～263

[19] 张义，杨其长，方慧．日光温室水幕帘蓄放热系统增温效应试验研究．农业工程学报，2012，28（4）：188～193

太阳能蓄热系统在日光温室中的应用*

阳　萍**，马承伟，陈　亮，金文卿，徐　凡
（中国农业大学 农业部设施农业工程重点实验室，北京　100083）

摘要：针对日光温室凌晨时段以及遇连续恶劣天气时室内温度过低，难以满足作物正常生长需求这一问题，研究者们利用将日光温室内白天多余的太阳能移至夜间使用的“削峰填谷”的原理，并结合太阳能的高效利用技术，设计了多种的蓄热系统。本文旨在对近年来研究设计的日光温室太阳能蓄热系统进行梳理分析，探讨其存在的问题和未来发展的方向；在此基础上提出使用日光温室自身结构进行集热与散热的设想，并研究开发了一套日光温室结构兼用式太阳能蓄热加温装置。

关键词：日光温室；太阳能；蓄热系统；加温

The Applications of Solar Heat Storage System in Solar Greenhouse

Yang Ping，Ma Chengwei，Chen Liang，Jin Wenqing，Xu Fan
(*China Agricultural University*，*Key Laboratory of Agricultural Engineering in Structure and Environment*，*Ministry of Agriculture*，*Beijing* 100083，*China*)

Abstract：The indoor temperature of the solar greenhouse is too law to meet the requirements for crop growing normally in the morning or encountering continuous bad weather. Focusing on this problem，the researchers combined the principle of collecting the excess solar energy in the day time to use at night and the efficient use of solar technology，and then designed a variety of heat storage systems. This paper aims at sorting out and summarizing the solar heat storage systems in the solar greenhouse researched and designed in recent years. On this basis，propose the conception of using the structure of solar greenhouse to storage and dissipate the heat，and develop a solar greenhouse's structure used along with solar heat storage device.

Key words：Solar greenhouse；Solar energy；Heat storage system；Rising temperature

1　引言

能源问题一直是制约世界各国经济发展的重要因素之一，2010 年，一次能源消费总量为 120.02 亿 t 油当量，中国的能源消费量占全球的 20.3%，已赶超美国成为世界最大的能源消费国。中国国家能源局公布的《中国可再生能源发展“十二五”规划目标》指出，到 2015 年，中国将努力建立有竞争性的可再生能源产业体系，风能、太阳能、生物质能及核电等非化石能源开发总量将达到 4.8 亿 t 标准煤。

日光温室是具有中国特色的作物栽培设施，以太阳能为主要能源。白天，太阳辐射进入

* 基金项目：现代农业产业技术体系建设专项资金（CARS—25），公益性行业（农业）科研专项（201203002），“十一五”国家科技支撑计划项目（2009BADA4B04—01）资助

** 作者简介：阳　萍（1988—），女，广西桂林人，在读硕士研究生。中国农业大学农业部设施农业工程重点实验室，北京 100083。E-mail：yangping882007@163.com

温室后，以热量的形式存储在温室后墙和土壤中；夜晚，当温室内气温下降时，墙体和土壤中蓄积的热量会缓慢释放到温室中，因而具有一定的保温节能性，在我国北方地区广泛用于冬季作物生产。但在冬季凌晨时段或遇到连续恶劣天气时，日光温室内的低温环境不能维持作物的正常生长，必要时需要进行辅助加温。目前的供暖系统主要使用煤、天然气等化石燃料，这样既耗费了大量能源、提高了生产成本，还会对环境造成污染。太阳能是一种清洁能源，量大而分布广泛，应用于温室加温，替代不可再生的化石燃料加热，是最有前景的可再生能源之一。因此，大力发展并更加充分有效利用清洁可再生的太阳能技术是我国温室产业现今所要研究的重要课题。

2　日光温室太阳能蓄热系统

太阳辐射具有间断性，为了实现“日间蓄夜间用，晴天蓄阴天用”，首先必须对日光温室结构进行优化，使其具有更充分利用太阳能资源的能力，因此，日光温室除了采取良好的保温措施以外，温室的朝向设计与坡向设计也至关重要。姜晨光等[1]根据天文学、气象学的基本原理，通过试验提出了不同地理位置日光温室（包括塑料大棚）朝向与坡向的设计方法。亢树华等[2~11]对日光温室结构的优化及围护结构材料的优化进行了广泛研究，提出了提高日光温室采光性和保温性措施的意见。日光温室由于具有良好的采光性和保温性，在天气晴好的中午，日光温室室内温度较高，一般需要放风进行降温，但通过放风就会浪费大量的热量。为了将白天这些多余的热量蓄积起来，并在夜间温室内温度降低时释放出来，提高夜间（尤其是凌晨）室内温度，专家学者们研究设计了种类繁多的太阳能蓄热系统。

2.1　以水作为媒介的蓄热系统

研究表明[12~14]，日光温室墙体和地面的吸、放热是维持日光温室室内温度的关键部分作用，但是在夜间（特别是凌晨时段）以及遇连续恶劣天气，室内温度过低时，它们的蓄热能力仍然不能满足需求；而水容易获得、价格便宜，使用时不污染环境、可循环利用，并且具有较大的比热，因此许多学者研究以水作为蓄热媒介，通过太阳能真空管、平板集热器、黑色 PE 管或黑色塑料薄膜等材料蓄积太阳能，通过地埋管的方式将热水循环至地下并将热量蓄积到土壤中，或者将热能蓄积到水箱中，夜间使水循环释放热量。

刘圣勇等[15]采用太阳能真空管集热器、保温蓄热水箱、循环水泵和地下散热器等部件构成了太阳能地下加热系统用以提高地温。王奉钦等[16]设计了一套全玻璃真空管太阳能集热器辅助加热系统，在室外平均温度为 -1℃ 的情况下，该系统能够提高根区土壤温度 2.2℃。王顺生等[17]在后墙内侧安装一套自制的太阳辐射集热调温装置，在冬季晴天能将集热器中的水温提高 20℃，并利用该热水加热土壤，可使地温提高 3.2 ~ 3.8℃。李炳海等[18~19]采用自主研发的太阳能地热加温系统，研究结果表明，日光温室内 15cm 深处土温在晴天时平均比不加温的高 2.94℃，阴天提高 2.56℃，15cm 深处最低土温由 11.0℃提高到 13.9℃。方慧等[20]在后墙表面平行安装 16 组并联的黑色 PE 管，白天通过水的循环将热量收集并储存到温室浅层土壤中，夜间通过土壤的自然放热将热量释放到温室中，提高温室夜间温度。此蓄热方法增加了温室的蓄热量，在盖上保温被以后，试验温室与对照温室的温度出现明显差异，平均气温差为 4.0℃。张义等[21]设计了一种水幕帘放热系统，白天利用水循环通过外层为透明薄膜，中间为黑色薄膜组成的水幕帘吸收太阳能，同时将能量储存在水池中，夜晚利用水循环通过水幕帘释放热量，应用该水幕帘蓄放热系统可将温室内夜间温度提高5.4℃以上，可将作物根际温度提高1.6℃以上。为了降低成本、利于推广，研究者们较多地使用自制的太阳能集热器，而非购买现成的太阳能集热器。

这类系统贮存热量的方法有两种：其一是将能量蓄积到保温水箱（水池）中；其二是使热水通入地下将热量蓄积到土壤中。以水为介质的太阳能加温系统进行土壤加温时，土壤温度得到了显著提高，使土壤夜间的持续放热能力加强，温室内空气温度得到提升，而且能提高作物根系温度，对作物的生产非常有利。此外，白天经过加温后的水还可以与温室内的灌溉系统连通，作为灌溉用水，减少冬季灌溉时对作物的刺激。由于水具有许多优良特性，以水作为蓄热媒介的太阳能系统得到了广泛地研究，但是这种蓄热方式为显热蓄热，蓄热容量较低。

2.2 以相变材料作为媒介的蓄热系统

相变材料蓄热是利用材料在相变（固-气、固-固、固-液、）过程中能够吸收或放出大量的潜热能，而温度变化很小的特点来进行蓄热。与显热蓄热相比，潜热蓄热具有蓄热密度大，温度变化相对稳定，设备体积小，使用方便等特点。由于固-气相变时体积变化过大，因此其应用受到一定限制；固-固、固-液相变时的体积变化小，使用方便，是相变材料研究的重点。目前，应用于温室中的相变材料分为有机和无机两类：有机相变材料以石蜡、硬脂酸正丁酯和聚乙二醇为主；无机相变材料主要是无机水合盐类，如 $CaCl_2 \cdot 6H_2O$、$Na_2SO_2 \cdot 10H_2O$[22]。邹志荣[23]、孙心心[24]等把相变储能材料经过封装后置于北墙中或是制成新型墙体材料用于日光温室中，试验温室内的温度波动幅度比对照温室小，具有明显的“削峰填谷”的作用。郭靖等[25]对日光温室中使用不同封装的相变材料蓄热效果进行了研究，试验表明，内渗型相变材料温室要优于外挂型相变材料温室。姚轩等[26]采用以玻化微珠为基材加入了相变材料的保温砂浆来代替传统的保温隔热层，大大提高了温室围护结构的蓄热能力。这种新型复合围护结构利用了玻化微珠保温砂浆绝热、保温的特点，可以有效的控制温室内的温度和湿度。

使用相变材料作日光温室中的太阳能蓄热介质，蓄热效率很高，蓄热效果很好，很有发展前景，但是其中的潜热能力只利用了一半左右；若使用过程中出现渗漏，还会对温室内环境造成污染。由于相变材料价格普遍较高，长期使用过程存在过冷现象和相分离，随着使用时间的延长，相变潜热值也会出现下降，因此其经济性和耐久性还有待深入研究。

2.3 以空气作为媒介的蓄热系统

目前，以空气作为媒介的蓄热系统研究较多的主要为地中热交换系统。温室地下蓄热系统由温室、土壤、轴流式风机、进气道、排气道等组成。白天，由于太阳辐射，温室内空气温度升高，此时土壤温度较低，因此进入地下管道的空气将与周围土壤发生热交换，而土壤与静止空气的换热较慢，安装轴流风机后可有效解决该问题，风机启动使室内热空气从地下管道流过，低温土壤蓄热，夜间启动风机使室内冷空气从地下管道流过，经过土壤加热后的空气重新循环至温室内，从而提高室内温度。

马承伟等[27]对连栋温室地中热交换系统进行了研究，系统在夜间加温能力可达加温热流量60 W/m²，可有效保证夜间室内气温高于室外11℃以上。白义奎等[28]对辽沈Ⅰ型日光温室地中热交换系统进行了研究，系统可提高地温和室内空气温度，并且在蓄热过程中能对日光温室进行除湿，使空气中的含水量降低，同时，空气流动使室内温度分布更均匀，更有利于作物生长。

地中热交换系统使用的设备较多，安装复杂，虽然增加了风机，增强了空气与土壤之间的传热，但是其相互间的换热量仍然有限，因此不能完全满足温室冬季加温的需要。

2.4 热泵系统

热泵技术最早起源于国外，在我国的发展起步较晚，但是发展速度很快，目前，已经广泛应用于工业生产、居民供暖、农业生产等领域中。热泵系统是以消耗一部分低品位能源为补偿，使热能从环境介质（地下水、地表水、土壤和空气等）向高温热源传递，并储存使用的系统。热泵系统能够提高热能的品位，因此提高了能源的有效利用率，从而可以解决显热蓄热太阳能加温系统的热能品位不高，效率较低的问题。热泵系统包括地源热泵系统、水源热泵系统、空气源热泵系统和混合热源热泵系统等。由于具有节能环保、效率高等优点，热泵技术应用于园艺设施工程的研究已成为热点，尤其是地源热泵。张晓慧等[29]对使用了地源热泵空调系统的日光温室的加温效果和经济性进行了分析，系统不仅明显提高了室内的气温和地温，还能有效降低室内的湿度，且其制热性能系数为4.16，说明系统的制热性能较好。热泵系统能实现“一机两用”，不仅可以用于冬季加温，而且可以用于夏季降温，柴立龙等[30]对使用地源热泵系统的日光温室的降温效果和性能进行了研究，系统有明显的降温和除湿效果，系统制冷性能系数平均为3.01。太阳能蓄热加温系统与热泵结合，可以提升热能的品位，从而提高热能利用效率。在日光温室中，与太阳能结合的热泵系统目前正在研究的有空气-水源热泵系统、太阳能联合双热泵系统等。热泵系统蓄热效率很高，对日光温室内加温降湿效果很明显，但限于其初次投资费用很高，目前，在普通农户中推广使用还有一定困难。

3 日光温室结构兼用式太阳能蓄热加温装置

在对以上太阳能蓄热系统分析的基础上，提出了使用日光温室自身结构来进行集热与散热的设想，以达到降低投资，同时能够有效利用太阳能的目的；并进一步研究开发了一套日光温室结构兼用式太阳能蓄热加温装置。该加温装置采用水作为蓄热介质，把日光温室钢骨架的上弦和下弦连通，构成水流回路，为使水流平衡，管道采用同程式布置，温室一侧地下建有保温水池，系统组成示意图如图所示。该加温装置采用自动控制运行的方式，白天天气晴好时，通过水流循环使用上下弦集热，保温水池贮热；夜间，温室内温度下降到一定程度时，再通过水流循环使用上下弦散热，以提高温室室内温度，改善温室内环境条件。

该蓄热装置的总投资不到2万元，创新点在于使用日光温室自身的结构作为集热器和散热器，屋架上下弦总面积约为56m^2，能接受太阳辐射总面积约为27m^2。晴天，可使水温提高的最大值为9.49℃，平均值为3.39℃，平均蓄热量约为106.86MJ；夜间放出热量后，水温降低的最大值为9.98℃，平均值为2.27℃，平均放热量为71.55MJ。与对照温室相比，使用屋架太阳能集热系统的温室在晴天中午时室内温度低3~5℃；遇恶劣天气时，夜间室内温度高1~3℃。由于试验的日光温室和屋架太阳能蓄热系统施工质量不甚理想等原因，系统的蓄热加温效果还有很大的提升空间，仍需进一步的优化和全面测试。

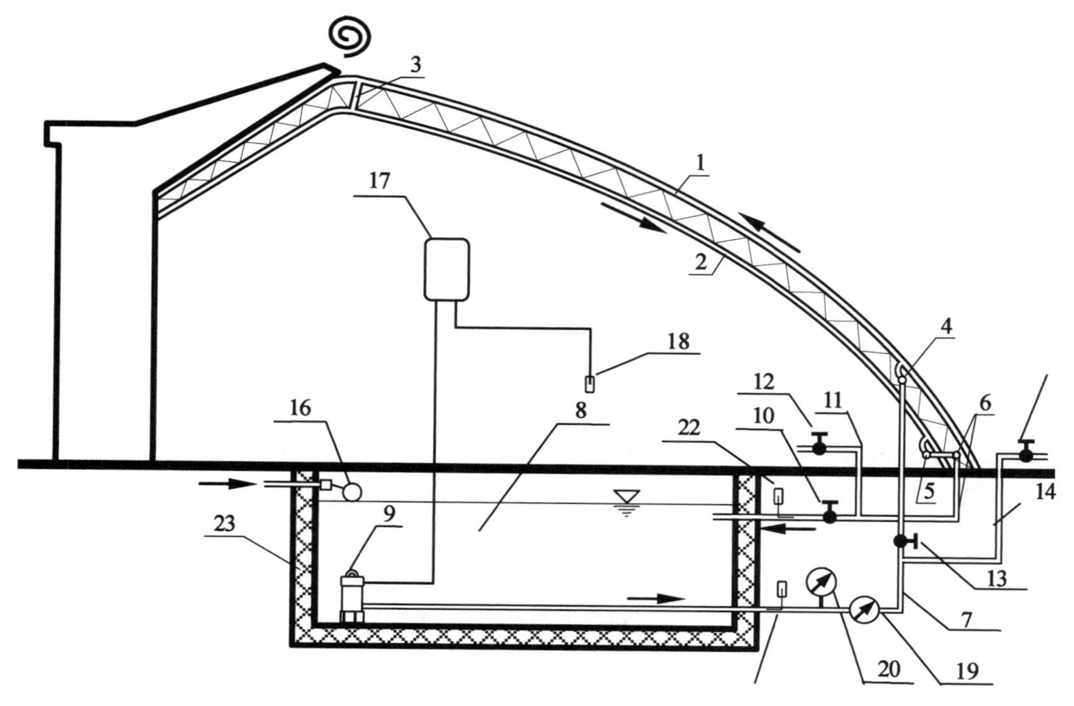

图　屋架太阳能蓄热系统组成示意图

（1）上弦，（2）下弦，（3）短管，（4）供水干管，（5）回水干管，（6）回水总管，（7）供水总管，（8）保温水池，（9）潜水泵，（10）回水阀，（11）热水支管（12）热水阀，（13）供水阀，（14）排污管，（15）排污阀，（16）补水浮球阀，（17）自动监测与控制箱，（18）室内气温传感器，（19）流量计，（20）水压力表，（21）水池温度计，（22）回水温度计，（23）聚苯泡沫板保温层

4　结论与讨论

在今后很长一段时间内，日光温室仍然会是我国北方地区冬季进行作物生产的最主要农业设施之一，温室夜间的加温问题也仍是专家学者们关注的焦点。为了不增加新的能源消耗，研究者们利用“削峰填谷”的原理充分利用太阳能资源，研究设计了各式各样的日光温室太阳能蓄热系统。这些系统可分为以水为媒介的太阳能蓄热系统，以相变材料为蓄热媒介的蓄热系统，以空气作为媒介的蓄热系统和热泵系统。日光温室蓄热系统还需进一步优化系统，以降低投资费用，提高其蓄热效率和夜间的加温效果，起到保证作物冬季的品质并最终提高农户的经济效益，这样才能让普通农户接受，真正达到经济实用的目的。

参考文献

[1] 姜晨光，吕中谦．提高日光温室太阳能利用率的措施．新能源．2000，22（12）：91～93

[2] 亢树华，房思强，戴雅东等．节能型日光温室墙体材料及结构的研究．中国蔬菜．1992（6）：1～5

[3] 郭慧卿，李振海，张振武等．日光温室北墙构造与室内温度环境的关系．沈阳农业大学学报．1995（2）：193～199

[4] 佟国红，许晓辉，白义奎．辽宁省日光温室结构现状及其改进途径．沈阳农业大学学报．1996（5）：26～29

[5] 佟国红．日光温室前屋面优化及结构分析．辽宁：沈阳农业大学（博士论文），1997

[6] 周长吉．“西北型”日光温室优化结构的研究．新疆农机化．2005（6）：37～38
[7] 孟力力．基于VB和MATLAB的日光温室热环境模型构建与结构优化．北京：中国农业科学院（博士论文），2008
[8] 武敬岩，刘荣厚．日光温室结构参数的优化设计-以北方农村能源生态模式为例．农机化研究．2008（2）：80～83
[9] 王晓冬，马彩雯，吴乐天等．日光温室墙体特性及性能优化研究．新疆农业科学．2009（5）：1 016～1 021
[10] 钟改荣，侯有良，郭盛等．对我国现代设施农业发展的思考．农业科技管理．2009，28（5）：30～32
[11] 刘在民．节能日光温室温光性能优化及其应用效果研究．黑龙江：东北农业大学（博士论文），2007
[12] 孟力力，杨其长，宋明军．北京地区日光温室温光及蓄热性能的试验研究．陕西农业科学．2008（4）：61～64
[13] 马承伟，卜云龙，籍秀红等．日光温室墙体夜间放热量计算与保温蓄热评价方法的研究．上海交通大学学报（农业科学版）．2008，26（5）：411～415
[14] 李建设，白青，张红亚．日光温室墙体与地面吸放热量测定分析．农业工程学报．2010，26（4）：231～236
[15] 刘圣勇，张杰，张百良等．太阳能蓄热系统提高温室地温的试验研究．太阳能学报．2003，24（4）：461～465
[16] 王奉钦．太阳能集热器辅助提高日光温室地温的应用研究．北京：中国农业大学（博士论文），2004
[17] 王顺生，马承伟，柴立龙等．日光温室内置式太阳能集热调温装置试验研究．农机化研究．2007（3）：130～133
[18] 李炳海，须晖，李天来．日光温室太阳能地热加温系统应用效果研究．沈阳农业大学学报．2009，40（2）：152～155
[19] 于威，王铁良，刘文合等．太阳能土壤加温系统在日光温室土壤加温中的应用效果研究．沈阳农业大学学报．2010，41（2）：190～194
[20] 方慧，杨其长，梁浩等．日光温室浅层土壤水媒蓄放热增温效果．农业工程学报．2011，27（5）：258～263
[21] 张义，杨其长，方慧．日光温室水幕帘蓄放热系统增温效应试验研究．农业工程学报．2012（4）：188～193
[22] 王宏丽，邹志荣，陈红武等．温室中应用相变储热技术研究进展．农业工程学报．2008，24（6）：304～307
[23] 张勇，邹志荣，李建明等．日光温室相变空心砌块的制备及功效．农业工程学报．2010，26（2）：263～267
[24] 孙心心．日光温室新型保温墙体材料的制备及应用效果的研究．杨凌：西北农林科技大学（博士论文），2010
[25] 郭靖，邹志荣，刘玉凤．不同方式封装的相变材料蓄热效果研究——基于日光温室．农机化研究．2012，34（2）：137～140
[26] 姚轩，王蕊，李珠．玻化微珠及珍珠岩在低能耗日光温室中的应用．山西建筑．2010，36（7）：224～225
[27] 马承伟，黄之栋，穆丽君．连栋温室地中热交换系统贮热加温的试验．农业工程学报．1999（2）：160～164
[28] 白义奎，王铁良．辽沈Ⅰ型日光温室配套设施（1）——地中热交换系统．农村实用工程技术．2002（1）：8
[29] 张晓慧，陈青云，曲梅等．地源热泵空调系统在日光温室中的加温效果．上海交通大学学报（农业科学版）．2008，26（5）：436～439
[30] 柴立龙，马承伟，张晓慧等．地源热泵温室降温系统的试验研究与性能分析．农业工程学报．2008，24（12）：150～154

新疆戈壁地区日光温室冬季环境测试与分析*

蒋程瑶[1**]，邹　平[2]，赵淑梅[1***]，马彩雯[2]，宋　羽[3]，滕光辉[1]，史慧锋[2]

（1. 中国农业大学 农业部设施农业工程重点实验室，北京　100083；2. 新疆农业科学院农业机械化研究所，乌鲁木齐　830091；3 新疆农业科学院品种资源研究所，乌鲁木齐　830000）

摘要：为考察新疆克州地区现有日光温室冬季生产环境性能，并为进一步优化非耕地条件下日光温室墙体及温室设计方案提供理论依据，论文通过智能环境监测仪等设备，对位于克州阿图什市的典型砖墙温室和戈壁石墙日光温室冬季生产环境进行了连续监测。测试结果表明，戈壁石墙温室相对于砖墙温室，在晴天具有升温更快的特性，不通风的情况下，白天最高气温可高于砖墙温室8℃以上；另一方面，墙体热通量的测试结果表明，白天两种墙体蓄热性能相近，而夜间戈壁石墙的放热效果优于砖墙；但是夜间无论是空气温度还是土壤温度，戈壁石墙温室内最低温度均显著低于粘土砖墙温室，甚至出现了零度以下的情况，昼夜温度波动很大，说明该温室保温性能及管理水平有待于进一步提高。

关键词：日光温室；非耕地；环境测试

The Test and Analysis of Solar Greenhouse Winter Environment Performance in Gobi District, Xin Jiang

Jiang Chengyao[1], Zou Pin[2], Zhao Shumei[1], Ma Caiwen[2], Song Yu[3], Teng Gguanghui[1], Shi Huifeng[2]

(1. *Key Laboratory of Agricultural Engineering in Structure and Environment*, *China Agricultural University*, *Beijing*, 100083, *China*; 2. *Institute of Agricultural Mechanization*, *Xinjiang Academy of Agricultural Science*, *Urumqi* 830091, *China*; 3. *Institute of Germplasm Resources*, *Xinjiang Academy of Agricultural Science*, *Urumqi* 830000, *China*)

Abstract: To evaluate the wall thermal insulation performance of solar greenhouse in Kezhou District, Xinjiang, and provide the basis for optimizing wall construction of solar greenhouse under the condition of non-arable land, two solar greenhouses, respectively with fired brick wall and mortar Gobi stone wall, which were completed in Artux City, 2010 were selected as a representative of the geographical features. The thermal environment of tested solar-greenhouses inwinter production were whole day monitored, and the results showed that the highest temperature in greenhouse of Gobi stone wall was higher than it in the fired brick wall one. The Gobi stone wall warmmed up faster in the sunny day, and without ventilation, the greenhouse highest temperature could be 8℃ higher than it fired brick wall. Meanwhile the test of wall heat flow showed such two kinds of wall shared a similar heat storage capacity in the

---

* 基金项目：适合西北非耕地园艺作物栽培的温室结构和建造技术研究与产业化示范（201203002）；“十一五”国家科技支撑计划项目（2009BADA4B04）

** 第一作者简介：蒋程瑶（1988—），女，在读硕士，研究方向为设施园艺。E-mail：catherinejiang@126.com

*** 通讯作者：赵淑梅（1967—），女，博士，副教授，现主要从事设施农业生物环境工程研究工作。E-mail：zhaoshum@cau.edu.cn

day time while the exothermic effect of Gobi stone wall was better in the night. However, in the Gobi Stone wall greenhouse there was a great day and night temperature fluctuations, and both the air and soil temperature were significantly lower than fired brick walls greenhouse, even below 0℃. The greenhouse insulation performance and management levels should be further improved

Key words: Solar-greenhouse; None-arable Land; Environment Test

1 引言

克州地区位于新疆西南部，是新疆人均耕地面积最少的地州之一，全州土地山地面积占90%，戈壁荒滩近1 000万亩[1]。近年来日光温室产业发展迅速，截至2010年8月，已建成日光温室3 469座，其中，戈壁温室1 289座，占温室总座数的37.2%；已完成定植投入生产2 805座，占温室总座数的80.9%[2]。由于克州独特的地理位置和气候条件，日光温室在设计及选材上应有别于其他地区。关于该地区日光温室构造参数的研究，梁建龙等[3]根据经验提出了相应的构造参数，但是，缺乏理论依据；裴先文等[4]对针对南疆巴州地区日光温室探讨了前屋面形状优化设计的方法；马彩雯等[5]于2010年对喀什地区的日光温室构造参数做了进一步优化设计，并且初步测试结果显示，经过优化设计的温室在冬至日前后室内最低气温可达10.7℃、最低地温可达12.1℃[6]。

但是，实际上，经过标准化设计的温室在建设过程中根据当地情况或多或少做过调整，另外，经过两年的使用，该地区标准化日光温室的环境性能是否还能维持在较好的水平，尚需要进一步测试检验。因此，于2012年冬至前后的近一个月的时间，对该地区2010年建设的典型日光温室的冬季生产环境进行了详细测试，以期为现有日光温室的管理、维护以及新建温室的改进提供理论依据。

2 材料与方法

2.1 测试温室概况

测试地点选择在克州阿图什市阿湖乡阿尔赛小区的温室园区。由于当地为戈壁荒滩，墙体采用隔壁石砌筑的温室较多，同时也有传统的土墙温室和粘土砖墙温室。为了测试方便，选择了构造参数相近的粘土砖墙温室和戈壁石墙温室各1栋作为测试温室。测试温室方位为南偏西5°，其他主要构造参数如表1所示。

冬季晴天保温被的卷放时间通常为北京时间10：00以后和18：00以前，根据天气情况灵活调整。

测试期间砖墙温室为处于定植准备期，戈壁石墙温室处于收获结束后的拉秧阶段。

表1 试验温室墙体材料及构造尺寸 （单位：m）

温室类型	墙体构造	东西跨度	南北净跨	下沉深度	屋脊高度	北墙高度	后墙厚
砖墙	0.12mm 黏土砖 + 沙土	100	8.5	0.85	3.7	2.7	1.5
戈壁石墙	戈壁石 + 砼抹面	80	8.5	0.85	3.69	2.26	1.5

2.2 测试方法

测试内容包括室内空气温度、相对湿度，土壤温度分布、墙体热通量，以及室外空气温度和相对湿度。其中，室内空气温度测点设置于温室中部的跨中位置，沿温室长度方向间隔20m共设3个测点，测点距地面1.0m；相对湿度测点设置在前述中间温度测点处。土壤温室在温室中部，沿温室跨度方向距南墙2m、4m、6m处、且距地面0m、0.1m、0.2m、0.3m深度处各设1个测点。墙体热通量在温室中部距地面0.9m和1.8m处各设一个点。在室外没有遮阴处设置一点测试室外空气温度和相对湿度。测点布置图如图1所示，所用测试仪器如表2所示。

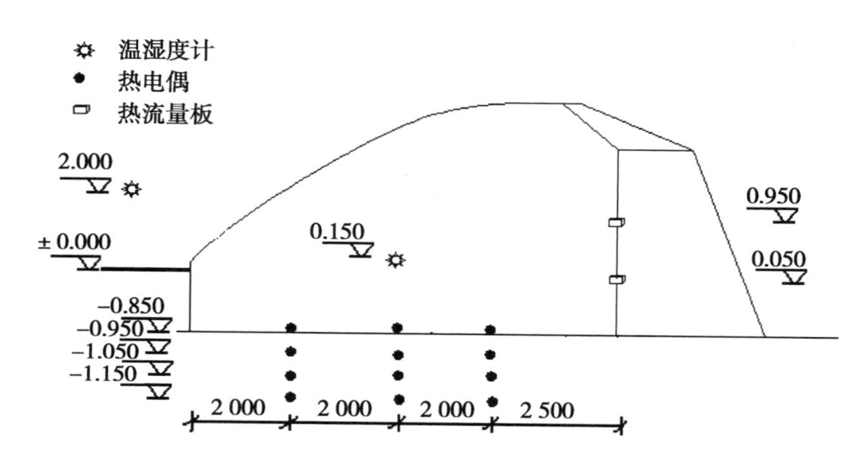

图1　温室内外测点分布图

表2　测试仪器一览表

仪器名称	测量内容	精度	生产厂家
RS-13、RS-13H	空气温度及相对湿度	±0.3℃、±5%	日本ESPECMIC公司
RT-13	空气温度	±0.3℃	日本ESPECMIC公司
T型热电偶	土壤温度	±0.15℃	国产
HFP01热流量板	墙体热流量	$50mV/W.m^{-2}$	荷兰HUKSEFLUX公司
34970A数据采集仪	温度及热流量	—	美国ANGILENT公司

3　结果与分析

3.1　典型天气条件下的室内空气温度

一般冬至日前后，室外光照条件较弱，气温较低，不利于日光温室的生产。特别是在阴雪等不利天气状况时，更是对日光温室的考验。测试期间的2012年12月28～31日，室外天气分别为雪、阴、晴/风和多云，其中，12月28日未揭保温被，具有一定的典型性，因此，以下将着重针对这段时间的测试结果进行分析。

图2为2012年12月28～31日的期间室内外空气温度测试结果。从室外温度来看，雪天的气温在－3.4～－10.4℃，不算太低，但是在晴天时，气温在1.1～－22℃，日变化较大，且最低温度显著降低。从室内气温来看，两种温室温度变化曲线规律相近，但温度值的差异显著。其中，在未揭帘的雪天和夜间，戈壁石墙温室的温度显著低于砖墙温室，最低出

现在12月30日早上11：00，为-6.9℃，二者最大气温差出现在12月31日10：00揭帘前，达8.4℃。但是在晴天的白天，戈壁石墙温室则显示出了快速的增温效果，在天气较好的情况下，最高温度达34.2℃，高出砖墙温室8.7℃。

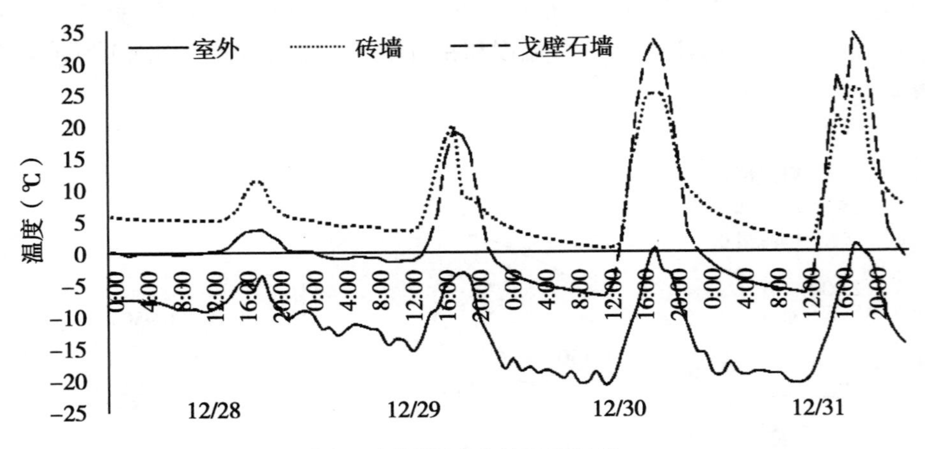

图2　测试温室内外空气温度

3.2　室内土壤温度

土壤的蓄放热作用对日光温室夜间热环境的维持具有极其重要的作用，因此，有必要考察土壤温度分布，特别是不利天气情况下的土壤温度变化情况。故而仍选择2012年12月28～30日的测试数据进行分析，对室内跨度方向同一深度的3个测点值取平均，作为该深度土壤温度的数值进行比较。不同天气条件下两种温室不同位置土壤温度分布，如图3所示。

从温室内垂直方向土壤温度分布情况看出，两个温室的地表温度日变化规律与空气温度相似，地表温度波动最大，数值与室内气温相近，而随着土壤深度的增加，温度的日变化逐渐缩小，在0.3m伸出基本接近稳定，这说明其受太阳辐射和室内气温影响较大。从定量数值上看，晴天条件下戈壁石墙温室的地表温度从半夜至中午揭帘前均低于0℃，最低温度达-4.7℃，与砖墙温室相差6℃以上；砖墙温室虽然地表温度高于戈壁石墙温室，但最低也仅为1.5℃。0.3m深度的温度比较稳定，砖墙温室和戈壁石墙温室的平稳度分别为9.3℃和7.9℃，砖墙温室略高。

3.3　墙体热通量

在不加温的条件下日光温室热环境得以维持的最重要的因素就是温室后墙这一构造所拥有的蓄热放热特性极其良好的保温作用。同样选择2012年12月28～30日的测试数据进行分析，测试结果如图4所示，其中，正值表示墙体蓄热、负值表示墙体放热。

从图4可以看出，有太阳辐射的条件下，两种材料的墙体均显示出了良好的蓄热特性，瞬时最大蓄热能力基本相同，但是，与砖墙相比，戈壁石墙体对太阳辐射变化的响应时间相对延迟，受其影响较小，整体的热稳定性较强，累计蓄热量略多于砖墙。但是在阴雪或者夜间，戈壁石墙体的放热能力却明显高于砖墙，这说明戈壁石墙具有优于砖墙的蓄、放热能力，更适用于日光温室。

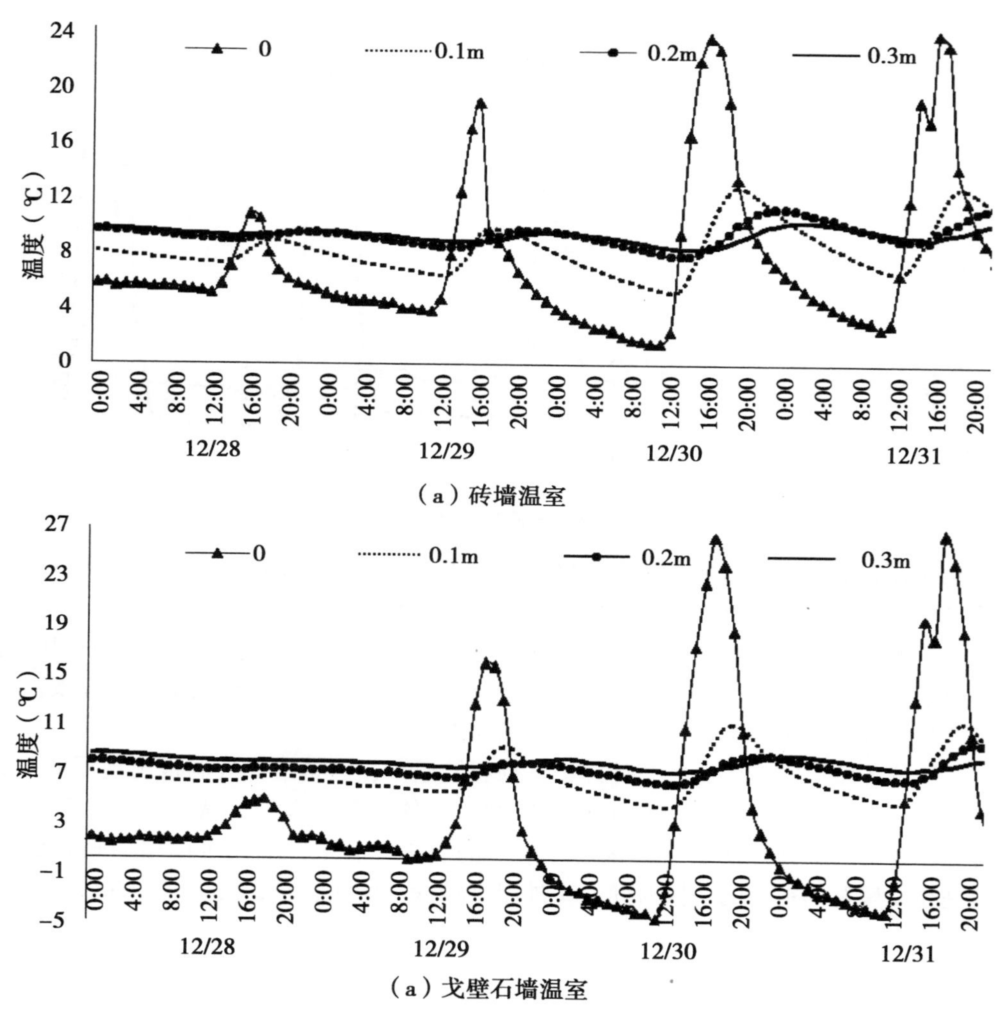

图3　测试温室土壤温度

4　讨论与结论

第一，在完全不加温的条件下，室外气温最低降到－20.5℃时，砖墙温室和戈壁石墙温室分别能维持23.7℃和15.3℃室内外空气温度差，再次证明了日光温室的良好保温蓄热性能。但从数值上也可以看出，测试的砖墙温室其整体保温性优于戈壁石墙温室。

第二，从两种温室墙体构造本身的蓄放热特性来看，戈壁石墙的白天累计蓄热能力和夜间放热能力更好，表明戈壁石墙构造更适合于日光温室。特别是从墙体材料的就地取材特性和经济性上来看，在新疆戈壁地区，戈壁石墙温室具有良好应用前景，应在新建温室中广泛应用。

第三，虽然戈壁石墙具有良好的蓄放热特性，但所测试的温室空气温度和土壤温度温度条件均低于砖墙温室，甚至出现了负值，说明温室的保温管理亟需加强。从现场的情况也可

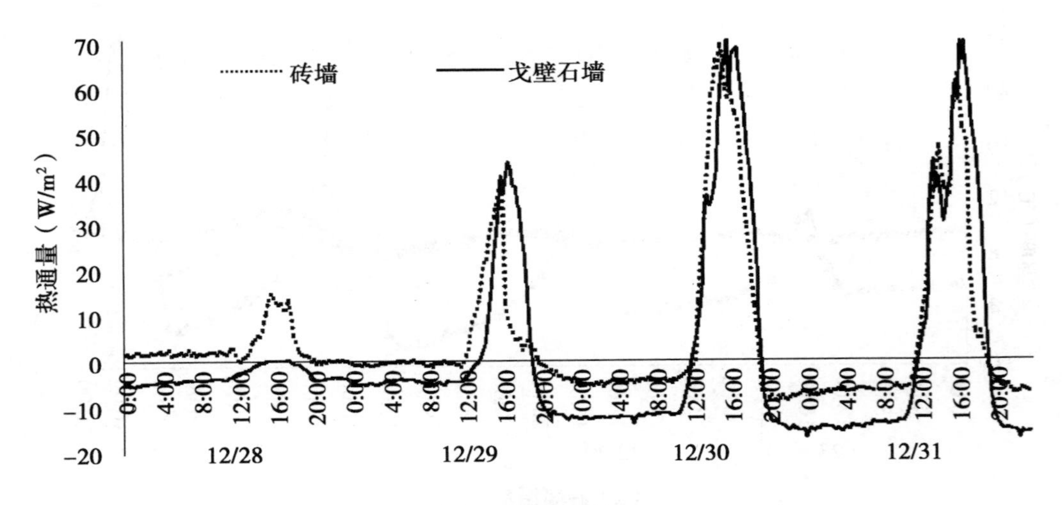

图 4　墙体热通量

以看出，戈壁石墙温室的保温被存在破损严重的现象，加之处于收获后的拉秧期，日常管理上也不是很精细，这应是导致该温室热环境条件明显不如砖墙温室的主要原因。

第四，两个测试温室在墙体构造上的差异，也导致了戈壁石墙温室单位土壤面积所对应的墙体面积小于砖墙温室，进而导致了温室总蓄热、放热量上少于砖墙温室。这应该也是戈壁石墙温室的温度环境条件不如所测砖墙温室的原因之一。

第五，尽管砖墙温室整体温度条件好于戈壁石墙温室，但是最低气温也出现了低于 5℃ 的情况，远低于番茄、黄瓜等常见设施栽培作物生长的最低要求[7]，从这一角度讲，两个温室现有温热环境并不完全满足该地区越冬生产的要求，有条件的话可以增加临时加温设备，特别是在凌晨至揭帘前的一段时间，室外温度较低，应注意防止发生冻害现象。

参考文献

[1] 努尔比亚・库尔班．克州戈壁产业的现状及解决对策．边疆经济与文化，2012（7）：16～18

[2] 克州党委办．克州设施农业经营模式及发展建议．http：//www. xjnb. gov. cn/news/Show. asp？ id = 11081. 2010 － 08 －10

[3] 梁建龙，杨保存．南疆地区节能日光温室结构参数的选择与设计．新疆农机化，2002（2）：43～45

[4] 裴先文，史为民，曲良举等．南疆巴州地区日光温室前屋面优化设计研究．北方园艺，2010（16）：63～66

[5] 马彩雯，王晓冬，邹平等．新疆新型高效节能日光温室标准化设计探讨．中国农机化，2010（2）：47～51

[6] 马彩雯，吴乐天，王晓冬等．喀什地区日光温室小气候试验研究．农机化研究，2010（9）：179～183

[7] 马承伟，苗香雯．农业生物环境工程．北京：中国农业出版社，2005

设施栽培理论技术

Effects of Light Intensity and Nutrient Addition on Growth, Photosynthetic Pigments and Nutritional Quality of Pea Seedlings

Liu Wenke*, Yang Qichang, Qiu Zhiping, Zhao Jiaojiao

(*Environment and Sustainable Development in Agriculture, Chinese Academy of Agricultural Sciences, Key Lab. for Energy Saving and Waste Disposal of Protected Agriculture, Ministry of Agriculture, Beijing* 100081, *China*)

Abstract: A glasshouse experiment was carried out to investigate the effects of shading and nutrient addition on growth, photosynthetic pigment content and nutritional quality of substrate-cultivated pea seedlings. The results showed that shading and nutrient addition significantly altered the growth, photosynthetic pigment, vitamin C and nitrate contents of pea seedlings. Nutrient addition improved shoot, root and total biomass significantly of pea seedling under non-shading condition, whereas nutrient addition only increased shoot biomass of pea seedling under shading condition. However, nutrient addition decreased root and total biomass of pea seedling under shading condition. Contents of chlorophyll a in stem and chlorophyll b in leaves were lowered by nutrient addition under non-shading condition, but carotenoid content in stem was improved. Additionally, contents of chlorophyll a and chlorophyll b in stem or leaves were similar under shading condition, and carotenoid content in stem was decreased under shading condition. Shading and nutrient addition markedly decreased vitamin C content in shoot of pea seedlings. Shading did not alter the nitrate content in shoot of pea seedlings, but nutrient addition increased nitrate content significantly. The results suggest that non-shading with nitrogen supply is beneficial conditions for commercial production of pea seedlings in terms of high yield, while non-shading without nitrogen supply is best for good nutritional quality. Reasonable matching of light intensity and nitrogen supply level may contribute the yield on the basis of high nutritional quality of pea seedlings.

Key words: Light intensity; Nitrogen fertilizer; Vitamin C; Carotenoids; Chlorophyll

1 Introduction

Nowadays, consumption of sprouts has become increasingly pupular worldwide because they are often perceived as part of a healthy diet. Sprouts are believed to be rich in health-promoting phytochemicals, particularly sprouts of cruciferous plants and legumes. For example, green sprouts of radish, broccoli, pea, alfalfa are known for their health-promoting phytochemicals with antioxidant properties (Hesterman *et al.*, 1981; Plum *et al.*, 1997; Takaya *et al.*, 2003; Zielinski *et al.*, 2007), such as vitamin C, flavonoids, carotenoid etc.. Additionally, research has showed that plant-based diets rich in phytochemicals with antioxidant properties were beneficial for decreasing the incidence of chronic and degenerative diseases, including cardiovascular diseases and several types

* Corresponding author: E-mail: liuwke@163.com

of cancer (Birt *et al.*, 2001; Hu, 2003). Generally, cultivation of green sprouts undergos a short growth period, usually less half a month, even several days for some species. Thus, as protected vegetables, sprouts are good candidates for promoting nutritional quality through regulation of environment and nutritients in facilities.

Environmental condition and nutrient supply are crucial factors that determine the yield and nutritional quality of sprouts during cultivation. In the past decade, some studies had been conducted to investigate the beneficial effects of light condition and environmental shocks on nutritional quality of green sprouts (Wu *et al.*, 2007; Xu *et al.*, 2005; Zhang *et al.*, 2010a, b). Light condition affects growth of green sprouts through light intensity and light quality. However, the optimal light intensity and spectral requirement for sprouts is not well understood. Therefore, many studies have attempted to declare the optimal light quality for various sprouts with high nutritional quality and yield, such as pea seedling, soybean sprouts, radish sprouts and toona sinensis seedling, and these literatures indicated that optimal light quality could improve the growth and nutritional quality of green sprouts (Wu *et al.*, 2007; Xu *et al.*, 2005; Zhang *et al.*, 2010a, b).

Previous reports showed that exposure of sprouts to high light or chilling resulted in higher total phenolic content and antioxidant capacity, and during recovery following shock treatments, high light produced a stronger response in increasing the levels of individual phenolic compounds in alfalfa sprouts without dry biomass loss (Oh and Rajashekar, 2009). In addition, after being exposed to heat shock (40°C for 10min), chilling (4°C for 1d) or high light intensity ($800\mu mol \cdot m^{-2} \cdot s^{-1}$ for 1d), a two to threefold increase in the total phenolic content and a significant increase in the antioxidant capacity was found in lettuce (Oh *et al.*, 2009). So far, however, very little is known about the effects of light condition and nutrient supply on growth and nutritional quality of green sprouts. In this study, the effects of shading and nutrient addition on growth and nutritional quality pea seedlings in glasshouse was conducted to clarify the roles of light intensity and nutrient supply in regulating nutritional quality of pea seedlings.

2 Materials and methods

Cutivar of pea used in experiment is 'Shenchun'. The experiment started on 23 August, 2011, all plants were harvested on 5 September, 2011. Four treatments are designed, including shading without nutrient addition (S), shading with nutrient addition (S + N), non-shading without nutrient addition (NonS) and non-shading with nutrient addition (NonS + N). Each treatment replicated three times. Uniform pea seeds were soaked in tap water for one hour, and then sown in plug tray with six holes. Three seeds were planted in each plug tray hole. Plug trays were placed in plastic box (length × width × height = 38cm × 18cm × 10cm) and immersed into distilled water or 1/2 strength nutrient solution for germination and growh. Nutrient solution used is the same as the full nutrient solution in Liu *et al.* (2012). 1/2 strength nutrient solution contained 5mmol/L nitrate nitrogen.

During the study period, the photosynthetic photon flux (PPF) of photosynthetic active radiation (PAR) between 400 ~ 700nm were determined with AvaSpec-2048 fibre spectrometer, made in the Netherlands. PPF of daytime ranged from 200 to 1400μmol · m^{-2} · s^{-1}. The shading treatment with aluminized screens decreased an average 54% ± 0.046% ($n = 20$) photosynthetically active radiation against the light intensity under the non-shading condition (Liu *et al.*, 2012).

After ten-day treatment, nine pea seedlings per pot were sampled and weighed for shoot and root biomass, then shoot was used to determine chlorophyll a, chlorophyll b, carotenoid (stem and leaf were determined separately), nitrate and vitamin C contents in shoot. Chlorophyll and carotenoid content of leaves was examined by colorimetric method extracted with 80% acetone (Li *et al.*, 2000), and nitrate and vitamin C contents in leaves and stems were determined by colorimetric method (Liu *et al.*, 2012) and titrimetry method respectively (Cao *et al.*, 2007).

3 Shoot and root biomass of pea seedlings

Shading and nutrient addition significantly altered the shoot, root and total biomass of pea seedlings (Table 1). Non-shading with nutrient addition treatment presented the highest shoot, root and total biomass, while non-shading without nutrient addition treatment showed the smaller shoot and total biomass in four treatments. Nutrient addition improved shoot, root and total biomass significantly of pea seedling under non-shading condition, whereas nutrient addition only increased shoot biomass of pea seedling under shading condition. Moreover, nutrient addition decreased root and total biomass of pea seedling under shading condition.

Table 1 Shoot and root biomass of pea seedlings treated with or without shading and nutrients addition

Treatments	Shoot biomass (g/pot)	Root biomass (g/pot)	Total biomass (g/pot)
NonS	5.50c	4.59ab	10.09b
NonS + N	7.55a	5.16a	12.71a
S	6.21bc	4.87a	11.08ab
S + N	6.77ab	3.85b	10.62b

Note: In the same column, the different letters following the averagesrepresent significant difference at $P < 0.05$.

4 Photosynthetic pigment contents in leaves and stems of pea seedlings

Shading and nutrient addition significantly affected the chlorophyll a, b and carotenoid contents in stem or shoot of pea seedling (Table 2). Contents of chlorophyll a in stem and chlorophyll b in leaves were lowered by nutrient addition under non-shading condition, but carotenoid content in stem was improved. However, contents of chlorophyll a and chlorophyll b in stem or leaves were similar under shading condition. Additionally, carotenoid content in stem was decreased under shading condition. Non-shading without nutrient addition treatment showed highest chlorophyll a and chlorophyll b content in stem and leaves respectively among four treatments, while non-shading with nutrient addition treatment presented the highest carotenoid content in stem.

Table 2 Photosynthetic pigments in leaves and stems of pea seedlings treated with or without shading and nutrients

Treatments	Chlorophyll a content (mg/g)		Chlorophyll b content (mg/g)		Carotenoid content (mg/g)	
	Stem	Leaves	Stem	Leaves	Stem	Leaves
NonS	0. 084a	1. 06a	0. 040a	0. 61a	0. 018b	0. 17a
NonS + N	0. 064b	0. 94a	0. 040a	0. 47b	0. 023a	0. 17a
S	0. 074ab	0. 94a	0. 038a	0. 47b	0. 017b	0. 17a
S + N	0. 076ab	0. 92a	0. 036a	0. 42b	0. 019ab	0. 14a

Note: In the same column, the different letters following the averagesrepresent significant difference at $P < 0.05$.

5 Vitamin C content in shoot of pea seedlings

Shading and nutrient addition markedly decreased vitamin C content in shoot of pea seedlings (Figure 1). Non-shading without nutrient addition treatment showed highest vitamin C content in shoot of pea seedlings, while shading with nutrient addition treatment presented the lowerest vitamin C content. There were no difference in shoot vitamin C content between treatments of non-shading with nutrient addition and shading without nutrient addition.

Nitrate content in shoot of pea seedlings

Shading did not alter the nitrate content in shoot of pea seedlings, but nutrient addition increased nitrate content significantly (Figure 2). Contents in shoot of pea seedling of nutrient addition treatments under shading or non-shading are similar, which about four times higher than treatments without nutrient addition.

6 Discussion

Pea seedling is kind of sprouts, delicious and rich in nutritional substances. Generally, light and nutrient are necessary factors that related to growth and quality of pea seedlings in cultivation. The results showed that shading and nutrient addition significantly altered the shoot, root and total biomass, pigment contents, vitamin C and nitrate contents of pea seedlings. Non-shading without nutrient addition treatments showed smaller shoot and total biomass than shading treatment, while non-shading with nutrient addition treatment presented the highest shoot, root and total biomass. This indicated that high light intensity inhibited pea seedling growth without nutirient supply, thus we suggested pea seedling production is not suitable being conducted directly under cover without nutirient supply. In addition, nutrient supply and non-shading combination is the optimal conditions for growth of pea seedlings. This may be attributed to the shortage of nutrients under high light intensity, particularly nitrogen during the rapid growth of pea seedlings. Furthermore, shading and nutrient addition significantly affected the chlorophyll a, b and carotenoid contents in stem and shoot of pea seedling differently, which also influences the growth of pea seedlings. Therefore, in order to obtain high productivity of pea seedlings, matchable light intensity and nitrogen supply is necessary.

Current data showed that shading and nutrient addition markedly decreased vitamin C content in shoot of pea seedlings. This suggested that certain level of light intensity is important for vitamin C accumulation since activity of key synthetic enzyme of vitamin C is promoted by elevated light intensity. Simultaneously, high light intensity is in favor of the decline of nitrate in pea seedlings. Therefore, higher light irradiance is beneficial for nutritional quality improvement of pea seedlings in terms of vitamin C and nitrate contents. However, as our data showed that over-high light intensity inhibited growth of pea seedlings due to shortage of nitrogen supply for photosynthesis. So, suitable nitrogen supply is helpful for high yield. Although shading did not altered the nitrate content in shoot of pea seedlings, nutrient addition increased nitrate content significantly. So in practical cultivation, much lower nitrogen concentration should be set for avoiding nitrate accumulation. These results suggest that high light intensity and low nitrogen solution may be the optimal conditions for commercial production of pea seedlings in terms of high yield and good nutritional quality. In conclusion, non-shading with low-level nitrogen supply is beneficial conditions for commercial production of pea seedlings in terms of high yield, while non-shading without nitrogen supply is best for good nutritional quality.

7 Acknowledgments

This study was supported by the National High Technology Research and Development Plan of China (863 Project, grant No. 2011AA03A114) and the Basic Scientific Research Fund of National Nonprofit Institutes (BSRF201004) and 2012 ~ 2013, Institute of Environment and Sustainable Development in Agriculture, CAAS.

References

[1] Birt, D. F., Hendrich, S. & Wang, W. 2001. Dietary agents in cancer prevention: flavonoids and isoflavonoids. Pharmacology & Therapeutics, 90: 157 ~ 177

[2] Cao, J. K., Jiang W. B. & Zhao, Y. M. 2007. Experimental guidance of postharvest physiology and biochemistry of fruits and vegetables. China Light Industry Press. Beijing

[3] Hesterman, O. O. B., Teuber, L. R. & Livingston, A. L. 1981. Effect of environment and genotype on alfalfa sprout production. Crop Science, 21: 720 ~ 726

[4] Hu, F. B. 2003. Plant-based foods and prevention of cardiovascular disease: an overview. The American Journal of Clinical Nutrition, 78: 544S ~ 551S

[5] Li, H. S. 2005. Principles and techniques of plant physiological biochemical experiment, Higher Education Press. Beijing

[6] Liu, W. K., Yang, Q. C. & Du, L. F. 2012. Effects of short-term treatment with various light intensities and hydroponic solutions before harvest on nitrate reduction in leaf and petiole of lettuce. Acta Agriculturae Scandinavica, Section B - Soil & Plant Science, 62 (2): 109 ~ 113

[7] Oh, M. M., Trick, H. N. & Rajashekar, C. B. 2008. Secondary metabolism and antioxidants are involved in environmental adaptation and stress tolerance in lettuce, Journal of Plant Physiology, 166: 180 ~ 191

[8] Oh, M. M., Carey, E. E. & Rajashekar, C. B. 2009. Environmental stresses induce health-promoting phytochemicals in let-

tuce. Plant Physiology and Biochemistry, 47: 578 ~583

[9] Oh, M. M. & Rajashekar, C. B. 2009. Antioxidant content of edible sprouts: effects of environmental shocks. Journal of the Science of Food and Agriculture, 89: 2 221 ~2 227

[10] Plum, G. W., Price, K. R., Rhodes, M. J. & Williamson, G.. 1997. Antioxidant properties of the major polyphenolic compounds in broccoli. Free Radical Research, 27: 429 ~435

[11] Takaya, Y., Kondo, Y., Furukawa, T. & Niwa, M. 2003. Antioxidant constituents of radish sprout (Kaiware-daikon), *Raphanus sativus* L. Journal of Agriculture and Food Chemistry, 51: 8 061 ~8 066

[12] Wu, M. C., Hou, C. Y., Jiang C. M., Wang Y. T., Wang, C. Y., Chen, H. H. & Chang, H. M. 2007. A novel approach of LED light radiation improves the antioxidant activity of pea seedlings. Food Chemistry, 101 (4): 1 753 ~1 758

[13] Xu, M. J., Dong, J. F. & Zhu, M. Y. 2005. Effects of germination conditions on ascorbic acid level and yield of soybean sprouts, Journal of the Science of Food and Agriculture, 85: 943 ~947

[14] Zieliński, H., Piskuta, M. K., Michalska, A. & Kozowska, H. 2007. Antioxidant capacity and its components of cruciferous sprouts. Poland Journal of Food Nutritional Science, 57: 315 ~322

不同樱桃番茄品种抗逆性评价试验*

高艳明**，周　筠，李建设

（宁夏大学农学院，宁夏　银川　750021）

摘要：综合各项逆境指标表明，3 种番茄在不同盐浓度胁迫、干旱胁迫和低温胁迫下，贝蒂叶片膜透性和 MDA 含量始终高于贝美和摩丝特，摩丝特又高于贝美；贝美叶片 SOD 酶活、POD 酶活始终高于贝蒂和摩丝特，摩丝特酶活性又高于贝蒂。因此，3 种番茄在逆境胁迫下表现出的抗逆性强弱为贝美 > 摩丝特 > 贝蒂。

关键词：樱桃番茄；抗逆性；生理指标

The Test of Resistance Evaluation on Different Cherry Tomato

Gao Yanming, Zhou Yun, Li Jianshe

(*Agricultural college of Ningxia University, Yinchuan Ningxia* 750021, *China*)

Abstract: All the adversity index showed that the comprehensive, three kinds of tomatoes in different salt concentration stress, drought stress and low temperature stress, the Bei Mei's leaf membrane permeability and MDA content always higher than the Bei Di and Mo Si Te, Mo Si Te is higher than Bei Di; The Bei Mei's leaf SOD, POD enzyme live always higher than the Bei Di and Mo Si Te, Mo Si Te is higher than Bei Di. Therefore, the three kinds of tomatoes in adversity stress resistance performance of strength for the Bei Mei > Mo Si Te > Bei Di.

Key words: Tomato; Stress resistance; Physiological index

盐渍化土壤在世界上分布很广，面积占世界陆地总面积的 1/3[1]。近年来我区设施栽培给农民和社会带来了显著的经济效益，与此同时，农业生产中化肥施用量高，且长期处于封闭环境，无自然降雨的淋洗，温度高，蒸发量大，因此，土壤盐分经过积累聚集在土壤表层，从而导致栽培作物品质下降。

在番茄的早春育苗期间常会因为低温的影响而使生长发育受阻，严重影响到其成株期的生长发育及早期产量和果实的商品性[2]。因此，培育耐低温的品种已成为国内外育种家们的主要目标之[3~4]，同时，对植物逆境生理进行研究也显得尤为重要。

水是农业的命脉，也是整个国民经济和人类生活的命脉。我国干旱、半干旱面积占全国土地面积的一半以上（52.5%），约有耕地面积 5.7 亿亩。全国人均水资源不足世界人均的 1/4，排在世界第 109 位，属 13 个贫水国之一[5]。因此，系统开展优异抗旱番茄品种的筛选

---

* 基金项目：国家星火计划重大项目“宁夏设施园艺优质高效安全生产关键技术集成示范与推广（2011GA880001）”宁夏科技攻关项目“宁夏设施园艺产业发展关键技术研究与示范”

** 作者简介：高艳明（1963—），女，宁夏石嘴山人，硕士，教授。研究方向为园艺植物营养与施肥，E-mail：myangao2@yahoo.com.cn

及其抗旱性鉴定与评价的研究，对指导我国干旱半干旱地区番茄产业化稳步发展尤为必要，且具有现实意义。

1 试验材料

试验在宁夏永宁领鲜设施果蔬标准化生产示范基地（永宁县杨和镇纳家户村）苗棚内进行。试验日光温室为钢架结构，外覆盖保温材料草帘。供试番茄品种为贝美（荷兰瑞克斯旺种苗公司）、贝蒂（荷兰瑞克斯旺种苗公司）、摩丝特（104，荷兰瑞克斯旺种苗公司）。

2 试验设计

试验于2011年5月5日将供试番茄品种播种于穴盘中进行育苗，育苗期间按常规方法管理。待番茄苗长到两叶一心时移栽到直径10cm、高10cm的塑料营养钵中，基质栽培，待番茄幼苗3片真叶完全展开时，选取长势一致的幼苗进行抗逆性试验。

2.1 耐盐性试验

盐胁迫分6组处理：①纯净水作为对照；②25mmol/L NaCl溶液；③50mmol/L NaCl溶液；④100mmol/L NaCl溶液；⑤150mmol/L NaCl溶液；⑥200mmol/L NaCl溶液。每处理100株，3次重复，在温室内随机排列，用各处理配液进行浇灌，15d后进行生理指标的测定。

2.2 耐寒性试验

选取长势一致的幼苗，将植株放入ZRX-300D智能人工气候培养箱内进行5℃低温胁迫处理7d。光照强度设为5 000lx，光照时间为12h（6：00～18：00），每品种株数为30株。

2.3 耐旱性试验

当植株长至3～4片真叶时进行极度干旱处理：每个营养钵浇30ml水，当营养钵中基质见干发白时，再连续干旱2～3 d，至60 %的材料有半数以上植株出现极度干旱症状时，调查各材料单株干旱症状。然后每个营养钵再浇30ml水，2d后观察植株恢复生长情况，统计各材料单株恢复情况。

旱情分级标准参照张丽英等番茄苗期抗旱性鉴定及其评价方法的研究[6]。

苗期旱情分级标准：

0级—幼苗正常，无任何症状；

1级—新叶暗淡，外叶明显萎蔫；

3级—新叶明显萎蔫，外叶严重萎蔫，茎直立；

5级—叶片严重萎蔫，茎倒伏或半倒伏；

7级—植株严重萎蔫失水，茎倒伏；

其中5级、7级为极度干旱症状。

极度干旱后植株恢复生长分级标准：

0级—幼苗完全恢复正常，无任何干旱伤害症状；

1级—幼苗基本恢复正常，下部1～2片叶明显萎蔫；

3级—茎直立，生长点有活力，叶片严重萎蔫难以恢复正常，植株受到严重伤害；

5级—植株严重萎蔫死亡，不可恢复。

3 测定项目

（1）膜透性的测定（电导仪法）；

(2) MDA 含量的测定（双组分分光光度计法）；
(3) 脯氨酸（Pro）含量的测定（酸性茚三酮法）；
(4) 超氧化物歧化酶（SOD）活性的测定［氮蓝四唑（NBT）法］；
(5) 过氧化物酶（POD）活性的测定（愈创木酚法）。

4 结果与分析

4.1 不同浓度 NaCl 对樱桃番茄生理指标的影响

4.1.1 对细胞膜相对透性和丙二醛含量的影响

植物在受到各种逆境（如干旱、低温、高温、盐渍和大气污染等）危害时，细胞膜的结构和功能首先受到伤害，导致膜透性增大。细胞膜在逆境下的稳定性反映了植物抗逆性的高低。丙二醛（MDA）是膜脂过氧化的产物，其含量可用来间接表示细胞膜的受伤程度。

由表 1 可知，在 NaCl 胁迫 15 d 后，随着盐浓度的增高 3 个品种膜透性、丙二醛含量呈升高趋势。与对照相比，贝美的膜透性分别提高 29.95%、35.49%、59.83%、100.19% 和 120.99%，丙二醛含量分别提高 21.15%、39.27%、50.15%、58.80% 和 84.29%；摩丝特膜透性提高 11.36%、26.58%、40.48%、67.97% 和 82.48%，丙二醛含量提高 3.69%、10.81%、19.16%、33.42% 和 55.53%；而贝蒂的膜透性分别提高 7.59%、20.31%、41.72%、63.05% 和 82.59%，丙二醛含量分别提高 1.27%、7.22%、15.92%、27.81% 和 67.30%。可见，盐胁迫对耐盐性较弱的贝蒂叶片细胞质膜的伤害程度始终高于耐盐性较强的贝美和摩丝特。

表 1 不同浓度 NaCl 对番茄叶片膜透性和丙二醛含量的影响

Table 1 Effect of NaCl stress on the relative electric conductivity and content of MDA in cherry tomato seedling leaves

NaCl 浓度（mmol/L）	膜透性（%）			丙二醛（mmol/g）		
	贝美	摩丝特	贝蒂	贝美	摩丝特	贝蒂
0（CK）	26.91dC	33.00eD	35.16eE	3.31 dC	4.07 dC	4.71 bB
25	34.97cBC	36.75eCD	37.83eDE	4.01 cdBC	4.22 dC	4.77 bB
50	36.46bcBC	41.77dBC	42.30dD	4.61bcABC	4.51 cdC	5.05 bB
100	43.01bB	46.36cB	49.83cC	4.97 bcAB	4.85 bcBC	5.46 bB
150	53.87aA	55.43bA	57.33bB	5.19 abAB	5.43 bB	6.02 bAB
200	59.47aA	60.22aA	64.20aA	6.10 aA	6.33 aA	7.88 aA

4.1.2 对叶片 POD、SOD 酶活性的影响

POD 作为植物内源的活性氧清除剂，逆境中维持较高的酶活性，才能有效地清除活性氧使之保持较低水平，减少其对膜结构和功能的破坏。SOD 是生物体内普遍存在的金属酶，在酶促保护系统中，它与 POD 酶、CAT 酶等酶协同作用，防御活性氧或其他过氧化物自由基对细胞膜系统的伤害，处于核心地位。SOD 酶活性可以间接反映植物抗逆性的高低。

由表 2 可以看出，随着盐浓度的增高，贝美、摩丝特的 POD 酶活性呈升高趋势，贝蒂 POD 酶活性则表现先升高后下降的趋势，在 150mmol/L 的盐胁迫下达到最大值，处理间差

异显著。在200mmol/L的盐胁迫下，贝美POD酶活比对照增加307.14%，摩丝特增加273.69%，贝蒂增加218.63%。结果表明，耐盐性不同的樱桃番茄品种幼苗叶片中POD活性表现出一定差异，在200mmol/L的盐浓度下胁迫15d后，幼苗叶片中POD活性能更好区分出樱桃番茄品种间的耐盐性。

3个品种的SOD酶活均表现为随着盐浓度的增大而逐渐升高，且在150mmol/L时达到最大值，以后随着盐浓度的加大又逐渐下降的趋势。达到最大值后，贝美SOD酶活下降1.15%，摩丝特SOD酶活下降2.77%，贝蒂SOD酶活下降5.25%。可以看出，在NaCl胁迫15 d后，盐胁迫对耐盐性较强的贝美叶片SOD酶活始终高于耐盐性较弱的贝蒂和摩丝特。

表2　不同浓度NaCl对番茄叶片POD和SOD酶活性的影响

Table 2　Effect of NaCl stress on activities of POD、SOD in cherry tamoto leaves

NaCl浓度（mmol/L）	POD酶（μg·g^{-1}FW·min^{-1}）			SOD酶（μg·g^{-1}FW·h^{-1}）		
	贝蒂	摩丝特	贝美	贝蒂	摩丝特	贝美
0（CK）	4.24dC	5.55dD	6.86dD	429.0dC	536.00dD	614.33eD
25	4.49dC	8.04cdCD	8.64dCD	556.33cB	636.33cC	737.00dC
50	6.60cB	10.02Cc	12.55cC	695.78bA	772.89bB	830.83cB
100	12.95bA	14.28bB	17.83bB	784.85aA	844.31aA	898.85bA
150	14.91aA	17.91aAB	19.37bB	783.01aA	887.93aA	955.16aA
200	13.51abA	20.74aA	27.93aA	741.93abA	863.34aA	944.16aA

4.1.3　对叶片脯氨酸含量的影响

由表3可以看出，随着盐浓度的增大，3个品种幼苗叶片中脯氨酸含量呈先升高后降低的趋势，贝蒂在100mmol/L时达到最大值，贝美、摩丝特在150mmol/L时达到最大值。达到最大值后，贝美脯氨酸含量下降4.65%，摩丝特下降13.35%，贝蒂下降16.55%。可以看出，在NaCl胁迫15 d后，耐盐性不同的3个樱桃番茄品种叶片的脯氨酸含量存在着差异，盐胁迫对耐盐性较强的贝美叶片脯氨酸含量始终高于耐盐性较弱的贝蒂和摩丝特。

表3　不同浓度NaCl对番茄叶片脯氨酸含量的影响

Table 3　Effect of NaCl stress on content of proline in cherry tamoto leaves

番茄品种	NaCl浓度（mmol/L）					
	0（CK）	25	50	100	150	200
贝蒂	207.21dD	229.60dD	654.44cC	913.49aA	780.01bB	650.92eE
摩丝特	145.23dD	324.42cC	1 000.07bB	1 000.88bB	1 071.06aA	928.09bB
贝美	478.66cC	592.05cC	1 063.95bB	1 115.62bB	1 506.32aA	1 436.52bAB

4.2　低温胁迫对樱桃番茄生理指标的影响

由表4可以看出，3种番茄进行5℃低温胁迫7d后，各个逆境指标均表现出不同程度的差异。膜透性差异不显著，其值在90%左右上下波动。脯氨酸含量、POD酶活、SOD酶活与植物的抗寒性呈正比，3个番茄品种脯氨酸含量、POD酶活、SOD酶活始终表现为贝美>

摩丝特 > 贝蒂。贝美的脯氨酸含量与摩丝特、贝蒂达到5%显著水平；贝蒂的SOD酶活与摩丝特、贝美达到5%显著水平；SOD酶活3个品种间均达到了1%极显著水平。

表4　番茄抗寒性测定

Table 4　Determination of cold resistance in different tomatoes

番茄品种	膜透性 (%)	丙二醛 (mmol/g)	脯氨酸含量 (μg/g FW)	POD酶 (μg · g^{-1}Fw/min)	SOD酶 (μg · g^{-1}FW · h^{-1})
贝美	88.77 aA	3.49 bB	429.91 aA	14.81 aA	1036.18 aA
摩丝特	89.09 aA	4.49 aAB	319.62 abAB	14.16 aA	879.91 bB
贝蒂	90.34 aA	4.61 aA	248.07 bAB	6.88 bB	591.60 cC

4.3　干旱胁迫对樱桃番茄生理指标的影响

4.3.1　不同樱桃番茄幼苗耐旱性调查

试验与2011年6月19日进行干旱胁迫处理，6月24日基质开始见干发白，6月27日统计干旱等级，结果如表5所示。可知，对3种番茄进行干旱胁迫后，均出现不同程度的干旱症状，其中5级、7级为极度干旱症状。贝美、贝蒂、摩丝特出现极度干旱症状的株数分别占总数的6.7%、66.7%、23.3%，未表现干旱症状的株数为0株。

表5　不同樱桃番茄幼苗干旱症状统计

Table 5　Statistik drought symptoms in different tomatoes

品种	0级	1级	3级	5级	7级	总株数
贝美	0	16	12	1	1	30
贝蒂	0	0	10	16	4	30
摩丝特	0	8	15	7	0	30

4.3.2　不同樱桃番茄幼苗干旱恢复调查

2011年6月27日进行干旱胁迫等级统计后，进行干旱胁迫恢复试验，6月29日观察统计不同品种樱桃番茄恢复状况如表6所示。3个樱桃番茄品种的恢复能力存在差异，贝美、贝蒂、摩丝特恢复为0级的株数分别占总数的66.7%、16.7%和50%；贝美、贝蒂、摩丝特恢复为1级的株数分别占总数的33.3%、30%和26.7%；贝蒂、摩丝特恢复为3级的株数分别占总数的33.3%和23.3%；贝蒂恢复为5级的株数占总数的20%。可见，贝美所有的材料均恢复为0级和1级，表明贝美耐旱性强于摩丝特和贝蒂。

表6　不同樱桃番茄幼苗干旱恢复状况

Table 6　Investigation drought recovery in different tomatoes

品种	0级	1级	3级	5级	总株数
贝美	20	10	0	0	30
贝蒂	5	9	10	6	30
摩丝特	15	8	7	0	30

4.3.3 干旱胁迫对不同樱桃番茄品种生理指标的影响

2011 年 6 月 27 日进行干旱胁迫等级统计后，于试验室测定番茄幼苗的各项生理指标，测定结果如表 7 所示。可以看出，3 个品种在干旱胁迫下，膜透性的差异均达到 5% 显著水平，且以贝蒂的膜透性为最高，贝美的膜透性为最低；贝蒂丙二醛含量与贝美、摩丝特品种间差异达到 5% 显著水平。脯氨酸与植物的抗旱性呈正比，3 个番茄品种脯氨酸含量从高到低依次为贝美 > 摩丝特 > 贝蒂，贝蒂与摩丝特、贝美达到 5% 显著水平；3 种番茄的 POD 酶活、SOD 酶活在逆境胁迫下存在差异，酶活性从高到低依次为贝美 > 摩丝特 > 贝蒂。POD 酶 3 个品种差异不显著；贝美品种的 SOD 酶活与贝蒂、摩丝特间差异达到 5% 显著水平。

表 7 干旱胁迫对不同樱桃番茄品种生理指标的影响

Table 7 Effect of watter stress on physiological indices in different tomatoes

番茄品种	膜透性 (%)	丙二醛 (mmol/g)	脯氨酸含量 (μg/g FW)	POD 酶 (μg · g^{-1}FW/min)	SOD 酶 (μg · g^{-1}FW · h^{-1})
贝美	81.38 cB	3.86 bB	337.34 aA	8.50 aA	854.39 aA
摩丝特	85.45 bAB	3.89 bB	229.70 bB	6.77 aA	680.41 bAB
贝蒂	90.12 aA	7.67 aA	180.41 cB	6.69 aA	627.61 bB

5 结论

3 个樱桃番茄品种抗逆性评价试验结果表明：

在 NaCl 胁迫 15 d 后，盐胁迫对耐盐性较弱的贝蒂叶片细胞质膜和细胞膜的伤害程度始终高于耐盐性较强的贝美和摩丝特；盐胁迫对耐盐性较强的贝美叶片 SOD 酶、POD 酶活、脯氨酸含量始终高于耐盐性较弱的贝蒂和摩丝特。

5℃低温胁迫 7 d 后，膜透性值在 90% 左右上下波动，表现为贝蒂 > 摩丝特 > 贝美；脯氨酸含量、POD 酶、SOD 酶 3 个指标在逆境胁迫下存在差异，3 个指标从高到低依次为贝美 > 摩丝特 > 贝蒂，表明贝美的耐寒性强于贝蒂和摩丝特。

干旱胁迫后，贝美、贝蒂、摩丝特出现极度干旱症状的株数分别占总数的 6.7%、66.7%、23.3%，恢复为 0 级的株数分别占总数的 66.7%、16.7% 和 50%；贝蒂膜透性、丙二醛含量显著高于贝美和摩丝特，而脯氨酸含量、POD 酶活和 SOD 酶低于贝美和摩丝特，表明的贝美的耐旱性强于贝蒂和摩丝特。

参考文献

[1] 翟凤林，曹鸣庆．植物的耐盐性及其改良．北京：农业出版社，1989
[2] 林多等．低温对番茄幼苗叶片活性氧代谢的影响．辽宁农业科学，2001 (5)：1 ~ 4
[3] 王孝宜等．番茄品种耐寒性与 ABA 和可溶性糖含量的关系．园艺学报，1998，25 (1)：56 ~ 60
[4] 吴晓雷等．番茄品种耐弱光性的综合评价．华北农学报，1997，12 (2)：97 ~ 101
[5] 孙景生．我国水资源利用现状与节水灌溉发展对策．农业工程学报，2000，16 (2)：1 ~ 4
[6] 张丽英，柴敏等．番茄苗期抗旱性鉴定及其评价方法的研究．中国蔬菜，2008 (2)：15 ~ 20

LED 光质对黄瓜幼苗生长及光合特性的影响*

邬 奇**，苏娜娜，崔 瑾
（南京农业大学生命科学学院，南京 210095）

摘要：采用发光二极管（Light emitting diode，LED）作为光源精确调制光质组成，以白光为对照，研究不同光质对黄瓜幼苗生长及光合特性的影响。结果表明，与对照相比，蓝光引起叶片的增厚和卷曲，促进了黄瓜幼苗叶片中 Rubisco 的合成和 *rca*、*rbcS* 和 *rbcL* 基因的表达，有利于 Fv/Fm、Φ_{PSII} 的提高。各单色光质之间相比显示，蓝光有利于叶片光合效率的提高，红光引起了幼苗真叶数的增多，提高了叶绿素相对含量。黄光和绿光处理后黄瓜幼苗矮小、新生叶片黄化，Rubisco 含量及相关基因表达受到抑制，不利于幼苗光合作用的进行。绿光和黄光的作用机理不同，绿光处理下幼苗的 Fv/Fm 与对照没有差异，说明绿光处理下幼苗仍具有较高的光合潜能，在调控植物的光合特性方面具有其独特作用。
关键词：LED；光质；黄瓜幼苗；生长；光合特性

Effects of LED Light Quality on the Development and Photosynthetic Characteristics of Cucumber Seedlings

Wu Qi，Su Nana，Cui Jin
（*College of Life Sciences*，*Nanjing Agricultural University*，*Nanjing* 210095，*China*）

Abstract： Light emitting diode（LED）was applied to accurately modulate the composition of light quality. With white light as control treatment，effects of LED light quality on the development and photosynthetic characteristics of cucumber seedlings were studied. The experimental results showed that，compared with the control treatment，blue light thickened and curled the seedling leaves，increased the Rubisco biosynthesis，upregulated the expression of *rca*，*rbcS* and *rbcL*，enhanced the Fv/Fm，Φ_{PSII}. Compared with other monochromatic lights，blue light increased the photosynthetic rate and red light increased the leaf number and Chl content. The seedlings under yellow and green light were dwarf and the new leaves were etiolated. Moreover，yellow and green light repressed the Rubisco biosynthesis and its gene expression，were bad for photosynthesis. Different from yellow light，compared with the control treatment，green light did not lowered the Fv/Fm and seedlings had high potential photosynthetic capacity. It is tempting to speculate that green light has unique role in regulating the photosynthetic characteristics.
Key words： LED；Light quality；Cucumber seedling；Development；Photosynthetic properties

光作为调节植物形态发育的关键环境因子，是植物光合作用的主要能量来源，光通过调控植物叶绿体的发育[1~2]，达尔文循环过程中的关键酶及光合相关基因的表达量[3]等影响植物的光合特性，进而调控植物的形态建成和生理发育。

---

* 基金项目：国家自然科学基金项目（31171998）；江苏省自然科学基金项目（BK2010439）
** 作者简介：邬 奇，硕士研究生；崔瑾，博士，副教授，主要从事植物光生物学方向的研究，E-mail：cuijin@ njau. edu. cn

光主要通过光强和光质来影响植物光合作用的高低，在光饱和点以内，随着光强的增加，光合效率不断增强，除光强外，光质也部分控制植物的光合机制[4]。并通过隐花色素、光敏色素和向光蛋白对不同光质的差异吸收，调节植物的生长发育[5]。研究表明不同波段的光质对植物叶绿素含量、Rubisco含量、光合特性及ROS清除系统的影响存在显著差异[4]。如蓝光可以增加叶片中光合色素的含量[6]，提高叶片的气孔导度[7]，红蓝组合光能增加叶片中氮素的积累量[8]，从而提高光合速率[9]等。

光质对植物光合特性及基因表达的研究已有很多[3,10]，且多集中在红、蓝光[6~8]的研究，对黄、绿光研究较少，有研究证明绿光能显著提高叶用莴苣的净光合速率（Pn），利于叶用莴苣的生长[11]，促进早期拟南芥（*Arabidopsis*）下胚轴的伸长[12]；黄光能增加大豆芽苗菜中异黄酮的含量[13]，证明了黄绿光在调控植物生长和代谢中的特殊作用。

新型半导体光源发光二极管（LED）具有的体积小、寿命长、耗能低、低发热并能精确配置光谱的组成等特点，使光质对植物生长发育的影响的试验研究变得方便准确[14]，本试验以黄瓜（*Cucumis sativus* L.）幼苗为试验材料，系统的研究不同单色光质在植物幼苗期形态建成中的不同效应，以及对光合效率和光合相关基因表达的影响，为LED光质补光育苗技术在设施栽培领域的应用提供科学依据。

1　材料与方法

1.1　试验材料和方法

供试材料为黄瓜（*Cucumis sativus* L.）幼苗，品种为“津春4号”，种子购于南京市神州种业。试验于2012年3月至2013年1月在南京农业大学生命科学学院植物光生物学实验室进行。黄瓜种子浸泡4 h后置于25℃培养箱中催芽，出芽后的种子播种于营养钵（10cm×15cm）中育苗，基质为蛭石：草炭（1：1），5d后洗根移入装满Hoagland's营养液的水培锥形瓶中，然后将水培的黄瓜幼苗置于不同光质的LED冷光源培养箱中培养，培养箱内相对湿度为（75 ± 5）%，温度为光照（28 ± 1）℃，黑暗（23 ± 1）℃，培养周期为10d，光照16 h·d^{-1}。

1.2　光质处理

使用LED冷光源培养箱（宁波海曙赛福实验仪器厂）培育幼苗，调节电流以及光源与植株的距离，照度计（Hansatech，UK）测量，使光照强度为（100 ± 3）μmol·m^{-2}·s^{-1}。试验以白光为对照（CK），单色光质分别为红光（R：658nm）、蓝光（B：460nm）、黄光（Y：585nm）、绿光（G：530nm）。

1.3　指标测定

（1）用直尺测定株高、根长，用游标卡尺测定茎基部作为茎粗；叶面积用便携式叶面积仪Li-3000C（Gene company limited）测量；干鲜重用电子天平测定。测定时对于各种处理的幼苗均随机取样。叶绿素相对含量用SPAD502叶绿素仪进行测量。叶绿素荧光参数用FMS-2叶绿素荧光仪（Hansatech，UK）测定。光合速率用Li-6400光合仪（美国LI-COR）进行测量。黄瓜幼苗叶片中Rubisco的SDS-PAGE分析采Power-Pac3000（美国）垂直电泳仪进行不连续双垂直板（22.0cm×14.5cm×0.75mm）聚丙烯酰胺凝胶电泳（PAGE）。脱色后凝胶用GE Healthcare Image Scanner Ⅲ（日本）扫描仪扫描。光合基因（*rbcL*、*rbcS*、*rca*）的相对表达量测定使用 ABI7500 荧光定量 PCR 仪进行检测。

（2）试验数据采用 Excel 2003 软件进行数据整理，SPSS 19.0 软件进行方差分析，显著性由 Duncan's 新复极差法检验，$P<0.05$。

2 结果与分析

2.1 不同光质对黄瓜幼苗生长发育的影响

不同光质对黄瓜幼苗形态发育的影响显著（图 1 和表）。随着处理时间的增加，蓝光下的幼苗叶片变厚且逐渐卷曲；红光有利于真叶数的增加，但单片叶的叶面积减小；黄光和绿光下的幼苗植株矮小、新生叶片黄化。培养至 10d 后，单色光下处理的黄瓜幼苗的形态指标除红光下的真叶数和根长与对照无显著性差异外，其他均显著低于对照，黄光和绿光处理下的叶面积和鲜重显著低于红光和蓝光处理（表）。

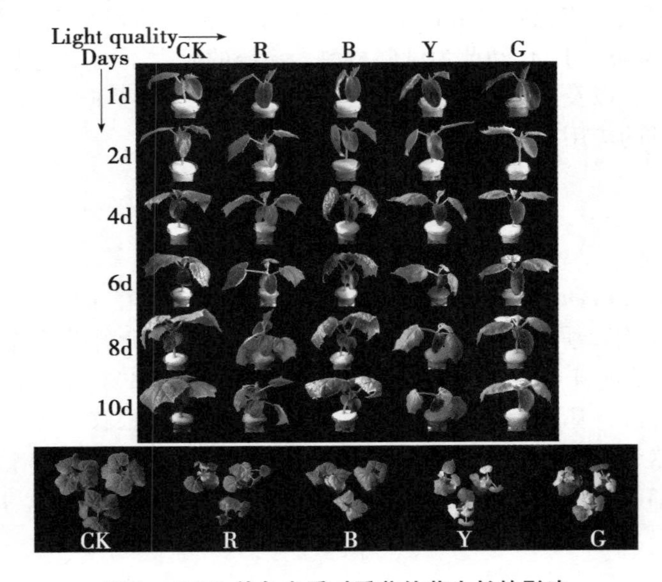

图 1　LED 单色光质对番茄幼苗生长的影响

Figure 1　Effects of light quality on the morphology of tomato seedlings

表　光质对黄瓜幼苗生长的影响

Table　Effects of light quality on the morphology of cucumber seedlings

光质 Light quality	真叶数	株高/cm Plant height	叶面积/cm^2 Leaf area	根长/cm Root length	根鲜重/g Root weight	地上鲜重/g Upper ground weight
CK	5.3 ±0.6a	10.87 ±0.35a	171.68 ±2.87a	17.6 ±2.5a	1.70 ±0.21a	9.06 ±2.50a
R	5.0 ±1.0ab	8.43 ±0.55b	107.26 ±2.88b	16.1 ±0.9ab	0.91 ±0.06b	6.05 ±0.90b
B	4.0 ±0.0b	7.47 ±0.49b	79.26 ±0.96c	13.6 ±0.9bc	0.75 ±0.11b	5.70 ±0.90b
Y	4.0 ±0.0b	8.13 ±0.75b	58.77 ±2.34d	12.5 ±1.5c	0.51 ±0.04c	3.48 ±1.49c
G	4.0 ±0.0b	7.40 ±1.01b	56.33 ±2.45d	14.7 ±0.6bc	0.52 ±0.07c	3.25 ±0.57c

2.2 不同光质对黄瓜幼苗叶绿素含量和荧光参数的影响

如图 2A 所示，随着不同光质处理天数的增加，各处理组黄瓜幼苗叶片中的叶绿素含量均呈现逐渐降低的趋势，且下降速率有差异。红光与对照处理相比无显著差异，蓝光、黄光和绿光随着天数增加与对照叶绿素含量差异越来越显著，其中黄光下降最快，呈直线下降。

随着处理天数的增加，黄瓜幼苗叶片的 $\Phi_{PS\,II}$ 值均呈现下降趋势，对照、红光和蓝光的 $\Phi_{PS\,II}$ 值在 0 ~ 10d 内降低不显著，且第 10d 蓝光处理下的 $\Phi_{PS\,II}$ 值高于对照。黄光和绿光的 $\Phi_{PS\,II}$ 值在 0 ~ 10d 内下降显著，至第 10d 显著低于对照，其中，绿光处理 10d 后 $\Phi_{PS\,II}$ 值下降一半左右，黄光处理下的 $\Phi_{PS\,II}$ 值在接第 10d 近于 0（图 2B）。

图 2C 表明，随着培养天数的增加，除黄光外，黄瓜叶片 Fv/Fm 值受光质影响不明显，红光和绿光处理下的第 10d 的 Fv/Fm 值与第 1d 相比略有降低，蓝光和对照在 0 ~ 10d 内无显著变化。黄光处理下 Fv/Fm 值从第 4d 开始呈直线下降趋势，第 10d 时接近于 0。

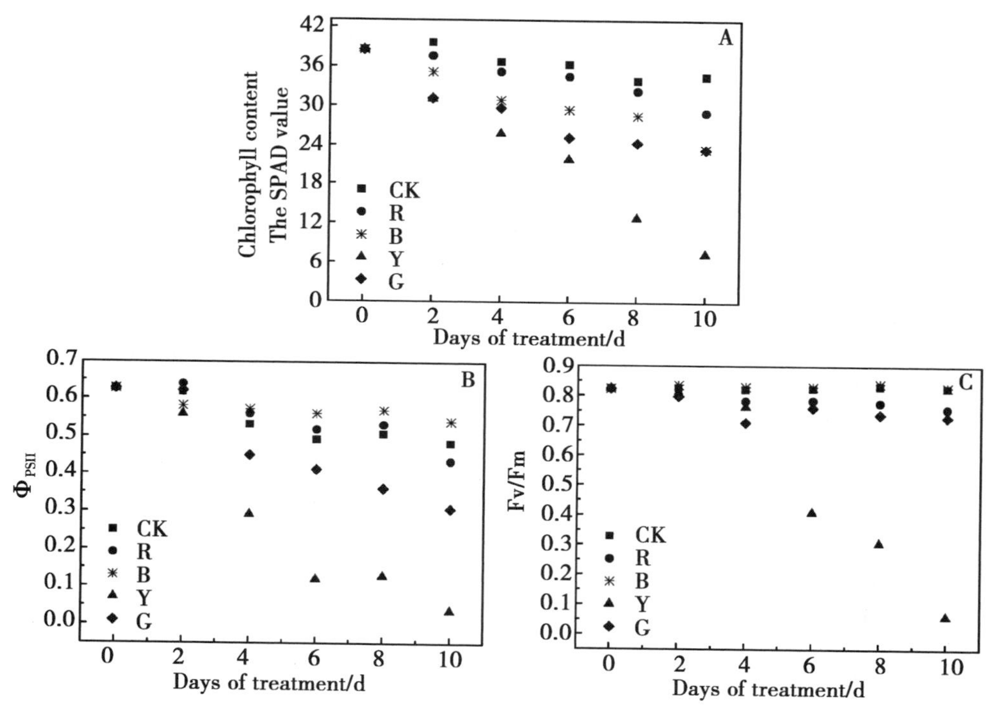

图 2 不同光质处理下番茄幼苗叶片 0 ~ 10d 内叶绿素含量（A）、Fv/Fm（C）和 ΦPS Ⅱ（B）的变化

Figure 2 The changes of chlorophyll content（C）, Fv/Fm（B）andΦPS Ⅱ（B）in cucumber seedlings leaves from 0 ~ 10 days under different light quality

2.3 不同光质对黄瓜幼苗叶片光合速率、Rubisco 含量及相关基因表达的影响

不同光质处理 10d 后，各单色光质处理下黄瓜幼苗叶片的净光合速率与对照组相比均显著降低，蓝光处理下的叶片净光合速率显著高于其他单色光质处理，黄光和绿光最低，其净光合速率值约为蓝光的 1/3，约为红光的 1/2。

提取黄瓜幼苗叶片总蛋白并进行 SDS-PAGE 分析后发现，与对照相比，蓝光处理下叶片中 Rubisco 大小亚基蛋白含量差异不显著，红光处理下略有降低，黄光和绿光处理下叶片中 Rubisco 大小亚基蛋白含量显著下降。对光合相关基因的 RT-PCR 分析表明，不同光质处理的黄瓜幼苗叶片中 *rca*、*rbcS* 和 *rbcL*3 种基因的表达量差异显著，与对照相比，蓝光下 3 种基因的表达量均显著升高，红光下除 *rbcL* 表达量显著升高，*rca* 和 *rbcS* 无显著变化，黄光和绿光下 3 种基因的表达量均显著降低，如图 3 所示。

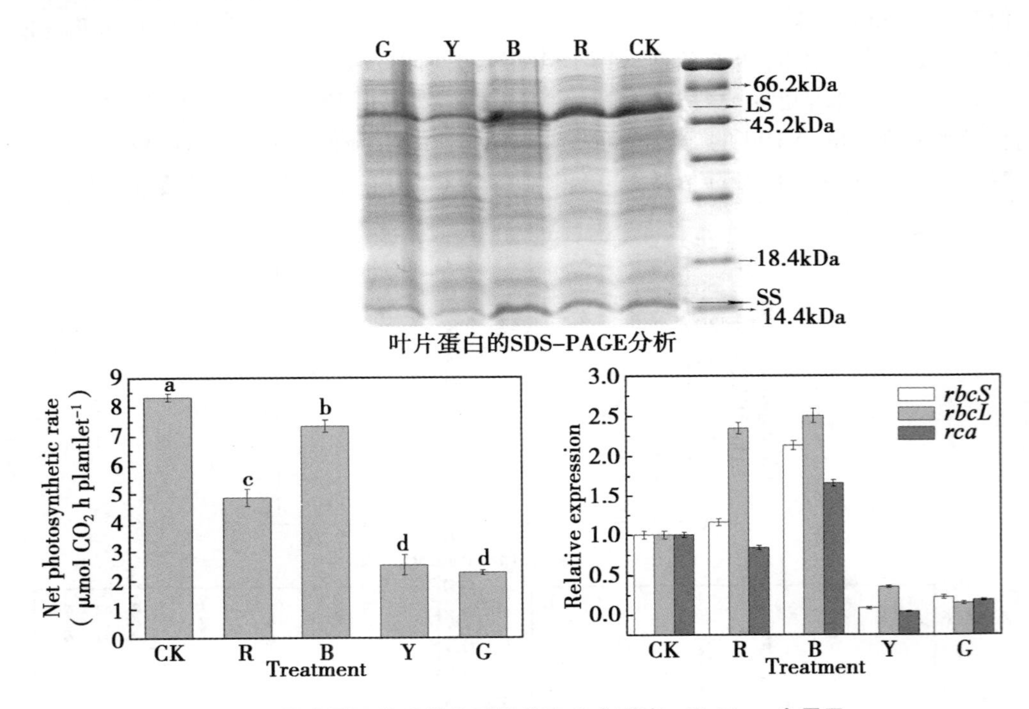

图 3　不同光质下黄瓜幼苗叶片的净光合速率、Rubisco 含量及 *rbcS*、*rbcL*、*rca* mRNA 相对表达差异

Figure 3　Effect of light quality on net photosynthetic rate, Rubisco content and relative mRNA expression of *rbcS*、*rbcL*、*rca* in tomato seedlings leaves

3　讨论

不同光质对黄瓜幼苗的形态发育和光合特性的影响差异显著。蓝光在控制叶片形态发育方面起着最重要的作用，其中，蓝光处理下的幼苗叶片变厚且逐渐卷曲（图 2），这可能是蓝光照射促进了叶片正面表皮细胞伸长或者抑制了叶片背面表皮细胞的伸长；增加栅栏组织的厚度，从而引起叶片加厚[15]。Fukuda 等[15]使用顶置光源照射天竺葵叶片正面的结果表明，红光使叶角（叶片与叶柄的夹角）变小，蓝光对叶角大小的影响与红光作用相反。而本试验中蓝光照射后叶角变小，红光照射后叶角增大，与 Fukuda 的结论相反，这种不同可能与本试验中培养光源为三面，叶片正反面均能感受光照有关。

不同单色光质下黄瓜幼苗叶片中的叶绿素含量均呈现逐渐降低的趋势（图 2A），这与

Wang[3]等（2009）的结果相似。但不同单色光处理下叶绿素含量下降速率不同，与对照相比，红光处理下无显著差异；蓝光、绿光和黄光处理的叶绿素含量随天数增加下降显著，表现为蓝光下叶片失绿，绿光和黄光下新生叶片黄化（图 1）。光在原叶绿素酸酯（Pchlide）还原为脱植基叶绿素（Chlide）的过程中发挥重要作用，此过程需要光依赖的 NADP 原叶绿素酸酯氧化还原酶（LPOR）的催化[16]。本试验的结果说明不同光质可能对 LPOR 的合成产生重要影响，特别是红光可能在 LPOR 的合成过程中起关键作用。

植物细胞中光合电子传递途径中产生的 ROS 能攻击细胞分子，例如，DNA、Rubisco、叶绿素等，从而加速叶片的衰老过程[17]。光敏色素（Phytochrome）通过调节基因 *APX*1 和 *APX*3 的表达，影响 APX 的合成，清除 ROS[18]。试验中红光处理组叶绿素和 Rubisco 含量与对照相比均无降低，而蓝光、绿光和黄光处理下的叶绿素含量显著降低，叶色失绿，表现衰老现象（图 1 和图 2A），表明红光通过光敏色素的吸收，促进 APX 的合成，抑制叶绿素的降解和叶片的衰老。

与对照相比，单色光下的净光合速率均显著降低，其中黄绿光降低最显著（图 4）。这与 Wang 等[3]在黄瓜和 Yu[19]等在黑荆树的研究结果一致。本试验结果表明，净光合速率的高低与叶绿素含量相关性较小，与叶片中的 Rubisco 酶含量相关性较高，表现为净光合速率的高低与 Rubisco 酶含量高低变化相一致。气孔也是影响净光合速率的重要因素，蓝光下净光合速率显著高于红光、黄光和绿光处理，可能与蓝光利于气孔发育及开放[3,7]及叶片中氮素的积累[8]有关。

本试验中，黄光处理下的黄瓜幼苗叶片中叶绿素含量、Φ_{PSII}和 Fv/Fm 值均呈显著下降趋势，但下降开始的时间不同，光质处理后的叶绿素含量、Φ_{PSII}和 Fv/Fm 值分别在第 1d、第 2d 和第 4d 开始下降（图 2）。说明光质处理首先引起了叶绿素含量的变化，再引起叶绿体光系统Ⅱ中电子传递速率的变化，最终反映为光合功能特性的改变。

光是影响植物发育相关基因表达的重要环境因素之一[3]，关于光质调节光合相关基因的表达的研究已有报道[20]。本试验中，光质引起了光合相关基因表达的显著差异。与对照相比，红光下除 *rbcL* 表达量显著升高，*rca* 和 *rbcS* 无显著变化，蓝光下 3 种基因的表达量均显著升高，黄光和绿光下 3 种基因的表达量均显著降低，与不同光质下叶片中 Rubisco 含量高低相一致（图 3）。试验同时表明同种光质下 *rbcL* 和 *rbcS* 基因表达的不一致性，红光下 *rbcL* 表达量显著升高，但 *rbcS* 无显著变化；蓝光下 *rbcL* 和 *rbcS* 表达均显著升高，但 *rbcS* 升高 70% 左右，*rbcL* 却升高 150% 左右（图 3），Wang[3]等（2009）也得出相似的结论，这可能与光敏色素调控植物核基因的表达差异有关[10]。

植物呈现绿色是因为反射绿光，因此，绿光被认为不能被植物利用，尤其是在植物的光形态发生和光合方面，植物只能在日光或者混合红蓝光下正常生长[21]，但有研究表明绿光能引起早期茎的伸长[12]，且生长在大于 300μmol · m^{-2} · s^{-1} PPF 绿光下的叶用莴苣形态与白色荧光灯下无差异，光合速率显著升高[11]。本试验中，黄光和绿光下净光合速率值相近，且显著低于对照，但绿光处理下 Fv/Fm 值相对稳定，Fv/Fm 值和叶绿素含量在第 10d 显著高于黄光处理，反映了绿光在调控植物生长和光合特性方面的独特作用。

参考文献

[1] Hudák J, Gálová E, Zemanová L. Plastid morphogenesis. In: Pessárakli M (ed) Handbook of photosynthesis. CRC Press/

Taylor and Francis Group, Boca Raton/London, 2005, 221 ~ 245
[2] Jiao Y, Lau OS, Deng XW. Light-regulated transcriptional net-works in higher plants. Nat Rev Genet. 2007, 8: 217 ~ 230
[3] Wang H, Gu M, Cui J, Shi K, Zhou Y, Yu J. Effects of light quality on CO_2 assimilation, chlorophyll-fluorescence quenching, expression of Calvin cycle genes and carbohydrate accumulation in *Cucumis sativus*. J Photochem Photobiol B. 2009, 96: 30 ~ 37
[4] Yamazaki J, Kamimura Y. Relationship between photosystem stoichiometries and changes in active oxygen scavenging enzymes in natural grown rice seedlings. Plant Growth Regul. 2002, 36: 113 ~ 120
[5] Whitelam G, Halliday K. Light and plant development. Oxford: Blackwell Publishing. . 2007
[6] Johkan M, Shoji K, Goto F, Hashida S, Yoshihara T. Blue Light-emitting Diode Light Irradiation of Seedlings Improves Seedling Quality and Growth after Transplanting in Red Leaf Lettuce. Hort Science. 2010, 45 (12): 1 809 ~ 1 814
[7] Hogewoning S W, Trouwborst G, Maljaars H, Poorte H, Ieperen W, Harbinson J. Blue light dose-responses of leaf photosynthesis, morphology, and chemical composition of *Cucumis sativus* grown under different combinations of red and blue light. Journal of Experimental Botany. 2010, 8: 3 107 ~ 3 117
[8] Ohashi-Kaneko K, Matsuda R, Goto E, Fujiwara K, Kurata K. . Growth of rice plants under red light with or without supplemental blue light. Soi l Science a nd Plant Nutrition. 2006, 52: 444 ~ 452
[9] Matsuda R, Ohashi-Kaneko K, Fujiwara K, Goto E, Kurata K. Photosynthetic characteristics of rice leaves grown under red light with or without supplemental blue light. Plant Cell Physiol. 2004, 45: 1 870 ~ 1 874
[10] Kuno N, Furuya M. Phytochrome regulation of nuclear gene expression in plants. Seminars in Cell and Developmental Biology. 2000, 11: 485 ~ 493
[11] Johkan M, Shoji K, Goto F, Hahida S, Yoshihara T. Effect of green light wavelength and intensity on photomorphogenesis and photosynthesis in *Lactuca sativa*. Environmental and Experimental Botany. 2012, 75 : 128 ~ 133
[12] Folta K M. Green light stimulates early stem elongation, antagonizing light-mediated growth inhibition. Plant Physiol. 2004, 135: 1 407 ~ 1 416
[13] Lee SJ, Ahn JK, Khanh TD, Chun SC, Kim SL, Ro HM, Song HK, Chung IM. Comparison of isoflavone concentrations in soybean [*Glycine max* (L.) Merrill] sprouts grown under two different light conditions. J Agric Food Chem. 2007, 55 (23): 9415 ~ 21
[14] Morrow R C. LED lighting in horticulture. Hortscience. 2008, 43 (7): 1 947 ~ 1 950
[15] Fukuda. N, Fujita. M, Ohta. Y, Sase S, Nishimura. S, Ezura H. . Directional blue light irradiation triggers epidermal cell elongation of abaxial side resulting in inhibition of leaf epinasty in geranium under red light condition. Scientia Horticulturae 2008, 115: 176 ~ 182
[16] Masuda T, Takamiya K. Novel insights into the enzymology, regulation and physiological functions of light-dependent protochlorophyllide oxidoreductase in angiosperms. Photosynth Res. 2004, 81: 1 ~ 29
[17] Yamazaki J. Is light quality involved in the regulation of the photosynthetic apparatus in attached rice leaves. Photosynth Res. 2010, 105: 63 ~ 71
[18] Mullineaux P, Ball L, Escobar C, Karpinska B, Creissen G, Karpinski S. Are diverse signalling pathways integrated in the regulation of *Arabidopsis* antioxidant defence gene expression in response to excess excitation energy. Phil Trans R Soc Lond B. 2000, 355: 1 531 ~ 1 540
[19] Yu H, Ong B L. Effect of radiation quality on growth and photosynthesis of Acacia mangium seedlings. Photosynthetica. 2003, 41: 349 ~ 355
[20] Tyagi A K, Gaur T. Light regulation of nuclear photosynthetic genes in higher plants. Critical Reviews in Plant Sciences. 2003, 22: 417 ~ 452
[21] Ohasi-Kaneko K, Takase M, Kon N, Fujiwara K, Kurata K. Effect of light quality on growth and vegetable quality in leaf lettuce, spinach and komatsuna. Environ. Cont. Biol. 2007, 45: 189 ~ 198

不同光质 LED 光源对金线莲组培苗生长的影响*

周锦业[1**]，康俊勇[2]，丁国昌[3,4]，吴志明[2]，曹光球[3,4]，
许珊珊[3,4]，刘　丽[4]，马志慧[4]，林思祖[3,4***]
（1. 福建农林大学园林学院，福州　350002；2. 厦门大学物理系，厦门　361005；
3. 福建农林大学药用植物研究所，福州　350002；4. 福建农林大学林学院，福州　350002）

摘要： 为了探讨不同光质光强处理对金线莲组培苗生长的影响，采用 LED 作为人工光源对漳州金线莲进行继代和生根培养，结果显示：70μmol · m^{-2} · s^{-1}的红光处理有利于株高生长，但不利于金线莲生根；绿光对金线莲各项指标均无促进作用，且不同强度处理之间差异不显著；70μmol · m^{-2} · s^{-1}的蓝光处理对金线莲组培苗叶片生长、生物量积累以及根系生长均有促进作用，但对株高有一定的抑制作用。结论：红光有利于植物地上部分生长，绿光对植物光合作用影响较小，蓝光对组培苗有一定的矮化壮苗作用。

关键词： LED；光质；金线莲；组培；生长量

Effect of Light Quality on the Growth of Tissue Culture Seedling in *Anoectochilus roxburghii* by Using LED

Zhou Jinye[1], Kang Junyong[2], Ding Guochang[3,4], Wu Zhiming[2], Cao Guangqiu[3,4],
Xu Shanshan[3,4], Liu Li[4], Ma Zhihui[4], Lin Sizu[3,4]
(1. *College of Landscape Architecture*, *Fujian Agriculture and Forestry University*, *Fuzhou* 350002, *China*;
2. *Department of Physics*, *Xiamen University*, *Xiamen* 361005, *China*;
3. *Institute of Medicinal Plant*, *Fujian Agriculture and Forestry University*, *Fuzhou* 350002, *China*;
4. *Forestry College*, *Fujian Agriculture and Forestry University*, *Fuzhou* 350002, *China*)

Abstract: To discuss the effect of different light quality and the light intensity treatment on the growth of *Anoectochilus roxburghii* tissue culture seedling, the LED light source was used as an artificial light source for *Anoectochilus roxburghii* in Zhang Zhou to subculture and rooting culture, the results showed that, 70μmol · m^{-2} · s^{-1}red light treatment is to the benefit of the growth of *Anoectochilus roxburghii* height, but go against to take root; it is not obvious that the effect of Green light for the indicators of *Anoectochilus roxburghii*, and no significant difference between different intensity; a promoting effect on leaf growth of tissue culture seedling and accumulation of biomass and root growth of *Anoectochilus roxburghii* under the 70μmol · m^{-2} · s^{-1} blue light treatment, but certain inhibitory effect on plant height. Conclusion: the red light is advantageous to the growth of aerial parts of plants, the green light have little impact on plant photosynthesis, and certain effect on dwarfed and strong seedling of tissue culture seedling under

---

* 基金项目：福建省农业高校产学合作科技重大项目（2012N5003）资助

** 作者简介：周锦业（1987—），男，硕士研究生，研究方向：园林植物生物技术，E-mail：P_ bearily@126. com

*** 通信作者：林思祖（1953—），男，教授，博士生导师，研究方向：森林培育理论与技术，E-mail：Szlin53@126. com

the blue light.

Key words：LED；Light quality；*Anoectochilus roxburghii*；Tissue culture；Growth

光能是植物光合作用重要基础，不同光质在植物生长发育过程中作用不同。植物一方面通过以吸收红光及远红光为主的光敏色素参与光形态建成的过程；另一方面通过感受外界蓝光信号的变化，使植物改变其生长发育过程（如向光性、抑制茎伸长、气孔运动、激活基因、合成色素等）以适应外界环境的变化[1]。植物对可见光的吸收主要集中于610～720nm的红橙区以及400～500nm的蓝紫区，LED光源具有体积小、发热少、能耗低和光色可选择等优点，选择不同光质LED作为人工光源，可以避免全光谱照射的光能浪费[2]。

金线莲（*Anoectochilus roxburghii*）为兰科开唇兰属多年生草本植物，别名金线兰、金丝草，是我国传统名贵中药材，全草可入药，具有清热凉血、除湿解毒的功效，对肺结核、糖尿病、肾炎、膀胱炎、风湿性关节炎以及毒蛇咬伤等症均有显著疗效[3]。除了药用价值之外，由于其株形优美、叶色独特且叶片具特殊金线网纹结构，因此具有很高的盆栽观赏价值。近些年金线莲的市场需求量越来越大，野生资源已经采收殆尽，人工种子及扦插繁殖的周期长且繁殖率低。组织培养技术能够在短期内大量扩繁，因此近些年许多学者对金线莲组培快繁技术进行了一系列的研究[4~6]，并取得了丰硕的成果，但研究主要集中于外植体诱导、培养基优化以及移栽方式研究等，关于人工微环境对金线莲组培影响的研究较少，特别是不同光质在金线莲组培中的应用研究还未见报道。本研究选择光色光强可调LED作为人工光源，研究了不同光色光强对金线莲组培苗生长及生根的影响，为寻求金线莲组培中新的人工光源提供理论依据。

1　材料与方法

1.1　试验地点与材料

试验于2011年10月至2012年1月在福建农林大学药用植物研究所组培室进行，所选用的材料为福建农林大学药用植物研究所内的漳州种源金线莲组培苗，光源采用福建农林大学与福建省半导体材料及应用重点实验室共同研发的光色光强可控的组合式LED光源。

1.2　试验方法

各试验组光强情况如表1所示，CK为荧光灯对照组，组培苗分别接种于对应的继代和生根培养基中，每瓶4株，每个处理3瓶作为3次重复，光源距离组培苗20～25cm，光照周期为12h·d^{-1}，分别继代和生根培养30d后统计相应的测试指标。

1.3　指标测定

继代30d后统计各试验组金线莲组培苗株高、叶片数、叶长、叶宽、平均鲜重以及平均干重；生根处理30d后统计各试验组金线莲组培苗的生根条数和根系长度。

1.4　数据处理

实验所的数据经Excel整理后，采用DPS7.05进行方差分析和多重比较分析。

表1　各处理组基本情况

Table 1　The basic situation in each treatment group

处理	光质	光强/μmol·m^{-2}·s^{-1}	波长/nm	功率/W
R1	红	70	634	0.945
R2	红	50	634	0.752
R3	红	30	634	0.549
G1	绿	70	502	0.675
G2	绿	50	502	0.536
G3	绿	30	502	0.392
B1	蓝	70	460	0.405
B2	蓝	50	460	0.320
B3	蓝	30	460	0.234
CK	荧光灯	——	——	28

2　实验结果与分析

2.1　不同光质光强对金线莲组培苗形态生长的影响

分析不同光质光强对金线莲组培苗形态生长的结果显示（表2），就株高生长量而言，同一光色不同处理间高强度红光处理株高生长量最高，与对照组相比提高了19.92%，绿光组株高生长量与对照相当，蓝光处理均显著低于对照；红蓝光处理株高生长量随着光照强度的增加分别表现为逐渐增加和逐渐减小的趋势，而绿光处理规律不明显。同一光强不同光色间差异明显，70μmol·m^{-2}·s^{-1}光强时株高生长量大小顺序为R>G>B，50μmol·m^{-2}·s^{-1}和30μmol·m^{-2}·s^{-1}光强时其顺序均表现为G>R>B。

分析不同光质光强对金线莲组培苗叶片数量的影响显示（表2），中高强度红光处理及高强度蓝光处理叶片数均高于对照，其中，以70μmol·m^{-2}·s^{-1}红光处理叶片数量最多，与对照组相比提高了40.20%。同一光色不同光强间叶片数量变化趋势一致，随着光强增加均表现为逐渐增加的趋势，红绿蓝3种光色叶片数最大最小值间分别相差52.48%、24.30%和20.86%；同一光强不同光色处理下叶片数量变化情况均表现为R>B>G，高中低光强处理下叶片数量最大最小值间差异分别为61.65%、46.15%和31.78%。

分析不同光质光强处理对金线莲组培苗叶长和叶宽影响的结果显示（表2），蓝光处理下金线莲叶片的叶长叶宽均高于对照，其中以70μmol·m^{-2}·s^{-1}光强处理时最大，叶长叶宽与对照相比分别提高了5.81%和13.16%。就同一光色不同光强间差异而言，红光和绿光处理下，金线莲叶长叶宽随着光强增加逐渐减小，红光处理叶长叶宽最大最小值分别相差6.67%和5.09%，绿光处理叶长叶宽最大最小值分别相差3.95%和12.00%，而蓝光处理下金线莲叶长叶宽随着光强增加均表现为逐渐增大的变化趋势，最大最小值间差异分别为4.37%和10.76%；同一光强不同光色间金线莲叶长叶宽变化趋势相似，3种光强处理下叶长叶宽值均表现为B>R>G，高中低3种光强下叶长最大最小值间分别相差53.77%、46.43%和41.73%，叶宽最大最小值间差异分别为56.90%、40.60%和26.47%。

表 2 不同光质光强对金线莲组培苗生长的影响

Table 2 Effect of different light quality and light intensity treatment on the growth of tissue culture seedling of *Anoectochilus roxburghii*

指标	光强/ $\mu mol \cdot m^{-2} \cdot s^{-1}$	光色		
		红	绿	蓝
高度/cm	70	6.260aA	5.500aB	4.560aC
	50	5.060bAB	5.240aA	4.660aB
	30	4.860bAB	5.420aA	4.680aB
	CK	5.2201b	5.220a	5.220b
叶片数/个	70	4.300aA	2.660aB	3.360aC
	50	3.420bA	2.340abB	2.800bC
	30	2.820cA	2.140bB	2.780bA
	CK	3.067bc	3.067ac	3.067ab
叶长/cm	70	1.469aA	1.088aB	1.673aC
	50	1.487aA	1.120aB	1.640abC
	30	1.567bA	1.131aB	1.603abA
	CK	1.581b	1.581b	1.581b
叶宽/cm	70	1.060aA	0.833aB	1.307aC
	50	1.107aA	0.867aB	1.219abA
	30	1.114aA	0.933aA	1.180abA
	CK	1.155a	1.155b	1.155b

注：纵向小写字母不同表示在5%水平差异显著，横向大写字母不同表示在5%水平差异显著，下同

2.2 不同光质对金线莲组培苗生物量积累的影响

不同光质光强处理30d后，测量各处理单株金线莲干鲜重值的结果表明（表3），70$\mu mol \cdot m^{-2} \cdot s^{-1}$红光及蓝光处理下，金线莲组培苗的鲜重和干重均高于对照，鲜重与对照相比红蓝光分别提高了17.81%和0.89%，干重与对照相比红蓝光分别提高了4.02%和9.77%；而绿光处理干鲜重值均显著低于对照。分析同一光色不同光强间的差异，红光蓝光处理下，金线莲干鲜重随着光强增加呈现逐渐增加的变化趋势，红光处理下鲜重和干重最大最小值间差异分别为43.47%和47.15%，蓝光处理下鲜重和干重最大最小值间差异分别为6.07%%和14.37%，绿光处理下变化规律不明显。就同一光强不同光色间差异而言，鲜重间差异表现为，70$\mu mol \cdot m^{-2} \cdot s^{-1}$和50$\mu mol \cdot m^{-2} \cdot s^{-1}$光强处理下鲜重由大到小依次为R>B>G，30$\mu mol \cdot m^{-2} \cdot s^{-1}$光强处理下鲜重由大到小依次为B>R>G，高中低光强处理下金线莲组培苗鲜重最大最小值间差异分别为64.55%、53.43%和22.07%；不同处理间干

重的差异表现为，3 种光强处理干重变化趋势均为 B > R > G，高中低光强处理下干重最大最小值间差异分别为 122.09%、134.62% 和 83.52%。

表 3 不同光质光强对金线莲组培苗生物量积累的影响

Table 3 Effect of different light quality and light intensity treatment on accumulation of biomass of tissue culture seedling of *Anoectochilus roxburghii*

指标	光强/ $\mu mol \cdot m^{-2} \cdot s^{-1}$	光色		
		红	绿	蓝
鲜重/g	70	1.713aA	1.041aB	1.467aAB
	50	1.453abA	0.947aB	1.406aA
	30	1.194bA	1.133aA	1.383aA
	CK	1.454ab	1.454b	1.454a
干重/g	70	0.181aA	0.086aB	0.191aA
	50	0.147bcA	0.078aB	0.183aC
	30	0.123cAB	0.091aA	0.167bB
	CK	0.174ab	0.174b	0.174ab
干重/鲜重	70	0.106aAB	0.083aA	0.130aB
	50	0.101aAB	0.082aA	0.130aB
	30	0.103aAB	0.080aA	0.121aB
	CK	0.120b	0.120b	0.120a

通过计算不同光质光强处理下干鲜重比值（表 3），可以得出不同处理对金线莲组培苗干物质积累率的影响，结果表明，不同光强蓝光处理干鲜重比值均高于对照，其中，$70\mu mol \cdot m^{-2} \cdot s^{-1}$和 $50\mu mol \cdot m^{-2} \cdot s^{-1}$光强处理下干鲜重比值相同，与对照相比均提高了 8.33%。就同一光色不同光强处理间干鲜重比值差异性而言，各光色处理随着光强的增加干鲜重比值变化较小，整体体现为高光强处理干鲜重比值要稍大。同一光强不同光色处理间变化趋势一致，金线莲干鲜重比值均表现为 B > R > G，高中低光强处理下其最大最小值间差异分别为 56.63%、58.54% 和 51.25%。

2.3 不同光质对金线莲组培苗生根的影响

生根培养 30d 后，统计金线莲组培苗生根情况如表 4 所示，$70\mu mol \cdot m^{-2} \cdot s^{-1}$和 $50\mu mol \cdot m^{-2} \cdot s^{-1}$光强蓝光处理下生根条数及根长均高于对照，其中又以 $70\mu mol \cdot m^{-2} \cdot s^{-1}$处理最高，与对照相比生根天数和根长分别提高了 12.83% 和 51.22%。同一光色不同光强处理下，随着光强增加金线莲生根条数和根长均表现为逐渐增加的变化趋势，红绿蓝 3 种光色处理下根条数最大最小值间分别相差 46.15%、15.52% 和 18.79%，红绿蓝 3 种光色处理下根长最大最小值间差异分别为 68.82%、18.97% 和 64.45%。而就同一光强不同光色处理间差异而言，$70\mu mol \cdot m^{-2} \cdot s^{-1}$和 $50\mu mol \cdot m^{-2} \cdot s^{-1}$光强处理下，生根条数和根长变化均表现为 B > R > G；$30\mu mol \cdot m^{-2} \cdot s^{-1}$光强处理下，生根条数和根长变化均表现为 B > G > R；高中低光强处理生根条数最大最小值间分别相差 19.02%、19.72% 和 42.46%，根长最大最小值间差异分别为 74.09%、54.57% 和 53.83%。

表 4　不同光质光强对金线莲组培苗生根的影响

Table 4　Effect of different light quality and light intensity treatment on root growth of tissue culture seedling of *Anoectochilus roxburghii*

指标	光强/μmol · m^{-2} · s^{-1}	光色		
		红	绿	蓝
生根条数/条	70	3. 167aA	3. 081abA	3. 667aA
	50	2. 833aA	2. 784aA	3. 333abB
	30	2. 167bA	2. 667aAB	3. 087bB
	CK	3. 250a	3. 250b	3. 250ab
生根长度/cm	70	0. 991aA	0. 853aA	1. 485aB
	50	0. 803bA	0. 733aA	1. 133bB
	30	0. 587cA	0. 717aA	0. 903bB
	CK	0. 982a	0. 982b	0. 982b

3　结论与讨论

3. 1　光质光强对金线莲组培苗形态生长的影响

实验结果表明，单色红光处理对金线莲株高生长有促进作用，这与王婷[7]等对不结球白菜的研究结果相同。绿光处理后的金线莲有一定程度的徒长，其叶片数量、叶长和叶宽值均较小，整体植株瘦弱，说明金线莲对绿光吸收较少。蓝光处理结果显示，金线莲组培苗株高较低，但是，叶片数量与对照相差较小，叶片长度及宽度值均较高，整体植株健壮，表明蓝光可以对植物有一定的矮化壮苗作用，这与张宇[8]的研究结果相一致。金线莲的药用成分主要集中于其叶片中，蓝光处理在不减少叶片数量的前提下可以增加叶片面积，这将有助于提高金线莲药用成分含量。

3. 2　光质光强对金线莲组培苗生物量积累的影响

研究结果显示，增加红光强度有利于提高组培苗生长速度和生物量积累。绿光处理金线莲组培苗生物量及干鲜重比值均较小，植株生长较弱，水分含量高，不同强度绿光处理之间差异较小，这与绿光对金线莲形态生长的影响一致，表明绿光很少参与植物光合作用。蓝光处理干鲜重值均较高，且随着光强增加生物量积累呈现一定上升趋势，同一光强下植株平均干重和干鲜重比值均要高于红光和绿光实验组，说明蓝光处理有利于植物生物量的积累，这与孙庆丽[9]的研究结果相同。

3. 3　光质光强对金线莲组培苗生根的影响

光强增加对组培苗生根条数和生根长度均有促进作用，低强度红光处理明显不利于金线莲组培苗生根。绿光处理对金线莲组培苗生根有轻度抑制效果，抑制作用随着光强的减弱而增强。增加蓝光强度蓝光有利于植物根系生长，这与柳金凤[10]的研究结果相似，但与王婷[7]的研究结果不一致，具体原因有待进一步研究。

整体而言，红光有利于植物的茎的伸长生长，但是会在一定程度上造成植物的徒长，且对植物根系生长有一定的抑制作用，单纯的红光照射难以得到生长状态良好的组培苗；绿光处理整体对金线莲组培苗生长没有促进作用；蓝光照射有利于得到健壮植株，植株地上部分节间变短、叶片面积增加、茎粗壮，地下部分根系数量和长度均有增加，但是植株纵向生长速度变慢。因此单色光源均不适合作为组培人工光源，但是可以通过周期性交替使用单色光照射来弥补上述不足。

参考文献

[1] 宋纯鹏，王学路．植物生理学（第四版）．北京：科学出版社，2009
[2] 何松林，闫新房，丁林波．LED 光源在植物组织培养中的应用．中国农学通报，2009（12）：42～45
[3] 毛碧增，娄沂春，蔡素琴等．金线莲的快速繁殖．浙江大学学报（农业与生命科学版），1999（5）：78～79
[4] 伍成厚，冯毅敏，贺漫媚等．金线莲种子培养的研究．中国野生植物资源，2008（1）：47～50
[5] 何云芳，杨霞，余有祥等．金线莲组培快繁技术．浙江林学院学报，1999（2）：64～68
[6] 肖木兴．台湾金线莲增殖培养试验．福建林业科技，2007（2）：124～126
[7] 王婷，李雯琳，巩芳娥等．LED 光源不同光质对不结球白菜生长及生理特性的影响．甘肃农业大学学报，2011（4）：69～73
[8] 张宇，杨晓建，刘世琦．不同 LED 光源对青蒜苗生长及叶绿素荧光特性的影响．中国蔬菜，2011（6）：62～67
[9] 孙庆丽，陈志，徐刚等．不同光质对水稻幼苗生长的影响．浙江农业学报，2010（3）：50
[10] 柳金凤，伍会萍，刘晓刚等．不同光质对枸杞试管苗生长的影响．中国农学通报，2011（22）：109～113

不同栽培容器对水果型黄瓜生长、产量和品质的影响

宋夏夏[*]，束 胜，郭世荣，张 钰，施 洋
（南京农业大学园艺学院，农业部南方蔬菜遗传改良重点开放实验室，南京 210095）

摘要：本试验在筛选出最佳醋糟混配基质的基础上，选择桶式栽培（T）、袋式栽培（D）和槽式栽培（C）3种栽培方式，研究不同栽培容器对水果型黄瓜生长、产量和果实品质的影响。结果表明，袋式栽培处理（D）水果型黄瓜植株生长健壮；与其他两个处理相比，袋式栽培（D）显著提高水果型黄瓜的单株产量和果实数，降低了畸形果率，并且显著增加果实可溶性总糖和可溶性蛋白含量。说明有机基质袋式栽培（D）有利于水果型黄瓜产量和品质的提高，栽培袋可作为水果型黄瓜有机基质栽培的适宜容器。
关键词：栽培容器；水果型黄瓜；生长；产量；品质

Effects of Different Cultivation Containers on the Growth, Yield and Quality of Mini-cucumber

Song Xiaxia, Shu Sheng, Guo Shirong, Zhang Yu, Shi Yang
(*College of Horticulture, Nanjing Agricultural University, Key Laboratory of Southern Vegetable Crop Genetic Improvement, Ministry of Agriculture, Nanjing* 210095, *China*)

Abstract: This test studied the effects of different cultivation containers on the growth, yield and quality of mini-cucumber ‘delstar’ by three cultivation methods—cultivation with bucket, plastic bag and groove, based on the best formula of vinegar residue mixed substrates. Results showed that cultivation with plastic bag made mini-cucumber plant haleness. Compared with the other two treatments, yield and number of fruit per plant of mini-cucumber of cultivation with plastic bag were significantly increased and the rate of malformed fruit was significantly decreased. Moreover, content of total soluble sugar and soluble protein was significantly increased. The conclusion was that organic substrate cultivation with plastic bag was beneficial to the increasing of yield and quality of mini-cucumber, so plastic bag can be made to be suitable container for organic substrate cultivation of mini-cucumber.
Key words: Cultivation containers; Mini-cucumber; Growth; Yield; Quality

选择合适容器是培育优质容器苗的关键，对作物的生长起着至关重要的作用（周跃华等，2005）。在实际生产中，基质栽培设施系统主要有槽式栽培、桶式栽培、袋式栽培和立体垂直栽培等。有机生态型无土栽培系统多采用基质槽式栽培的形式，在没有标准规格的成品槽供应时，可选用当地易得的材料建槽，如用木板、木条、竹竿、砖块等。建槽的基本要求为槽与土壤隔绝，槽框能保持基质不散落到走道上即可，而在实际栽培过程中，不同的栽培作物对栽培槽的宽度和深度都有不同的要求（杨波，2002）；桶式栽培一般选用适当体积的花盆或桶装容器为栽培容器，如NAU-G1型（卜崇兴和李式军，2003），或者选用一些专

* 第一作者简介：宋夏夏，女，江苏南京人，硕士在读，主要研究方向为设施高效栽培

用的无土栽培装置（郑奕等，2008a；郑奕等，2008b），；袋式栽培，分为地面栽培和立体栽培两种，而地面栽培又可分为筒式栽培和枕头式栽培，立体栽培包括柱状栽培和长袋栽培两种（李式军，2002），其分类标准主要根据栽培袋的外观形状和形式，一般常见的是枕头式栽培，在长 70cm、直径 30 ~ 35cm 的编织袋内装基质 20 ~ 30L，两端封严，在栽培袋上设定植孔，依次按照行距要求摆放到栽培温室内（李式军，2002）。

本文选取以木板槽为代表的槽式栽培，以 NAU-G1 型为代表的桶式栽培和以枕头式栽培为代表的袋式栽培，研究和比较 3 种不同基质栽培容器下水果型黄瓜植株生长状况、果实产量和品质，从而提出合适的有机生态型无土栽培容器。

1　材料与方法

1.1　材料与设计

试验栽培基质为预备试验筛选出的最适混配基质（醋糟、蛭石、草炭体积比 6 : 1 : 3），3 种基质均由镇江培蕾有机肥有限公司提供，其理化性状指标和营养元素含量如表 1 和表 2 所示，供试黄瓜为水果型黄瓜品种“戴多星”，由荷兰瑞克斯旺种苗集团公司生产。

表 1　醋糟栽培基质的物理性质

Table 1 Main physical properties of mixed vinegar residue cultivation substrates

指标 Parameters	容重 Bulk density ($g \cdot cm^{-3}$)	总孔隙 Total porosity (%)	通气孔隙 Aeration porosity (%)	持水孔隙 Water-Holding porosity（%）	水气比 WHP/AP
栽培基质 Cultivation substrate	0. 20	67. 4	38. 2	29. 3	0. 77

表 2　醋糟栽培基质的化学性质

Table 2 Main physical properties of mixed vinegar residue cultivation substrates

指标 Parameters	全氮 Nitrogen ($g \cdot kg^{-1}$)	全磷 Phosphorus ($g \cdot kg^{-1}$)	速效钾 Available potassium ($g \cdot kg^{-1}$)	Ca^{2+} Calcium ($g \cdot kg^{-1}$)	Mg^{2+} Magnesium ($g \cdot kg^{-1}$)	pH 值	电导率 EC（$mS \cdot cm^{-1}$）
栽培基质 Cultivation substrate	4. 65	2. 62	16. 22	5. 41	0. 71	6. 61	0. 86

试验设置 3 个处理：桶式栽培（T）处理，采用南京农业大学自行研制的 NAU-G1 型专无土栽培装置，其规格为下底直径 22. 5cm，上口直径 35cm，高 30cm，网芯盘直径 26. 5cm，网芯筒下口直径 11. 5cm，上口 13. 5cm，高 12cm；袋式栽培（D）处理，采用可装 30 L 基质的编织袋，制成长度为 50cm，宽度为 30cm，厚度为 20cm 的栽培袋；槽式栽培（C）处理，采用长度为 80cm，宽度为 50cm，深度为 20cm 的栽培槽。3 种栽培方式都保证每株水果型黄瓜占有基质 15 L。单因素随机区组设计，重复 3 次。

1.2 试验方法

1.2.1 水果型黄瓜栽培试验

试验在南京农业大学现代化温室中进行。于2012年9月3～27日进行水果型黄瓜育苗，待黄瓜幼苗三叶一心时移栽至不同的栽培容器中，株行距为30cm×60cm。桶式栽培处理20个栽培桶，每桶一株黄瓜；袋式栽培处理10个栽培袋，每袋两株黄瓜；槽式栽培处理5个栽培槽，每个栽培槽4株黄瓜。于2012年9月28日至2013年1月26日进行栽培管理，缓苗后，白天温度控制在28～30℃，夜间14～16℃。采用滴灌系统根据植株不同生育期定时定量浇灌Hoagland营养液。定植成活后，每株每天浇1/2S Hoagland营养液（200～500ml）；开花结果期，每株每天需浇1S Hoagland营养液（1.0～2.5 L）。为培育壮苗，一般五节以下不留瓜。植株调整、病虫害防治按正常栽培管理进行。

1.2.2 指标测定方法

在植株营养生长期，分别于10月5日、12日、19日和26日测定黄瓜生长势，第一次采样全株拔起，剩下3次定株观测株高、茎粗、叶片数和最大叶叶面积的生长动态，每个处理选定6株。用直尺测量株高；用游标卡尺测量茎粗；烘干法测定干重；完全展开叶及叶长超过3cm的叶数均为叶片数；选出最大叶，测量叶长（叶片基部至叶尖的距离）和叶宽（叶片上部肩宽测量值），运用公式 $y=0.7430x$（x 为叶长×叶宽）计算最大叶面积（裴孝伯等，2005）。

选取节位一致的黄瓜，用游标卡尺测量果实直径和果实长度，去除瓜瓤，测定果肉厚度和果肉质量。总酸度采用NaOH滴定法测定，维生素C含量用碘滴定法测定，可溶性总糖含量用蒽酮比色法测定，可溶性固形物含量取汁液用折射仪Pocket. PAL-1（日本株式会社）测定，可溶性蛋白用紫外吸收法测定。黄瓜单株产量从开始采收到花凋谢时所有采收瓜称重的质量，采收瓜长应达到13～16cm，直径达2～3cm，并记录每个处理的瓜条数以及畸形果数。

1.2.3 数据分析

试验数据采用Microsoft Excel 2003和SAS9.2软件进行数据处理和统计分析，在 $P<0.05$ 的条件下分析数据的差异显著性。

2 结果与分析

2.1 不同栽培容器对水果型黄瓜生长的影响

从表3中可以看出，定植后第7天，不同栽培容器对水果型黄瓜植株生长指标和壮苗指数均没有显著性的差异。从图1可以看出，10月12日，处理C株高显著大于其余两个处理（C为40.2cm，T为34.5cm，D为33.3cm）。10月19日与26日，处理C株高显著大于D，C与T、T与D之间没有显著性差异。

从图2可以看出，10月26日之前的3个时期，各处理间茎粗没有显著性差异。10月26日，处理D茎粗显著大于C和T（D、C、T分别为7.05mm、5.41mm、5.30mm）。从图3可以看出，营养生长期间各处理叶片数没有显著性差异。

表 3 不同栽培容器对水果型黄瓜植株生长的影响

Table 3 Effect of different cultivation containers on the plant growth of mini-cucumber

处理 Treatments	株高 Plant height (cm)	茎粗 Stalk width (mm)	根干重 Root dry biomass (g·plant^{-1})	地上干重 Shoot dry biomass (g·plant^{-1})	全株干质量 Plant dry biomass (g·plant^{-1})	壮苗指数 Seedling index
T	8.80a	4.52a	0.10a	0.67a	0.77a	0.54a
D	8.67a	4.74a	0.08a	0.64a	0.73a	0.50a
C	10.07a	4.34a	0.09a	0.57a	0.66a	0.39a

注：a、b、c 小写字母表示 5% 显著差异下同

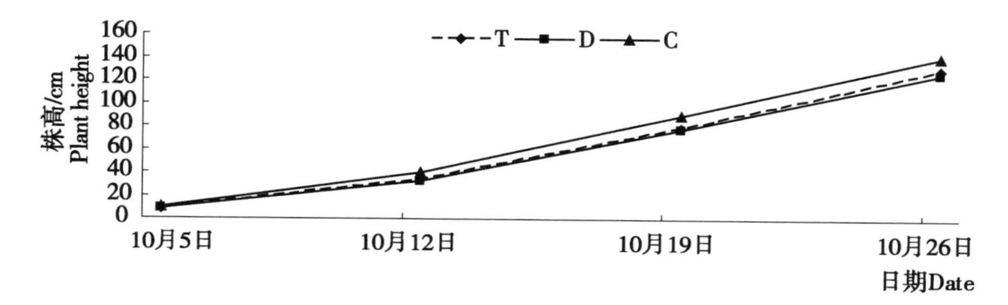

图 1 不同栽培容器对水果型黄瓜植株株高的影响

Figure 1 Effect of different cultivation containers on the plant height of minicucumber

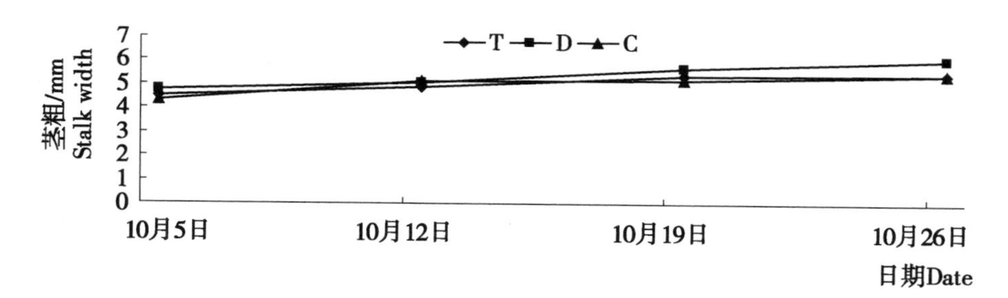

图 2 不同栽培容器对水果型黄瓜植株茎粗的影响

Figure 2 Effect of different cultivation containers on the stalk width of minicucumber

从图 4 可以看出，10 月 5 日与 12 日，各处理之间最大叶片叶面积没有显著性差异。10 月 19 日，处理 D 最大叶片叶面积显著大于处理 T，D 与 C、C 与 T 之间没有显著性差异（D、C、T 最大叶片叶面积分别是 614.0cm^2、504.8cm^2、450.2cm^2），增加幅度分别为 50.6%、34.6%、30.9%。10 月 26 日，各处理之间最大叶片叶面积没有显著性差异。

2.2 不同栽培容器对水果型黄瓜果实性状及果实品质的影响

从表 4 中可以看出，处理 D 的果实质量、果实直径、果肉质量显著大于处理 T 和 C，而处理 T 与 C 之间差异不显著；各处理之间的果实长度、果肉厚度、果肉质量/果实质量和果肉厚度/果实直径均没有显著性差异。

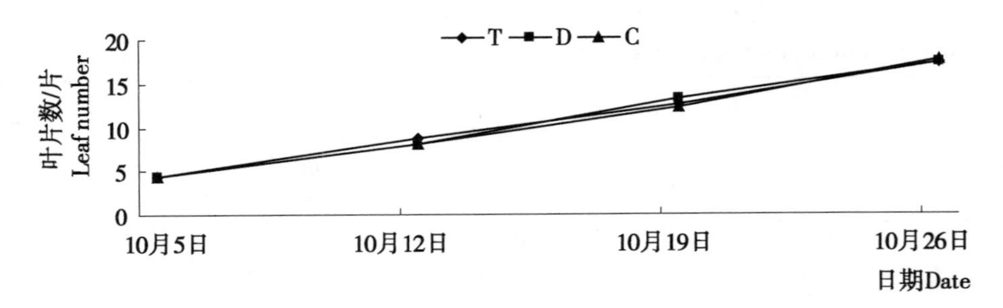

图3 不同栽培容器对水果型黄瓜植株叶片数的影响

Figure 3 Effect of different cultivation containers on the leaf number of minicucumber

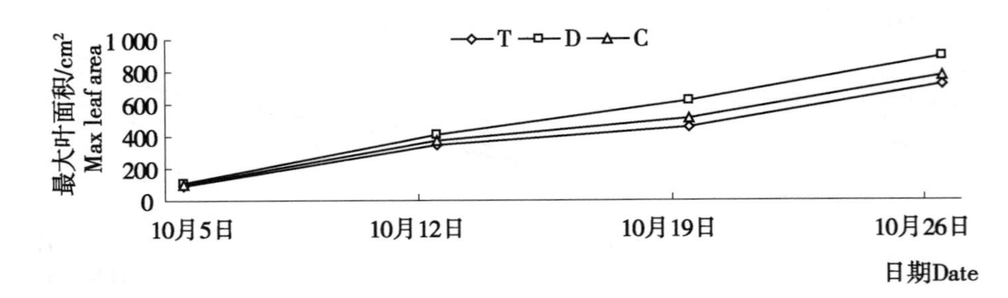

图4 不同栽培容器对水果型黄瓜最大叶面积的影响

Figure 4 Effect of different cultivation containers on the max leaf area of minicucumber

表4 不同栽培容器对水果型黄瓜果实性状的影响

Table 4 Effect of different cultivation containers on the fruit characters of mini-cucumber

处理 Treatments	果实质量 Fruit weight (g・fruit^{-1})	果实直径 Fruit diameter (cm)	果实长度 Fruit length (cm)	果肉厚度 Flesh thickness (mm)	果肉质量 Flesh weight (g・fruit^{-1})	果肉质量/果实质量 FlW/FrW	果肉厚度/果实直径 FT/FD
T	77. 34b	2. 93b	12. 99a	6. 42a	57. 06b	0. 74a	0. 22a
D	83. 43a	3. 04a	13. 34a	6. 85a	62. 50a	0. 75a	0. 23a
C	76. 54b	2. 90b	13. 06a	6. 34a	56. 34b	0. 74a	0. 22a

从表5可以看出，D和C的总酸度最高，显著大于处理T；处理D的可溶性糖含量最高，显著大于处理C，但与处理T之间没有显著性差异，处理T与处理C之间差异不显著；处理D的可溶性蛋白含量最高，显著大于处理T，但和处理C之间差异不显著，处理T与处理C之间差异不显著。各处理的还原型Vc和可溶性固形物没有显著性差异。

2.3 不同栽培容器对水果型黄瓜产量构成的影响

从表6中可以看出，各处理的单果重在61.13～66.54g之间，差异显著。处理D的单株果数和单株总产量分别为26.10个和1.69kg，显著高于其他处理，而处理T的单株果数最少，因此，单株总产量也最低。各处理之间畸形果率差异显著，处理T的畸形果率最高，这可能与营养液流失过多有关。

表 5　不同栽培容器对水果型黄瓜果实品质的影响

Table 5　Effect of different cultivation containers on the fruit quality of mini-cucumber

处理 Treatments	总酸度 Total acidity (%)	还原型维生素 C Antitype vitamin c ($mg \cdot kg^{-1}$)	可溶性总糖 Total soluble sugar (%)	可溶性固形物 Soluble solids (%)	可溶性蛋白 Soluble protein ($mg \cdot g^{-1}$)
T	0. 09b	860. 52a	1. 69ab	4. 30a	0. 27b
D	0. 10a	866. 58a	1. 99a	4. 34a	0. 33a
C	0. 10a	884. 76a	1. 15b	4. 14a	0. 30ab

表 6　不同栽培容器对水果型黄瓜产量构成的影响

Table 6　Effect of different cultivation containers on the yields of mini-cucumber

处理 Treatments	单株总产量 Yield ($kg \cdot plant^{-1}$)	单果重 Fruits weight (g)	单株果数 Fruits number ($No \cdot plant^{-1}$)	畸形果率 Irregular fruit rate (%)
T	1. 12c	66. 54a	16. 85c	7. 16a
D	1. 69a	64. 92b	26. 10a	1. 32c
C	1. 54b	61. 13c	25. 25b	1. 75b

3　讨论

植物根系的生长发育及对养分和水分的吸收，受到植物根际固、液、气三相比例的直接影响，并影响其地上部的生长发育（赵旭 等，2010）。基质栽培条件下，常因基质中气态比例低或气体中 CO_2 浓度高、O_2 浓度低而影响植物根系的吸收和生长发育，进而影响地上部分的生长（Nakano，2007；Chong *et al.*，2004；Vartapetian and Jackson，1997）。与其他栽培容器相比，虽然 NAU-G1 型专用桶式栽培装置能够很大程度改善作物根际的通气环境、促进植株根系的生长，但在本试验中所采用的基质配方通气性良好，若在栽培装置上再改善通气状况的话，栽培过程中基质的液、气比例可能会减小至不适宜作物生长的值。在栽培管理过程中发现，桶式栽培淋失的营养液量多于其他栽培方式。由于栽培槽中基质表面完全与空气接触，故栽培槽基质表面水分蒸发量多于栽培袋，即营养液量耗损较多。而栽培袋中基质与空气接触面积较少，基质保水能力最强，固、液、气比例较佳，因而这种栽培方式下水果型黄瓜生长、产量和品质都较佳。

在生产成本方面，3 种栽培条件下，采用袋式栽培的成本最低，因为在购买基质时，可以将基质包装袋直接作为袋式栽培的袋子，不需要再购买额外的装置和设施；采用槽式栽培，可就地取材，甚至在地上挖出栽培槽，然后铺上塑料薄膜，实现与土壤隔离，成本亦较低；而采用桶式栽培，需要额外的投资和购买大量的设备，还需要根据当地实际选用合适的栽培方式。

4　结论

试验结果表明，采用枕头式袋式栽培，水果型黄瓜植株生长健壮，提高了水果型黄瓜单株果实数和单株总产量，降低了畸形果率，果实品质有所改善。而且从投资大小上来说，袋式栽培投资较低。说明有机基质袋式栽培（D）有利于水果型黄瓜产量和品质的提高，栽培袋可作为水果型黄瓜有机基质栽培的适宜容器。

参考文献

[1] 周跃华，聂艳丽，赵永红等．国内外固体基质研究概况．中国生态农业学报，2005（4）：40～43

[2] 杨波．有机生态型无土栽培．中国花卉园艺，2002（7）：23～25

[3] 卜崇兴，李式军．无土栽培新装置根区环境温、湿度及设施内地表湿度变化．上海农业学报，2003，19（3）：83～86

[4] 郑奕，卜崇兴，黄量等．一种桶式无土栽培新装置在番茄栽培上的应用研究．上海农业学报，2008a，24（3）：58～60

[5] 郑奕，卜崇兴，黄量等．一种无土栽培新装置根区温湿度及对番茄根系生长的影响．长江蔬菜（学术版），2008b，5b，60～62

[6] 李式军，郭世荣．设施园艺学（第二版）．北京：中国农业出版社，2011

[7] 裴孝伯，李世城，张福墁等．温室黄瓜叶面积计算及其与株高的相关性研究．中国农学通报，2005，8（21）：80～82

[8] 赵旭，李天来，孙周平．番茄基质通气栽培模式的效果．应用生态学报，2010，21（1）：74～78

[9] Nakano Y. Response of tomato root systems to environmental stress under soilless culture. Japan Agricultural Research Quarterly，2007（41）：7～15

[10] Chong S K，Boniak R，Indorante S，*et al.* Carbon dioxide content in golf green rhizosphere. Crop Science，2004（44）：1 337～1 340

[11] Vartapetian B B，Jackson M B. Plant adaptations to anaerobic stress. Annals of Botany，1997（79）：3～20

氮素形态和水平对生菜生长、叶片光合色素及营养液中根分泌物累积的影响*

邱志平[1,2]**，杨其长[1,2]，刘文科[1,2]***

（1. 中国农业科学院农业环境与可持续发展研究所，北京　100081；

2. 农业部设施农业节能与废弃物处理重点实验室，北京　100081）

摘要：采用温室盆栽试验的方法，研究了不同氮形态和水平处理，包括 10mmol/L NO_3^--N（N10），5mmol/L NO_3^--N + 5mmol/L NH_4^+-N（N5 + A5），5mmol/L NO_3^--N（N5）和 2.5mmol/L NO_3^--N + 2.5mmol/L NH_4^+-N（N2.5 + A2.5），对生菜生长、叶片中光合色素含量及营养液中根分泌物累积的影响。结果表明，N10 和 N5 处理生菜地上部和根系生物量显著高于复合氮形态处理 N5 + A5 和 N2.5 + A2.5。相同氮形态下，不同氮水平处理间在生菜地上部和根系生物量上差异不显著。N10 和 N5 的生菜叶绿素 a 含量、叶绿素 b 含量、类胡萝卜素含量、营养液中 TOC 浓度及 TOC 含量均显著低于复合氮形态处理 N5 + A5 和 N2.5 + A2.5。同一氮形态下，不同氮水平处理的叶绿素 a 含量、叶绿素 b 含量、类胡萝卜素含量、营养液中 TOC 浓度及 TOC 含量间差异不显著。可见，在本试验条件下，铵态氮虽不利于生菜的生长，而且还会增加营养液中 TOC 的含量，但却可以提高叶片中光合色素的含量。而相同氮形态下，不同氮水平对生菜生长，叶片光合色素及营养液中根分泌物的累积影响不大。

关键词：生菜；氮形态；氮水平；光合色素；根分泌物

Effects of Nitrogen Forms and Levels on Hydroponic Lettuce Growth, Content of Leaf Photosynthetic Pigments and Root Exudate Accumulation in Nutrient Solution

Qiu Zhiping[1,2], Yang Qichang[1,2], Liu Wenke[1,2]

(1. *Institute of Environment and Sustainable Development in Agriculture*, *Chinese Academy of Agricultural Science*, *Beijing* 100081, *China*; 2. *Key Lab. of Energy Conservation and Waste Management of Agricultural Structures*, *Ministry of Agriculture*, *Beijing* 100081, *China*)

Abstract: A pot experiment in greenhouse was carried out to examine the effects of different nitrogen forms and nitrogen levels, including 10mmol/L NO_3^--N (N10), 5mmol/L NO_3^--N + 5mmol/L NH_4^+-N (N5 + A5), 5mmol/L NO_3^--N (N5) 和 2.5mmol/L NO_3^--N + 2.5mmol/L NH_4^+-N (N2.5 + A2.5) (10mmol/L NO_3^--N, 5mmol/L NO_3^--N + 5mmol/L NH4$^+$-N, 5mmol/L NO_3^--N, 2.5mmol/L NO_3^--N + 2.5mmol/L NH4$^+$-N), on lettuce

* 基金项目：国家 863 计划重大项目（2011AA03A114）和国家科技支撑计划（2011BAE01B00）；中央级公益性科研院所基本科研业务费项目

** 作者简介：邱志平（1986—），女，硕士研究生，研究方向为设施蔬菜营养与品质调控。E-mail：youngapple1986@126.com

*** 通讯作者：刘文科（1974—），男，博士，研究员，硕士生导师，研究方向为设施蔬菜营养与品质调控。E-mail：liuwke@163.com

growth, leafy photosynthetic pigments contents and root exudate accumulation in nutrient solution. The results showed that lettuce shoot and root biomass with N10 and N5 treatments were higher than those of N5 + A5 and N2.5 + A2.5 treatments, and no difference was found between treatments of same nitrogen form and different nitrogen levels. On the contrary, the content of leafy chlorophyll a, chlorophyll b, carotenoid contents, and TOC concentration and content in nutrient solution of N5 + A5 and N2.5 + A2.5 treatments were higher than N10 and N5 treatments. Also, no difference was found between treatments of same nitrogen form and different nitrogen levels.

Key words: Lettuce; Nitrogen forms; Nitrogen levels; Photosynthetic pigments; Root exudates

蔬菜是人们日常生活饮食中不可缺少的农产品，含有丰富的维生素、蛋白质、纤维素、矿质元素等，对维持人们的身体健康起着重要的作用。氮素是蔬菜营养的第一大元素，对蔬菜的生长发育及营养品质的形成具有重要影响。关于氮素水平和氮素形态与蔬菜的生长、品质的关系研究较多，尤其是对提高维生素 C、可溶性蛋白、可溶性糖含量及降低硝酸盐的研究[1,2,3,4,5]，但是关于氮水平与形态对光合色素，尤其是对类胡萝卜的研究相对较少。

类胡萝卜素广泛存在于植物体内，不仅是一种光合辅助色素[6]，而且一些类胡萝卜素是维生素 A 的前体，是人饮食中必需的营养成分，可提高人体的免疫能力。同时，类胡萝卜素作为一种抗氧化剂，可以清除体内产生的自由基，具有抗癌作用[7]。因此，了解氮素水平及形态对蔬菜中类胡萝卜素的累积具有重要理论与现实意义。

蔬菜在生长发育过程中，根系会分泌有机物，根系分泌物的多少影响光合产物的分配与利用，并直接决定着作物的产量。研究证明，根分泌物的释放量相当可观，一年生植物净光合产物的 30% ~60% 被分配到根部，其中，4% ~70% 以有机碳的形式被释放到根际中[8]。在设施蔬菜无土栽培中，根系分泌物随着栽培时间延长累积于营养液中，甚至产生连作障碍。据报道，芦笋水培循环利用营养液中积累了大量的根分泌物，结果造成了芦笋的严重减产[9]。因此，探讨氮水平及形态对根分泌物累积的影响，有助于从氮素营养角度调控根分泌物的累积，降低连作障碍的发生几率，减少由于根系分泌有机物造成的光合产物的浪费。

本试验在温室条件下采用营养液水培盆栽的方法研究了不同氮水平及氮形态对生菜生长、叶片光合色素含量、营养液中根分泌物累积的影响，以期从氮素营养角度为设施蔬菜的优质高产提供理论依据。

1 材料与方法

1.1 试验材料

供试材料生菜为意大利品种。基础营养组成为（mmol/L）：0.75 K_2SO_4，0.5 KH_2PO_4，0.1 KCl，0.65 $MgSO_4$，1.0×10^{-3} H_3BO_3，1.0×10^{-3} $MnSO_4$，1.0×10^{-4} $CuSO_4$，1.0×10^{-3} $ZnSO_4$，5×10^{-6} $(NH_4)_6Mo_7O_{24}$ 和 0.1 EDTA-Fe。

1.2 试验设计

本试验设 4 个处理，包括 10mmol/L NO_3^- N（N10），5mmol/L NO_3^- N + 5mmol/L NH_4^+-N（N5 + A5），5mmol/L NO_3^- N（N5）和 2.5mmol/L NO_3^- N + 2.5mmol/L NH_4^+-N（N2.5 + A2.5），每个处理重复 3 次。采用栽培盆栽培生菜，4 株/盆，营养液 5L/盆，盆的规格：直径 18.5cm，高 20cm。采用充气泵补充营养液中氧气。本试验硝态氮由

Ca（NO_3）$_2$、KNO_3 提供，铵态氮由（NH_4）$_2SO_4$ 提供，利用增加 K_2SO_4 的方法来平衡各处理因 KNO_3 的加入所造成的的钾离子浓度差异。

各处理营养液中氮素构成：N10 由 0.5mmol/L 的 Ca（NO_3）$_2$ 和 9mmol/L KNO_3 组成；N5 + A5 由 0.5mmol/L Ca（NO_3）$_2$，4mmol/L KNO_3 和 2.5mmol/L 的（NH_4）$_2SO_4$ 组成；N5 由 0.5mmol/L Ca（NO_3）$_2$ 和 4mmol/L KNO_3 组成；N2.5 + A2.5 由 0.5mmol/L Ca（NO_3）$_2$，1.5mmol/L KNO_3 和 1.25mmol/L（NH_4）$_2SO_4$ 组成。

试验地点在中日中心楼楼顶温室。温室室内温度 28～35℃，2011 年 7 月 12 日育苗，育苗 10 天后定植，定植 40 天后采收，测定地上鲜重，根系鲜重，叶片中光合色素（叶绿素 a，叶绿素 b，类胡萝卜素）含量，营养液中 TOC 浓度及各盆中剩余营养液的体积。

1.3 测定方法

生菜的地上鲜重及根系鲜重采用称重法。叶片中光合色素测定，采用 80% 丙酮浸提比色法[10]，测定相应的吸光度值。营养液中 TOC 浓度测定，将剩余营养液混匀后，用量筒准确量取其体积，然后取样，用 TOC 分析仪（Apollo-9000，美国 Tekmar Dohrmann 公司产）分析营养液中 TOC 的浓度，计算出 TOC 的含量。

1.4 数据分析

各实验数据用 SAS8.2 进行统计分析

2 结果与分析

2.1 不同氮形态及水平对生菜生物量、生菜叶片中叶绿素 a 及叶绿素 b 含量影响

由表可知，全氮处理 N10 和 N5 的生菜地上部和根系生物量显著高于复合氮形态处理 N5 + A5 和 N2.5 + A2.5。这表明，生菜在全硝态氮供应条件下生长速率更快，产量更高。N10 根系鲜重略高于 N5 处理外，地上生物量无显著差异。N5 + A5 和 N2.5 + A2.5 之间根系和地上部生物量差异不明显。数据表明，硝态氮更有利于的生菜地上及地下部分的生长。

全氮处理 N10 和 N5 的生菜叶片中的叶绿素 a 和叶绿素 b 含量显著高于复合氮形态处理 N5 + A5 和 N2.5 + A2.5。全氮处理 N10 和 N5 的生菜叶片中的叶绿素 a 和叶绿素 b 含量之间无差异，而且复合氮形态处理 N5 + A5 和 N2.5 + A2.5 间在生菜叶片中的叶绿素 a 和叶绿素 b 含量上也无显著差异。可见，与全硝态氮供应相比，复合氮形态供应更能提高叶片光合色素含量。

表 不同处理生菜的生物量及叶片中叶绿素 a 及叶绿素 b 含量

Table Shoot and root biomass, leafy chlorophyll a and chlorophyll b contents of lettuce of different treatments

处理 Treatments	地上部生物量 Shoot biomass (g/pot)	根系鲜重 Root biomass (g/pot)	叶绿素 a 含量 Chlorophyll a content (mg/g)	叶绿素 b 含量 Chlorophyll b content (mg/g)
N10	56.56a	5.98a	0.33b	0.15b
N5 + A5	33.57b	4.51b	0.51a	0.21a
N5	50.39a	5.30ab	0.35b	0.15b
N2.5 + A2.5	44.83ab	4.71b	0.51a	0.17b

注：同列数据后不同字母表示差异达 5% 显著水平。(下同)

Note: Different letters in column data show significantly diference at 5% level. (The same below)

2.2 不同氮形态及水平对生菜叶片中类胡萝卜素含量影响

由图 1 可知，同一氮形态下，N10 处理与 N5 处理间生菜叶片中类胡萝卜素含量无显著差异，N5 + A5 处理与 N2. 5 + A2. 5 处理生菜叶片中类胡萝卜素含量也无显著差异，但在不同氮形态处理之间，生菜叶片中类胡萝卜素的含量存在显著差异，无论是否处于同一氮水平，加入铵盐的处理类胡萝卜素的含量都要高于全硝态氮处理，说明铵态氮对叶片中类胡萝卜素的增加有促进作用。

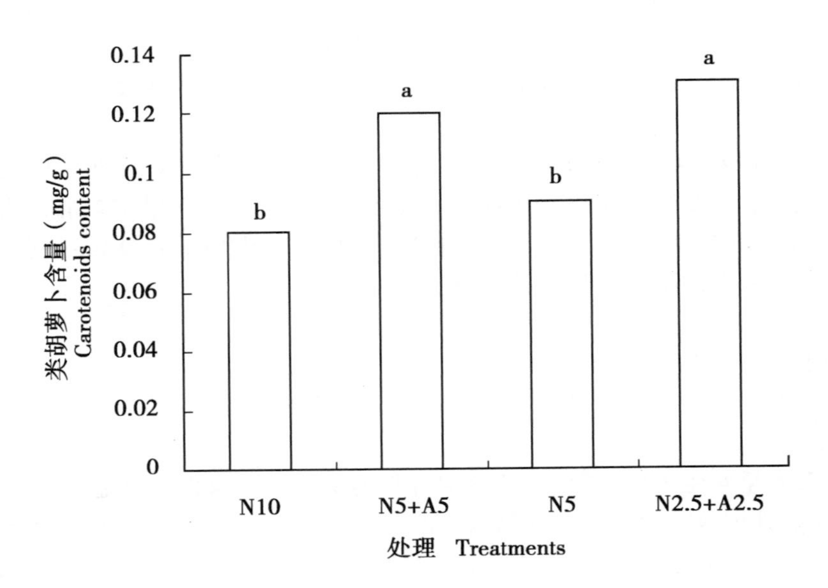

图 1 不同氮形态及水平对叶片中类胡萝卜素含量的影响

Figure 1 Effect of different nitrogen forms and levels on carotenoids content in leaves

2.3 不同氮形态及水平对营养液中 TOC 浓度的影响

由图 2 可知，营养液中 TOC 浓度，10mmol/L 氮水平不同氮形态处理，TOC 浓度存在显著差异，N10 处理的 TOC 浓度明显低于 N5 + A5，5mmol/L 氮水平不同氮形态处理，TOC 浓度存在显著差异，N5 处理的 TOC 浓度明显低于 N2. 5 + A2. 5。同一氮形态不同氮水平下，N10、N5 处理之间无显著差异，N5 + A5、N2. 5 + A2. 5 处理之间差异显著，且 N2. 5 + A2. 5 处理 TOC 浓度要高于 N5 + A5 处理。

2.4 不同氮形态及水平对营养液中 TOC 含量影响

由图 3 可知 10mmol/L 氮水平下，N10 处理与 N5 + A5 处理比较，营养液中 TOC 含量存在显著差异，且 TOC 含量前者低于后者，两者差值约为 16mg。5mmol/L 氮水平下，N5 处理与 N2. 5 + A2. 5 处理比较，营养液中 TOC 含量存在显著差异，且 TOC 含量前者低于后者，前者比后者低约 17mg；总体来看，无论是否处于同一氮水平，N5 + A5、N2. 5 + A2. 5 处理，TOC 含量要高于 N10、N5；而同一氮形态处理之间差异不显著，说明铵态氮不仅抑制生菜地上地下部的生长，而且会刺激根分泌物的分泌。

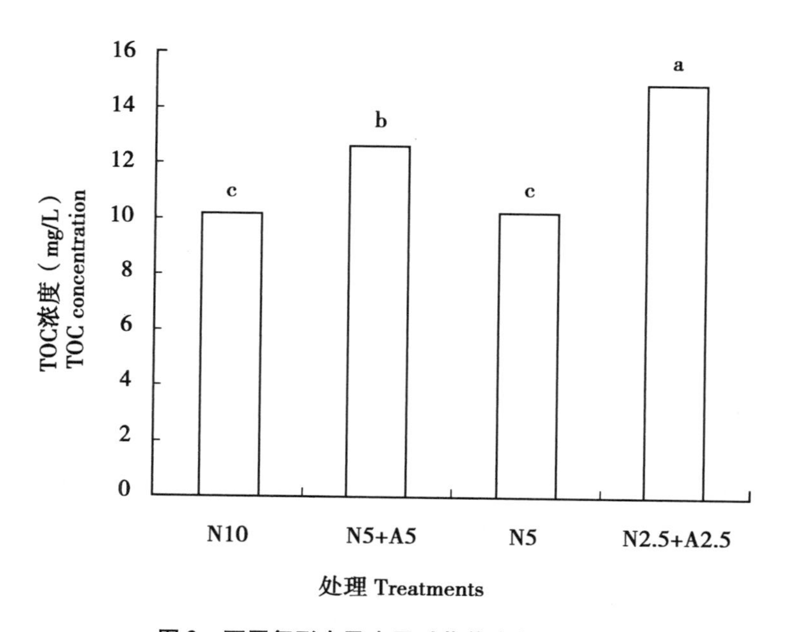

图 2　不同氮形态及水平对营养液中 TOC 浓度的影响

Figure 2　Effect of diferent nitrogen forms and level on TOC concentraton in nutrition

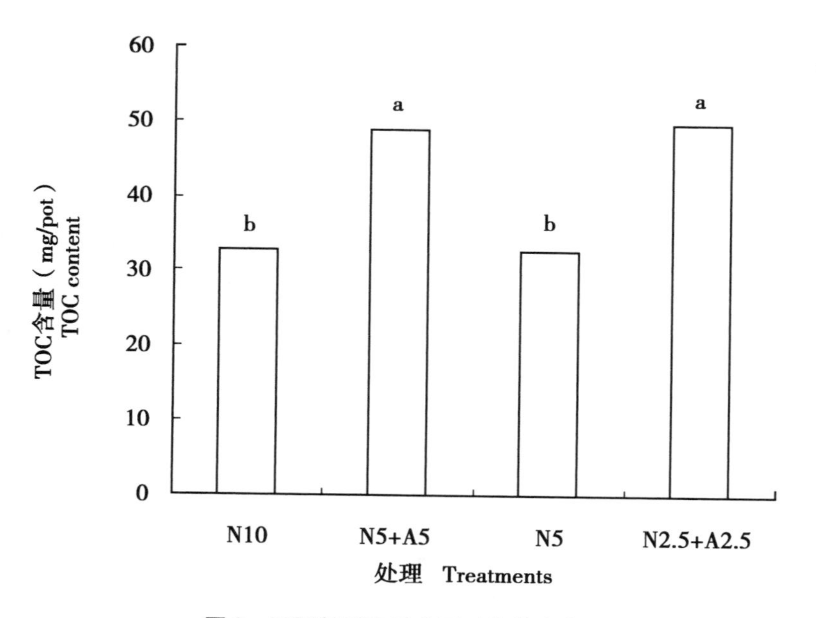

图 3　不同氮形态及水平对营养液中 TOC 含量的影响

Figure 3　Effect of different nitrogenforms and levels on TOC content in nutrition

3 讨论

本试验结果表明，NO_3^- N 处理生菜的根系及地上部分的生物量要高于 NO_3^- N + NH_4^+ – N 处理，且在 10mmol/L NO_3^- N 处理下生菜的地上部和根系生物量最大。关于不同氮形态对蔬菜的生长的影响已有大量报道，张英鹏等[11]研究了在供氮浓度 8mmol/L 不同氮形态对菠菜生长的影响，结果表明，硝铵比为 1∶1 时菠菜的生物量最高。王波[12]对生菜的研究结果表明，营养液中供氮浓度为 7.5mmol/L，硝铵比为 1∶3 时，生菜的地上部分生物量最大。本试验的结果与以上两篇文献的报道略有差异，一方面可能是不同的蔬菜品种对铵态氮的敏感性不同，另一方面，可能是本试验设置的 NO_3^- N + NH_4^+-N 比例，已经超出了生菜的耐受铵态氮的范围。

铵态氮可增加生菜叶片中叶绿素 a，叶绿素 b 及类胡萝卜素的含量。陈艳丽等[2]的研究结果表明铵态氮部分替代硝态氮利于小白菜叶绿素和类胡萝卜素含量的增加。荣秀莲等[13]两种白菜品种的研究结果表明，2 个白菜品种的叶绿素含量均在铵态氮与硝态氮比例为 1∶3 时含量最高。植物体内铵态氮含量增加，提高了叶绿素合成前提谷氨酸或 a-酮戊二酸的含量，促进了叶绿素的合成[14]。与全硝态氮处理相比，复合氮形态处理可提高叶片类胡萝卜素的含量，从而提高了生菜的营养品质和光合能力。已有很多研究表明，高水平硝态氮供应将显著增加地上部硝酸盐的累积，降低蔬菜的营养品质，因此适宜增铵被认为是降低硝酸盐含量的好方法[3,4]。

但是，本研究表明，增加铵态氮可导致生菜营养液中 TOC 的浓度及含量的提高，TOC 的浓度和含量的提高反映了营养液中根分泌物的累积程度加剧，造成这种影响可能是因为生菜对铵态氮的比较敏感，为了适应铵态氮的一种反应，其原因有待进一步研究。研究证明，植物在自身的生长发育过程中会分泌有机物，以适应周围不利环境因素的影响[15,16,17]。营养液中根分泌物的累积程度加剧将增加连作障碍的发生几率，导致营养液循环利用效益降低，无法获得高产。日本已有研究表明，根分泌物导致的 TOC 增加大幅降低了园艺作物的番茄和芦笋的产量[9,18]。综上所述，营养液中适当增加铵态氮可以提高生菜叶片中叶绿素的含量，增加类胡萝卜素含量，但导致营养液中 TOC 的浓度及含量的增加，因此，在设施无土栽培中，制定合理的氮营养供应策略对实现优质高产至关重要。

参考文献

[1] 田宵鸿，王朝辉，李生秀．不同氮素形态及配比对蔬菜生长和品质的影响．西北农业大学学报，1999，27（2）：6 ~ 10

[2] 陈艳丽，高新生，李绍鹏．不同形态氮素替代部分硝态氮对水培小白菜的生长和品质．中国农学通报，2010，26（15）：306 ~ 309

[3] 张攀伟，罗金葵，陈巍等．硝铵比例影响小白菜生长和叶绿素含量的原因探究．植物营养与肥料学报，2006，12（5）：711 ~ 716

[4] 孙园园，林咸永，金崇伟等．氮素形态对菠菜体内抗坏血酸含量及其代谢的影响．浙江大学学报，2009，35（3）：292 ~ 298

[5] 胡承孝，邓波儿，刘同仇．氮肥水平对蔬菜品质的影响．土壤肥料，1996（3）：34 ~ 36

[6] Goodwin T. W. The Biochemistry of the Carotenoids，Chapman and Hall，London，Ed2，1980（1）：377

[7] Olson J. A. , Provitamin-A Function of carotenoids：the conversion of β-carotenoid into vitamin-A，J. Nutr. , 1989，119：105～108
[8] 申建波，张福锁．根分泌物的生态效应．中国农业科技导报，1999，1（4）：21～27
[9] Kayano Sunada，Kazuhito Hashimoto *et al.* Detoxification of phytotoxic compounds by TiO_2 photocatalysis in a recycling hydroponic cultivation system of asparagus . Journal of Agriculture and Food Chemistry，2008（56）：4 819～4 824
[10] 李合生．植物生理生化实验原理和技术．北京：高等教育出版社，2000
[11] 张英鹏，林咸永，章永松等．不同氮形态对菠菜生长及体内抗氧化酶活性的影响．浙江大学学报，2006，32（2）：139～144
[12] 王波，王梅农，赖涛等．不同氮形态营养对生菜光合特性的影响．南京农业大学学报，2007，30（4）：74～77
[13] 荣秀莲，王梅农，宋采博等．不同铵态氮/硝态氮配比对白菜叶绿素含量的影响．江苏农业科学，2010（1）：298～300
[14] 潘瑞炽，董愚得．植物生理学．北京：高等教育出版社，1995
[15] Nambiar E K S. Uptake of ^{65}Zn from dry soil by plants. Plant Soil，1976，44：267～271
[16] Roemheld V. Different strategies for iron acquisition in higher plants. Physiol Plant，1987，70：231～234
[17] Dinkelaker B，Roemheld V，Marschner H . Citric acid secretion and precipitation of calcium cirrate in the rhizosphere of white lupin（*Lupines albus* L. ）. Plant Cell Environ，1989，12：285～292
[18] Miyama Y. , Sunada K. , Fujiwara S. , Hashimoto K. Photocatalytic treatment of waste nutrient solution from soil-less cultivation of tomatoes planted in rice hull substrate. Plant and Soil，2009，318：275～283

根际温度对高温下温室番茄根系显微结构的影响*

韩亚平**，李亚灵***，宋敏丽

（山西农业大学园艺学院，太谷 030801）

摘要：［目的］针对夏季高温对温室番茄栽培的限制作用，试图通过降低根际温度来缓解高气温所带来的危害。［方法］采用营养液循环栽培法，以28℃ ±1℃的自然根温为对照，分别对根系进行23℃ ±1℃和33℃ ±1℃的处理，采用石蜡切片法对植株根系结构进行观察测量。［结果］与28℃ ±1℃的自然根温相比，23℃ ±1℃处理根系导管直径增大12%、导管密度减小20%；33℃ ±1℃处理根系的导管直径减小17%、导管密度增大30%；导管总面积3个处理间差异不显著。［结论］降低根温至23℃左右，可增强根系的吸收能力，利于植株生长，可有效缓解夏季高温危害。

关键词：温室番茄；根际温度；高温；石蜡切片；根系结构

Influence of Root-zone Temperature on the Microstructure of Greenhouse Tomato Root under High Temperature

Han Yaping, Li Yaling, Song Minli

(*College of Horticulture*, *Shanxi Agricultural University*, *Taigu*, 030801, *China*)

Abstract: High temperature in summer is a main limit factor for greenhouse tomato production. Therefore, the experiment was carried out to reduce the root-zone temperature in order to relieve the harm by high temperature. Adopting DFT hydroponics method, the natural root-zone temperature of 28 ±1℃ was as control, and the root-zone temperature of 23 ±1℃ and 33 ±1℃, were respectively realized as two treatments. The root microstructure was observed and measured by the method of paraffin section. Compared with the natural root-zone temperature of 28 ±1℃, the diameter of root vessel was increased about 12%, and its density was reduced about 20% with the treatment of lower root-zone temperature of 23 ±1℃. On the opposite, the diameter of root vessel was decreased about 17%, and its density was increased 30% in the treatment of 33 ±1℃. The total area of root vessel had no significant difference among the three treatments. Therefore, lower root-zone temperature of 23 ±1℃, could enhance the root ability, be conducive to plant growth, and effectively relieve heat harm in summer. It is a good way for over-summer cultivation in greenhouse tomato.

Key words: Greenhouse tomatoes; Root-zone temperature; High temperature; Paraffin section; Root microstructure

1 引言

番茄植株生长发育的最适温度为昼温24～26℃，夜温18℃左右，超过30℃则温度过高，到40℃就停止生长。根际温度的最高界限为33℃。然而，在光热资源丰富的华北地区，

* 基金项目：中国高等学校博士学科点专项科研基金资助课题（项目编号：20091403110002）

** 作者简介：韩亚平（1985—），女，在读硕士，研究方向为蔬菜栽培生理。E-mail：liyapinglyp01@sina.com

*** 通信作者：李亚灵，博士，教授，研究方向为设施蔬菜栽培。E-mail：yalingli1988@yahoo.com

夏季温度偏高，温室6~8月份温度很容易上升到35℃甚至更高（田鹏，2004），这样的温度对番茄栽培而言是灾难性的。因此可以认为高温是温室番茄长季节栽培稳产、高产的主要限制因子（王冬梅等，2004）。

高根温往往伴随着高气温而产生。植物的生长发育可能对根温更为敏感。根温变化1℃就能引起植物生长的明显变化（Walker，1969）。植物体是内在统一的有机整体，根部和地上部存在着密切的联系，任一部分受到环境因子的影响，均会迅速涉及和传递到另一部分。根温直接影响根系的生长和根系对水肥及矿质元素的吸收，从而影响番茄植株的生长。因此本试验力图通过降低根温来缓解高气温所带来的危害，为温室番茄越夏栽培提供理论支持。

2 材料与方法

试验材料为番茄“新星101”。试验于2009年5~8月在山西太谷（北纬37°25′，东经112°25′）山西农业大学设施农业工程中心的非对称三连栋温室内进行。水培育苗，待植株现蕾后，移栽于水培槽内进行营养液循环栽培，缓苗一周后于每天8：00~18：00进行不同的根温处理：T1：23℃ ±1℃（番茄生长比较适宜的根温，通过向营养液中放冰桶来控温），T2：28℃ ±1℃（根际自然温度状态，不进行处理），T3：33℃ ±1℃（设施栽培中由高气温可能产生的高根温状态，通过电热线加热来控温）。

分别在处理10d、20d、30d、40d后，选择生长健壮植株主根的中部，切取5mm长的小段，迅速浸于FAA固定液中，真空抽气后固定48h以上。采用石蜡切片法对固定好的材料进行酒精梯度脱水、二甲苯梯度透明、透蜡、包埋后切片，切片厚度10μm，对切片再进行脱蜡、番红-固绿对染法染色、二甲苯透明后用中性树胶封片，干片后在Olympus显微镜下进行观察、拍照，并用台式测微尺测定3个温度下的导管直径、导管密度、导管总面积各10次，并进行差异显著性分析。

3 结果与分析

3.1 根系的导管直径

根系导管直径随着根系温度的增高而显著降低（图1和图2A），即 $T_1 > T_2 > T_3$。全期平均，T_1 为21.6μm，T_2 为19.1μm，T_3 为16.1μm 。试验结束时，导管直径 T_1 处理比 T_2 处理的大12.22%，T_3 处理比 T_2 处理小17.44%，导管直径 T_1 处理显著大于 T_2 处理，T_3 处理极显著小于 T_2 处理。

根系的导管直径随着处理时间的增加而减小（图1和图2A）。处理结束时，T_1、T_2、T_3 处理的根系导管直径分别由处理10d时的24.2μm、20.3μm、17.4μm减小到处理40d时的19.75μm、17.6μm、14.53μm，分别减小22.53%、15.34%、19.75%。

3.2 根系的导管密度

木质部导管的密度对其输导能力也有着显著的影响。单位面积上导管个数越多，其输导能力就越强。随着根系温度的增加，根系导管的密度显著增加（图1和图2B），T_1 处理比 T_2 处理小20%，T_3 处理比 T_2 处理大30%。

随着处理时间的延长，根系导管的密度在增高（图1和图2B）。处理结束时，T_1、T_2、T_3 处理的根系导管密度分别由处理10d时0.05个/mm^2、0.08个/mm^2、0.1个/mm^2 增加到

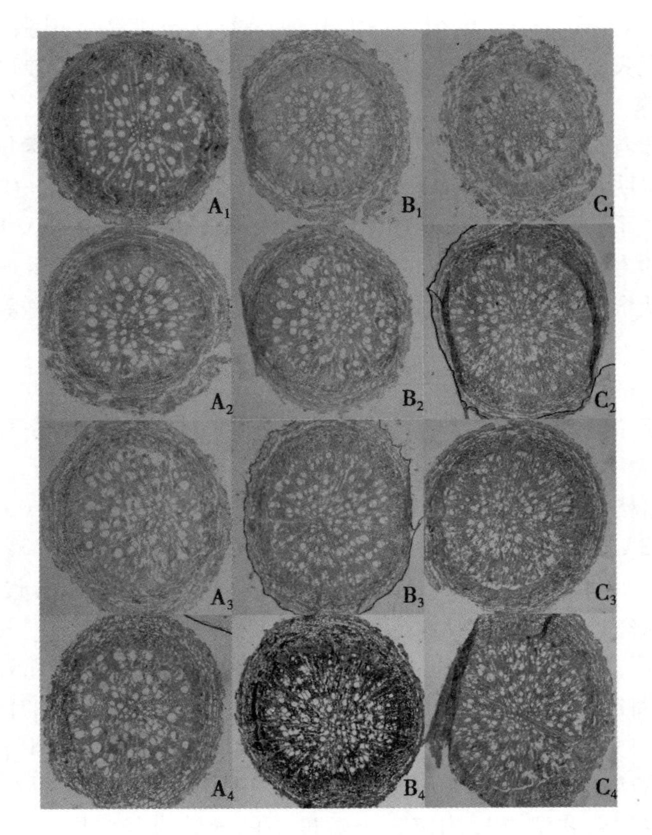

图 1　不同根温处理下的根系结构横切图（40 倍）

注：图中 A_1、B_1、C_1 分别为 T_1、T_2、T_3 处理 10d 的根结构横切图；A_2、B_2、C_2 分别为 T_1、T_2、T_3 处理 20d 的根结构横切图；A_3、B_3、C_3 分别为 T_1、T_2、T_3 处理 30d 的根结构横切图；A_4、B_4、C_4 分别为 T_1、T_2、T_3 处理 40d 的根结构横切图。

Figure 1　Transverse section of tomato root at different temperature treatments（40 times）

Note：A_1，B_1，C_1 were the root structure transverse chart of T_1，T_2，T_3 treatment for 10 days after treatment respectively；A_2，B_2，C_2 were T_1，T_2，T_3 treatment for 20 days after treatment respectively；A_3，B_3，C_3 were the root structure transverse of T_1，T_2，T_3 treatment for 30 days after treatment respectively；A_4，B_4，C_4 were T_1，T_2，T_3 treatment for 40 days after treatment respectively.

处理 40d 时 0.08 个/mm^2、0.1 个/mm^2、0.13 个/mm^2，分别增大 37.5%、20%、23.08%。

3.3　根系的导管总面积

随着处理时间的延长，根系导管的总面积也在增加（图 1 和图 2 C）。处理结束时，T_1、T_2、T_3 处理根系导管的总面积分别由处理 10d 时的 893.6mm^2、882.15mm^2、847.4mm^2 增加到处理 40d 时的 1231.8mm^2、1196mm^2、1104mm^2，分别增加了 27.24%、26.24%、23.24%，T_1 处理比 T_2 大 3%，T_3 处理比 T_2 处理小 7.7%，但 3 个处理间差异不显著。

4　讨论与小结

根系的解剖结构在一定程度上决定了它的吸收和输导能力（强盛，2006），从而影响地

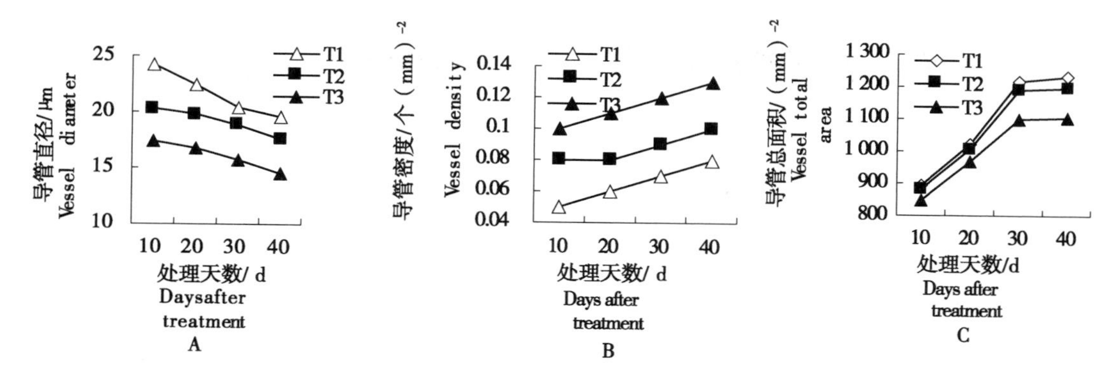

图 2　不同根温处理对番茄根系结构的影响

图 A 为根温对根系导管直径的影响；图 B 为根温对导管密度的影响；图 C 为根温对导管总面积的影响

Figure 2　The influence of different root temperature treatment on tomato root structure

Figure A the influence on the diameter of root vessel; Figure B the influence on the root vessel density; Figure C the influence on the total area of root vessel

上部的生长。本试验通过对不同根温处理后番茄主根解剖结构的观察，以及对其导管密度、导管直径和导管总面积的测量、统计，所得结果与曾文芳（2006）对枳根中维管组织的解剖结果一致。与夏季温室内的自然根温 28℃相比，降低根温至 23℃ ±1℃时，番茄根系导管分布分散，导管直径相对较大，导管数量少；这说明降低根温至 23℃ ±1℃可使其根系吸收能力增强，输送水分和无机营养的阻力减小，有利于植株的生长。当根温继续升高至 33℃ ±1℃，则植株根系导管直径减小，导管密度增加，即植株通过增加其根系横截面单位面积上导管个数多来增强其吸收和输导能力，以利于植株有效地利用水分，以致根温升高至 33℃ ±1℃时的导管密度比自然根温下的大 30%；这说明番茄植株在高温逆境下能不断地进行自我调节以适应高温逆境，保证了根系的吸收和输导功能，能较多地吸收和运输水和无机营养，来尽量满足植株的生存和生理需要。

综上所述，夏季高温下降低根温至 23℃ ±1℃可以增强番茄植株根系的吸收和输导能力，利于植株生长，可以有效地缓解高温带来的危害，从而为生产提供了新的越夏栽培技术。

参考文献

[1] 李景富，徐鹤林．中国番茄．北京：中国农业出版社，2007

[2] 李正理．植物制片技术．北京：科学出版社，1987

[3] 曾文芳．枳橙和枳根中维管组织的解剖学研究．武汉：华中农业大学（学士学位论文），2006

[4] 强盛．植物学．北京：高等教育出版社，2006

[5] 田鹏．太原地区温室番茄限产因素探讨——光照、温度对产量的影响．山西农业大学（硕士学位论文），2004

[6] 王冬梅，许向阳，李景富．番茄耐热性研究进展．中国蔬菜，2003（2）：58～60

[7] Walker J M. One degree increment in soil temperature affect maize seeding behavior. Pro. Soc. Soil Sci. Am，1969（33）：729～736

光调控在芽苗菜生产中的应用及前景*

魏圣军**，张晓燕，鲁燕舞，崔　瑾
（南京农业大学生命科学学院，南京　210095）

摘要：光调控技术作为调节芽苗菜生产的一种新技术，采用物理手段进行调控，符合绿色农业的要求，在芽苗菜生产中具有广阔的应用前景。本文从我国芽苗菜的发展现状、存在问题，光调控技术在芽苗菜生产中的研究及在芽苗菜生产中的应用前景进行了综述，为芽苗菜的工业化生产提供参考。
关键词：光调控；芽苗菜；发展；应用

Applications and Prospects of Light Control in Sprout Seedling Vegetable Cultivation

Wei Shengjun，Zhang Xiaoyan，Lu Yanwu，Cui Jin
（*College of life Sciences*，*Nangjing Agricultural University*，*Nangjing* 210095，*China*）

Abstract：Light control technology，a new technology for controlling sprouts' production，controls plant growth by physical means，which meets the requirements of green agriculture and has broad application prospects in sprout seedling vegetable cultivation. Here we review present status and existing problems of sprout seedling vegetable cultivation in China，and then discuss investigation and applications about light control in sprout seedling vegetable cultivation.
Key words：Light control；Sprout seedling vegetable；Development；Application

芽苗菜一般是指用植物种子或其他营养体，在一定条件下培育出可供食用的嫩芽、芽苗、芽球、幼梢或幼茎等芽苗类蔬菜[1]，也是一种新兴蔬菜。芽苗菜比一般蔬菜能更好地满足21世纪我国蔬菜消费——从数量消费型向质量消费型的转变，具有品种多样、优质、无污染、食用安全、方便、养生保健功能高等特点，适于设施栽培或工业化集约生产，能满足都市农业发展的需要，有很高的生产价值和经济效益。因此，芽苗菜作为新兴蔬菜的一种在21世纪将会得到前所未有的发展[2]。

植物幼苗个体发育对环境因素的变化十分敏感，特别是光对它的影响。传统栽培中，依据照光程度不同，将芽苗菜分为黄化型、半绿化型和全绿化型。如遮光栽培的大豆芽菜是传统的黄化型芽菜。虽然遮光或弱光有利于产品鲜嫩，但也有研究表明，适当照光下的大豆芽菜颜色翠绿，含有丰富膳食纤维及其他生物活性物质，如维生素C（VC）、叶绿素含量的增加，越来越受到消费者的喜爱[3]。近年来，芽苗蔬菜种类不断丰富，半绿化型和绿化型的芽苗蔬菜，如萝卜芽、豌豆、紫花苜蓿芽、油葵芽、香椿芽、芦丁甜荞芽等是风味独特、具

---

*　基金项目：国家自然科学基金项目（31171998）；江苏省自然科学基金项目（BK2010439）
**　魏圣军：硕士研究生；崔瑾：博士，副教授，主要从事植物光生物学方向的研究。E－mail：cuijin@njau.edu.cn

有药用和保健功能、营养价值较高的绿色食品，受到消费者青睐，越来越引起关注。

已有不少研究报道，芽苗菜作为植物的幼嫩组织，光对其生长和营养品质都有很大的影响[4]。芽苗类蔬菜生长期间如光照过弱或不足，则易引起下胚轴或茎叶柔长、细弱，并导致倒伏、腐烂和减产；若光照过强则将促使纤维提前形成，不利于优质产品的形成[5]。通过环境控制尤其是光环境调控是改善芽苗菜生长发育，提高产量和品质的重要措施，具有安全环保、经济有效的显著特点。而在芽苗菜工业化生产中，对于光调控技术的研究一直没有引起足够的重视，研究基础薄弱。因此，全面深入地研究芽苗菜工业化生产光调控技术及其应用机理，研发符合生产实践所需的光调控设施对于芽苗产业的发展具有重要意义。

1　芽苗菜的研究现状

1.1　芽苗菜栽培技术的研究

韩春梅等[6]对皮大麦和裸大麦的研究表明，23g · 100 cm^{-2}和18g · 100 cm^{-2}分别为皮大麦和裸大麦芽苗菜生产的最佳播种密度，在此密度下，大麦芽苗菜的经济产量最高，分别为1.51kg · m^{-2}和2.49kg · m^{-2}。刘乃森等[7]研究发现，播种密度对萝卜芽苗菜的产量和维生素 C 含量有着重要的影响，而对蛋白质含量影响较小。播种密度为850 g · m^{-2}时产量较高，且成本最低，同时维生素 C 含量达到最高（23.34mg · kg^{-1} · FW^{-1}）。张大龙等[8]发现，播种密度与生长时间对豌豆芽苗菜生长、产量和品质有显著影响。在播种密度 2.4kg · m^{-2}，第 8 天采收时为豌豆芽苗菜最优化栽培措施。

张余洋等[9]从浸种时间、栽培温度、播种密度和采收时间等方向对豌豆和萝卜芽苗菜的生长、产量以及营养品质进行了研究。结果表明，豌豆芽苗菜在播种密度为2.4kg · m^{-2}、浸种 24 h、温度 20 ~ 25 ℃左右、培养 8 d 后采收的经济产率和品质最佳，萝卜芽苗菜在播种密度为0.5kg · m^{-2}、浸种 6 h、温度 25 ℃左右、培养 7 d 后采收的经济产率和品质较高。杨秀坚和罗富英[10]研究表明，在萝卜芽苗菜生长期间喷施不同浓度的 GA_3 对萝卜芽苗菜生长均有不同程度的促进作用，其中，500mg · L^{-1}的 GA_3 极显著地提高了萝卜芽苗菜的高度，增产效果显著；喷施不同浓度的 6-BA，对萝卜芽苗菜的生长均有不同程度的抑制作用，萝卜芽苗菜的高度极显著地比对照和 GA_3 处理的低。张桂芬[11]的研究证明对香椿喷施赤霉素可促进顶芽提前萌发，100 ~ 150mg · L^{-1}的赤霉素能促进香椿顶芽迅速生长，单株产量显著增加，采收提前。6-BA 处理后能诱发侧枝大量萌发，6-BA 为 50mg · L^{-1}时最为显著。6-BA 为 75mg · L^{-1}时能够促进顶芽提前萌发、抑制顶芽的顶端优势，香椿芽生长速度快、粗壮，显著提高其单株产量。张静等[12]报道，500mg · L^{-1}赤霉素处理能较好地促进蕹菜芽苗菜生长，表现出经济产量最高、整齐度较好等优点。

1.2　芽苗菜营养品质的研究

刘懿等[13]研究表明，用不同质量浓度硫酸锌（$ZnSO_4$）在芽苗生长过程中对其进行叶面喷施，经观察，在各表观指标无明显差异情况下，$ZnSO_4$ 浓度在 0.1 ~ 0.5mg · L^{-1}时芽苗菜体内锌含量上升，当施加 $ZnSO_4$ 在 0.6mg · L^{-1}及以上时萝卜芽苗菜体内锌含量下降。刘雁丽等[14]发现，用 37.5 μmol · L^{-1}亚硒酸钠溶液对绿豆浸种 12 h，对绿豆芽的伸长生长有促进作用，收获的绿豆芽硒干重含量为 4.43 μg · g^{-1}，达到了富硒效果。

Kim 等[15~16]研究发现锗处理的豆芽含水量、灰分、膳食纤维、锗、钙和铁含量高于对照，产量可增加10% ~20%，胚轴较粗，颜色较深，但胚轴长度和子叶粗壮度与对照无明显差异。从感官品质上看，这种豆芽外观、风味、口感都较好。史铀等[17]用含铁锌的溶液培养豆芽，可以形成高铁和高锌豆芽。张兴志等[18]研究表明，在不影响芽苗菜生长的情况下，锌离子浓度在0~0.5mg·L^{-1}范围内，苜蓿芽苗菜中的锌随着浓度的增加，锌积累量呈增加趋势；锌离子浓度超过一定限度（0.5mg·L^{-1}）后，苜蓿芽菜生长受阻，而且对锌的积累也受到一定的影响。铁离子浓度在0~35mg·L^{-1}范围内，芽菜对铁的积累量随着铁离子浓度的增加而增大。

1.3 芽苗菜生产过程中存在的问题

随着人们对芽苗菜需求量的增加，芽苗菜生产发展迅速，但多为小作坊式生产，工厂化、集约化生产较少，此外，为了提高芽苗菜的生物产量和经济产量，现在生产中通常使用一些生长调节剂或微量元素溶液进行浸种或喷洒。虽然在芽苗菜栽培过程使用植物生长调节剂或微量元素溶液能够起到一定程度的调控生长、富集矿质元素、增强豆芽的保健功能，然而由于生长迅速，生产周期短，这些措施也易造成化学物质在芽苗菜内的积累，从而引发安全与卫生隐患。现在人们逐渐认识到化学调控手段存在环境污染和食品安全等问题，逐渐在很多国家和地区被禁止使用。美国有相关法令禁止蔬菜种苗公司使用任何化学生长调节剂来调控幼苗生长。因此，研究开发安全环保、经济有效的芽苗菜生产技术迫在眉睫。光环境调控技术采用物理手段调控植物生长，符合绿色农业的要求，在芽苗菜生产中具有广阔的应用前景[19]。

2 光调控在芽苗菜生产的研究

光在植物生长发育中具有特殊重要的地位，因为它不仅通过光合作用为植物提供能量，同时也几乎影响着植物发育的所有阶段，光调节的发育过程包括发芽、茎的生长、叶和根的发育、向光性、叶绿素的合成、分枝以及花的诱导等[20]。已有多篇论文报道芽[21~34]，芽苗菜作为植物幼苗，光不仅可以调控其形态建成、生长发育，对其营养品质改善也有显著影响。生产上可以通过光调控技术来促进芽苗菜生长，提高品质。光调控技术在芽苗菜工业化生产中具有广阔应用空间。

2.1 光质对芽苗菜生长和营养品质的影响

邢泽南等[21]对油葵芽苗菜的研究发现，红光显著提高了叶绿素 a、叶绿素总量和类胡萝卜素含量；蓝光显著提高芽苗菜干物质重量、可溶性蛋白含量及 POD、CAT 活性；黄光对根长有抑制效果，而对游离氨基酸的积累有促进作用；UV-B 显著提高下胚轴长度和 SOD、CAT 的活性。Amal[22]对鹰嘴豆芽的研究结果表明，绿光显著提高抗坏血酸含量，荧光和 γ 射线对豆芽生长和一定程度上 Vc 生物合成有促进作用。Mahmoud[23]报道，UV-A 或 UV-C 能显著的增加小扁豆芽苗菜总酚类物质含量。Wu 等[4]研究表明，LED 红光显著增加了豌豆幼苗的茎长和叶面积，显著提高 β-胡萝卜素含量和抗氧化酶活性，LED 蓝光显著提高了幼苗的质量及叶绿素含量。Li[24]报道，与在白光下培养的生菜相比，在 UV-A 和蓝色（B）补充光质下花青素的含量分别增加了11%和31%；在 B 补充光质下类胡萝卜素增加12%；多酚类物质在红色（R）补充光下增加6%；在远红外（FR）补充光质下花青素、类胡萝卜素和叶绿素含量分别减少了40%、11%、14%；FR补充光下鲜重、干重、茎长、叶长、叶宽显著比白光增加28%、15%、14%、44%。

张立伟[25]发现，红蓝混合光、蓝光处理下萝卜芽苗菜的维生素C、蛋白质含量显著高于白光和红光处理，但是蓝光显著提高了硝酸盐含量。Lin[26]等研究表明，与红蓝组合光相比，红、蓝、白组合光显著的增加可溶性糖和硝态氮的含量；然而，叶绿素、类胡萝卜素和可溶性蛋白质含量在红蓝、红蓝白和白光处理间没有显著差异。刘文科等[27]发现，蓝光和红蓝光可促进豌豆苗地上部生长，增加叶片叶绿素a、b含量，红蓝光处理可提高豌豆苗叶片维生素C含量；白光和红蓝光处理下豌豆茎叶中类胡萝卜素含量较高，白光处理的豌豆苗茎叶中花青素含量最高。蓝光和红蓝光有利于增加豌豆苗菜产量，而白光和红蓝光有利于提高豌豆苗的营养品质。也有报道表明，红光偏向于形成更多的叶绿素a，蓝光促进形成更多的叶绿素b，生产上以红光蓝光比为5∶1或3∶1为最好。蓝光使芽苗菜更脆嫩，红光使芽苗菜产量更高、颜色更浓绿，两者科学结合为最佳光质搭配模式[28]。

2.2 光周期和光照时间对芽苗菜生长和营养品质的影响

张欢等[29]对油葵芽苗菜生长和品质的研究表明，红光光周期0～12 h·d^{-1}，下胚轴长显著降低，子叶面积显著增加，与Yusuke[30]在野生型拟南芥中观察到的现象一致，光周期0～16 h·d^{-1}，油葵芽苗菜子叶颜色由浅黄变成深绿，叶绿素和类胡萝卜素含量显著提高，光周期到达16 h·d^{-1}后含量无显著变化，随着光周期延长，油葵芽苗菜蔗糖的含量无显著变化，淀粉含量有所提高，在16h·d^{-1}时达到最高值，维生素C的含量随光周期延长呈现逐渐提高的趋势，在光周期为12 h·d^{-1}时显著提高。

马超[31]发现延长光周期显著抑制大豆芽苗菜下胚轴和根的伸长，增加下胚轴的直径。光照时间大于8h·d^{-1}时显著抑制大豆芽苗菜的地上部分和根鲜重，但能显著提高地上部和根的干重。延长光周期可显著提高叶绿素、维生素C和异黄酮的含量，这与Gibon[32]在拟南芥中和刘磊等[33]对洋葱幼苗的研究一致。王成章和史莹华[34-35]研究发现，SOD和POD的活性与光照周期有关，短日照可提高SOD和POD的活性，增强抗逆性。文明玲等[36]研究发现，8h光照处理总体上使过氧化物酶活性升高，过氧化氢酶活性降低；16h光照处理使过氧化物酶活性降低，过氧化氢酶活性升高。

Michał[37]结果表明：连续光照可以刺激多酚类物质生成，发芽培养3～4d后观察羟基苯甲酸、苯甲酸、咖啡酸含量显著增加，芽的抗氧化活性与酚类物质含量和种植条件紧密相关。

2.3 光强对芽苗菜生长和营养品质的影响

光强对植物幼苗的生长发育具有显著的生物学效应[38]。植物幼苗部分机制不健全，对阳光的耐受能力较差，在高光强下生长会受到抑制[39]。赤霉素对作物节间伸长和细胞扩大起作用，在弱光下，作物大量产生赤霉素，促进节间伸长[40]，这与杨兴有和杨盛昌[41～42]的发现相似，在弱光下，幼苗茎的生长高度会增加。

光强对作物次生代谢产物的合成和积累过程其重要作用，这已被不同试验结果所证实，如Graham[43]发现黑暗中生长的大豆秧苗根尖部异黄酮仅为光照下的1/3，说明照光有利于异黄酮的生物合成。Sun[44]等对大豆芽苗菜的研究发现，黑暗培养的黄化型芽苗菜和光照培养的绿化型芽苗菜在异黄酮的总量上没有显著差异，均是种子中的4倍，但总异黄酮含量在芽苗菜的不同器官差异性显著，黄化型芽苗菜染料木苷的含量最高，绿化型芽苗菜根中的异黄酮含量最高，子叶下胚轴中的黄豆黄苷含量高于子叶和根。Siviengkhek[45]研究发现，在高光强下，大豆芽苗能合成较高的大豆苷和染料木苷，在低光强下，合成的大豆素和染料木素较高，而染料木素和大豆素为游离型，有利于人体吸收，弱光下培养大豆芽苗菜较好。

郭银生[46]研究表明，黑暗处理和低于 9 μmol · m^{-2} · s^{-1}的光强处理都能显著提高黑豆芽苗菜的鲜质量，过高的光强不利于芽苗菜鲜重量的积累，这与高光强对滇青冈幼苗的处理处理结果一致，表现出鲜质量显著的抑制作用[47]；在 0 ~ 9 μmol · m^{-2} · s^{-1}的光强范围内，随着光强的增加，黑豆芽苗菜的叶绿素 a、叶绿素 b、叶绿素总含量和类胡萝卜素含量都呈现显著增加的趋势，随着光强的继续增大色素含量没有显著的变化，这与刘国顺[48]在烤烟幼苗上的研究成果一致；光强处理的黑豆芽苗菜蛋白质、可溶性糖和蔗糖含量显著高于黑暗处理，其中光强为 9μmol · m^{-2} · s^{-1}处理的蔗糖含量显著高于其他光强处理。

3　光调控技术在芽苗菜生产中的应用前景分析

芽苗菜是利用植物的种子或其他营养器官，直接生长出可供食用的芽、芽苗、幼稍或幼茎等幼嫩组织，光对其生长和营养品质有很大影响。利用光调控技术培育芽苗菜是一项环保、经济有效且简便易行的新方法，它能有效的改善或提高芽苗菜的产量和营养品质，增加芽苗菜生产的经济效益，具有巨大的生产应用价值。

光调控技术的应用研究逐渐成为当前芽苗菜生产研究的热点之一。然而，迄今为止，芽苗菜生产光调控的应用尚缺乏成熟的技术和深入、全面的理论基础。为了实现光调控在芽苗菜生产中的广泛应用，需要加强以下 3 个方面的研究：①光调控在芽苗菜生产中的应用基础研究。深入研究光质、光强及光周期与 CO_2 浓度、温湿度和培养基质等环境因子综合调控对芽苗菜生长发育、物质代谢和营养品质的影响机理，为不断优化栽培条件，实现芽苗菜的工业化生产提供理论依据。②光调控技术及调控系统和设施的研究。研究和应用环境监测和调控系统，对栽培环境内光因子进行实时监测和直接、精准调控，满足芽苗菜在不同生长发育阶段对光环境的需求，提高资源利用效率和工作效率，降低生产成本。目前，国内对该项技术的研究还处于起步阶段，急需生物与工程等学科深入合作，研究符合我国芽苗菜生产实际的环境调控技术和相关配套设施。③新型环保节能光源的研究。目前，自然光和荧光灯是规模化芽苗菜生产常用光源。然而，自然光光强、光照时间不易人工控制；荧光灯只能发射固定的光谱波长，光效低、发热量大、能耗成本占其运行费用的 40% ~50%[49]。随着光电技术的发展，具有光谱能量调制便捷、发热低、节能环保等重要特点的发光二极管（LED）、激光二极管（LD），在芽苗菜光调控生产领域具有良好的前景。

参考文献

[1] 张德纯．体芽菜及其营养．中国食物与营养，2006（2）：48 ~50

[2] 王德槟，张德纯．新兴蔬菜的兴起和发展．中国蔬菜，2000（1）：5

[3] 张颖．绿瓣大豆芽菜生产技术研究．南京：南京农业大学，2006

[4] Wu M C, Hou C Y, Jiang C M, *et al.* A novel approach of LED light radiation improves the antioxidant activity of pea seedlings. Food Chemistry, 2007, 101 (4): 1 753 ~1 758

[5] 王德槟，张德纯，王小琴等．芽苗蔬菜工业化栽培技术．中国蔬菜，1996（3）：17 ~22

[6] 韩春梅，李春龙，叶少平等．播种密度对不同品种大麦芽苗菜生产的影响．北方园艺，2011（19）：34～35
[7] 刘乃森，刘福霞，胡群等．播种密度对萝卜芽苗菜产量及品质的影响．北方园艺，2009（7）：84～85
[8] 张大龙，张军良，黄东等．播种密度与生长时间对豌豆芽苗菜产量及品质的影响．西北农业学报，2012，21（7）：123～126
[9] 张余洋，胡全凌，李汉霞．不同处理对豌豆和萝卜芽苗菜生长、产量及品质的影响．华中农业大学学报，2008（27）：289～293
[10] 杨秀坚，罗富英．不同浓度 GA3、6-BA 对萝卜芽苗菜产量影响的研究．北方园艺，2006（4）：22～23
[11] 张桂芬．不同浓度赤霉素和 6-BA 对香椿芽萌发及产量的影响．甘肃农业，2005（231）：181
[12] 张静，杜庆平，汤鹏先等．赤霉素对蕹菜芽苗菜生长影响．北方园艺，2011（22）：33～35
[13] 刘懿，窦羿，张平．不同质量浓度锌对萝卜芽苗菜生长及锌含量的影响．科技信息，2008（29）：327
[14] 刘雁丽，吴峰，宗昆等．富硒芽苗菜的培育及几种大众蔬菜硒含量分析．江苏农业科学，2010（3）：204～206
[15] Kim E J，Kyeoung I L，Kun Y P. Effects of germanium treatment during cultivation of soybean sprouts. Journal of the Korean. Society ofFood Science and Nutrition，2002a，31（4）：615～620
[16] Kim E J，Kyeoung I L，Kun Y P. Quantity analysis nutrients in soybean sprouts cultured with germanium. Journal of the Korean . Societyof Food Science and Nutrition，2002b. 31（6）：1 150～1 154
[17] 史铀，汪红，尤康．高微量元素豆芽研究．成都大学学报，2007（26）：101～102
[18] 张兴志，张菊平，成玉梅等．锌铁配施对苜蓿芽菜锌铁富集的影响．陕西农业科学，2007（1）：48～49
[19] 马超，张欢，郭银生等．LED 在芽苗菜生产中的应用及前景．中国蔬菜，2010（20）：9～13
[20] 廖祥儒，张蕾，徐景智等．光在植物生长发育中的作用．河北大学学报，2001（9）：341～347
[21] 邢泽南，张丹，李薇等．光质对油葵芽苗菜生长和品质的影响．南京农业大学学报，2012，35（3）：47～51
[22] Amal B K，Aurang Z，Maazuulah K. Influence of germination techniques on sprout yield，biosynthesis of ascorbic acid and cooking ability，in chickpea（*Cicer arietinum* L.）Food Chemistry 2007，103：115～120
[23] Mahmoud E Y，Mohammed N ，Heba M M. An enhancing effect of visible light and UV radiation on phenolic compounds and various antioxidants in broad bean seedlings Plant Signaling & Behavior. Landes Bioscience 2010
[24] Qian Li，Chieri K. Effects of supplemental light quality on growth and phytochemicals of baby leaf lettuce. Environmental and Experimental Botany 2009. 67：59～64
[25] 张立伟，刘世琦，张自坤等．光质对萝卜芽苗菜营养品质的影响．营养学报，2010，32（4）：390～392，396
[26] Kuan-Hung Lin，Meng-Yuan Huang，Wen-Dar Huang The effects of red，blue，and white light-emitting diodes on the growth，development，and edible quality of hydroponically grown lettuce（*Lactuca sativa* L. var. *capitata*）Scientia Horticulturae，2013：86～91
[27] 刘文科，杨其长，邱志平等．LED 光质对豌豆苗生长、光合色素和营养品质的影响．中国农业气象，2012，33（4）：500～504
[28] 池田彰．光ランプを光源とした人工光型植物工场の开-近接照射による照明电力の低．植物工场学会，1992，3（2）：111～123
[29] 张欢，章丽丽，李薇等．不同光周期红光对油葵芽苗菜生长和品质的影响．园艺学报，2012，39（2）：297～304
[30] Yusuke Niwa，Takafumi Yamashino，Takeshino Mizuno. The Circadian Clock Regulates the Photoperiodic Response of Hypocotyl Elongation through a Coincidence Mechanism in Arabidopsis thaliana Plant Cell Physiol. 2009，50（4）：838～854
[31] 马超．光环境调控对绿瓣型大豆芽苗菜生长和品质的影响．南京：南京农业大学，2012
[32] Gibon Y，Pyl ET ，Sulpice R，Lunn JE，Hohne M，Gunther M，Stitt M. Adjustment of growth，starch turnover，protion content and central metabolism to a decrease of the carbon supply when Arabidopsis is grown in very short photoperiods. Plant，Cell and Environment，2009（32）：859～874
[33] 刘磊，刘世琦，徐莉等．光周期及春化处理对洋葱蛋白质合成代谢与 POD 活性的影响．西北农业学报，2005，14（6）：90～95
[34] 王成章，李建华，郭玉霞等．光周期对不同秋眠型苜蓿 SOD、POD 活性的影响．草地学报，2007（13）：5
[35] 史莹华，张伟毅，于晓丹等．光周期对紫花苜蓿 SOD、POD 活性的影响．草原与草坪，2009（1）：74～77

[36] 文明玲，牛应泽，刘小俊等．光周期处理对黄瓜花性分化及其氧化酶活性影响．长江蔬菜，2007（4）：46～48
[37] Michał S，Urszula G D，Dariusz K，Urszula Z. Impact of germination time and type of illumination on the antioxidant compounds and antioxidant capacity of Lens culinaris sprouts Scientia Horticulturae. 2012，140：87～95
[38] Gavagnaroa JB，Trioneb SO. Physiological and biochemical responses to shade of Trichloris crinita，a forage grass from the arid zone of Aregentina . Arid Environ. 2007，68：337～347
[39] He S Q，Wang F Q. Relationship between the seedings growth of Quercuse liaotungensisand light. Forest Research，2001，14：697～700
[40] Dudley SA，Schmit TJ. Genetic differentiation in morphological responses to simulated of liage shade between populations of impatiens-capensis from open and woodland sites. Functional Ecology. 1995，9（4）：655～666
[41] 杨兴有，叶协锋，刘国顺等．光强对烟草幼苗形态和生理指标的影响．应用生态学报，2007（18）：11
[42] 杨盛昌，林鹏．光强对秋茄幼苗的生长和光合特性的影响．厦门大学学报，2003（42）：2
[43] Graham LT. Flavonoid and isoflavonoid distribution in developing soybean seeding tissues and in seed and root exudates. Plant Physiol，1991，95：594～603
[44] Lee SJ，Ahn JK，Khanh TD，Chun SC，Kim SL，Ro HM，Song HK，Chung IM.．Comparison of isoflavone concentrations in soybean［*Glycine max*（L.）Merrill］sprouts grown under two different light conditions. Food Chemistry. 2007，55（23）：94，15～21
[45] Siviengkhek Phommalth，Yeon-Shin Jeong，Yong-Hoon Kim，Krishna Hari Dhakal and Young-Hyun Hwang. Effects of Light Treatment on Isoflavone Content of Germinated Soybean Seeds. Plant Biochemistry 2008，56（21）：10 123～10 128
[46] 郭银生．光环境调控对水稻幼苗和黑豆芽苗菜生长发育的影响．南京：南京农业大学（博士生论文），2011
[47] 曹建新，张光飞，张磊等．滇青冈幼苗的光合和生长对不同生长光强的适应．广西植物，2008，28（1）：126～398
[48] 刘国顺，杨兴有，叶协锋等．不同生育期弱光胁迫对烤烟生长和品质的影响．中国农学通报，2006（22）
[49] Kim H H，Goins G D，Wheeler R A，*et al.* Green-light supplementation for enhanced lettuce growth under red-and-blue light-emitting diodes. Hort Science 2004，39（7）：1

光环境人工调控技术在设施栽培中的研究与应用*

孙俪娜**，马月虹，刘　霞，姜鲁艳，张彩虹，邹　平，马彩雯
（新疆农业科学院农业机械化研究所、新疆设施农业工程与装备
工程技术研究中心，乌鲁木齐　830091）

摘要：近年来，我国农业设施工程水平有了明显提高，以塑料大棚和日光温室为主的设施农业逐步趋向于大型化。很多学者在日光温室光环境调控方面进行了大量、深入的理论和实验研究。本文对常用的光环境人工调控技术在设施栽培中的研究与应用现状进行了简要阐述。

关键词：光环境；反光膜；反光幕；LED；设施栽培

Research and Utilization of the Light-environment Artificial Control Technology in Facility Culture

Sun Lina，Ma Yuehong，Liu Xia，Jiang Luyan，Zhang Caihong，Zou Ping，Ma Caiwen*
(*Institute of Agricultural Mechanization*，*Xinjiang Academy of Agricultural Science Urumqi* 830091，*China*)

Abstract：Recently，facility agriculture engineer level in our country had been significantly improved. Facility-based plastic tunnel and greenhouse agriculture are the main facilities，and the facility agriculture had tend to be large scale. Relative researchers had carried out many theory and practice study deeply and intensively. This article briefly elaborated the current situation of research and utilization of the light-environment artificial control technology in facility culture

Key words：Light-environments；Reflecting film；Reflecting curtain；LED；Facility culture

设施农业是科技含量高、高投入、高产出、高效益的集约化生产方式。20 世纪 70 年代末到 80 年代初，我国进入了设施农业的高速发展时期。为了提高设施栽培的生产效益，温室环境调控技术已经越来越受到国内外学者的关注，信息采集系统、遥感监测系统等现代信息化技术的应用使得设施农业逐步向自动化、智能化、机械化发展。光照是作物生长发育的基本要素，光环境调控是最重要的温室环境调控技术之一。

目前常用的温室光环境调控技术有：铺设反光膜、悬挂反光幕、人工光源补光等。

1　铺设反光膜

该项技术主要是在温室栽培的植株下方铺设不同光质的反光膜，提高叶幕下方的光照强度，从而影响作物的生长发育。

---

*　基金项目：新疆维吾尔自治区科技计划 201130104，PT1003

**　孙俪娜，女，1984 年 9 月出生于河北，硕士研究生，助理研究员，主要从事设施农业工程与技术的研究工作

李靖等[1]研究不同覆盖材料对设施枇杷果实品质的影响。结果表明，白膜、银灰色反光膜、稻草均提高了叶幕下方光强，以银灰色反光膜的反射率最高，为9.3%。银灰色反光膜处理所得果实的可溶性固形物和可溶性糖含量最高。梁浩等[2]通过铺设不同反光膜增加设施内的光照度，均提高了葡萄果实的大小、含糖量。金国强[3]等对巨峰葡萄进入着色期后铺反光膜，结果表明，露地巨峰葡萄铺反光膜主要有3方面作用：①促进浆果着色以实现提早上市。②明显减少裂果 铺设地膜的地块连续2年几乎无裂果。③提高果实糖含量。常美花等[4]研究了在桃、杏树冠下铺设反光膜对光合速率及果实品质的影响，结果表明温室内铺设反光膜、吊反光幕，光照增强，行间的光照强度提高56.4%～150.9%，株间提高30.9%～53.1%。早红霞桃光合速率日平均增长17.8%，凯特杏增长19%。所以，温室内铺设反光膜、吊反光幕都能不同程度提高桃、杏的光合速率，从而对促进树体健壮、提高果实品质具有重要的意义。

生产上通过铺设不同类型的反光膜或其他覆盖材料来改善植物生长的光环境，这在一年生作物和果树种植上应用比较广泛，但该方法存在着一定的环境污染问题，因此推广起来势必受此影响。

2 悬挂反光幕

20世纪80年代，国内就有相关学者对反光幕在改变温室光环境方面的作用进行了研究。吴国兴等[5]利用银色聚酯膜做反光幕挂与温室北侧，通过实验研究表明：利用反光幕可增加温室内光照强度30%以上，温室后部的增光率大于前部。有利于白天温室内小区气温及低温的少量提高。王冰亚[6]研究了反光幕对越冬茄子的影响，悬挂聚酯镀铝反光幕后，光照强度增加，茄面着色率增加29.4%，气温、地温提高，茄子越冬生长限制因子作用时间缩短，光合效率增加，茄株生长健壮，叶片数、茎粗、株高、抗病性等都有明显改善，开花期提前，座果率，成果率增加。崔庆法等[7]研究了温室后墙悬挂反光膜进行补光对日光温室内光照环境与温度的综合影响以及反光膜悬挂位置对日光温室内光照的影响，结果显示反光膜不仅通过直接反射太阳光提高温室内一定范围内光照，且这些反射光从温室后方照向前方，较好地改善了温室内总体光环境。但反光膜减弱了后墙的蓄热能力，降低了温室内凌晨温度。吕桂云等[8]在河北几个试验示范点通过张挂反光幕，变温管理，换头整枝等一系列技术集成使冬春茬番茄种植取得了优质高产和高效益。孔云等[9]在日光温室内，以不挂反光幕区域为对照，调查了反光幕处理后室内光照强度的时空分布和番茄植株形态。结果表明：温室内北部光照强度日变化趋势基本没有改变，但中午光照强度明显增加，南北光照强度差异减小，垂直光照梯度增加。增光幅度受离反光幕距离、离地面的高度、天气状况和番茄植株冠层的影响。靠近反光幕的番茄植株形态特征也发生了明显的改变，如株高减小、茎节间变短；朝南的叶片数减少，叶片明显变厚，小叶面积减小；果实纵、横径明显增大，但番茄植株茎粗和节间数没有明显的变化。张利华等[10]计算了反射光在床面上的落点，推算出全国各主要纬带、主要使用期日光温室反光幕的最优悬挂角度。王静等[11]在温室内后墙悬挂反光膜增加了后墙对光的反射，但却减弱了后墙热量缓冲的作用，影响了温室内夜间温度。

3 人工光源补光技术

在现代农业生产系统中，补充人工光照已经成为高效生产的重要手段，目前国内大部分的人工光源仍然比较简单，多采用白炽灯、荧光灯、高压钠灯、高压汞灯等传统光源作为直接补光措施。发达国家在这方面的研究工作起步较早，美国、俄罗斯、日本、加拿大、荷兰等发达国家已经走在了该领域的前列。美国 LEDTRONICS 公司已经研发出效果较好的植物生长灯，广泛用于温室花卉栽培方面，但价格昂贵[12]。日本三菱公司开发出利用发光二极管栽培蔬菜的技术，该发光二极管能产生一种特定波长的红光，可以有效促进植物光合，缩短生长周期，降低生产成本[13]。荷兰针对花卉种植专门研制出一种特殊的光合作用照明系统，该系统采用的飞利浦农艺钠灯的发射光谱与植物生长所需的光谱相符合，为植物生长提供所需的蓝光和红光组分，有效增加作物产量与品质[14]。

3.1 传统光源

具有很大的局限性，设计简单，控制功能少，耗电量大，发热快，存在重金属污染的隐患，大大阻碍了国内农业发展的现代化进程。

作为第四代新型照明光源，LED 具有节能环保、光电转换效率高、寿命长、发热低、冷却负荷小、光量与光质可调节、易于分散或组合控制等许多不同于其他电光源的重要特点[15~16]。由于这些显著的特征，LED 适合应用于可控环境中的植物培养或栽培，如植物组织培养、设施园艺与工厂化育苗和航天生态生保系统等[17~18]。

3.2 LED 光源用于设施栽培的研究

刘文科等[19]探讨了采收前不同光质 LED 连续光照处理对温室土培生菜生长及营养品质的影响。结果表明，采收前以 LED 红光或红蓝光进行连续光照处理，对提高设施土培生菜的营养品质的效果较好。这与陈文昊等[20]的研究结果相似。闻婧等[21]认为红蓝光可以有效提高植物的光能利用率，促进光合作用；叶绿素 A 和 B 会随着红光的减少与蓝光的增加而减少；适宜的红蓝光比例还能有效增加植物维生素 C 含量，并降低硝酸盐含量。邢泽南等[22]采用 LED 调制光谱能量分布，研究光质对油葵芽苗菜生长和品质的影响。发现红光照射促进油葵芽苗菜色素形成。蓝光下油葵芽苗菜可溶性蛋白含量显著提高，可食干质量、全干质量均达到最大值且显著高于对照。陈娴等[23]认为红/蓝（7：1）处理下韭菜的营养品质最好，蓝光最差。齐连东等[24]发现在抑制菠菜叶柄伸长方面，蓝光具有显著效果。红光处理可明显促进萝卜苗、香椿苗等芽苗菜茎的伸长，蓝光抑制[25]。闻婧[26~27]、李雯琳[28]等研究了 LED 光源对叶用莴苣及其幼苗光合特性、生理性状和品质的影响，结果显示 LED 光源在 630nm + 460nm 的波长组合和 R/B =8 的条件下，提高光合速率，更适宜叶用莴苣的生长发育和品质提高。刘晓英等[29~30]研究了不同光质 LED 照射对樱桃番茄植株形态和光合性能以及樱桃番茄果实品质的影响。各种光质对植株生长发育的影响具有相互协同、相互制约的关系。不同比例的蓝光和红光显著地调控樱桃番茄果实的品质。

4 小结

设施内的光环境对作物的生长有重大影响，调节好光环境是实现高产优质的首要条件。我国对设施光环境调控的研究明显落后于美国、日本、荷兰等发达国家，近年来国内学者对光环境调控的研究主要集中于LED光源，且多为不同光质处理的研究，关于照射周期和时间的研究较少，且研究对象多为设施蔬菜类，对水果的研究较少。LED技术以其显著的优势势必会在设施栽培中得到广泛应用，但关于LED智能温室照射装备的研究较少，这个难关还需要相关专业学者共同攻克，相信在不久的将来我国会迎来设施光调控技术的新发展。

参考文献

[1] 李靖，孙淑霞，陈栋等．不同覆盖材料对设施大五星枇杷果实品质的影响．西南农业学报，2011，24（2）：695～698

[2] 梁浩．不同反光材料对设施栽培葡萄生长和果实品质的效应．北京：中国农业大学（硕士学位论文），2005

[3] 金国强．葡萄品种巨峰露地栽培铺反光膜效果试验．中国果树，2011（3）：75

[4] 常美花，张小红，师占君等．反光膜对温室桃杏光合速率及品质的影响．江苏农业科学，2006（4）：79～81

[5] 吴国兴，陈巧芬．利用聚酯膜反光幕改善温室光照效应初探．蔬菜，1987（12）：10～12

[6] 王冰亚．反光幕对越冬茄子的影响．长江蔬菜，1995（6）：30～31

[7] 崔庆法，曹春晖，胡小进等．日光温室内反光膜补光的实践与理论探析．中国生态农业学报，2005，13（1）：82～83

[8] 吕桂云，乜兰春，胡淑明等．日光温室冬春茬番茄优质高效关键技术．北方园艺，2012（14）：43～44

[9] 孔云，孟利云．反光幕对日光温室光照分布和番茄形态特征的影响．北方园艺，2004（5）：10～11

[10] 张利华，张永强，张仁祖等．日光温室反光幕悬挂角度的研究．江西农业学报，2010，22（6）：160～161

[11] 王静，崔庆法，林茂兹等．不同结构日光温室光环境及补光研究．农业工程学报，2002，18（4）：86～89

[12] Pfundel E，Baake E. Aquantitative description of fluorescence excitation spectra in intact bean leaves greened under intermittent light． Photosynthesis Research，1990，26：19～28

[13] Yanagi T.，K. Okamoto and S. Takita. Super-bright light emitting diodes as an artificial light source for plant growth. Abstract of 3rd International Symposium on Artificial Lighting in Horticulture，1994，19

[14] Folta K M. Green light stimulates early stem elongation，antagonizing light — mediated Growth inhibition． Plant Physiology，2004，135：407～1416

[15] Brown C S，Schuerger A C，Sager JC. Growth and photom orphogenesis of pepper plants under red light-emitting diodes with supple-mental blue or far-red lighting. Am Soc Hort Sci，1995，120：808～813

[16] Guo S，Liu X，AiW，TangY，Zhu J，WangX，WeiM，Qin L，YangY. Development of an improved ground-based prototype of space plant-growing facility. Advances in Space Research，2008（41）：736～741

[17] 饶瑞佶，方炜，李登华．超高亮度发光二极体作为组培苗栽培人工光源之灯具制作与应用．中国园艺，2001，47（3）：301～312

[18] Bula R J，Morrow R C，Tibbits T W，Barta D J. Light-emitting diodes as a radiation source for plants． Hort Science，1991，26（2）：203～205

[19] 刘文科，杨其长，邱志平等．不同LED光质对生菜生长和营养品质的影响．蔬菜，2012（11）：63～65

[20] 陈文昊，徐志刚，刘晓英等．LED光源对不同品种生菜生长和品质的影响．西北植物学报，2011，31（7）：1 434～1 440

[21] 闻婧，杨其长，魏灵玲等．不同红蓝LED对生菜形态与生理品质的影响．设施园艺创新与进展——2011第二届中国·寿光国际设施园艺高层学术论坛论文集，2011

[22] 邢泽南，张丹，李薇等．光质对油葵芽苗菜生长和品质的影响．南京农业大学学报，2012，35（3）：47～51

[23] 陈娴，刘世琦，孟凡鲁等．不同光质对韭菜营养品质的影响．山东农业大学学报（自然科学版），2012，43（3）：361～366

[24] 齐连东．光质对菠菜生理特性及其品质的影响．山东农业大学，2007

[25] 张立伟，刘世琦，张自坤等．不同光质下香椿苗的生长动态．西北农业学报，2010，19（6）：115～119

[26] 闻婧，杨其长，魏灵玲等．不同红蓝LED组合光源对叶用莴苣光合特性和品质的影响及节能评价．园艺学报，

2011，38（4）：761～769
[27] 闻婧，鲍顺淑，杨其长等. LED 光源 R/B 对叶用莴苣生理性状及品质的影响. 中国农业气象，2011（3）：303～309
[28] 李雯琳，郁继华，张国斌，杨其长. LED 光源不同光质对叶用莴苣幼苗叶片气体参数和叶绿素荧光参数的影响. 甘肃农业大学学报，2010，45（1）：47～51
[29] 刘晓英，常涛涛，郭世荣等. 红蓝 LED 光全生育期照射对樱桃番茄果实品质的影响. 中国蔬菜，2010（22）：21～27
[30] 刘晓英，徐志刚，常涛涛等. 不同光质 LED 弱光对樱桃番茄植株形态和光合性能的影响. 西北植物学报，2010，30（4）：645～651

浇灌微/纳米气泡氧气水对番茄幼苗生长的影响*

何华名[1**]，郑　亮[1]，栗亚飞[1]，邢文鑫[1]，宋卫堂[1,2***]
（1. 中国农业大学水利与土木工程学院，北京　100083；
2. 农业部设施农业工程重点（综合）实验室，北京　100083）

摘要：采用微/纳米气泡发生技术，以番茄幼苗为试验材料，研究了浇灌3种不同溶氧浓度的营养液对番茄幼苗生长的影响。结果表明，营养液中溶氧浓度的提高对番茄幼苗的生长量有一定的促进作用，溶氧量为19.4mg/L的处理组与不充氧的对照组相比，根长增加了71.8%，地下干重增加了108.3%，明显提高了植株的壮苗指数；另外，充氧处理对叶片内SOD酶的活性以及MDA、总酚、类黄酮的含量影响不大，而低浓度的充氧处理促进了花青素的合成，高浓度的充氧处理表现为对花青素含量的抑制。综上所述，利用微/纳米气泡氧气水浇灌番茄幼苗，有利于培育壮苗。

关键词：微/纳米气泡；番茄幼苗；溶氧

Effects of Irrigation of Micro/Nano Bubble Oxygen Water on Tomato Seedlings Growth

He Huaming[1], Zheng Liang[1], Li Yafei[1], Xing Wenxin[1], Song Weitang[1,2]
(1. *College of Water Resources and Civil Engineering*, *China Agricultural University*, *Beijing* 100083, *China*;
2. *Key Laboratory of Agricultural Engineering in Structure and Environment*, *Beijing* 100083, *China*)

Abstract: Effects of irrigation of micro/nano bubble oxygen water on the growth and physiological characters of tomato seedlings were investigated. In this study, the concentrations of dissolved oxygen were from 6.2mg/L up to 19.4mg/L. For the purpose of comparison, the effect of normal nutrient solution was also evaluated. The results indicated that insufflating oxygen in the nutrient solution could promote the growth of tomato seedlings. With insufflating oxygen for 9minutes and DO = 19.4mg/L, the root length and dry matter accumulation were 71.8% and 108.3% higher than the treatment without insufflating oxygen. Also aeration treatment had remarkable influences on the content of anthocyanin. Proper concentration dissolved oxygen can help increase the level of anthocyanin. However, aeration treatment had no impacts on the content of MDA and the activities of root and SOD compared to the contrast. Overall, the micro/nano bubble water is beneficial for culturing tomato seedlings.

Key words: Micro/nano bubble water; Tomato seedling; Dissolved oxygen

氧气是维持植物正常呼吸的重要因子，缺氧会造成根系呼吸受阻，根系生长缓慢，对水

---

* 基金项目：“十二五”国家科技支撑计划项目——封闭式无土栽培营养液灭菌技术研究与示范（2011BAD12B01—A）

** 作者简介：何华名（1990—），女，中国农业大学农业生物环境与能源工程专业2012级在读硕士

*** 通讯作者：宋卫堂（1968—），男，博士，教授，主要从事设施园艺栽培技术与设备研究。北京 中国农业大学水利与土木工程学院，100083。E-mail：songchali@cau.edu.cn

分和养分的吸收减少，光合作用受到抑制[1]，影响植物的生长发育[2]。此外，处于缺氧胁迫的根系抗性差，容易成为微生物或病原菌侵染的目标[3]。因此，充足的氧气是维持植物良好生长的必要条件。有研究表明：对番茄根部进行通气处理可以提高番茄的生物量[2,4]；水培中改善黄瓜幼苗的供气条件可以提高植株根系活力和生长量[5]；增氧条件可以使水稻根系变长，提高根系活力和产量[6]；改善根际通气条件可以促进马铃薯光合作用与光合代谢产物的转运和积累[7]；向营养液中通气可以提高生菜的生物产量[8]。

微/纳米气泡发生技术是日本2006年研发的一项新技术，该技术是通过特定的装置，使气体以微/纳米级直径的气泡溶解于水中，可以促进气体在水中的溶解，使气体在水中分布均匀，存续时间延长。

本试验以番茄幼苗为试材，通过灌溉含有不同浓度微/纳米气泡溶解氧的营养液，研究其对番茄幼苗生长发育的影响。

1 材料与方法

1.1 材料

试验于2012年8月至12月，分两次在中国农业大学水利与土木工程学院密闭式育苗工厂内进行。供试番茄（*Lycopersicon esculentum* Mill.）品种为佳红五号。采用标准200孔穴盘育苗，育苗基质为草炭：蛭石：珍珠岩=3V：1V：1V。播种后将穴盘放入催芽室进行催芽，出芽后移入密闭植物工厂中。待子叶展开后，开始浇灌；浇灌方式为潮汐式灌溉。

1.2 试验设计

试验中所用营养液为山崎营养液配方。试验设置3个处理组和一个对照组，每组4个穴盘，共16穴盘。处理组采用微/纳米气泡发生器对待浇灌的营养液进行曝气处理，灌溉开始时营养液的3种溶氧浓度分别为A（DO = 14.4mg/L）、B（DO = 16.8mg/L）、C（DO = 19.4mg/L），对照组营养液不充氧（DO = 6.2mg/L）。其后，不再对营养液中的溶氧浓度进行控制，让其自然衰减。各组采用相同的灌溉制度，每隔2 d灌溉一次。

1.3 测定方法

1.3.1 幼苗生长指标

播种后24d为番茄的苗期。苗期结束时，从各组随机选取10株幼苗，用直尺测定株高、根长，用游标卡尺测定茎粗，用电子天平测定地上部和地下部的鲜、干质量。采用下述算式计算番茄幼苗的壮苗指数、根冠比。

$$\text{根冠比} = \text{地下部鲜重}/\text{地上部鲜重} \tag{1}$$

$$\text{壮苗指数} = (\text{茎粗}/\text{株高} + \text{地下部干重}/\text{地上部干重}) \times \text{全株干重} \tag{2}$$

1.3.2 生理指标

才用TTC染色法测定幼苗的根系活力；用叶绿素测定仪SPAD测定叶片的叶绿素含量，以SPAD值表示；用紫外吸收法测定叶片中的过氧化氢酶活性；用氮蓝四唑法测定叶片中超氧化物歧化酶的活性；用盐酸甲醇浸提比色的方法测定叶片中的总酚物质、类黄酮与花青素的含量；用过氧化脂质硫代巴比妥酸分光光度法测定叶片中的MDA含量。

1.4 数据分析

采用Excel软件进行数据整理，采用SPSS 19.0软件进行方差分析，显著性由Duncan's

新复极差法检验。

2 结果与分析

2.1 不同处理对番茄幼苗生长的影响

株高和茎粗是植株长势强弱的重要指标，尤其茎粗在一定程度上反映幼苗的健壮程度。由表1可以看出：不同处理对番茄幼苗的株高和茎粗有一定的影响，处理C的幼苗最高，株高为119mm；处理A居中，株高为109.5mm；处理B和对照组较矮，分别为97.65mm和100.5mm，但各组之间株高差异不显著。各处理中，茎粗以对照组最细，其次为处理A和B；处理C为最粗，相比对照组增加了17.8%，但各组之间茎粗没有明显差异。根长方面，处理A和B的根长较对照组有所增加，但无明显差异，处理C的根长明显高于对照组，增加了71.8%，说明充氧处理可以促进根的伸长。在干重方面，各处理组的地上干重没有明显的差异，处理C组的地下干重明显高于其他组，说明充氧处理促进了根部的生长，增加了地下部的干物质积累。

表1 不同处理对番茄幼苗生长的影响

Table 1 Effect of different treatments on the growth of tomato seedlings

处理	株高/mm	茎粗/mm	根长/mm	地上干重/g	地下干重/g
CK	100.00 ±6.93a	2.56 ±0.33a	83.50 ±8.20b	0.180 ±0.058a	0.012 ±0.003b
A	109.50 ±8.65a	2.86 ±0.14a	107.75 ±14.25ab	0.203 ±0.054a	0.018 ±0.005b
B	97.65 ±2.79a	2.63 ±0.15a	103.00 ±26.28ab	0.168 ±0.030a	0.019 ±0.004b
C	119.00 ±21.25a	3.01 ±0.33a	143.50 ±30.87a	0.236 ±0.012a	0.026 ±0.004a

注：Duncan's Multiple Range Test（$P=0.05$），下同。

2.2 充氧处理对番茄幼苗素质的影响

由表2可见：各处理的根系活力和叶绿素含量方面没有明显差异；处理A、B和C的根冠比与对照组差异显著，处理A、B、C之间无明显差异，对照组根冠比为0.08，而处理A、B、C分别为0.13、0.14和0.14，相较对照组均增长了近50%；从植株的壮苗指数来看，处理A、B较对照组没有明显变化，处理C壮苗指数较对照组有显著增加。说明充氧处理可以促进番茄幼苗地下部的生长，与对照组相比，表现为根冠比和壮苗指数增加，提高了番茄幼苗的素质。

表2 不同处理对番茄幼苗素质的影响

Table 2 Effect of different treatments on the quality of tomato seedlings

处理	根系活力/（mg TTF·g^{-1}·h^{-1}）	叶绿素/*SPAD*值	根冠比	壮苗指数
CK	0.45 ±0.13a	32.55 ±2.42a	0.08 ±0.01b	0.0187 ±0.0035b
A	0.33 ±0.18a	31.15 ±2.09a	0.13 ±0.03a	0.0258 ±0.0047b
B	0.37 ±0.05a	32.95 ±3.05a	0.14 ±0.02a	0.0259 ±0.0051b
C	0.26 ±0.08a	34.83 ±0.98a	0.14 ±0.03a	0.0362 ±0.0044a

2.3 充氧处理对番茄幼苗抗氧化系统的影响

由图所示，处理B的SOD酶活性最高，为84.11 U/g，比对照组高21.9%，处理A和C与对照组差异不明显。在MDA含量方面，处理A组番茄幼苗的MDA含量最低，与处理C差异不明显，明显低于对照组和处理B组，处理C与对照组差异不明显，与处理B组有明显差异，处理B组的MDA含量明显高于其他处理。

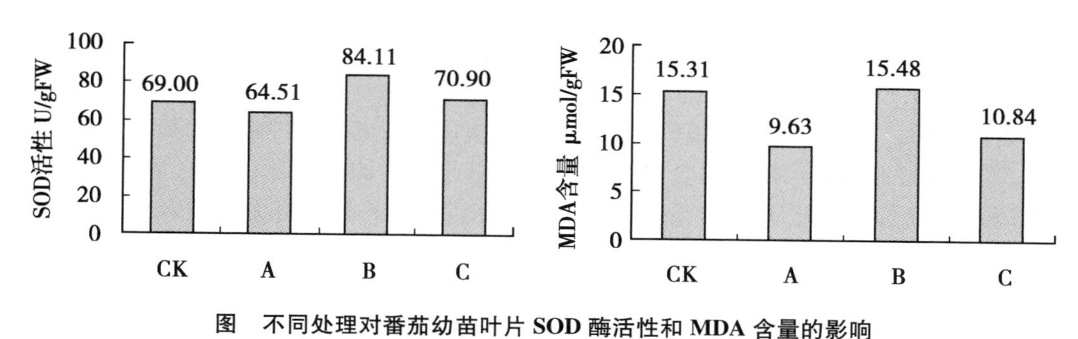

图　不同处理对番茄幼苗叶片SOD酶活性和MDA含量的影响

Figure　Effect of different treatments on the SOD activity and MDA content

2.4 充氧处理对番茄叶片内类黄酮、花青素等物质含量的影响

由表3可以看出，各充氧处理组的番茄叶片内总酚类物质较对照组有所增加，但没有显著差异；类黄酮含量方面，各充氧处理组比对照组略有下降，但差异同样不显著；花青素含量方面，各处理组之间差异明显，由多到少依次为处理C > 对照组 > 处理组A > 处理B。总体看来，充氧处理不影响叶片内总酚和类黄酮含量，低浓度的充氧处理促进了花青素的合成，高浓度的充氧处理表现为对花青素含量的抑制。

表3　不同处理组番茄叶片内总酚、类黄酮和花青素的含量

Table 3　Effect of different treatments on the content of total phenol, flavonoid, anthocyanin

处理	总酚/（$OD_{280}\cdot g^{-1}$）	类黄酮/（$OD_{325}\cdot g^{-1}$）	花青素/［（$OD_{530}-OD_{600}$）$\cdot g^{-1}$］
CK	0.164 ±0.062a	0.148 ±0.019a	0.281 ±0.023c
A	0.199 ±0.063a	0.108 ±0.054a	0.405 ±0.011b
B	0.203 ±0.014a	0.138 ±0.078a	0.475 ±0.011a
C	0.193 ±0.032a	0.125 ±0.043a	0.139 ±0.015d

3　讨论

育苗是蔬菜生产的重要环节，由于幼苗的形态建成是一个不可逆转的过程，培育出的幼苗的健壮程度将直接影响植株的生长发育，并与作物的产量和品质密切相关[9]。本试验通过微/纳米气泡充氧处理灌溉番茄所用的营养液，研究其对番茄幼苗生长发育的影响。结果表明，在育苗期间，通过微/纳米气泡发生器对番茄育苗期间的营养液进行增氧处理，可以促进根的伸长，增加根系的干物质积累，提高了植株的根冠比和壮苗指数，有利于培育壮苗。这与前人的研究结果一致，其中，各充氧处理中，以充氧浓度最高的处理C组对番茄

幼苗生长的促进作用最大。

根系活力的大小反映了根系代谢能力的强弱，与环境氧含量有密切的关系。各充氧处理与对照组没有明显差异，但是稍低于对照组，这与黄金秋[5]等人的研究结果有相同之处，其原因有待进一步研究。当植物受到胁迫时，会诱导植物体的氧化代谢[10]，超氧化物歧化酶（SOD 酶）是一种重要的抗氧化酶，可以减少活性氧对机体的毒害作用。当植物处于逆境或衰老时，SOD 酶的活性下降。由试验结果可以看出，充氧处理可以在一定程度上提高番茄叶片的 SOD 酶活性，因此，充氧可以保持功能叶片活力，但是，处理 C 组与处理 B 组相比 SOD 酶活力反而下降。丙二醛（MDA）是过氧化产物的一种，它能强烈地与细胞内各种成分发生反应，因而引起酶和膜的严重损伤，膜电阻及膜的流动性降低，最终导致膜的结构及生理完整性的破坏。MDA 的含量积累，可以引起植物的衰老。由试验结果可以看出，充氧处理对于番茄叶片内 MDA 含量影响不明显。花青素是一种强有力的抗氧化剂，能够保护机体免受自由基的损伤，本试验中，低浓度的充氧浓度可以显著提高番茄叶片的花青素含量，提高植株的抗氧化能力，但是，高浓度的充氧浓度使花青素含量显著降低。综合所测番茄叶片的生理指标，发现并不是充氧浓度越高越好，这与前人的研究有相同之处：孟彩霞[2]等人发现，对樱桃番茄进行通气处理，结果表明，生长前期并不是充氧浓度最高的最利于植株生长，营养液中氧浓度过高，在一定程度上可以抑制幼苗的生长，但是，具体机理尚待考察；王中梅[3]等人研究发现，生菜生长初期，氧需求量和吸收量均较少，只需要补充少量的氧就可满足正常生长的需要；韩勃[6]发现，饱和氧气环境处理的水稻根系反而不发达。

综上所述，育苗期间对番茄幼苗灌溉微/纳米气泡充氧营养液，可以促进番茄幼苗的生长，对培育壮苗有一定的生产意义，至于营养液的溶氧量应达到什么水平对番茄幼苗的生长最有利，有待进一步研究。

参考文献

[1] Mustroph A, Albrecht G. Tolerance of crop plants to oxygen deficiency stress: fermentative activity and photosynthetic capacity of entire seedlings under hypoxia and anoxia. Physiologia Plantarum. 2003, 117 (4): 508~520

[2] 孟彩霞，王合理．不同通气处理对樱桃番茄的影响．江西农业学报．2011，23（6）：68~70

[3] 王中梅，李胜利，王永华．秋茬生菜深水栽培氧的需求规律．中国农学通报．2007，23（3）：483~485

[4] Gislerod H R, Kempton R J. The oxygen content of flowing nutrient solutions used for cucumber and tomato culture. Scientia Horticulturae. 1983, 20 (1): 23~33

[5] 黄金秋，王秀峰，宋述尧等．不同供气条件对水培黄瓜幼苗生长的影响．山东农业科学，2009（5）：41~44

[6] 韩勃．增氧条件下水稻根系及地上部生长特性研究．扬州：扬州大学（博士论文），2007

[7] 孙周平，郭志敏，刘义玲．不同通气方式对马铃薯根际通气状况和生长的影响．西北农业学报，2008（4）：125~128

[8] 张殿高．非循环水培设施改进及活性炭再生研究．武汉：华中农业大学（博士论文），2005

[9] 汪俏梅．设施栽培中培育壮苗的一些技术措施．沈阳农业大学学报，2000（1）：120~123

[10] Queiroz C, Alonso A, Mares-Guia M, *et al.* Chilling-induced changes in membrane fluidity and antioxidant enzyme activities in Coffea arabica L. roots. Biologia plantarum. 1998, 41 (3): 403~413

日光温室番茄植株叶温与气温差异研究*

谭 敏**，李亚灵***，温祥珍

（山西农业大学园艺学院，太谷 030801）

摘要：试验利用热电偶测温仪测定了日光温室番茄植株叶温与同高度空气温度差异的日变化，并对植株不同密度、不同高度、不同天气条件的叶－气温差进行了测定。试验结果表明，日光温室番茄植株叶温与同高度空气温度有明显的差异。夜间叶温与气温的平均差值为－2℃，白天叶温随气温变化而变化，基本同步；行距75cm（密度为6株/m^2）时，叶温为上部＞中部＞下部，行距150cm（密度为3株/m^2）时，叶温为上部小于中下部；晴天叶－气温差变化幅度很大，阴天几乎没有大的波动；当气温在32～35℃之间时，由于植株的蒸腾作用，叶－气温差变化比较剧烈，约在－3～－5℃，当气温超过35℃，叶－气温差可以达到－5～－7℃。

关键词：番茄；叶温；气温；叶－气温差

Study of the Difference between Leaf Temperature and Air Temperature of Tomato Plant in Solar Greenhouse

Tan Min, Li Yaling, Wen Xiangzhen

(*Horticultural college*, *Shanxi Agricultural University*, *Taigu* 030801, *China*)

Abstract: In this experiment, the thermocouple temperature instrument was adopted for measuring diurnal variation of leaf temperature, leaf-air temperature differences for tomato plant. The result showed that the leaf temperature was clear different with air temperature on the same level. The leaf-air temperature difference during night was maintained around -2℃, but it was varied with the change of air temperature synchronously during the day. The leaf temperature for plant with plant space 75cm between rows (density of 6 plants/m^2) was higher for upper leaf, next for central, then for lower part. However, the leaf temperature for upper part with plant space of 150cm (density of 3 plants/m^2) was smaller than that of the central and lower. On sunny day, the leaf-air temperature difference changed greatly; but, on cloudy day, it was almost no fluctuation. When the air temperature in the greenhouse was between 32 ~ 35℃, the leaf-air temperature difference was －3 ~ －5℃ with the great fluctuation due to the transpiration of plants; When the air temperature was over 35℃, the leaf-air temperature difference was up to －5 ~ －7℃.

Key words: Tomato; Leaf temperature; Air temperature; Leaf-air temperature difference

1 引言

作物是变温有机体，其体温是同环境进行能量交换的结果。这种能量交换使得作物地上

* 基金项目：高等学校博士点科研基金项目（20091403110002），山西省回国留学人员科研资助项目（Research Project Supported by Shanxi Scholarship Council of China, 2012－051）

** 谭敏（1987—），女，在读硕士，研究方向为蔬菜栽培与栽培生理

*** 通讯作者：李亚灵，教授，博士生导师，主要从事设施园艺作物生长与环境调控方面的研究，E-mail：yalingli1988@yahoo.com

部分的温度显著偏离于气温[1]。精确农业的发展，要求日光温室中对环境因子的调控应尽可能准确地反映作物本身的温度，所以，有必要研究作物体温。作物的叶温是环境和植物内部因素共同影响叶片能量平衡的结果，是研究作物与其环境进行物质和能量交换的重要参数[2]。方学敏等[3]指出气温对叶温起着支配作用。用叶温减去气温所得到的叶-气温差，联系着作物叶片水平的水分与能量平衡[4]。番茄作为我国温室主栽作物，其高效栽培，需要精细化管理，由此可见，研究番茄叶温有重要的意义。本文用热电偶测温仪测定日光温室番茄植株叶温的变化，研究叶温与气温的差异，从而依据植株自身温度有效调节温室温度，为温室番茄高效生产提供理论依据。

2　材料与方法

试验在山西农业大学设施农业工程中心的日光温室内进行，供试品种为以色列石头番茄“NS2258”，无限生长型，定植时间为 2012 年 4 月 20 日，共设两个密度处理，分别为行距 75cm（密度为 6 株/m^2），行距 150cm（密度为 3 株/m^2）。

试验采用的测温仪器为 CID（北京）生态科学仪器有限公司生产的 CB—0232 热电偶测温仪。将探针插在测定叶片主脉内部的中间，测温基点一端放在冰水混合物中，测得二者温差，每 5min 记录一次数据。空气温度用 HOBO 测定，测定位点与热电偶测温探针同高度，每 5min 记录一次数据。叶-气温差是植株的叶片温度减去空气温度的差值。选取植株距地 150cm 左右高处的叶子挂牌标记，测定叶温、气温、叶-气温，并进行不同天气的比较；同一植株上、中、下部气温和叶温变化比较的测定位点，分别距离地面 150cm、100cm 和 50cm。

3　结果与分析

3.1　植株叶温、气温及叶-气温差日变化

试验中经多次测定，番茄植株气温、叶温合叶-气温差日变化规律基本相同，如图 1 所示。从图 1 可以看出，同时期的叶温和气温的变化趋势基本一致，一昼夜的时间内，同时期的叶温低于气温，叶-气温差为负值。早晨番茄植株叶温与周围气温差异较小，随着太阳辐射的增强，气温和叶温都升高，但是，气温比叶温上升的幅度大，叶-气温差增大；中午 12：00左右，气温、叶温同步达到峰值，图 1 显示的测定日内峰值分别为 32.4℃、26.2℃；之后随着太阳辐射的降低，气温、叶温均降低，但是，气温比叶温下降的幅度大，叶-气温差缩小。夜晚，气温、叶温几乎保持恒定，叶-气温差值保持在-2℃左右。

3.2　同一植株上中下部叶温与气温变化比较

图 2 中，75cm 处理植株上、中、下 3 部分叶温与气温的变化趋势一致，但是，上部比中下部变化快，变化幅度大。这是由于植株上部叶片截获的光能较多，与周围空气交换热量大，而植株中下部由于种植密度过大，株间郁闭，因而获得的光能较少，变化比较缓慢。测定日内上部叶温最大值出现在 13：25 时，为 33.9℃，中部和下部最大值的出现时间、温度分别为，13：25 时、32.1℃和 14：20 时、31.8℃，方学敏[5]认为，最大叶温发生在中午或午后，植株上部的叶温一般高于下部，与本研究结果一致。

从图 2 整体看，株距 150cm 处理同株距 75cm 处理一样，植株上、中、下 3 部分叶温与

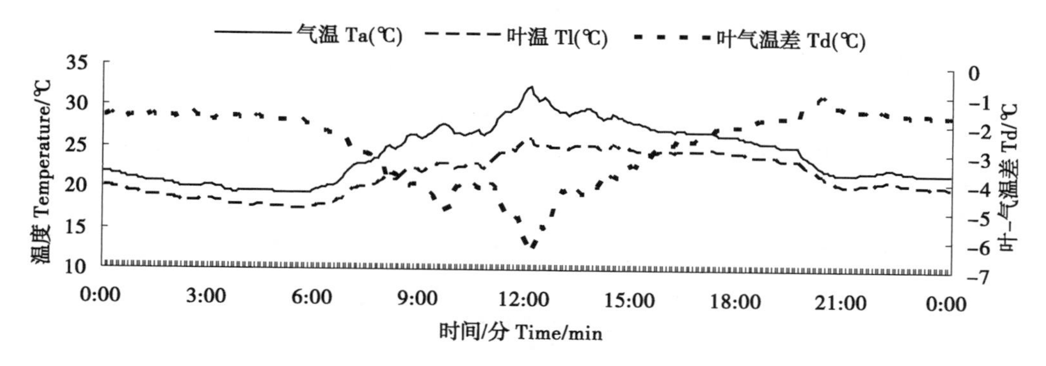

图1　番茄气温、叶温、叶－气温差的日变化

Figure 1　Diurnal course of air temperature（Ta）, leaf temperature（Tl）and the difference temperature between leaf and air（Td）in tomato

注：图中数据为2012年7月4日（晴）测得的数据，测定时间为24h

Note：The data of the plot was measured on the July 4，2012（sunny）.

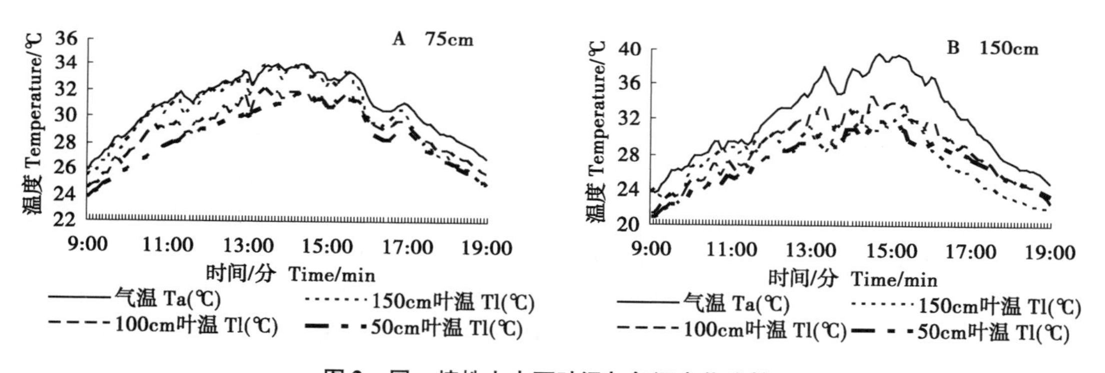

图2　同一植株上中下叶温与气温变化比较

Figure 2　Comparison of leaf temperature and air temperature in the different parts of the same plant

注：A图数据为株距75cm，晴天【2012年8月1日（50cm）、2日（150cm、100cm）2d】测得的数据；B图数据株距150cm，晴天【8月19日（100cm、50cm）、20日（150cm）2d】测得的数据；测定时间均为每日的9：00～19：00

Note：The data in the panel A was 75cm between the rows，measured on the August 1（50cm），2（150cm and 100cm）in 2012. And the data in the panel B was 150cm between the rows，measured on the August 19（100cm and 50cm），the 20th（150cm）respectively. The measuring time was from 9：00 to 19：00 on sunny day.

气温的变化趋势相一致，但并不像75cm处理的变化有规律，而是上部变化幅度小于中、下部。这可能是由于150cm大行距处理，冠层内各层的光能分布相对均匀，使得植株上、中、下3部分与周围空气热量交换相差不多，并且由于阳光可直射于地面，使得地面辐射加强，从而中、下层气温、叶温变化要比上层剧烈。

3.3　不同天气下植株叶－气温差变化比较

图3是晴天、阴天不同天气状况下测得的叶－气温差变化。晴天，随着一天中太阳辐射先增大后减弱的变化，气温先升高后降低，叶－气温差呈先增大再减小的变化。叶－气温差最小值在10：45时出现，约为－1.5℃；叶－气温差最大值在13：20时出现，约为－4℃。

阴天太阳辐射小，气温变化不大，叶－气温差的变化几乎没有大波动，日均值为－1.9℃。从图3可知，晴天叶－气温差变化大于阴天。

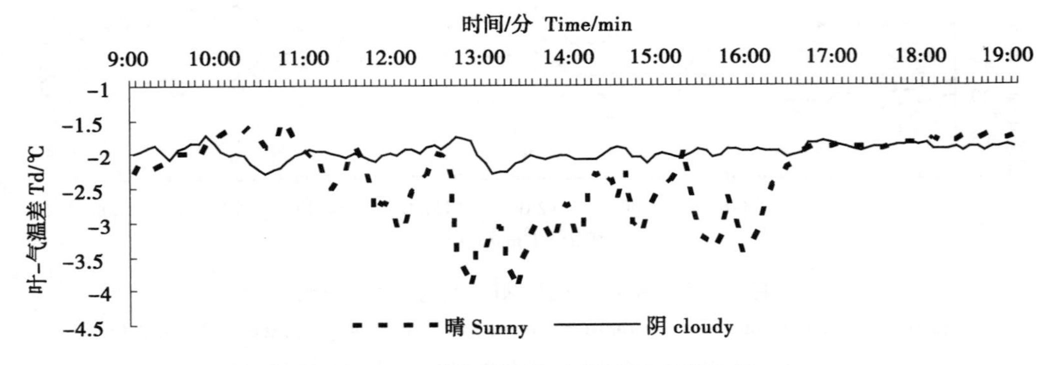

图3　不同天气番茄叶－气温差变化的比较

Figure 3　Comparison of the leaf-air temperature difference（Td）on sunny and cloudy day

注：图中为2012年8月13日（晴）、17日（阴），每天9：00～19：00测得的数据。

Note：The data in the plot was measured from 9：00 to 19：00 on the August 13（sunny） and 17（cloudy） respectively.

3.4　叶－气温差随气温的变化

以气温为横坐标，叶气温差为纵坐标，把测定日几个典型晴天的数据做成图4。由图4可以看出，叶－气温差随气温变化规律比较明显，表现为叶气温差为负值，即叶温始终低于气温。在测定期间，叶气温差基本集中在－2～－7℃，这可能是因为植株在不断通过蒸腾进行自我调节，使得自身的温度不至于太高而影响正常的生命活动。累计这4d的叶－气温差测定值，发现当气温在25～32℃之间时，叶－气温差小于3℃所占比例为89.9%；当气温超过32℃时，叶－气温差大于3℃所占的比例高达78.3%，当气温在35℃以上时，叶－气温

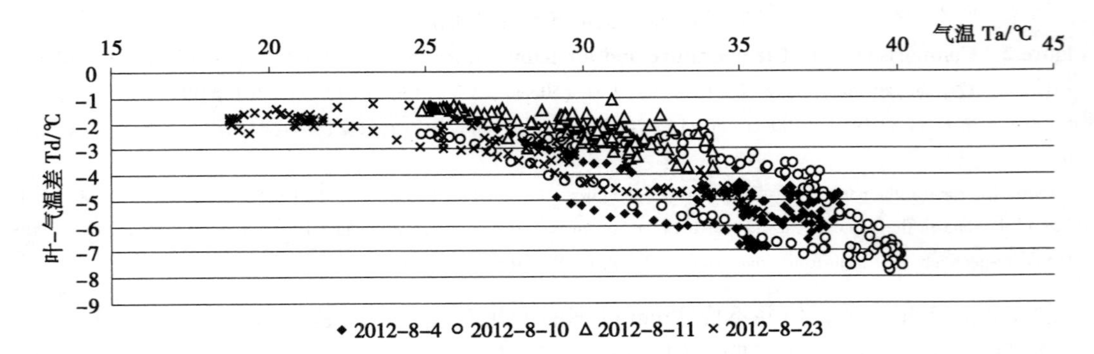

图4　番茄叶－气温差随气温的变化

Figure 4　Variation of the leaf-air temperature difference（Td）with the changes of the air temperature（Ta）

注：图中为2012年8月4日、10日、11日及23日四个典型晴天，每天9：00～19：00测得的数据。

Note：The data in the plot were measured daily from 9：00 to 19：00 on the August 4，10，11，23 in 2012 on sunny days.

差大于5℃所占比例为67%。当气温达到40℃时，叶－气温差为－7℃或接近－8℃。李利平[6]认为当叶气温差超过3℃时，植株蒸腾加大，耗能太多，因此植株虚弱，生长不良，这也就是说，当气温长时间高于32℃，不适合番茄的生长。

4 小结

第一，观测表明，日光温室番茄植株一天中植株叶温与同高度空气温度变化有着相同的趋势，即叶温随气温的变化而变化，且二者变化基本同步。测定日内叶－气温差始终为负值。

第二，日光温室番茄种植密度、植株部位可以影响叶温的变化。株距75cm（密度6株/m^2）植株上部叶温比中下部变化快，变幅大；株距150cm（密度3株/m^2）植株上部叶温变化幅度小于中部、下部。

第三，晴天太阳辐射强，叶－气温差变幅较大；阴天几乎没有大波动，维持在－2℃左右。

第四，番茄叶－气温差随气温变化明显。气温在25～32℃之间时，叶－气温差基本集中在0～－3℃；气温在32～35℃时，叶－气温差为－3～－5℃；当气温达到40℃时，叶气温差接近－8℃。叶气温差大于3℃，植株因蒸腾耗能，生长衰弱，因此当气温高于32℃，不适合番茄的生长。

参考文献

[1] 韩湘玲．作物生态学．北京：气象出版社，1991

[2] 李国臣，于海业，赵红霞等．基于叶片空气温差的温室黄瓜水分胁迫指数的应用分析．吉林农业大学学报．2006，28（4）：469～472

[3] 方学敏．叶面温度变化规律的试验研究．武汉水利电力学院学报．1989，22（6）：105～110

[4] 彭世彰，徐俊增，丁加丽．节水灌溉条件下水稻叶－气温差变化规律与水分亏缺诊断试验研究．水利学报，2006，37（12）：1 503～1 508

[5] 方学敏，李恩羊．以叶温为参数的作物缺水指标的研究方法．水利水电技术，1991（8）：56～61

[6] 李利平．不同根际温度对番茄植株体温及生长的影响．太谷：山西农业大学（硕士学位论文），2011

设施无土基质栽培辣椒品种比较试验*

孙晓军[1**]，王　强[2]，李翠梅[3]，贝丽柯孜·阿西木[3]，依米提·米吉提[3]，李鹏发[3]

（1. 新疆农业科学院 植物保护研究所，乌鲁木齐　830091；2. 新疆农业科学院园艺作物研究所，乌鲁木齐　830091；3. 疏勒县农业技术推广中心，疏勒　844200）

摘要：［目的］为了筛选出适合喀什日光温室无土基质栽培的辣椒品种。［方法］引进4个日本系列辣椒品种进行品种比较试验，通过植物学性状、果实商品性状、果实品质及产量对品种进行综合评价。［结果］试验品种阪田316、大板金秀、川野品种植株生长旺盛，植株直立，果实为长羊角形，纵径长达到20cm以上，果型指数达到7以上，果实皮薄质脆、微辣，其产量、品质和抗病性表现较好。［结论］阪田316、大板金秀、川野等3个品种均适应喀什地区栽培气候和消费习惯，可在秋冬茬生产中作为无土基质栽培品种引进并推广。

关键词： 辣椒；无土基质栽培 ；品种比较

Facility Soilless Media Cultivated Trials in Pepper Varieties Comparison

Sun Xiaojun[1], Wang Qiang [2], Li Cuimei [3], Beilikezi · aximu[3], Yi Miti · mijiti[3], Li Penfa[3]

(1. *Institute of Plant Protection*, *Xinjiang Academy of Agricultural Sciences*, *Urumqi* 830091, *China*; 2. *Institute of Horticulture*, *Xinjiang Academy of Agricultural Sciences*, *Urumqi* 830091, *China*; 3. *Centre of Agricultural Techniques Extension of Shule County*, *Shule Xinjiang* 844200, *China*)

Abstract: [Objective] To select pepper varieties were suitable for soilless cultivation in Kashi greenhouse. [Method] The introduction of four Series of different Japan pepper varieties were compared, as a comprehensive evaluation by botanical characters, trade traits, fruit quality and yield of varieties. [Result] The results showed that Ban tian316、Da ban jin xiu、Chuan ye varieties of grew well 、upright, the fruit of long claw-shaped longitudinal diameter reached 20cm or more, had thin crisp、spicy, the type of fruit index reached more than 7, its production, quality and disease resistance showed better performance. [Conclusion] Ban tian316、Da ban jin xiu、Chuan ye varieties adapted to the climate and consumption habits, as in the Fall Winter crop production the soilless substrate cultivars to introduce and promote in Kashi area.

Key words: Peppers; Soilless substrate culture; Variety comparison

采用无土基质栽培，可以有效克服土壤盐渍化、土传病虫害等连作障碍问题，并可减少农药用量，有效提高单位面积产量和产品品质[1~4]，用于绿色蔬菜的生产。而以有机废弃物菌糠为主的无土基质栽培技术因其取材容易，成本低廉，处理简单，操作方便，产品品质优，成为喀什地区非耕地设施蔬菜栽培示范推广主要的形式。其次，辣椒是喀什地区各族群众喜爱的蔬菜种类之一，但无土基质栽培技术的推广应用在当地起步晚，缺乏适宜的辣椒种

---

* 基金项目：国家“十一五”科技支撑计划项目（2009BADA4B04）；新疆“十二五”科技支撑计划项目（201130104）

** 作者简介：孙晓军（1962—），男，副研究员，主要从事设施蔬菜栽培技术研究与推广

植品种。试验通过对引进的辣椒新品种在基质栽培条件下生长发育和产量等方面的综合评价，筛选出适合于喀什地区设施无土基质栽培的辣椒品种，为无土基质栽培技术的推广和生产应用中品种选择提供参考依据。

1 试验材料与方法

1.1 试验材料

供试辣椒品种为阪田316、大板金秀、川野、日本长川和正佳1号（对照），均为鲜食品种。试验温室为土墙钢架无立柱式，长80m，内径宽8.5m。有机废弃物菌糠采用堆积高温发酵处理50d，炉渣过筛、水洗。栽培槽采用下挖式，槽内径50cm，深30cm，长6.7m，槽间距70cm，槽底铺黑塑料与土壤隔离，槽内先铺5cm厚的粗炉渣作渗水层，用剪开的废旧编制袋将炉渣全覆盖，再用配方基质装满栽培槽。栽培基质采用喀什地区推广使用的基质配方：菌糠：河沙：炉渣=5：3：2（体积比），定植前每立方基质施入腐熟鸡粪20kg，硫酸钾复合肥（N：P_2O_5：K_2O=15：15：15）2kg做基肥。灌溉采用重力自压式滴灌施肥系统，每槽铺设两条滴灌管，地膜覆盖栽培槽。

1.2 试验方法

试验在喀什地区疏勒县疏勒镇2村9组1号温室进行，采取无土基质秋冬茬栽培，2011年7月14日穴盘基质育苗，9月11日定植，11月11日采收，各品种生长期间环境保持一致。试验在同一温室内进行，随机区组排列，每个处理3次重复，定植株距40cm，双行双株品字型定植，每667m^2保留株数为5 600株，2012年3月25日拉秧。追肥在门椒长10cm左右时开始，隔10～15d一次。

1.3 测定方法

维生素C含量采用2，6-二氯靛酚滴定法测定；可溶性有机酸的测定采用NaOH滴定法，可溶性糖含量采用蒽酮法测定；植株性状、果实性状、产量等采用记载表记载。

2 结果与分析

表1 供试品种植株性状表

Table 1 The plant characters of tested cultivars

	株高（cm）	茎粗（mm）	幅宽（cm）	首花节位高（cm）	叶色	株型
阪田316	45.5ab	6.6c	39.6a	21.0a	绿色	直立型
大板金秀	45.1ab	7.3ab	36.6ab	19.5a	深绿色	直立型
川野	46.5a	7.1bc	39.1a	21.1a	绿色	直立型
日本长川	47.2a	8.0a	38.8ab	20.4a	绿色	直立型
正佳1号（CK）	43.1b	6.5c	35.1b	19.6a	深绿色	直立型

由表1可知，参试的4个日本辣椒品种株高、幅宽显著于正佳1号（CK）、首花节位高差异不显著。在茎粗方面，日本长川、大板金秀显著于正佳1号（CK），阪田316相对较弱。叶色绿色或深绿色，株型为直立型。结果表明，4个日本辣椒品种在良好的基质栽培环

境中，除阪田316茎秆相对稍弱外，试验品种各性状间差异不大，表现出植株生长健壮。

表2 供试品种果实特征表

Table 2 The fruit characteristics of tested varieties

	肉厚（mm）	果横径（cm）	果纵径（cm）	果型指数	果型	口感
阪田316	2ab	2.7ab	23.4a	8.9a	长羊角形	微辣
大板金秀	1.9ab	3a	23a	7.7b	长羊角形	微辣
川野	1.8b	2.8ab	23.1a	8.1ab	长羊角形	微辣
日本长川	2.1a	2.6b	20.3b	8ab	长羊角形	微辣
正佳（CK）	1.8b	2.7ab	20.2ab	7.6b	长羊角形	微辣

由表2可知，从横径来看，大板金秀显著于对照，其他各处理差异不显著。参试品种的果实纵径都在20cm以上，4个引进品种与对照相比差异不显著。果型指数都在7以上，阪田316显著于对照，其他各处理与对照相比较差异不显著，表明4个品种均为长羊角形果实，长而大。从果实的果肉厚度性状看，日本长川显著于对照，说明肉质稍厚，其他品种间没有表现出较大差异，均表现为皮薄质脆、微辣。

表3 供试品种果实品质性状

Table 3 The fruit quality traits of tested varieties

	还原糖（%）	VC（mg/100g）	干物质（%）
阪田316	2.8	3.56	6.9
大板金秀	2.3	4.78	6.8
川野	2.4	3.18	6.8
日本长川	2.6	2.44	6.8
正佳1号（CK）	2.8	2.74	7.0

由表3可以看出，3个辣椒品种维生素C含量明显高于正佳1号（CK），大板金秀最高，阪田316、川野次之，其含量分别比对照高74.5%、29.9%、16.1%，而日本长川维生素C含量低于对照10.9%。干物质在品种间含量相差不大；还原糖含量除阪田316与对照相同外，其他各处理都低于对照。

表4 不同品种产量及抗病性差异

Table 4 Comparison of yield and disease resistance in different varieties

	单果重（g）	早期产量（kg/666.7m²）	产量（kg/666.7m²）	病情指数	
				白粉病	疫霉病
阪田316	70.5	1 392.5	4 495.3	4.2	4.6
大板金秀	60.2	1 243.1	3 868.1	5.3	1.9
川野	70.3	1 246.5	4 023.5	6.9	4.5
日本长川	51.0	1 124.4	2 940.0	8.9	4.0
正佳（CK）	55.4	1 041.1	3 294.8	3.7	5.3

由表 4 可知，阪田 316、大板金秀、川野单果重分别比对照高 27.2%、8.7%、26.9%，而日本长川比对照低 7.9%。早期产量试验品种均高于对照，其中，以阪田 316 最高、其次是川野和大板金秀，最低是日本长川，总产量以阪田 316 最高、其次是川野和大板金秀，分别比对照高 36.4%、22.1%和 17.4%，日本长川则低于对照 10.8%。结果表明，阪田 316、川野、大板金秀表现出较高的前期产量和总产量水平。通过对生育期辣椒白粉病和疫霉病调查结果表明，参试品种白粉病病情指数均高于正佳（CK），但相差不大，发病较轻，病指均未超过 9%；试验品种大板金秀、日本长川、川野和阪田 316 的疫霉病发病率比正佳（CK）分别低 3.4%、1.3%、0.8%和 0.7%。

3 讨论

第一，辣椒是喀什地区设施蔬菜种植的主要种类之一，其产量的高低直接影响到当地种植户的经济效益，而品种是影响产量的决定性因素[5]。前人的研究主要针对设施蔬菜土壤栽培适宜品种的筛选，而设施无土基质栽培适宜辣椒品种的筛选报道较少。随着喀什地区设施农业规模的不断扩大，有限的耕地资源使得粮、棉、果、菜间争地矛盾不断加剧。无土基质栽培技术可在不适宜耕种的土地上进行蔬菜种植，理化性状良好的栽培基质，能有效地解决水分、空气和养分供应的矛盾，充分发挥作物的增产潜力[6]，是今后设施农业发展的主要方向之一。因此，开展无土基质栽培技术的应用研究，筛选适宜当地无土基质栽培的辣椒品种就显得尤为迫切。

第二，白粉病是辣椒的主要病害之一，发生严重会造成叶片、花、蕾大量脱落，植株早衰，导致减产，调查参试品种在基质栽培条件下白粉病病情指数相差不大，发生较轻，但在生产上仍要注意早防早治。辣椒疫霉病是新疆辣椒种植中威胁最大的病害，药剂防治效果不佳，土壤栽培的死秧率达 20% ~90%[7]，参试品种在基质栽培条件下辣椒疫病发病率较常规土壤栽培明显降低，表明无土基质栽培对克服作物土传病害方面具有明显的效果。

第三，试验引进的辣椒品种株型均为直立型，品种各性状间差异不大，植株生长旺盛，阪田 316 茎秆相对稍弱，在管理上要合理控制水肥以防生长过旺、落花落蕾，对植株要吊绳扶蔓以防倒伏。

第四，由于设施辣椒栽培的茬口模式多样，对品种特性的要求有差异。因此，对不同茬口模式适宜辣椒品种的筛选研究有待进一步开展。

4 结论

通过对引进品种植株性状、果实性状分析表明，4 个参试日本辣椒品种植株生长旺盛，植株直立，果实为长羊角形，纵径长达到 20cm 以上，果型指数达到 7 以上，果实皮薄质脆、微辣，均适应当地气候和消费习惯，但从品质、抗病性、前期产量和产量等方面综合考虑，阪田 316、大板金秀和川野表现较好，说明阪田 316、大板金秀、川野等 3 个品种均可作为喀什日光温室秋冬茬辣椒无土基质栽培的适宜品种。

参考文献

[1] 段崇香，于贤昌．有机基质栽培黄瓜化肥施用技术的研究．植物营养与肥料学报，2003，9（2）：238 ~241

[2] 段崇香，于贤昌．日光温室黄瓜有机基质型无土栽培基质配方的研究．农业工程学报，2002（增刊）：193～196
[3] 蒋卫杰，郑光华，汪浩等，有机生态型无土栽培技术及其营养生理基础．园艺学报，1996，23（2）：139～144
[4] 白纲义．有机生态型无土栽培营养特点及其生态意义．中国蔬菜，2000（增刊）：40～45
[5] 王浩，买合木提·肉孜，艾斯卡尔·吾守尔等．日光温室进口番茄品种生长发育和产量性状分析．北方园艺，2009（10）：155～157
[6] 张广楠．基质栽培技术研究的现状与发展前景．甘肃农业科技，2004（2）：6～8
[7] 杨华等，王志田，崔元玗，新疆辣椒疫霉病发生规律及综合防治研究．新疆农业科学，1997（2）：70～72

设施园艺固碳减排工业源 CO_2 原理与优势分析

刘文科*，刘喜明，杨其长，程瑞锋

（中国农业科学院农业环境与可持续发展研究所/农业部设施农业节能与废弃物处理重点实验室，北京 100081）

摘要：温室气体减排是当前重要研究课题。为了实现环境友好、彻底性减排工业源温室气体，本文基于设施园艺系统内 CO_2 浓度的日变化特征和园艺作物处于碳饥饿急需 CO_2 气肥的科学事实，提出了利用设施园艺固碳减排工业源 CO_2 的方法，并分析了其应用潜力与优势，明确了现有的技术支撑。总之，设施园艺固碳减排工业源 CO_2 是有效的方法，该方法具有固碳量大，周年运行，减排彻底无二次释放，可增加设施园艺产量等多重优点，可实现减排温室气体和增加设施园艺收益等双重效益。

关键词：设施园艺；固碳；减排；高效栽培；周年生产

Principle and Advantage Analysis of Method to Decrease Environmental Release of Industry-produced CO_2 by Protected Horticultural Systems

Liu Wenke, LiuXiming, Yang Qichang, Cheng Ruifeng

(*Institute of Environment and Sustainable Development in Agriculture*, *Chinese Academy of Agricultural Sciences*; *Key Laboratoryof Energy Conservation and Waste Management of Agricultural Structures*, *Minstry of Agriculture*, *Beijing* 100081, *China*)

Abstract: In order to decrease environmental release of industry-produced CO_2, a method of fixation CO_2 through protected horticultural systems was put forward based on diurnal change of CO_2 concentration in protected horticultural systems and the facts of protected horticultural plants severely lacking of CO_2 fertilizer as carbon material. Then, application potential, advantages and supporting techniques of this method were collected and pointed out. To conclude, fixing industry-source CO_2 through protected horticultural systems to reduce environmental emission is an effective method with advantages of in large quantity without secondary release, also with double benefits being achieved.

Key words: Protected horticulture; CO_2 fixation; emission reduction; High-efficient culture; Year-round production

工业革命以来，随着全球工业化的蓬勃发展和化石燃料的大量消耗，以 CO_2 为主体的温室气体大量排放（甲烷、氧化亚氮、氢氟碳化物、臭氧等）。温室气体对太阳短波辐射很少具有较强辐射吸收带，而在长波段则具有强烈的辐射吸收带。这样造成太阳短波辐射段可以透过大气使地球表面升温，而地球向宇宙空间发射的长波辐射则被大气吸收，使大气升温，这就是温室气体产生温室效应的基本原理[1]。二氧化碳的全球排放在 1751 年仅为 300

* 刘文科（1974—），男，博士，硕士生导师，研究员，主要从事设施园艺环境工程方面的研究工作。E-mail：liuwke@ 163. com

万 t，经过近两百年的缓慢增长后，自 20 世纪 40 年代开始迅速攀升，到 2008 年，全球的二氧化碳排放已经达到 299 亿 t，增速为 3%，略低于世界实际 GDP 增速 3.78%。CO_2 累积在空气中，其吸热和隔热的功能使太阳辐射到地球上的热量无法向外层空间发散，并逐年增强，从而引发全球气候变暖等一系列严重问题。温室效应所导致的全球气候变化，全球变暖，极端天气事件频发，给人类带来的是一系列的气候灾难。为应对全球气候变化对人类经济和社会带来不利影响，在《联合国气候变化框架公约》和《京都议定书》确立的共同但有区别的责任的原则上基础上，达成的巴厘路线图以及哥本哈根协议，使减排温室气体成为全球关注的问题。

规模化的工业生产是产排 CO_2 的主要源泉，同时也是减排 CO_2 的重要领域。通过对废气中 CO_2 的分离与回收，以利于 CO_2 的封存和应用。如已发展较为成熟的新型高效的 CO_2 分离回收技术、CO_2 封存技术、固定源和流动源的减排 CO_2 工艺技术，提高能效和新型能源技术等[2~5]。但是，利用 CO_2 作为一种碳资源合成附加价值较高的化工产品或材料是最困难也是最有意义的长期减排目标。CO_2 资源化利用是工业减排 CO_2 主要技术模式，但现有的一些资源化利用方法存在技术难度大、利用率低、二次释放可能性高等问题，未从“碳固定”的角度根本上解决工业源 CO_2 减排问题。开发低成本、绿色环保、经济效益较高的工业 CO_2 固碳减排技术十分迫切。

1 设施园艺系统内 CO_2 浓度日变化特征

植物是生物固碳的主体，通过叶片的光合作用固定大气中的 CO_2，形成光合产物、生物量和农产品，符合生态学和生物学规律，绿色环保。但受低温制约，北方露地植物冬春季节无法进行光合固碳，使植物固碳效能受到限制。设施园艺是在温室等设施内进行园艺作物生产的现代农业方式，可周年生产，复种指数高，冬春季节仍具有较高固碳效能，周年固碳潜力巨大，目前已成为反季节蔬菜和花卉栽培与供应的主体，发挥着不可或缺的作用。在我国西北、华北和东北地区，设施园艺发展迅速，已具规模。设施园艺作物在冬春季节旺盛的固碳作用可弥补露地植物固碳活力的空白期，成为北方冬春季节的重要碳汇。

受日照条件和农艺措施的控制，设施园艺系统内 CO_2 浓度表现出明显的日变化特征，即上午持续降低，下午缓慢上升，晚上快速上升的变化趋势。从植物生理角度而言，设施园艺系统内上午 CO_2 浓度常低于大气浓度，植物处于碳饥饿状态，光合作用受到抑制，需要人工补充 CO_2 气肥。原因如下：①植物光合作用适宜的 CO_2 浓度为 600×10^{-6} 以上水平，而植物 CO_2 的饱和点一般为 $800 \sim 1\ 000 \times 10^{-6}$，但大气 CO_2 浓度仅为 $300 \sim 400 \times 10^{-6}$，相差很大。②设施内园艺作物受密闭设施控制和植物密度大光合作用碳需求强度大的影响，在上午太阳照射 1 ~ 2h 内出现了 CO_2 浓度急剧下降的现象，经常降至 100×10^{-6} 甚至更低，由于无法得到有效的补充，极大地限制了园艺作物的光合作用和优质高产。实际生产中，鉴于保温需要上午很少换气通风，无法补充设施外空气源获得 CO_2。③设施园艺人工施用 CO_2 气肥缺乏气源。众所周知，补充 CO_2 气肥是设施园艺作物优质高产的必须的农艺措施，人工施用 CO_2 气肥十分必要。但因净化罐装 CO_2 气体及运输成本较高等问题，园艺农户很难支付费用，放弃了人工施用 CO_2 气肥的举措。化学发生法和秸秆生物反应法等方法因 CO_2 释放不可控、成本高及负效应多等原因未能大规模推广应用。因此，将工业 CO_2 提纯净化后直接用作设施园艺生产气肥，既能提高设施园艺作物的产量，又为工业 CO_2 浓资源化利用、固碳减排提供了新方法，具有双重效益。

2 设施园艺固碳的潜力与优势

设施园艺具有规模化固碳减排工业源 CO_2 的潜力，原因如下。①我国设施园艺规模大，居世界首位，具有大规模利用固碳减排工业源 CO_2 的潜力。截至 2012 年底，我国设施园艺面积已达 350 万 hm^2 以上，日光温室面积达 125 万 hm^2，玻璃温室面积在 10 万 ~20 万 hm^2，设施蔬菜产量位列世界第一。按每公顷固碳 5t 计算，固碳量相当客观公顷。②设施园艺可周年生产，复种指数高，单位面积固碳能力强。

在应用实践时，通常需要将设施内的 CO_2 浓度从 100×10^{-6} 甚至更低提升到 800 ~ 1 000 $\times10^{-6}$，提升幅度大，固碳潜力大。设施园艺具有规模化固碳减排工业源 CO_2 的优势，原因如下。①设施园艺固碳减排工业源 CO_2 可产生双重效益。工业源 CO_2 用作设施园艺气肥，既可达到固碳减排工业源 CO_2，也可增加设施园艺产量和品质，具有双重效益优势。②设施园艺具有规模化固碳减排工业源 CO_2 可工程化实施，可行性高。设施园艺具有集中特点，可采用工程手段实现工业源 CO_2 的输送释放与运用，连接工业 CO_2 气源和温室群，可行性高。

3 设施园艺系统固碳技术问题

利用设施园艺具有规模化固碳减排工业源 CO_2 需要解决几个技术问题。首先，开发工业废气中 CO_2 的净化提纯技术，确定设施园艺作物杂质气体含量基准，将工业源 CO_2 气体转变为符合植物光合要求的 CO_2 气肥。在研究设施园艺作物杂质气体耐受能力基础上，开发工业废气 CO_2 的提纯净化技术，减少杂质气体含量水平，使之不对设施园艺作物造成生理伤害。其次，研发工程技术实现工业源 CO_2 管道运输到规模化温室群，建立气肥供给网络。再次，需要开发适合各种类型温室设施 CO_2 智能化释放技术，实现温室中 CO_2 浓度的智能控制，最大程度地提高 CO_2 气肥的利用效率，保证固碳效果。最后，筛选高效固碳植物品种和高产栽培技术提高 CO_2 气肥的利用效率，保证固碳效果。现今，园艺作物响应 CO_2 气肥的增产效果和品质效应不清等，有待解决。

4 设施园艺固碳的技术支撑

目前，较为成熟的工业废气 CO_2 净化提纯技术为通过设施园艺固碳减排工业源 CO_2 提供了技术支撑，需要根据设施园艺作物生物学性质和耐受性基准，建立合理经济的净化提纯技术。其次，完善的设施园艺环境控制技术和高效高产立体周年栽培技术发展成熟，为设施园艺固碳减排工业源 CO_2 提供了技术支撑。目前，设施园艺的 CO_2 浓度控制技术已经十分完善，基于 PLC 的 PWM 控制可实现 CO_2 浓度渐进式控制，在密闭空间内控制精度可达到 50 ~ 100×10^{-6}，完全能够满足设施园艺 CO_2 施肥的需求。近 10 年，基于无土栽培的设施园艺高产立体栽培技术发展迅速，设施蔬菜生物学产量得以大幅度提升，在某种意义上提高了设施园艺系统的固碳减排能力。设施蔬菜高产立体栽培技术是利用人工墙体、立柱、栽培床等装置和工程工艺，采用果叶菜可调控栽培方法，并通过营养液进行养分的供给，环境因子控制创造十分有利于园艺作物生长的环境条件，充分利用蔬菜无限生长的生物学特性，形成超常规栽培冠幅和寿命，实现超常规产量和固碳效果。目前，多种高效立体栽培技术已广泛应用于设施园艺生产、观光休闲农业中。其中，蔬菜树式栽培技术大规模应用。设施蔬菜立体栽培是利用营养液槽等装置和工程技术，采用果菜无土栽培技术，利用果菜（辣椒、茄子和番茄）无限生长的生物学特性，运用设施环境可控的优势，充分延长辣椒、茄子和番茄等的生育期，形成高大的树形体态，具有观赏点。例如，中国农业科学院农业环境与可持续发展研究所首创的甘薯根系功能分离栽培技术，采用无土栽培技术，使甘薯块根和营养根空间分离，在甘薯茎蔓叶柄处诱导出块根，由于这种栽培模式根系养分吸收力强，冠层光合作用强，光合产物运输距离短，可明显提高产量。试验证明，单株甘薯产量可达 1t 以上，生物量高达 5t 以上，固碳能力非常高。总之，设施园艺固碳减排大有可为，具有广阔的应用前景。

参考文献

[1] 胡庆东，余博鹏，陈京远．温室效应与全球变暖．科技创新导报，2012（23）：134～135
[2] 洪大剑，张德华．二氧化碳减排途径．电力环境保护，2006，22（6）：5～8
[3] 周韦慧，陈乐怡．国外二氧化碳减排技术措施的进展．中外能源，2008，13（3）：7～13
[4] 栾健，陈德珍．二氧化碳减排技术及趋势．能源研究与信息，2009，25（2）：88～93
[5] 邝生鲁．全球变暖与二氧化碳减排．现代化工，2007，27（8）：1～12

添加 EM 对柠条粉基质番茄苗生长发育的影响*

曲继松**，张丽娟，冯海萍，杨冬艳
（宁夏农林科学院种质资源研究所，银川 750002）

摘要：为了丰富柠条基质穴盘育苗理论，为生产实际提供优质健壮秧苗提供参考。研究通过对柠条基质番茄苗苗期采用 EM 浇根处理（以清水为对照），结果表明：添加 EM 使番茄幼苗株高、茎粗、根长、地上部鲜质量、地下部鲜质量、全株鲜质量、地上部干质量、地下部干质量、全株干质量、根冠比和壮苗指数等生长发育指标上均高于对照，其中根鲜质量增加 18.34%，根冠比增加 17.2%，壮苗指数增加 16.66%。因此，合理添加 EM 有助于柠条基质番茄幼苗的生长发育。

关键词：干旱区；柠条基质；育苗；EM；番茄

The Effect of Adding EM on the Growth and Development of Tomato Seedlings in Nursery Substrate of Caragana

Qu Jisong, Zhang Lijuan, Feng Haiping, Yang Dongyan
(*Institute of Germplasm Resources*, *Ningxia Academy of Agriculture and Forestry Science*, *Yinchuan* 750002, *China*)

Abstract: For rich of Caragana microphylla matrix plug seedlings theory, provide a reference for the actual production to provide quality robust seedlings. Research by Caragana intermedia matrix tomato Miao Miao period using EM pouring root as the control treatment (clean water), the results showed that: add EM tomato seedling height, stem diameter, root length, shoot fresh mass underground fresh mass, full plant fresh mass, shoot dry weight, root dry quality, total dry weight, root to shoot ratio and seedling index growth and development indicators are higher than that of the control, of which 18.34% increase in the quality of the fresh root, root to shoot ratio increased by 17.2% seedling index increased by 16.66%. Therefore, it is reasonable to add to the growth and development of the EM help to of Caragana microphylla matrix tomato seedlings.

Key words: Arid regions; Caragana substrates; Seedling; EM; Tomato

蔬菜育苗是栽培过程中的重要环节，培育优质壮苗的重要条件之一就是要具有优良的育苗基质。目前，国内外开发应用了不同形式的工厂化育苗模式和机械设备，根据当地资源也开发出了不同来源的育苗基质，为壮苗培育的研究和商业应用奠定了良好的基础。在我国，由于设施农业的快速发展，对优质壮苗的需求量日益增大，对基质的需求量逐年增加。开发性能优越，且应用稳定，来源充足的商业育苗基质一直是农业科技工作者和基质开发商的重

---

* 基金项目：宁夏自然科学基金资助项目（NZ12241）；宁夏重大科技攻关计划（2011ZDN0401）；中-日国际合作项目“近自然农法技术在宁夏蔬菜生产中的引进与研究”；国家星火项目（2011GA880001）

** 作者简介：曲继松（1980—），男，吉林永吉人，助理研究员，主要从事设施环境调控和蔬菜栽培生理研究。E-mail：qujs119@126.com

要研究开发内容。

柠条粉作为基质的探索性试验已经取得了初步成功，尤其是在西瓜[1]、甜瓜[2]、茄子[3]、辣椒[4]、黄瓜[5]等育苗上和樱桃番茄[6,7]、辣椒[8]、番茄[9]等作物基质栽培上均有较好表现。因此，本试验以柠条基质番茄穴盘苗为试材，通过添加 EM 培育优质瓜菜壮苗，为培育优质健壮秧苗及工厂化育苗提供理论依据和支持。

1 材料与方法

1.1 试验时间及地点

试验时间为 2012 年 3 月至 2012 年 5 月，地点位于宁夏同心县旱作节水高效农业科技园内，地处宁夏中部干旱带，位于东经 105°59′，北纬 36°51′，海拔 1 363m。

1.2 试验材料

供试番茄品种为“保罗塔”来自于先正达种业有限公司，供试柠条粉（颗粒粒径为 0.4 ~0.8mm）购自宁夏回族自治区盐池县源丰草产业有限公司，柠条粉中加入有机-无机肥料（1m^3 柠条粉加入 2.8kg 尿素、100kg 消毒鸡粪）腐熟发酵 90 d，加入珍珠岩和蛭石（柠条粉：珍珠岩：蛭石 =3：1：1，体积比）后作为育苗基质使用，育苗穴盘采用 72 穴标准苗盘。

1.3 试验方法

试验共计 2 个处理，每处理 10 盘，EM 菌液处理浓度（体积比）为 1%（厂家推荐使用量）对番茄幼苗进行浇灌，以清水为对照。出苗时间为自播种之日起到出苗数为 30%；齐苗时间为自播种之日起到出苗数为 80%；出苗后天数以出苗时间之日算起；出苗率 = 出苗株数/72；成苗率 = 成苗株数/72；根冠比 = 地下部干质量（g）/地上部干质量（g），壮苗指数 = ［茎粗（cm）/株高（cm）+地下部干质量（g）/地上部干质量（g）］×全株干质量（g）[10]，根系活力测定采用甲基蓝法[11]。

测定各项指标时每重复取样 10 株，所有数据均为 3 次重复的平均值。数据采用 Microsoft Excel 2003 和 DPS 3.01 进行处理和统计分析。

2 结果与分析

2.1 添加 EM 对番茄幼苗出苗状况的影响

从表 1 可以看出，添加 EM 和清水处理（CK）育苗的出苗天数均为 7d，齐苗时间均为 9d；在出苗率方面，添加 EM 比清水处理（CK）的略低，低 0.4 个百分点，差异不显著；在成苗率方面，无差异，均为 100%。

2.2 添加 EM 对番茄幼苗生长的影响

由于株高、茎粗和叶片数的变化是幼苗生长状况的综合体现，且叶片多少直接关系到植株光合同化能力；根系是作物吸收水分和养分的主要器官，又是许多物质同化、转移和合成的重要场所。本试验以株高、茎粗、叶片数、根长等作为衡量幼苗生长势的指标。由表 2 可知，添加 EM 对茎粗和根长两方面均有显著影响，分别增加了 9.1% 和 7.1%；但在出苗天数、株高、叶片数方面差异不显著。

表 1　添加 EM 对番茄幼苗出苗状况

Table 1　Emergence situation of tomato seedling from adding EM

处理	出苗时间 Seeding time (d)	齐苗时间 All seeding time (d)	出苗株数 Seeding number	成苗株数 Mature seedling number	出苗率 Seeding rate (%)	成苗率 Mature seedling rate (%)
CK	7a	9a	70.5a	70.5a	97.9a	100a
EM	7a	9a	70.2a	70.2a	97.5a	100a

注：同列不同小写字母表示差异显著（$P<0.05$）；同列不同大写字母表示差异极显著（$P<0.01$），下同

表 2　添加 EM 对番茄幼苗生长状况

Table 2　The growth situation of tomato seedling from adding EM

处理	出苗天数 (d)	株高 Plant height (cm)	茎粗 Stem diameter (cm)	叶片数 Leaf number	根长 Root length (cm)
CK	45	12.35a	0.307a	4a	9.45a
EM	45	12.55a	0.335b	4a	10.12b

2.3　添加 EM 对番茄幼苗根系生长的影响

从表 3 可以看出，添加 EM 番茄幼苗根体积和根重均大于 CK，其中根体积较 CK 增加 22.22% 倍、根鲜质量增加 18.34%；由于幼苗根体积大小关系，添加 EM 处理根系总吸收面积、活跃吸收面积（m^2）均大于 CK，且差异显著；添加 EM 处理比表面积略小于 CK，差异不显著。

表 3　添加 EM 对番茄幼苗根系状况

Table 3　The Seedling roots situation of tomato seedling from adding EM

处理	根体积 Root volume (ml)	根鲜质量 Root fresh weight (g)	总吸收面积 Total absorption area (m^2)	活跃吸收面积 Active absorption area (m^2)	比表面积 Specific surface area (cm^2/cm^3)
CK	0.9a	0.447A	1.0584a	0.6622a	12 693.22a
EM	1.1b	0.529B	1.1325b	0.6943b	12 529.43a

2.4　添加 EM 对番茄幼苗干物质积累的影响

从表 4 可以看出，添加 EM 使番茄幼苗地上部鲜质量、地下部鲜质量、全株鲜质量、地上部干质量、地下部干质量、全株干质量根冠比和壮苗指数均高于对照，其中地上部鲜质量和地上部干质量差异不显著，地下部干质量、全株鲜质量、全株干质量和壮苗指数差异显著，地下部鲜质量、根冠比差异极显著。

表 4　添加 EM 对番茄幼苗干物质积累的影响

Table 4　The influence on accumulation of dry matter of tomato seedling from adding EM

处理	地上部鲜质量 Shoot fresh weight（g）	地上部干质量 Shoot dry weight（g）	地下部鲜质量 Root fresh weight（g）	地下部干质量 Root dry weight（g）	全株鲜质量 Total fresh weight（g）	全株干质量 Total dry weight（g）	根冠比 Root/Shoot	壮苗指数 Seedling index（g）
CK	1.388a	0.113a	0.447A	0.043a	1.835a	0.155a	0.372A	0.0466a
EM	1.401a	0.117a	0.529B	0.051b	1.89b	0.168b	0.436B	0.0525b

3　讨论与结论

目前国内诸多学者对于替代草炭的新型育苗基质开发、利用方面进行了大量的研究工作，新型基质材料主要包括：花生壳[12]、锯末[13]、苇末[14]、作物秸秆[15]、树皮[16]、椰子壳纤维[17]、菇渣[18]、稻壳[19]以及下水道污泥[20]等，这些工业和农业生产中的废弃物都是很好的草炭替代材料，而且在试验中得到良好的结果。本试验针对西北内陆地区贮量极为丰富的沙生植物—柠条进行研究，丰富了柠条基质的理论和使用技术。

EM（Effective microorganisms）是一个微生物群的混合体，主要包括乳酸菌、酵母菌、光合细菌、放线菌、芽孢菌和其他一些类型，所有这些菌群在培养液中协调一致、相互依存。EM 常被用于接种制作有机堆肥，EM 被认为在提高作物生产和净化环境等多方面具有重要作用。

结果表明，添加 EM 使番茄幼苗株高、茎粗、根长、地上部鲜质量、地下部鲜质量、全株鲜质量、地上部干质量、地下部干质量、全株干质量、根冠比和壮苗指数等生长发育指标上均高于清水处理（CK）。因此，在柠条基质育番茄苗时使用 EM，有助于幼苗的生长发育。

参考文献

[1] 曲继松，郭文忠，张丽娟等．柠条粉作基质对西瓜幼苗生长发育及干物质积累的影响．农业工程学报．2010，26（8）：291～295

[2] 张丽娟，曲继松，冯海萍等．利用柠条发酵粉作育苗基质对甜瓜幼苗质量的影响．北方园艺，2010（15）：165～167

[3] 曲继松，张丽娟，冯海萍等．混配柠条粉基质对茄子幼苗生长发育的影响．西北农业学报，2012，21（11）：162～167

[4] 曲继松，张丽娟，冯海萍等．发酵柠条粉混配基质对辣椒幼苗生长发育的影响．江苏农业学报，2012，28（4）：846～850

[5] 孙婧，买买提吐逊·肉孜，曲梅等．柠条基质理化性质和育苗效果研究．中国蔬菜，2011，（22/24）：68～71

[6] 冯海萍，曲继松，郭文忠等．基于发酵柠条为栽培基质对樱桃番茄产量及品质的影响．北方园艺，2010，（3）：22～23

[7] 冯海萍，曲继松，郭文忠等．不同栽培方式下樱桃番茄基质栽培试验及效益分析．北方园艺，2010，（7）：38～39

[8] 冯海萍，郭文忠，曲继松等．不同营养液对辣椒柠条基质栽培产量和品质的影响．北方园艺，2010，（15）：153～155

[9] 冯海萍，曲继松，郭文忠等．栽培模式对柠条复合基质栽培有机番茄生长发育的影响．北方园艺，2012（18）：30～31

[10] 韩素芹，王秀峰．氮磷对甜椒穴盘苗壮苗指数的影响．西北农业学报，2004，132）：128～132

[11] 于小凤，李进前，田昊．影响粳稻品种吸氮能力的根系性状．中国农业科学，2011，44（21）：4 358～4 366
[12] 孙治强，赵永英，倪相娟．花生壳发酵基质对番茄幼苗质量的影响．华北农学报，2003，18（4）：86～90
[13] 籍秀梅，孙治强．锯末基质发酵腐熟的理化性质及对辣椒幼苗生长发育的影响．河南农业大学学报，2001，35（1）：66～69
[14] 李谦盛．芦苇末基质的应用基础研究及园艺基质质量标准的探讨．南京：南京农业大学，2003：90～94
[15] 高新昊，张志斌，郭世荣．玉米与小麦秸秆无土栽培基质的理化性状分析．南京农业大学学报，2006，29（4）：131～134
[16] Sánchez-Monedero M A，Roig A，Cegarra J，*et al.* Composts as media constituents for vegetable transplant production. Compost Science & Utilization，2004，12（2）：161～168
[17] 陈萍，郑中兵，王艳飞等．南方甜瓜育苗基质的研究．种子，2008，27（7）：63～64，66
[18] 时连辉，张志国，刘登民等．菇渣和泥炭基质理化特性比较及其调节．农业工程学报，2008，24（4）：199～203
[19] 孙向丽，张启翔混配基质在一品红无土栽培中的应用．园艺学报，2008，35（12）：1 831～1 836
[20] Medina E，Paredes C，Pérez-Murcia M D，*et al.* Spent mushroom substrates as component of growing media for germination and growth of horticultural plants. Bioresource Technology，2009，100（18）：4 227～4 232

微/纳米气泡臭氧水杀灭温室土壤根传病菌的研究*

齐太山[1**]，李兴隆[1]，高朝飞[1]，尹柏德[1]，宋卫堂[1,2***]
［1. 中国农业大学水利与土木工程学院，北京 100083；
2. 农业部设施农业工程重点（综合）实验室，北京 100083］

摘要：为了探索绿色、无残留、无污染的土壤灭菌新方法，研究了微/纳米气泡臭氧水对土壤根传病菌的杀灭效果。将番茄尖镰孢菌配成浓度为 2.4×10^4 CFU/ml 的菌液后，通入微/纳米气泡臭氧水，对目标菌进行灭菌试验。在此基础上，对温室土壤样品，每周浇灌一次微/纳米气泡臭氧水灭菌，连续进行 3 次。试验结果表明：当水中臭氧浓度为 5.28mg/L、接触时间 2min 时，对浓度为 2.4×10^4 CFU/ml 的番茄尖镰孢菌菌液的灭菌率接近 100%。采用浓度为 1.2～1.7mg/L 和 7.8～8.3mg/L 的臭氧水，对温室土壤样品进行 3 次连续灭菌后，对真菌的灭菌率分别达到 72.9% 和 91.8%；对细菌的灭菌率分别达到 63.2% 和 86.0%。因此，利用微/纳米气泡臭氧水进行温室土壤灭菌，是一种可行的灭菌新方法。

关键词：土壤灭菌；微/纳米气泡臭氧水；番茄尖镰孢菌；灭菌率

Germicidal Efficacy of Micro/nano Bubbles Ozone Water on Root-Infecting Pathogens in Greenhouse

Qi Taishan[1], Li Xinglong[1], Gao Zhaofei[1], Yin Bode[1], Song Weitang[1,2]
(1. *College of Water Resources & Civil Engineering*, *China Agricultural University*, *Beijing* 100083, *China*;
2. *Key Laboratory of Agricultural Engineering in Structure and Environment*, *Beijing* 100083, *China*)

Abstract: To explore an environmental friendly soil sterilization method with no residue and pollution, the effect of micro/nano bubbles ozone water on sterilization of soil pathogens was studied. Experiments were carried out in samples containing 2.4×10^4 CFU/ml of *Fusarium oxysporum* of tomato. Micro/nano bubble Ozone Water was added to the samples and thus the sterilization efficacy of the target bacteria was tested. Based on the results of vitro study, irrigation of micro/nano bubble ozone water to greenhouse soil samples once a week was conducted to investigate the germicidal efficacy. The results showed that when the concentration of ozone was 5.28mg/L and contact time was 2minutes, the killing rate was close to 100%. In the sterilization experiment of micro/nano bubbles ozone water on greenhouse soil samples, the concentration of ozone was 1.2～1.7mg/L and 7.8～8.3mg /L. After three consecutive irrigations, the fungal killing rates were 72.9% and 91.8% while the bacterial killing rates were 63.2% and 86.0%, respectively. Therefore, the use of micro/nano bubbles ozone water for greenhouse soil sterilization is a viable new sterilization method.

---

* 基金项目：“十二五”国家科技支撑计划项目——封闭式无土栽培营养液灭菌技术研究与示范（2011BAD12B01—A）；大学生创新实验计划项目“微/纳米气泡臭氧水灭菌温室土壤中根传病原菌的试验”

** 作者简介：齐太山（1991—），男，中国农业大学农业建筑环境与能源工程专业 2010 级学生

*** 通讯作者：宋卫堂（1968—），男，博士，教授，从事设施园艺中营养液的物理灭菌技术及设备研究。北京 中国农业大学水利与土木工程学院，100083。E-mail：songchali@cau.edu.cn

Key words: Soil sterilization; Micro/nano bubble ozone water; *Fusarium oxysporum*; Killing rate

土壤灭菌能有效杀灭土壤中的真菌、细菌、线虫、土传病毒、地下害虫等，能很好的解决高附加值作物连续种植中的重茬问题，并显著提高作物的产量和质量[1~2]。目前的土壤灭菌技术主要有物理、化学和生物熏蒸灭菌等[3]。物理灭菌主要包括太阳能灭菌、蒸汽灭菌等[4~5]。化学灭菌是指将氯化苦、威百亩、棉隆、碘甲烷和福尔马林等熏蒸剂注入到土壤中进行灭菌。生物熏蒸灭菌是利用有毒气体杀死土壤害虫和病菌[6]。但上述灭菌方法存在受环境影响大、操作繁琐、能耗大等缺点，很多化学灭菌法还对环境有污染，在土壤中有残留，对人类健康和自然生态环境造成危害[7]。正因为这些缺点，使得在生产实际中的推广受到限制。

臭氧是一种氧化性极强、功能多样化、极具开发价值的气体，用臭氧水进行温室土壤灭菌可以有效防护和控制土传病虫害的发生；臭氧无论在气体状态，还是溶于水的状态，都极易分解为氧气，不会对环境造成二次污染，因此臭氧被称为“理想的绿色灭菌剂”[8~10]。虽然人们应用臭氧对水体灭菌已有100多年的历史，但臭氧应用于农业用水灭菌的研究，是随着封闭式无土栽培（Closed Cultural System）的发展，在20世纪80年代末才开始的[11~13]。Runia[14]等将臭氧通入无土栽培灌溉后回收的营养液中，用来杀灭其中的植物根传病害，然后再将灭菌后的营养液重新利用。在荷兰、法国等，臭氧已经在营养液的灭菌上应用于生产[15]。喻景权[16]等用10 L/min的臭氧，对放有番茄青枯病病原菌和番茄枯萎病病原菌的营养液，分别曝气90min和120min，营养液中的菌落数分别从10^4 CFU/ml和10^5 CFU/ml降至0。

微/纳米气泡，是指直径等于或小于50μm的微小气泡，具有自增压、带负电荷和强氧化性的自由基等特性，因此农业行业和食品科学技术领域对微/纳米气泡技术的应用日益重视[17]。与普通微小气泡相比，微/纳米气泡在水中上升缓慢，能够在水中停留更长的时间，并且微/纳米气泡会自我压缩，溶解在水中，所以其在水中具有很高的溶解度。

本试验研究微/纳米气泡臭氧水进行土壤灭菌的效果，主要包括：测试试验条件下微/纳米气泡臭氧水中臭氧的溶解和衰减特性；在此基础上，研究微/纳米气泡臭氧水对目标病原菌—番茄尖镰孢菌的灭菌效果；进而，通过连续3次浇灌微/纳米气泡臭氧水，研究对温室土壤中真菌、细菌的灭菌效果，为采用微/纳米气泡臭氧水进行温室土壤的灭菌提供技术基础。

1 材料方法

试验于2012年5月至2013年2月在中国农业大学水利与土木工程学院农业部设施农业工程重点实验室进行。实验装置主要有氧气罐、臭氧发生器、微/纳米气泡发生器、无菌操作台等。试验水源为实验室自来水，pH值=7.42。土壤样品取自于北京市通州区潞城镇卜落垡村温室内。

1.1 试验装置

如图1所示，臭氧发生器与微/纳米气泡发生器相连，氧气罐为臭氧发生器提供气源，通过水冷方式为臭氧发生器降温；微/纳米气泡发生器主要包括气液混合泵和微/纳米气泡曝气头。通过气液混合泵将储液桶（试验时容积为35 L，盛水量为20 L）中的自来水，与臭氧发生器产生的臭氧进行水与臭氧混合，然后气液混合体在曝气头中以22 500～45 250rpm的速度高速旋回，使得臭氧被切分为直径5～30nm的气泡，形成微/纳米气泡臭氧水。

臭氧发生器（GQ-TGI）：为天津水产研究所研制[18]，属搪瓷单管式水冷臭氧发生器，臭氧产量为25 g/h。

微/纳米气泡发生器（定制）：为本洲（北京）纳米科技有限公司生产，处理水量2～3m³/h，功率0.37 kW。

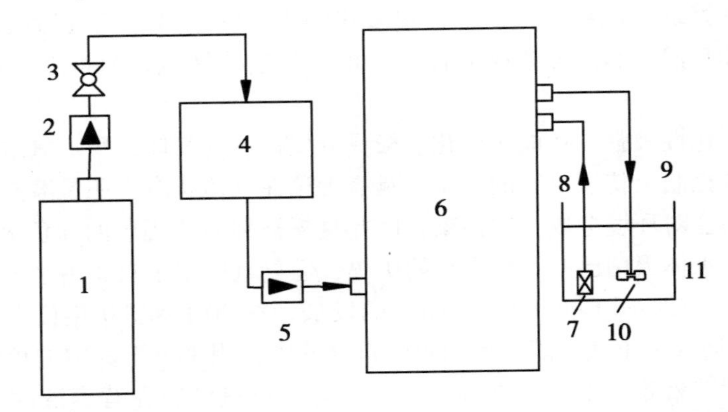

1. 氧气瓶 2. 流量计 3. 调节阀 4. 臭氧发生器 5. 流量计 6. 微纳米气泡发生器 7. 进水阀 8. 进水管 9. 出水管 10. 曝气头 11. 储液桶

图1 微/纳米气泡臭氧水发生装置示意图

Figure 1 Sketch of micro/nano bubbles ozone water generator

1.2 目标菌

半知菌亚门真菌—番茄尖镰孢菌番茄专化型（*Fusarium oxysporum* f. sp. *Lycopersici Snyderet* Hansen），可引起番茄枯萎病。

1.3 试验方法

1.3.1 微/纳米气泡臭氧水中臭氧浓度的升高与衰减规律

采用碘量法[19]测定微/纳米气泡臭氧水中的臭氧浓度。

试验时，先用温度计测量水温，然后用秒表计时，当曝气时间达到预设时间后，按照碘量法对微/纳米气泡臭氧水中的臭氧浓度进行测定。

曝气结束后，开始每隔5min对水中的臭氧浓度测定一次；浓度衰减变慢后，每隔10min或15min测定一次；待水中的臭氧浓度下降到起始浓度的一半以后，每隔30min或45min测定一次。

1.3.2 微/纳米气泡臭氧水对目标菌（番茄尖镰孢菌）的灭菌试验

试验步骤如下：①先将培养好的目标菌稀释成10^4数量级的菌液；②将曝气头放入配好的目标菌中，取一个空白样品，3次重复，作为对照（CK）；③开启试验装置，秒表开始计时，按设定的时间：1min、2min、4min和6min分别取样，每个样品3次重复；④用移液枪取0.5ml样品注入事先准备好的、含4.5ml中和试剂的试管中，摇匀。每次取样时（包括空白取样），都同时采用碘化钾进行滴定，以确定对应取样时间时水中臭氧的浓度；⑤取样结束后，在超净工作台中，将样品稀释10倍，然后把原样品和稀释10倍的样品涂板，每个培养皿所涂样品量为0.1ml。在28℃恒温箱中培养3d，然后计数。以实际空白样品（CK）长出的菌落数为起始菌落数进行计算灭菌率[20]：

灭菌率 =（对照存活菌落数 - 处理后存活菌落数）/对照存活菌落数 × 100%　　(1)

1.3.3　微/纳米气泡臭氧水对温室土壤的灭菌试验

试验步骤如下：①从温室中取回土壤样品，取样深度为35cm；②在3个大小15×25×15cm³的容器中，分别装入15kg的土壤样品，容器底部有6个排水孔；③取1 000ml自来水，作为对照（CK）缓慢浇灌于其中一个容器的土壤中；④取同时刻的自来水，进行曝气，分别取曝气时间为30 s、3min的微/纳米气泡臭氧水，同时测定其臭氧浓度。将两种浓度的臭氧水，各1 000L，分别浇灌于另外2个容器中的土壤中；⑤24 h后，分别从3个容器中取10 g土壤，溶于90ml无菌生理盐水中。曝气3min臭氧水处理的土壤溶液，以及稀释10倍后的土壤溶液分别涂布在PDA培养基上；曝气30 s臭氧水处理的土壤溶液和自来水浇灌的土壤溶液，分别稀释10倍和100倍涂布在PDA培养基上，培养72h，统计存活真菌菌落数，计算灭菌率；⑥曝气3min臭氧水处理的土壤溶液，稀释10倍和100倍后分别涂布在营养琼脂培养基上；曝气30 s臭氧水处理的土壤溶液和自来水浇灌的土壤溶液，分别稀释100倍和1 000倍涂布在营养琼脂培养基上，培养48 h，统计存活细菌菌落数，计算灭菌率；⑦第一次试验结束7d、14d后，对上述土壤样品，采用同样的方法，分别进行第二次、第三次灭菌试验。

2　结果与分析

2.1　微/纳米气泡臭氧水中臭氧浓度的升高与衰减规律

2.1.1　微/纳米气泡臭氧水中臭氧浓度的升高规律

在微/纳米气泡发生器的进氧气流量为0.4 L/min、水温20℃、自来水pH = 7.4、EC = 0.48mS/cm的条件下，检测了微/纳米气泡臭氧水中，臭氧浓度随曝气时间的变化规律如图2所示。

由图2可知，在自来水初始温度为20℃的条件下，微/纳米气泡臭氧水中臭氧浓度随曝气时间的增加而上升，上升速度逐渐减缓，直至趋于稳定。溶解规律曲线的回归方程为：$y = -0.21x^2 + 3.204x - 0.451$；回归系数$R^2 = 0.984$。曝气0～1min阶段，臭氧浓度增加较快，1min时臭氧浓度达到1.8mg/L；曝气1～4min这段时间是臭氧浓度增加最快的阶段，臭氧浓度大约增加了9.4mg/L，占最终浓度的87%；4min后就逐渐开始稳定，浓度稳定在10.8～11.1mg/L。因此，曝气1～4min这段时间是微/纳米气泡臭氧水中臭氧溶解最关键的时间段。

2.1.2　微/纳米气泡臭氧水中臭氧浓度的衰减

在微/纳米气泡发生器的进氧气流量为0.4 L/min、水温20℃、自来水pH = 7.4、EC = 0.48mS/cm的条件下，检测了微/纳米气泡臭氧水中，臭氧浓度随时间的衰减规律如图3所示。

由图3可知，曝气结束时臭氧水中的臭氧浓度为10.4～10.8mg/L。开始，臭氧浓度衰减的十分明显，停止曝气后18～24min，臭氧浓度衰减了一半；到停止曝气1 h时，臭氧浓度又衰减了一半；而后衰减速度变得比较缓慢，臭氧浓度能较长时间（大于3 h）稳定，稳定时的浓度约为1.3mg/L；臭氧衰减规律的回归曲线方程为：$y = -1.92\ln(x) + 10.77$，回归系数 $R^2 = 0.98$。因此，停止曝气后的60min内（若对臭氧浓度要求较高，则是20min内），应作为用于灭菌的主要时间段。

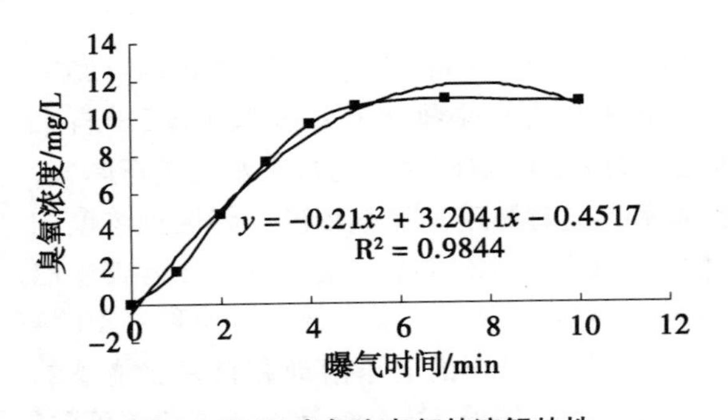

图2　微/纳米气泡臭氧的溶解特性

Figure 2　Dissolution properties of micro/nano bubbles ozone water

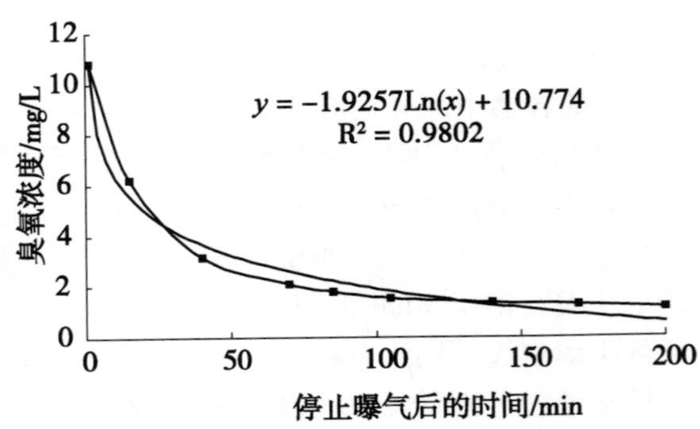

图3　微/纳米气泡臭氧的衰减特性

Figure 3　Attenuation properties of micro/nano bubbles ozone

2.2　微/纳米气泡臭氧水对目标菌（番茄尖镰孢菌）的灭菌试验

采用图1所示实验装置，所取自来水温度为16.2℃，在氧气气源、进口流量0.4 L/min的条件下，曝气0.5min、1min、2min、4min分别制备浓度为0.9mg/L、3.1mg/L、7.3mg/L、12.3mg/L的微/纳米气泡臭氧水，测定臭氧水对番茄尖镰孢菌的灭菌效果。病原菌孢子平均起始浓度为 2.4×10^4 CFU/ml。

由图4可以看出，在通气时间0.5min时，目标菌菌液中的臭氧浓度为0.9mg/L，臭氧对番茄尖镰孢菌的灭菌率平均已达96%；在通气时间1min时，臭氧浓度为3.1mg/L，灭菌率达到99%；在通气时间分别为2min和4min时，目标菌菌液中的臭氧浓度达到7.3和12.3mg/L，对 2.4×10^4 CFU/ml浓度的番茄尖镰孢菌病原菌灭菌率都达到100%。微/纳米气泡臭氧水对浓度为 10^4 CFU/ml数量级菌液中番茄尖镰孢菌病原菌的灭菌效果，是十分显著的。

2.3　微/纳米气泡臭氧水对温室土壤的灭菌试验

将含有根传病菌的土壤，分成3组进行灭菌试验。以曝气时所用的自来水做为对照，在微/纳米气泡发生器的进氧气流量为0.4 L/min、水温16℃、自来水pH=7.4、EC=0.48mS/cm的条件下，制备臭氧浓度分别为1.2~1.7mg/L和7.8~8.3mg/L的微/纳米气泡臭氧水。灭菌后取土样进行接种试验，所取土壤样品深度为5cm，统计存活菌落数量，计算灭菌率。3次灭菌试验的结果分别如图5至图7所示。

由图5可以看出，进行第一次灭菌，当微/纳米气泡臭氧水中的臭氧浓度为1.2~1.7mg/L时，真菌灭菌率为55.3%，细菌灭菌率为35.4%；当微/纳米气泡臭氧水中的臭氧浓度达到7.8~8.3mg/L时，真菌灭菌率达到89.5%，细菌灭菌率也达到了62.5%。因此，臭氧水中的臭氧浓度越高，杀菌效果越好。

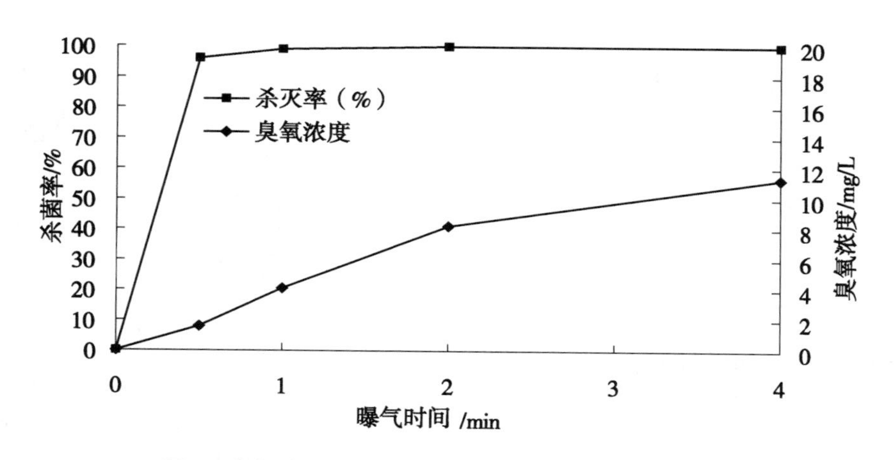

图 4　臭氧浓度及其与病原菌接触时间与灭菌率的关系

Figure 4　Relationship of sterilization rate with ozone concentration and contact time

(*Fusarium oxysporum* f. sp. *Lycopersici Snyderet* Hansen)

图 6 是在第一次灭菌的基础上，7 天后进行的第二次灭菌的试验结果。两次灭菌所用的臭氧水浓度相同。当微/纳米气泡臭氧水中的臭氧浓度为 1.2 ~ 1.7mg/L 时，对真菌的灭菌率为 60.6%，对细菌的菌灭菌率为 55.3%；当臭氧浓度达到 7.8 ~ 8.3mg/L 时，真菌灭菌率达到 78.8%，细菌灭菌率也达到了 77.9%。同样，臭氧水中的臭氧浓度越高，杀菌效果越好。

由图 7 可以看出，第三次灭菌后，当臭氧浓度为 1.2 ~ 1.7mg/L 时，对真菌的灭菌率为 37.9%，对细菌的灭菌率为 54.3%；当臭氧浓度达到 7.8 ~ 8.3mg/L 时，真菌灭菌率达到 85.7%，细菌灭菌率也达到了 85.6%。因此，臭氧水中的臭氧浓度越高，灭菌效果越好。

从 3 次连续灭菌效果来看，当臭氧水中的臭氧浓度较低（1.5mg/L 左右）时，对真菌和细菌的灭菌效果相差较大；当浓度较高（8mg/L 左右）时，对真菌和细菌的灭菌效果均较高，而且后两次的灭菌率也几乎相同。

试验土壤样本的总体灭菌效果如图 8 所示。灭菌前土壤中的菌落数为第一次的对照组菌落数，灭菌后的菌落数为第三次灭菌后的菌落数。由图可知，土壤经过三次微/纳米气泡臭氧水灭菌，当臭氧水中的臭氧浓度为 1.2 ~ 1.7mg/L 时，对土壤中真菌的灭菌率为 72.9%，对土壤中细菌的菌灭菌率为 63.2%；当臭氧浓度达到 7.8 ~ 8.3mg/L 时，对土壤中真菌灭菌率达到 91.8%，对土壤中细菌灭菌率也达到了 86.0%。

3　讨论与结论

臭氧以微/纳米气泡形式溶解于水中，相对于普通直接曝气方式，其溶解度有很大提高。在微/纳米气泡发生器的进气流量为 0.4 L/min、水温 20℃、自来水 pH 值 = 7.4、EC = 0.48mS/cm 的条件下，检测发现：在曝气 1 ~ 4min 时间段内臭氧浓度增加得最快，增加了约 9.4mg/L，占最终浓度（约 10.8mg/L）的 87%。这种条件下的臭氧衰减试验表明，臭氧的第一个半衰期为 18 ~ 24min；停止曝气 1 h 后，臭氧浓度又衰减了一半。而后衰减速度变得很慢，最终能较长时间地稳定在 1.3mg/L。因此，微/纳米气泡臭氧水中的臭氧溶解度很

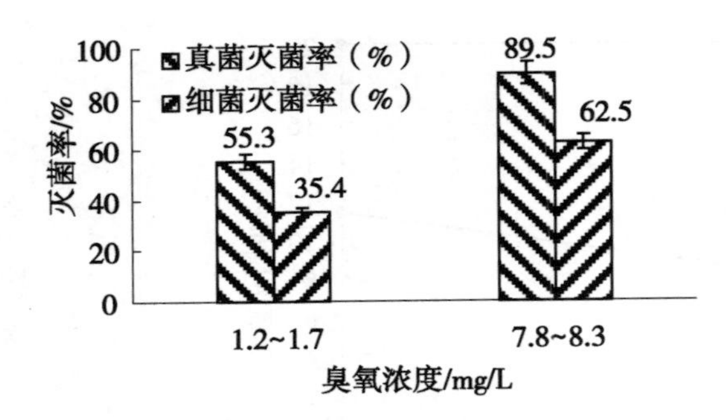

图 5　第一次土壤灭菌的效果

Figure 5　Results of soil sterilization for the first time

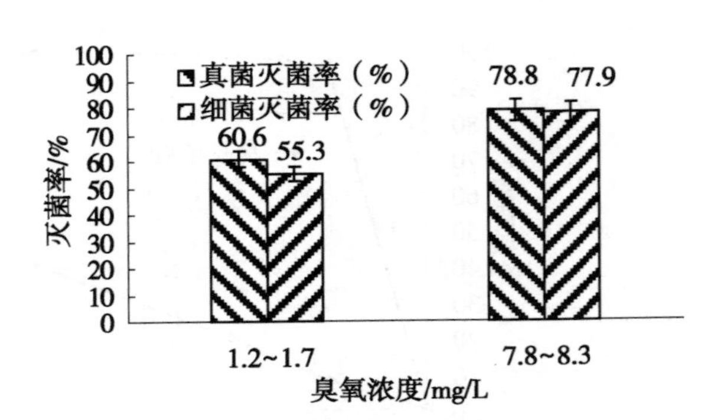

图 6　第二次土壤灭菌的效果

Figure 6　Results of soil sterilization for the second time

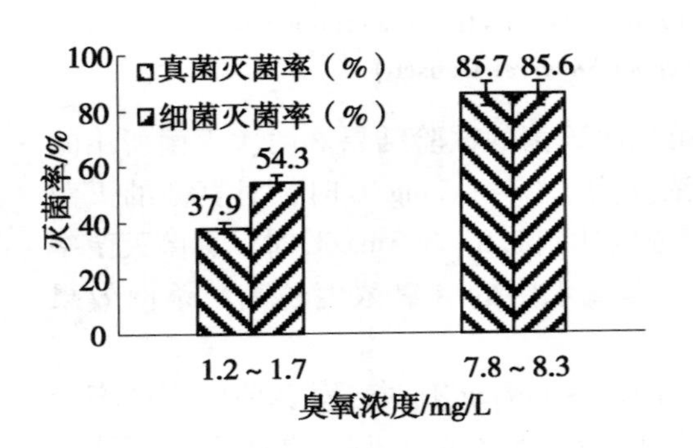

图 7　第三次土壤灭菌的效果

Figure 7　Results of soil sterilization for the third time

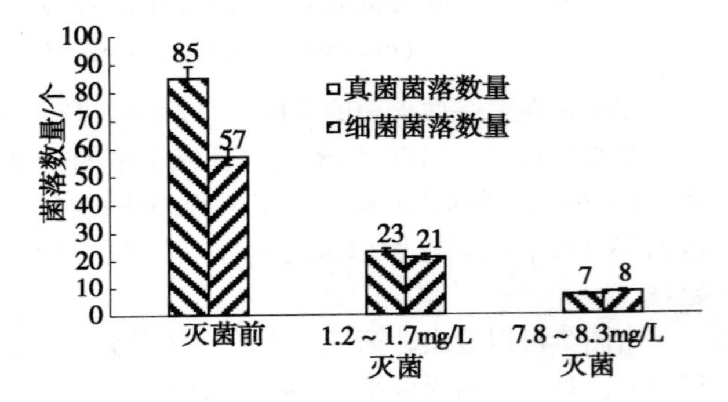

图 8　总体灭菌效果

Figure 8　Total sterilization effects

高，并且很稳定。

在实验室内用微/纳米臭氧水处理 2.4 CFU/ml 的番茄尖镰孢菌菌液，当臭氧浓度为 0.9mg/L、接触时间 0.5min 时，臭氧对番茄尖镰孢菌的灭菌率平均已达 96%；在臭氧浓度为 7.3mg/L、接触时间 2min 时，灭菌率达到 100%，灭菌速率和效率很高。

在杀菌试验中可以发现，对于特定目标菌配成的特定浓度菌液，很低浓度（约 1mg/L）的臭氧水就能达到很好的杀菌效果，杀菌率能达 95% 以上。但从土壤灭菌试验发现，即使浓度较高（7.8 ~ 8.3mg/L），杀菌率也达不到 90%。试验的结果是，臭氧浓度为 1.2 ~ 1.7mg/L 的臭氧水杀菌率只有 50% 左右，并且对真菌和细菌的杀菌效果存在差异。这说明土壤的复杂性要求更高浓度的臭氧水才能达到较好的灭菌效果。因此，微/纳米气泡臭氧水具有的、臭氧溶解度高且稳定性好的特性，对于采用臭氧水进行土壤灭菌意义重大。

本研究表明，用微/纳米气泡臭氧水对土壤灭菌时，可以采取较高浓度的臭氧水进行灭菌，如试验浓度 7.8 ~ 8.3mg/L，一次杀菌后，对真菌的灭菌率为 89.5%，细菌灭菌率为 62.5%。但较高的臭氧浓度，一般是在水温较低时才容易达到，如果采取对土壤多次灭菌的方式，即使使用浓度为1.2～1.7mg/L的臭氧水，经过三次灭菌后，对真菌的灭菌率可达到72.9%，对细菌的菌灭菌率可达到63.2%。而当采用较高的臭氧浓度、多次灭菌时，如臭氧浓度7.8～8.3mg/L，对真菌灭菌率可达91.8%，对细菌灭菌率也达到了86.0%。因此，在采用适当臭氧浓度、适宜灭菌制度时，温室土壤用微/纳米气泡臭氧水进行灭菌，是可行的。当然，多次灭菌之间的间隔时间、用不同无机酸调节水的pH值后再制备弱酸性臭氧水对土壤的灭菌效果，以及臭氧水灭菌对土壤的理化性质和肥力的影响等方面，有待开展进一步的试验研究。

参考文献

[1] 曹坳程，郭美霞，王秋霞．土壤灭菌技术．世界农药，2010，32（2）：10～12

[2] 张利英，李贺年，翟姗姗等．太阳能土壤灭菌在草莓保护地栽培中的应用效果．北方园艺，2010（14）：67～68

[3] H. Gómez-Cousoa，M. Fontán-Sainza. P. Fernández-Ibánez，*et al.* Speeding up the solar water disinfection process（SODIS）against Cryptosporidium parvum by using 2.5 l static solar reactors fitted with compound parabolic concentrators（CPCs）. Acta Tropica，2012，124，235～242

[4] P. D. Kapagiannia，G. Boutsisb，M. D. Argyropoulou，*et al.* The network of interactions among soil quality variables and nematodes：short-term responses to disturbances induced by chemical and organic disinfection. Applied Soil Ecology，2010，44（1）：67～74

[5] A. Peruzzi，M. Raffaelli，M. Ginanni，*et al.* An innovative self-propelled machine for soil disinfection using steam and chemicals in an exothermic reaction. Biosystems Engineering，2011，110（4）：434～442

[6] H. E. Dickens，J. M. Anderson. Manipulation of soil microbial community structure in bog and forest soils using chloroform fumigation. Soil Biology and Biochemistry，1999，31（14）：2 049～2 058

[7] Susanne Klosea，Veronica Acosta-Martínezb，Husein A. Ajwaa. Microbial community composition and enzyme activities in a sandy loam soil after fumigation with methyl bromide or alternative biocides. Soil Biology and Biochemistry，2006，38（6）：1 243～1 254

[8] 曹志平，陈国康，郑长英等．五种甲基溴土壤灭菌替代技术比较研究．农业工程学报，2004，20（5）：250～253

[9] 郑建秋．利用臭氧灭菌园艺设施中有害生物的装置．中国专利：CN1860868，2006

[10] Mujeebur Rahman Khan a，M. Wajid Khan. Effects of intermittent ozone exposures on powdery mildew of cucumber. Environmental and Experimental Botany，1999（42）：163～171

[11] 薛广波．灭菌·灭菌·防腐·保藏．北京：人民卫生出版社，1993

[12] 王芳，刘育京，张文福．臭氧水稳定性及杀菌性能的试验观察．中国灭菌学杂志，1999，16（2）：69～73

[13] Runia WT，Vanachter A. A review of possibilities for disinfection of recirculation water from soilless cultures. Acta Horticulturae，1995，382：221～229

[14] Van Os E A，Runia W T，Van Buclr ent J，*et al.* Prospects of slow sand filtration to eliminate pathogens from recirculating nutrient solution. Acta Horticultrate，1998，458：377～382

[15] M. Takayamaa，K. Ebiharaa，H. Stryczewska. Ozone generation by dielectric barrier discharge for soil sterilization. Thin Solid Films，2006，6（26）：396～399

[16] 喻景权，驹田旦．臭氧对培养液中两种植物病原菌的杀菌效果．园艺学报，1998，25（1）：96～98

[17] Takahashi，M. 2005. Potential of microbubbles in aqueous solutions：Electrical properties of the 146 Biological Engineering Transactions gas-water interface. J. Phys. Chem. B 109（46）：21 858～21 864

[18] 宋卫堂，刘伟，赵淑梅．一种土壤的灭菌方法．中国专利：CN102668932A，2012

[19] 中国建设部，《臭氧发生器臭氧浓度、产量、电耗的测量》标准 CJ/T 3028.2～94，1994

[20] 宋卫堂，孙广明，刘芬．臭氧灭菌循环营养液中三种土传病原菌的试验．农业工程学报，2007，23（6）：189～192

微生物菌剂在牛粪好氧堆肥中的应用研究*

李　杰**，郁继华***，冯　致，颉建明，张国斌，李雯琳，吕　剑
（甘肃农业大学农学院　兰州　730070）

摘要：以牛粪和玉米秸秆为原料进行牛粪高温好氧堆肥，研究微生物菌剂在牛粪堆肥发酵中的应用。结果表明，在整个堆制过程中，2 个处理的堆体温度呈现升温期、高温期和降温期 3 个阶段，添加微生物菌剂可迅速提高堆肥初期的发酵温度，较不添加菌剂处理提前 4d 进入高温期，最高温度为 58℃。发酵 30d 种子发芽指数大于 80%；最终物料含水率比对照低 5.35%，全氮、全磷和全钾分别高 0.06%、0.05% 和 0.07%。综合多项腐熟指标分析，添加微生物菌剂可使牛粪堆肥时间比对照提前 7d 腐熟，可降低对种子发芽的抑制作用，缩短发酵时间，加快物料水分蒸发，使堆肥中无机营养成分含量相对增加，从而提高牛粪堆肥的质量。

关键词：微生物菌剂；牛粪；高温堆肥；腐熟

Effect of Microbial Inoculants on Aerobic Composting of Cow Manure

Li Jie, Yu Jihua, Feng Zhi, Xie Jianming, Zhang Guobin, Li Wenlin, Lv Jian
(*College of Agronomy, Gansu Agricultural University Lanzhou*, 730070, *China*)

Abstract: Aerobium were added to the compost using cow manure and corn stalk as raw materials. Effects of inoculating microbes on the cow manure compost fermentation were investigated. The results show that temperature of caused by the two treatments defined three stages including a warming period, high temperature period and cooling period for the compost pile. The temperature of the compost is increased at the early composting stage by adding the inoculating microbes, and temperature of adding microbe treatment is relatively high, less to no adding microbial inoculants processing advance four days reaches the highest temperature, reach to 58℃. Seed germination index (GI) of the treatment with inoculating microbes is about 80% after composting. Compared to the control, there is a 5.35% decrease of water content of the compost, and a 0.06%, 0.05% and 0.07% increase of total nitrogen, total phosphorus, and total potassium content. With regarding to the maturity parameters, there is more than 7 day composting time reduction in the treatments with inoculating microbes. Our study suggests that the addition can reduce the adverse effects of material on GI, shorten composting time significantly, and further accelerate water volatilization of compost material. Meanwhile, contents of inorganic nutrients are relatively increased; Therefore, the quality of cow manure compost is improved.

---

* 基金项目：农业部公益性行业（农业）专项（201203001）；现代农业产业技术体系专项资金资助（CARS-25-C-07）；甘肃省重大专项（1002FKDA038）

** 作者简介：李杰（1987—），男，甘肃省甘谷人，硕士研究生，主要从事设施蔬菜栽培基质生产的研究. E-mail：gsau23@126.com

*** 通讯作者：郁继华，男，教授，博士生导师，主要从事蔬菜栽培生理及设施作物生产的教学和研究. E-mail：yujihua@gsau.edu.cn

Key words: Inoculating microbes; Dairy manure; Compost; Maturity

研究畜禽粪便循环利用技术，既可以提高资源利用效率，保护和改善环境，又能为非耕地设施农业生物质基质栽培提供原料[1~3]。畜禽粪便的无害化处理技术是生物质栽培基质生产的有效途径，高温堆肥是实现畜禽粪便减量化、无害化处理和资源化利用的有效措施[4~5]。为提高堆肥效率，缩短发酵时间，不少研究者采用添加菌剂的方法来达到这一目的。研究认为，添加菌剂能够使堆温最高温度升高2~10℃，较不添加菌剂提前10d左右腐熟[3,5]。近几年，西北荒漠区非耕地设施农业已逐步发展，已进行规模化生产，对生物质栽培基质需用量很大，当地资源也非常丰富，但是，生物质基质生产技术较为落后，因此，本试验以生物质栽培基质牛粪为试验原料，采用高温好氧堆肥技术，研究添加微生物菌剂对牛粪高温堆肥的影响，旨在为非耕地设施农业生物质栽培基质生产及微生物菌剂规模使用提供依据。

1 材料与方法

1.1 试验区概况与试验材料

所有堆肥试验均在甘肃省酒泉市总寨镇沙河村荒漠区设施农业生产基地进行，试区年平均气温约7.9℃，6~7月平均气温32.2℃，昼夜温差大，风沙大，≥8级大风日数超过30d，年日照数为2 800h，年降水量为84mm，蒸发量2 141mm，超过降雨量27.3倍，年相对湿度为46%，属典型的半沙漠干旱性气候。

供试新鲜牛粪采自试区周边奶牛养殖场，玉米秸秆从周边农户购买。微生物菌剂由山东农业科学院农业资源与环境研究所研发，是枯草芽孢杆菌（*Bacillus subtilis*）、唐德链霉菌（*Streptomyces tendae*）、白浅灰链霉菌（*Streptomyces albogriseolus*）、黑曲霉（*Aspergillus niger*）、里氏木霉（*Trichoderma reesei*）几种菌剂的复合菌系，有效活菌数>80亿CUF/g。

1.2 试验设计

堆肥自2012年4月26日开始，于2012年6月14日结束。试验设添加微生物菌剂（T）和不添加菌剂（CK）2个处理。将新鲜牛粪平摊于地上晾晒，根据秸秆和牛粪的全碳氮含量，将牛粪与秸秆混合并调整物料C/N值为20：1，多点监测物料含水率，直到降至65%左右时加入微生物菌剂，添加量为物料重的0.3%，均匀撒入摊薄的牛粪后翻混3~4次，将其堆制成高为1.5m，长约8m、宽约6m的条形垛式，每10d翻堆1次。

1.3 采样与样品测定

采样：每次翻堆时按多点采样法从堆体10个点分层取样约500g，混匀，一部分冷藏，部分经风干后带回实验室进行相关指标测定。

温度测定：每天上午10：00测定所有堆肥的温度，以堆体东、南、西、北和中心5点温度的平均值作为堆体的发酵温度，测量时温度计插入堆体距表面25cm以下，同时记录周围环境温度。

含水率测定：将样品放入铝盒，在105℃下烘至恒重，冷却至室温后用电子天平称重。

pH值测定：取混合后的样品，用去离子水按粪水比1：10（W：V）混合，振荡30min，过滤后取上清液，用便携式PHSJ-3F型数显pH计测定pH值。

种子发芽指数（GI）：将滤纸放入干净无菌直径为9cm的培养皿中，整齐摆放20粒饱满青菜种子，吸入5ml堆肥浸提液（同pH测定所用浸提方法）于培养皿中，同时用蒸馏水作为对照，每个处理3次重复，在28℃恒温培养箱、黑暗条件下培养48h，用游标卡尺测定根长，计算种子发芽率，根据下列公式计算种子发芽指数。

种子发芽指数（%）＝（堆肥处理的种子发芽率×种子根长）/（对照的种子发芽率×对照种子根长）

有机质、全氮、全磷、全钾的测定参照国家农业标准NY525-2011[6]。

1.4 数据分析

采用SPSS 17.0软件和Excel 2003对试验数据进行统计分析。

2 结果与分析

2.1 堆肥过程中温度的变化

从图1可以看出，各处理堆肥温度的变化呈现先升高后降低的趋势，在整个发酵过程经历了快速升温期、保温期和后熟降温期的阶段性变化规律。初堆肥时，各处理物料温度与环境温度基本一致，在18～23℃。两个处理进入高温发酵阶段（>50℃）持续时间为18d，符合中华人民共和国粪便无害化标准。添加微生物菌剂处理比对照提前4d升至50℃以上，可见CK升温最慢；迅速增温后进入高温阶段，物料温度维持在50～58℃，添加菌剂处理和CK堆肥温度保持在50℃以上的时间依次为18d和13d；前12d添加菌剂处理比CK温度高，但后期降温快。说明添加菌剂能缩短前期升温时间，提高发酵温度，缩短发酵时间。

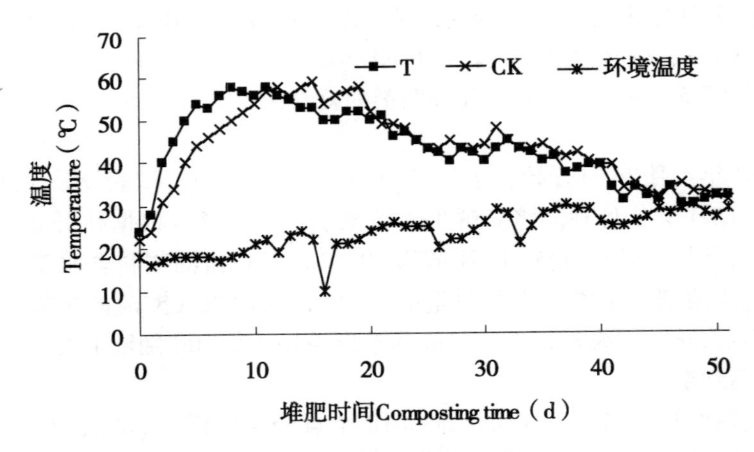

图1 不同处理堆肥温度的变化

Figure 1 Changes of temperature by different treatments during composting of cow manure

2.2 堆肥过程中水分的变化

在发酵过程中（图2），随着堆料温度上升，各处理含水率因大量蒸发而快速下降。初堆时，堆肥含水率为65.1%～66.2%，在第7d，各处理含水率因微生物的活动而快速下降，添加菌剂处理和CK含水率分别比初堆时含水率降低了3.40%和1.66%。由于添加菌剂作用于物料时其微生物活动较 CK 剧烈，致使其产生大量热，蒸发作用增强使堆体水分随之减少；到堆肥结束时，堆体的含水率维持在 16.7% ~22% 左右，比 CK 低 5.3%。

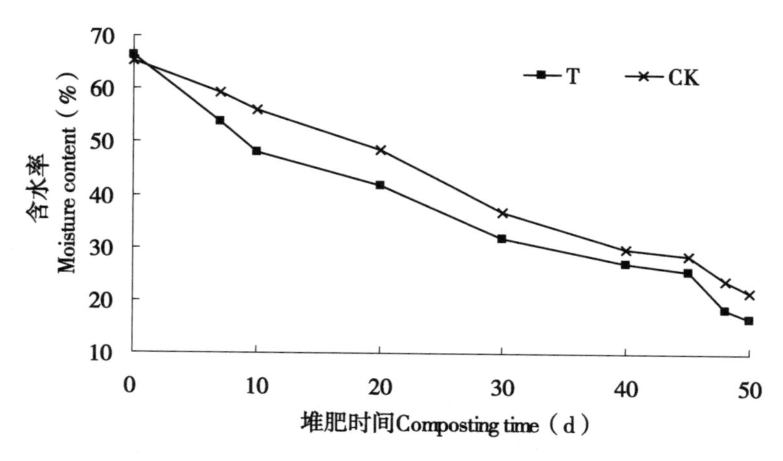

图 2　不同处理堆肥物料含水率的影响

Figure 2　Changes of moisture content by different treatment during composting of cow manure

2.3　堆肥过程中 pH 值的变化

从图 3 可以看出，堆肥初期由于物料中微生物大量繁殖，分解蛋白质等有机物产生大量的 NH_4^+-N，导致 pH 值有所上升；随着堆肥的进程，NH_4^+-N 转化为氨挥发、蛋白质有机物的彻底降解及硝化作用使 NH_4^+-N 含量大大降低，使物料的 pH 值下降，另外，微生物活动产生的大量有机酸也会引起堆肥后阶段 pH 值的降低。添加菌剂处理 pH 值在 7 ~40d 一直高于 CK，而在发酵后期明显低于 CK，其 pH 值升高和下降的速度均大于 CK，说明添加菌剂处理在整个发酵过程中微生物活动和代谢比 CK 剧烈，且对堆料氨气释放的具有抑制作用，即具保氮效果。最后各处理 pH 值均稳定在 8.01 ~8.31，符合腐熟堆肥 pH 值指标标准[5]，其变化均在好氧微生物要求的适宜范围内，有利于堆肥的进行。

2.4　不同微生物菌剂对牛粪堆肥总有机碳、全氮、全磷和全钾的影响

有机质是微生物活动和代谢的能量来源，一部分微生物分解成 CO_2 和 H_2O，释放出能量被微生物所利用，而另外一部分则以稳定的腐殖质形式存在于物料中[7]。用有机碳来反映有机质的变化趋势。从表中可以看出，在堆肥过程中有机质含量呈下降趋势。与堆制前相比，腐熟后添加菌剂处理和 CK 分别降低了 37.90% 和 29.56%，添加菌剂处理下降幅度大于 CK。发酵结束时，添加菌剂处理和 CK 的全氮含量较初堆期分别增加 20.41% 和 15.46%。

在堆肥过程中，P 元素为比较稳定的营养成分，其绝对含量在发酵过程中通常不会随着发酵的进行出现明显的变化；钾元素是极易迁移的元素，它的变化趋势极易受堆肥系统和堆肥条件的影响[7]。牛粪物料始末全磷和全钾的含量均有所增加，添加微生物菌剂与 CK 相比，前者的增幅较后者大，在发酵结束时各处理的全磷（P_2O_5）含量分别为 0.556% 和 0.506%，添加菌剂处理全磷含量的增幅较 CK 大。在发酵结束后物料全钾（K_2O）含量比堆制前分别增加了 37.06% 和 27.97%。可见添加菌剂处理比 CK 提前腐熟。

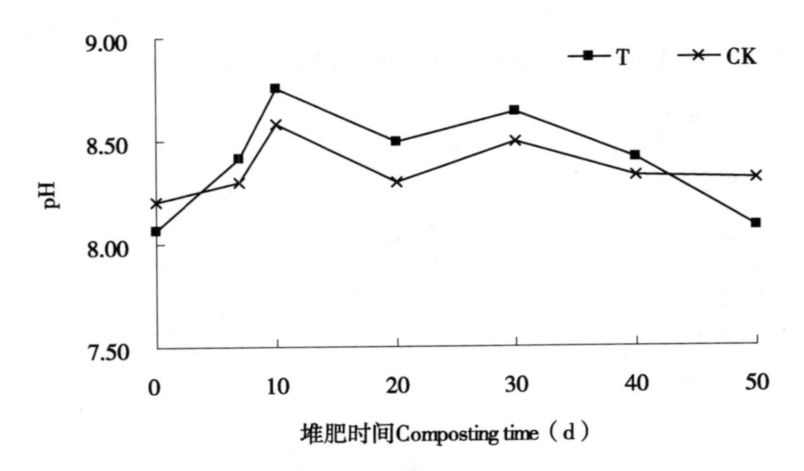

图3　不同处理堆肥物料 pH 值的变化

Figure 3　Changes of pH by different treatments during composting of cow manure

表　不同处理堆肥物料总有机碳、全氮、全磷和全钾的变化

Table　Changes of the total organic carbon, total nitrogen, total phosphorus and total potassium by different treatments during composting of cow manure

处理	总有机碳 TOC（$g \cdot kg^{-1}$）			全氮 TN（%）			全磷 TP（%）			全钾 TK（%）		
	始	末	减少（%）	始	末	增加（%）	始	末	增加（%）	始	末	增加（%）
T	265.23	164.7	37.90	1.352	1.628	20.41	0.4129	0.556	34.66	0.8593	1.1763	37.06
CK	267.36	188.34	29.56	1.358	1.568	15.46	0.4131	0.506	22.49	0.8612	1.1021	27.97

2.5　堆肥物料对种子发芽指数的影响

用生物学方法测定堆肥的植物毒性是检验堆肥腐熟度的有效方法[5]，种子发芽指数（GI）是判断堆肥的植物生物毒性和腐熟度的重要参数[8]。从图4可以看出，随着堆肥进程的推移，堆肥种子发芽指数呈升高趋势，即对种子发芽抑制作用逐渐减弱。在堆制初期，各处理种子发芽指数均在40%左右，在堆制7d时，由于物料pH值升高，多酚、胺类等有害物质增加，对种子发芽指数的抑制也随之增强。堆肥10～30d时，添加菌剂处理种子发芽指数的增幅较CK高，且差异明显。堆肥进行至20d时，添加菌剂处理GI为78.04%，接近80%，而CK的GI为58.16%，从GI大于80%这一标准判断，添加菌剂处理需30d达到堆肥无害化腐熟标准，而CK需要40d。在堆肥结束时，各处理的种子发芽指数分别为90.10%和84.92%，均大于80%。

3　讨论

堆肥是一个微生物参与和作用的过程，微生物活动对堆肥物料的分解起重要的作用[9]。调节物料各因素在最佳范围，加入适当的外源功能微生物是加快腐熟的重要手段。近年来关于畜禽粪便好氧堆肥过程中添加微生物菌剂的研究都证实了微生物菌剂用于堆肥的可行性及

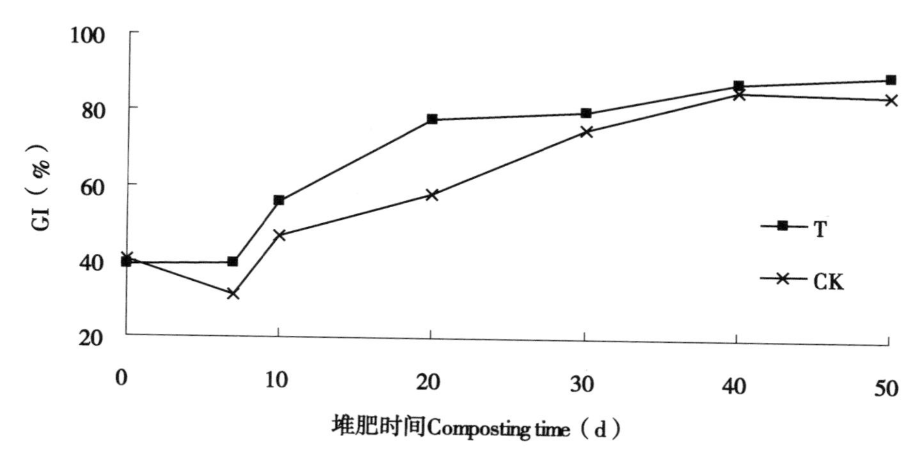

图 4　不同处理堆肥物料对 GI 的影响

Figure 4　Effects of the germination index by different treatment during composting of cow manure

有效性[3,10]。添加了微生物菌剂，增加了堆料中微生物数量，不仅使堆肥初期升温加快，高温时间延长，而且促进了有机碳的分解和降低了堆肥物料的 pH 值，缩短了堆腐时间，提高了腐熟效率。

温度是堆肥过程重要影响因素之一，同时也是堆肥腐熟度的一个重要指标。王岩等[11]研究表明，高温 50～60℃有利于杀死病原菌，但温度过高会抑制并杀死部分有益微生物，温度过低不利于有机物降解和无害化处理。本研究中，堆肥通气较好，气温高，但翻堆间隔时间太长，导致堆肥降温期延长，堆肥效率降低。

水分是影响堆肥效果的重要参数，含水率过高或过低，都将影响微生物的生长及活性。张建华等[7]研究表明，堆肥结束后堆料含水率为 20.76%，李琬等[12]研究发现，腐熟后含水率为 29.6%。本研究表明，新鲜牛粪初堆时含水率在 65.1%～66.2%，随着堆肥的进行，堆料含水率也随之下降，在堆肥第 40d 由于降雨致使物料的含水率略有升高，与张建华等[7]和李婉等[12]研究不同的是本试验中物料含水率各处理下降均较快，这可能是由于本试验所处环境日蒸发量较大，而且堆置时间正逢夏季，与该地区半沙漠干旱性气候条件有关。

前人许多研究[1,10,13]认为，堆肥实际上代表氮素损失的一条潜在途径。本研究中，在堆肥结束时，堆体含水率较堆肥前下降了 23.2%～28.4%，且物料总有机碳下降了 29.6%～40.1%，堆肥总有机碳的降解程度和堆体质量下降程度均较大；堆肥过程中有机氮的矿化、氨的挥发及硝态氮的反硝化作用均引起堆肥过程中氮素的损失，虽然物料中硝化细菌通过硝化作用，使其中氮素得以固定，氮素的损失逐渐变小，但在发酵过程中，堆体质量和水分的散失以及降解有机质等因素，物料中干物质总量下降幅度远大于氮素损失，因此氮素的含量较初堆时其含量相对增加，这与王晓娟[10]和王小琳等[13]的研究结论一致。

综合堆体温度、物料含水率、种子发芽指数等各项指标，添加微生物菌剂使牛粪堆肥比对照提前 7d 腐熟，物料含水率低 5.35%，全氮、全磷和全钾分别高 0.06%、0.05% 和 0.07%。各种微生物菌剂有较强的专一性，本试验所用微生物菌剂对牛粪高温好氧发酵具有良好的促进作用，但对其他畜禽粪便的是否有促进作用，还需进一步研究。

参考文献

[1] 魏彦红，郁继华，冯致等．不同添加剂对牛粪高温堆肥的影响．甘肃农业大学学报，2012，47（3）：52～56

[2] 马丽红，黄懿梅，李学章等．两种添加剂对牛粪堆肥中氮转化及相关微生物的影响．干旱地区农业研究，2010，1（1）：76～80

[3] 沈根祥，尉良，钱晓雍等．微生物菌剂对农牧业废弃物堆肥快速腐熟的效果及其经济性评价．农业环境科学学报，2009，28（5）：1 048～1 052

[4] Bustamante MA，Paredes C，Marhuenda-Egea FC，*et al.* Co-composting of distillery wastes with animal manures：Carbon and nitrogen transformations in the evaluation of compost stability. Chemosphere，2008，72：551～557

[5] Wei Shi，Jeanette M Norton，Bruce E Miller，*et al.* Effects of aeration and moisture during windrow composting on the nitrogen fertilizer values of dairy waste composts． Applied Soil Ecology，1999（11）：17～28

[6] 中华人民共和国农业行业标准 NY525-2011，有机肥料

[7] 张建华，田光明，姚静华等．不同调理剂对猪粪好氧堆肥效果的影响．水土保持学报，2012，26（3）：131～135

[8] Tiquia S M，Tam N F Y. Composting of spent pig litter in turned and forced-aerated piles． Environmental Pollution，1998，99：329～337

[9] 龚建英，李国学，李彦明等．微生物菌剂和鸡粪对蔬菜废弃物堆肥化处理的影响．环境工程学报，2012，6（8）：2 813～2 817

[10] 王晓娟，李博文，刘微等．不同微生物菌剂对鸡粪高温堆腐的影响．土壤通报，2012，43（3）：637～642

[11] 王岩，李玉红．添加微生物菌剂对牛粪高温堆肥腐熟的影响．农业工程学报，2006，22（10）：220～223

[12] 李琬，许修宏．外源微生物对堆肥理化性质及酶活影响的研究．农业环境科学学报，2010，29（3）：592～596

[13] 王小琳，陈世昌，袁国锋等．促腐剂在鸡粪堆肥发酵中的应用研究．植物营养与肥料学报，2009，15（5）：1 210～1 214

新疆喀什地区设施番茄晚疫病调查及品种抗病性鉴定*

何　伟**，崔元玗***，杨　华，张　升，孙晓军
（农业部西北荒漠作物有害生物综合治理重点实验室，新疆农业科学院植物保护研究所，乌鲁木齐　830091）

摘要：为明确喀什地区设施番茄晚疫病发生情况并筛选出适宜本地设施栽培的抗番茄晚疫病的品种，在晚疫病发生时期，调查喀什地区疏勒县、英吉沙县、莎车县和叶城县温室番茄晚疫病发生情况并采用室内苗期人工接种的方法鉴定新引进品种对番茄晚疫病的抗性。调查结果表明，番茄晚疫病病叶率为15.6%～100%；病果率为3.4%～57.3%；病情指数为14.1～100。不同番茄品种抗病性鉴定中，909、新粉7号、DBE-1、KY-2、新粉8号、ZBO-2、8500-4和俄毕德病级指数分别为1.09、3.4、1.24、4.5、5.25、5.04、4.17和4.43。9月至翌年3月，番茄晚疫病可在喀什地区温室中发生流行。品种909、新粉7号和俄毕德表现为抗番茄晚疫病，其他5个供试品种表现为感病。
关键词：设施；番茄晚疫病；调查；抗性鉴定

Survey of Tomato Late Blight and Identification of Resistance of Cultivars in Greenhouse in Kashi Region of Xinjiang

He Wei, Cui Yuanyu, Yang Hua, Zhang Sheng, Sun Xiaojun
(*Key Laboratory of Integrated Pest Management on Crops in Northwestern Oasis Ministry of Agriculture, P. R. China, Institute of Plant Protection, Xinjiang Academy of Agriculturel Sciences, Urumqi* 830091, *China*)

Abstract: In order to know occurrence of tomato late blight in greenhouse in Kashi area of Xinjiang and screen resistant varieties, In the late blight occurred period, surveyed occurrence of tomato late blight in greenhouse in Shule, Yingjisha, Shache and Yecheng of kashi and identificated resistance of new varieties to tomato late blight by indoor inoculation in seedling stage. The results show that disease rate of leaf was 15.6% ~100%, disease rate of fruit was 3.4% ~57.3%, disease index was 14.1 ~100. Identification of disease resistance of different tomato varieties, disease index of 909, Xinfen 7, DBE-1, KY-2, Xinfen 8, ZBO-2 and 8500-4 were respectively 1.09, 3.4, 1.24, 4.5, 5.25, 5.04, 4.17 and 4.43. Occurrence and prevalence of tomato late blight from September to the second year in March in greenhouse in kashi. 909, Xinfen 7 and Ebide were resistance to tomato late blight, but other five varieties were susceptibility.
Key word: Facility; Tomato late blight; Survey; Identification of resistance

1　引言

番茄晚疫病俗称番茄疫病，又称黑秆病，由致病疫霉菌 *Phytophthora infestans* (Mont.)

---

* 基金项目：国家科技支撑计划项目“设施蔬菜高效种植技术研究与示范”；新疆维吾尔自治区“十二五”重大专项（201130104－3）

** 作者：何伟（1980—），男，安徽阜阳，助理研究员，硕士，研究方向为蔬菜病虫害防治，E-mail：hewei_888y@yahoo.com.cn

*** 通讯作者：崔元玗（1955—），女，研究员，研究方向为蔬菜病虫害，E-mail：cuiyuanyu@126.com

De Bary 侵染所致。番茄晚疫病是一种全世界番茄生产上的重大病害之一，给世界番茄生产造成重大损失。法国、美国、加拿大、墨西哥、澳大利亚和瑞士等国家均报道该病的严重危害[1~4]。早在1912年，我国河北和北京一带就有晚疫病发生，随后山西、贵州、江苏、河南、甘肃、福建、湖北、黑龙江、陕西和山东等省市均有番茄晚疫病发生，并给当地的番茄生产造成重大损失[5~9]。近年来，新疆设施农业发展迅速，温室中蔬菜新病害不断出现，造成设施蔬菜产量和品质下降。2011年，喀什温室番茄发生番茄晚疫病，调查番茄晚疫病的发生情况并对新品种进行抗病性鉴定，旨在为喀什地区番茄晚疫病的防治提供科学依据。

2　材料与方法

2.1　温室番茄晚疫病发生情况调查

调查温室发生番茄晚疫病的情况，在温室两端和中间分别随机选取两垄，共调查六垄，调查其病株率和病果率，根据番茄晚疫病分级情况，在温室中两端和中间各随机选取10株，每株调查上、中、下3片叶，共调查90片叶，计算其病情指数。番茄品种试验棚在苗期发生番茄晚疫病，调查不同品种病株数，计算其病株率。

2.2　番茄品种抗性鉴定

2.2.1　供试菌种及番茄品种

供试菌种为分离自新疆喀什地区疏勒县温室中发病番茄植株，经形态学鉴定为 *Phytophthora infestans*（Mont）de Bary。供试番茄品种为新疆农业科学院园艺研究所引进的7个新品种和疏勒县一个常规栽培品种。引进品种分别为909、新粉7号、DBE-1、KY-2、新粉8号、ZBO-2和8500-4，常规栽培品种为俄毕德。

2.2.2　接种方法及病情调查

接种工作在光照培养箱内进行，当鉴别寄主幼苗7~8叶时接种晚疫病菌菌株，接种浓度为 5×10^4 孢子囊/ml，在12℃恒温箱中放置2~3 h，以释放的游动孢子。采用喷雾接种法，每一番茄品种32株，4次重复，接种后24 h内，室温为（20 ±2）℃、黑暗、100 % HR，以后每天光照12h。

接种后第7天调查每一植株叶面积的被害百分率，进而确定各个鉴别寄主每一植株的病害严重度，即单株病害等级。

2.2.3　鉴别标准

单株病害等级：0级：无病症；1级：病斑细小，叶面积被害率≤5%；2级：限制型病斑，5% <叶面积被害率≤15%；3级：叶部有病斑，茎部无病斑，15% <叶面积被害率≤30%；4级：茎部病斑少量，30% <叶面积被害率≤60%；5级：茎部病斑扩展型，60% <叶面积被害率≤90%；6级：茎部严重受害，叶面积被害率 >90%，甚至植株死亡。

群体抗性等级：根据每一鉴别寄主每一植株的单株病害等级，按如下方法计算出每一鉴别寄主4次重复的病级指数平均值。病级指数 =Σ 每个病级的植株数 × 级别数）/ 调查总植株数。依据每一鉴别寄主4次重复的病级指数平均值，再按下面3个抗性等级将不同的鉴别寄主划分成免疫、抗性和感性等不同的反应类型。免疫（I）：病级指数 =0；抗病（R）：0 <病级指数≤4.0；感病（S）：4.0 <病级指数≤6.0。

3 结果与分析

2011~2013 年，新疆喀什地区 4 个县温室中发生番茄晚疫病，笔者调查该病害在温室中番茄上发生情况。

3.1 疏勒县与英吉沙县温室番茄晚疫病发生情况

2012 年 2 月中下旬，喀什地区疏勒县疏勒镇 2 村 9 组温室发生番茄晚疫病，2 月 23 日，对全部发病温室进行调查。

9 个番茄温室全部发生番茄晚疫病，温室栽培品种为 1887 和 6629，番茄生长时期为座果期，其中，3 座温室番茄发生番茄晚疫病后，种植户将病叶和病果全部摘除，无法调查其发病真实情况。另外 6 个温室中，番茄晚疫病发生较为严重且两个品种均有发病较重的温室，病株率最低为 41%，最高达 91.7%；病叶率最低为 15.6%，最高达 82.2%；病果率最低为 3.4%，最高达 23.1%；病情指数最低为 14.1，最高达 77.8（表 1）。

2011 年 12 月，英吉沙县 3 个温室发生番茄晚疫病，发病率 100%，3~5d，3 个温室番茄病指都达 100，全部枯死。

3.2 叶城县和莎车县温室番茄晚疫病发生情况

2012 年 9 月中上旬，叶城县依提木孔乡 12 个温室发生番茄晚疫病，番茄品种为福瑞特，番茄晚疫病发生在座果期。9 月 14 日调查番茄晚疫病发病情况并计算病叶率、病果率和病情指数。

12 个温室中番茄为同一品种，由于管理水平存在差异，晚疫病发生严重度不同，不同温室番茄晚疫病病情指数存在差异。番茄晚疫病病株率为 100%，病叶率从 46.67%~100%，病果率从 0.3%~10.3%，病情指数从 37.53~100（表 3）。

9 月中上旬，叶城县温室未覆膜，笔者调查并记录了 9 月份气象条件和番茄晚疫病发生情况。9 月 1 日至 10 日，叶城县平均气温 22.8~25.5℃，白天气温在 27℃以上，夜间气温在 17.3~20.8℃，相对湿度为 31%~48%，早晚结露，此条件适宜番茄晚疫病的发生与流行。9 月 17 日至 19 日，温室番茄晚疫病病叶率、病果率和病情指数迅速增加，病叶率从最低 46.67%增加到 98.75%，病果率从 0.3%~10.3%增加到 23.1%~57.3%，病情指数从最低 37.53 增加到 81.98（表 3）。原因分析：气象资料表明，9 月 17~19 日，叶城县降雨，相对湿度 57%~75%，早晚结露持续时间长，易于番茄晚疫病的快速流行。

2012 年 3 月上旬，喀什地区疏勒县香妃湖庄园一个定植 10d 的品种试验温室中发生番茄晚疫病，调查不同品种番茄晚疫病发生的情况。

表 2 表明，品种 909、新粉 7 号、俄毕德和 DBE-1 病株率最低，4 个番茄品种晚疫病病株率没有显著差异，品种 8500-4 病株率最高，达到 48%，与其他番茄品种晚疫病病株率存在极显著差异。

2013 年 1 月上旬，莎车县英吾斯塘乡 1 村一个温室发生番茄晚疫病，1 月 6 日调查其发病情况，病叶率达 88.3%，病情指数为 54.63。

表1　疏勒县温室番茄晚疫病发生情况

Table 1　The occurrence of tomato late blight in greenhouse in Shule county

品种 (variety)	病株率 (%) (diseased plant rate)	病叶率 (%) (diseased leaf rate)	病果率 (%) (diseased fruit rate)	病情指数 (disease index)
1887	81. 40	50	10. 10	42. 3
1887	96. 10	82. 20	23. 10	77. 8
1887	42. 20	15. 60	5	14. 1
6629	78. 10	27. 80	5. 50	20. 1
6629	91. 70	68. 90	21. 90	68. 9
6629	41	18. 90	3. 40	16. 4

表2　不同番茄品种苗期晚疫病病株率比较

Table 2 The comparison of disease rate of plant of different varieties in seedling stage

品种 (variety)	病株率 (%) (diseased plant rate)	显著差异性 (significance of difference)	
		5%	1%
909	10	a	A
新粉7号	13	a	A
俄毕德	14	a	A
DBE-1	16	a	A
新粉8号	26	ab	AB
KY-2	31	abc	AB
ZBO-2	36	bc	AB
8500-4	48	c	B

表3　叶城县温室番茄晚疫病发生情况

Table 3　The occurrence of tomato late blight in greenhouse in Yecheng county

温室编号 (number of greenhouse)	病叶率 (%) (diseased leaf rate)		病果率 (%) (diseased fruit rate)		病情指数 (disease index)	
	9月14日	9月19日	9月14日	9月19日	9月14日	9月19日
1号棚	86. 67	100	5. 3	57. 3	68. 89	98. 02
2号棚	46. 67	98. 75	0. 3	25. 3	37. 53	90. 61
3号棚	67. 78	100	4	46. 7	43. 33	81. 98
4号棚	84. 44	100	4. 3	50. 1	60. 49	95. 8
5号棚	58. 89	98. 89	0. 7	23. 1	32. 47	82. 96
6号棚	64. 44	100	2	30. 1	48. 15	88. 39
7号棚	95. 55	100	5	50. 3	68. 4	95. 31
8号棚	80	100	3. 7	37. 8	62. 22	96. 3
9号棚	100	100	10. 3	57. 3	100	100
10号棚	95. 56	100	4. 7	41. 2	85. 93	100
11号棚	82. 22	100	2	26. 7	58. 77	95. 06
12号棚	85. 56	100	4	47. 8	66. 54	97. 04

3.3 品种抗病性鉴定

调查结果表明，室内接种后番茄品种909、新粉7号和俄毕德表现抗病，其他品种表现感病。番茄品种909与俄毕德病级指数分别为1.09和1.24，表现最抗病。新粉8号和KY-2病级指数分别为5.04和5.25，表现最感病（表4）。田间调查情况表明，番茄品种909、新粉7号和俄毕德病株率最低，这与室内接种试验发病情况相同；品种DBE-1田间发病率为16%，与番茄品种909、新粉7号和俄毕德发病率无显著差异，但室内接种试验表明，DBE-1表现感病。

表4 不同番茄品种抗病性比较

Table 4 The comparison of disease resistance of different tomato varieties

品种（variety）	病情指数（disease index）	抗病类型（resistant type）
909	1.09	R
新粉7号	3.4	R
俄毕德	1.24	R
DBE-1	4.5	S
新粉8号	5.04	S
KY-2	5.25	S
ZBO-2	4.17	S
8500-4	4.43	S

4 结论与讨论

4.1 番茄晚疫病发生情况调查

在9月上旬至第二年3月中下旬，新疆喀什地区温室中温湿度环境适宜番茄晚疫病的发生与流行。番加晚疫病菌主要以分生孢子借气流传播，从气孔或表皮直接侵入，在田间形成中心病株，病菌丝营养菌丝在寄主细胞间或细胞内扩展蔓延，经3~4d潜育，病部长出菌丝和孢子囊，借气流传播蔓延，进行多次重复侵染，引起该病流行。此病发生见于白天气温24℃以上，夜间10℃以上，相对湿度75%~100%。相对湿度高于85%，孢囊梗从气孔中伸出，相对湿度高于95%，孢子囊形成，当寄主表面有水膜时，孢子囊才能产生游动孢子或休眠孢子萌发并产生芽管。温度条件容易满足，能否发病或流行取决于有无饱和的相对湿度或水滴，因此，棚内结露持续时间长短是决定该病发生和流行的重要条件。近两年，喀什地区在9~10月份，降雨量比往年偏多。此时温室尚未覆膜，早晚有结露，温度和湿度适宜番茄晚疫病的发生与流行。11月至翌年3月，温室中温度较低，且多采用沟灌模式种植，棚内早晚湿度较大，早晚形成结露时间较长，易于番茄晚疫病的发生与流行。

4.2 品种抗病性鉴定

新引进栽培的品种中，909和新粉7号表现为抗病，其他品种表现为感病。常规栽培品种俄毕德表现为抗病。番茄晚疫病的常用防治方法包括栽培管理技术、喷施杀菌剂和栽培抗病品种[1]。*Phytophthora infestans* 侵染番茄，如果环境适宜，喷施化学药剂可能没有效果，栽培管理措施在生产中因管理水平低限制了其防治晚疫病的作用，因此，栽培抗病品种是防治番茄晚疫病最有效的措施。

参考文献

[1] Nowicki M, Foolad M R, Nowakowska M, *et al.* Potato and tomato late blight caused by *Phytophthora infestans*: an overview of pathology and resistance breeding. Plant Disease, 2012, 96 (1): 4 ~ 17

[2] Peters R D, Hwbud, Plattrhall. Hypotheses for the inter-reginonal movement of new genotypes of *Phytophthora infestans* in Canada. Cana Plant Pathol, 1999, 21: 132 ~ 136

[3] Erinile I D, Quimv. An Epiphytotic of late blight of tomatos in Nigeria. Plant Disease, 1998, 64 (7): 701 ~ 702

[4] Hartangl Y, Huang H. Characteristics of *Phytophthora infestans* Isolates and development of late blight on tomato in Taiwan. Plant Disease, 1995, 79 (8): 849 ~ 850

[5] 黄河，魏世义，孟晓云. 北京番茄晚疫病的发生规律和预测. 中国农业科学，1984 (4): 85 ~ 89

[6] 钟仕田，刘志涛，肖述炳. 大棚番茄晚疫病的发生及防治. 湖北植保，1992 (3): 8

[7] 冯兰香，杨宇红，谢丙炎等. 中国 18 省市番茄晚疫病生理小种的鉴定. 园艺学报，2004，16 (3): 758 ~ 761

[8] 李新智. 平凉市露地番茄晚疫病发生原因及防治措施. 甘肃农业科技，2004 (9): 51 ~ 52

[9] 陈加福. 厦门地区番茄晚疫病的发生原因与综合防治措施. 中国植保导刊，2004 (2): 20 ~ 22

有机复合肥对温室郁金香生长发育的影响

张彩虹[*]，邹　平，马彩雯
（新疆农业科学院农业机械化研究所，乌鲁木齐　830091）

摘要：［目的］以“巴塞罗娜”郁金香品种为试材，在定植前对其施入两种有机复合肥（F60、F35），研究选用的两种有机复合肥在设施郁金香栽培中对郁金香生长发育的的影响。［方法］对施入有机复合肥的不同处理的郁金香株高、茎粗、花苞大小及发芽率和开花率的数据采集，与空白对照进行对比分析。［结果］选用的两种有机复合肥对郁金香株高、茎粗、花冠大小较对照有显著增大。其中 F60 有机复合肥作用效果较 F35 作用效果好。［结论］不同浓度有机复合肥施入量对郁金香生长特性影响不同，其中以 200kg/667m^2 的 F60 有机复合肥和 300kg/667m^2 的 F35 有机复合肥的作用效果最好。
关键词：有机复合肥；郁金香；促生栽培；生长特性

Effect of Organic Compound Fertilizer on Promoting Cultivation of Tulip

Zhang Caihong，Zou Ping，Ma Caiwen
（*Institute of Agricultural Machinery Research Xinjiang Academy of Agricultural Sciences*，*Urumqi* 830091，*China*）

Abstract：［Object］This paper use tulip varieties Barcelona as researching materials. Applied two organic omplex fertilizers（F60，F35）before field planting was done. To heck the influence of soil improvements and nutrition supply of plants under forcing ulture in greenhouse，after two organic omplex fertilizers were applied.［Method］acquisition the data of plant height，stem diameter，buds and flower size and germination percentage rates of the tulips after applied different organic omplex fertilizers，and comparative with the comparison.［Results］tulips which were applied organic omplex fertilizers was bigger and stronger in height，stem diameter and flower size than the heck which were not applied. ompare the two fertilizers，plants which were applied with F60 organic omplex fertilizer was better.［Conclusion］fertilizer rate is important for plants，this paper shows 200kg/667m^2 for F60 fertilizer and 300kg/667m^2 for F35 fertilizer which have best effects for plants growing.
Key words：Organic omplex fertilizer；Tulip；Forcing ulture；Growth haracteristics

1　前言

花卉种植业是我国的一个新型产业，尤其近几年花卉生产业在我国也得到了大力发展。新疆鲜切花生产业正处于逐渐形成一个区域优势明显的特色产业的时期[1]。由于化肥超常规使用和设施栽培，造成土地板结，“透支”的土地盐碱化严重，已不利于花卉种植生产[2]。造成个别地区大幅度减产减收，虽经多方努力，问题仍得不到解决，成为困扰制种单位、种植户的一大难题。开展在设施花卉促生栽培中日光温室土壤改良、养分供给效果的

---

* 作者简介：张彩虹，女，1982 年 6 月生，新疆农业科学院农业机械化研究所助理研究员，主要研究方向为温室蔬菜栽培

有机复合肥试验，检验有机复合肥（F60、F35）在设施农业花卉生产中的肥效，更好的解决郁金香生产中无机化肥使用不合理造成烧根和生长不良的现象，同时保护土地资源，是本实验研究的目的所在。

2　材料与方法

2.1　试验设计

试验于2012年12月至2013年2月在乌鲁木齐水西沟东湾镇日光温室内进行。选用“巴塞罗娜”郁金香品种为试材。种球定植前施入两种不同复合有机肥，在标准常规有机肥施入水平（$6\sim8m^2/667m^2$）的基础上，每种有机复合肥试验设置4个施肥量处理（$100kg/667m^2$、$200kg/667m^2$、$300kg/667m^2$、$400kg/667m^2$）、一个常规对照处理，整个试验共计9个处理，每个处理3次重复，分别为：CK、F 60-1、F 60-2、F 60-3、F 60-4、F 35-1、F 35-2、F 35-3、F 35-4。于采花时期测定郁金香植株的株高、茎粗、花冠等生长参数，以此衡量选用有机复合肥对郁金香生长的影响。

2.2　测定项目与方法

株高、茎粗、花冠测量：采用直尺、游标卡尺测量[3]。

始花期、盛花期统计：始花期为郁金香开出第一朵花的时期，盛花期开始时间为花朵开放数达到该处理总株数的50%[4]。

温室内温湿度测定：采用国家农业信息化工程技术研究中心研制的温室环境参数自动监测记录仪器——“温室娃娃”测量，精确度已经通过国家检测中心检测。

2.3　数据处理

采用Microsoft Excel和SPSS进行数据处理，并对差异显著指标进行Duncan多重比较。

3　结果与分析

3.1　不用处理对郁金香生育期的影响

表1　不同肥料及浓度对郁金香开花期的影响

Table 1 Effect of different fertilizer and density on the flowering of Tulip

处理	定植时期/（月-日）	初花期/（月-日）	盛花始期/（月-日）	盛花持续时间/d
F60-1	12-12	2-12	2-16	7.5a
F35-1	12-12	2-12	2-17	7.3a
F60-2	12-12	2-13	2-14	8.9b
F35-2	12-12	2-12	2-16	7.9c
F60-3	12-12	2-13	2-14	9.1b
F35-3	12-12	2-13	2-14	8.6b
F60-4	12-12	2-13	2-15	8.6b
F35-4	12-12	2-13	2-14	8.2c
CK	12-12	2-12	2-17	7.1a

由表1可见，与对照相比，不同浓度有机复合肥的使用对郁金香进入初花期的时间没有显

著影响，其中 F60-2、F60-3、F60-4、F35-3、F35-4 等 5 个处理进入初花期的时间较对照提前 1d，进入盛花期的时间提前 3d，F35-1 处理与对照进入初花期及盛花期的时间基本一致。从单花盛花持续时间来看，F60-2、F60-3、F60-4、F35-3 均显著延长单花盛花期，而 F35-1 处理与对照无明显差异。因此，与对照生育期为 60d 相比较，通过施用合适浓度（F60-2、F35-3）的复合有机肥，使郁金香生育期均能提前 3d，这对于郁金香提前前上市、取得较好经济效益和降低市场风险具有重要的现实意义，同时，也为茬口安排和花期调控提供了理论依据。

3.2　不用处理对郁金香切花品质的影响

表 2　不同肥料及浓度对郁金香切花品质的影响

Table 2　Effect of different fertilizer and density on the cut quality of Tulip

处理	株高（cm）	茎粗（cm）	花苞大小（cm×cm）
F60-1	37.6c	1.03ab	1.76×5.06
F60-2	37.33c	1.13b	1.86×5.11
F60-3	38.4b	1.14b	1.86×5.07
F60-4	39.4b	1.09b	1.91×5.18
F35-1	41.17a	0.90a	1.64×5.01
F35-2	40.71a	0.99a	1.69×5.06
F35-3	37.6c	1.09b	1.88×5.12
F35-4	37.9c	1.07b	1.84×5.17
CK	41.67a	0.88a	1.72×5.01

如表 2 所示，与对照相比，F35-1、F35-2 对郁金香株高的影响不明显，而 F60-1、F60-2、F60-3、F60-4、F35-3、F35-4 这六个处理对郁金香的茎高影响较大，茎高分别为 37.6cm、37.33cm、38.4cm、39.4cm、37.6cm 和 37.9cm，比对照分别下降了 4.07cm、4.34cm、3.27cm、2.27cm、4.07cm 和 3.77cm，统计分析表明，其差异达显著水平。与对照相比，所有处理对郁金香花苞大小的影响不明显，F60-2、F60-3、F60-4、F35-3、F35-4 这五个处理的花苞较其他四个处理稍大。

3.3　不用处理对郁金香开花率及发芽率的影响

如表 3 所示，与对照相比，施用有机复合肥的所有处理对郁金香的发芽率和开花率影响都较明显，8 个处理的发芽率和开花率都较对照要高。

3.4　试验期间温室内温度和湿度的变化

郁金香最适温度为 15～18℃，最高不超过 25℃。定植后的前 2 周内土壤温度应保持在 12℃以下。整个生长期温室内夜间温度控制在 6℃以上。温室内相对湿度不可高于 80% 或低于 60%，湿度大小可通过浇水和放风来调节[5,6]。白天温室内温度超过 25℃或湿度超过 80% 便要注意用通风来降温排湿。如图所示，温室内郁金香生长期间温度范围在 6～26℃，湿度范围在 50%～70% 之间，可见，本次试验设置地点新疆乌鲁木齐市的冬季气候是非常适合郁金香栽培的，无需增加其他温度调控设施。

表 3　不同肥料及浓度对郁金香切花品质的影响

Table 3　Effect of different fertilizer and density on the rate of germination and flowering of Tulip

处理	发芽率（%）	开花率（%）
F60-1	98.6	98.6
F60-2	99.4	99.4
F60-3	99.6	99.6
F60-4	98.9	98.4
F35-1	97.8	97.6
F35-2	98.8	98.8
F35-3	97.9	98.8
F35-4	98.2	98.3
CK	95.4	95.4

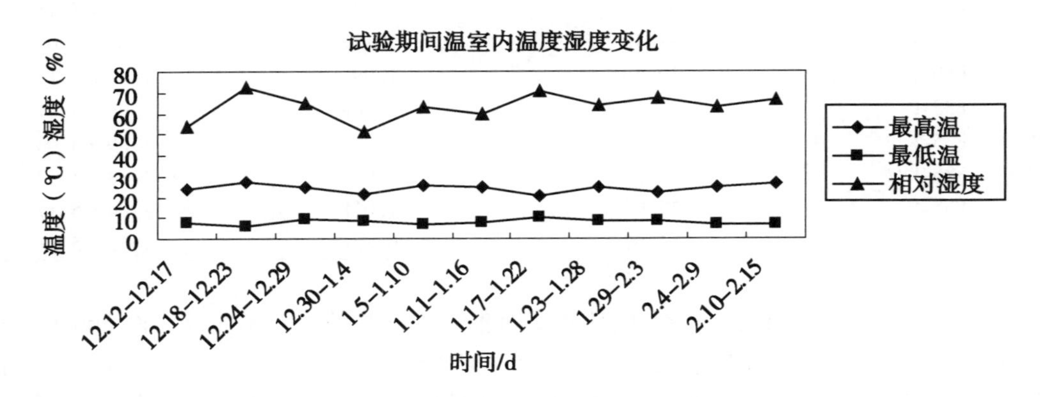

图　试验期间温室内温度和湿度变化

Figure　The change of temperature and relative humidity during the test

4　小结和讨论

在郁金香生育期内，施入的两种有机复合肥 F60（有机质含量 60%）、F35（有机质含量 35%），均能促进郁金香生长，并显著缩短其生育期，其中，以 F60-2（200kg/667m^2 使用量）。F35-3（300kg/667m^2 使用量）最为显著。分析认为，施入一定量的有机复合肥可增强养分，促进郁金香生长发育，一方面保证了郁金香在较低温度条件下良好发育，提高了郁金香切花品质，另一方面缩短郁金香生育期，保证郁金香能在春节期间上市，提高了种植效益，降低市场风险。

与对照相比，施用有机复合肥的所有处理对郁金香的发芽率和开花率影响都较明显且处理间差异不显著，这可能与郁金香这一类球根花卉的特殊营养代谢有关，具体原因还有待进一步探讨。

通过试验表明，两种肥料 F60-2、F60-3、F60-4、F35-3、F35-4 五个处理对郁金香株高、茎粗、花苞大小、生育期等能产生较对照显著的作用。其中，F60 有机复合肥的作用效果较 F35 的显著，说明有机复合肥的施入能提高郁金香的生长发育能力，但是，考虑生产成本，通过试验选择出 F60-2 作为郁金香栽培过程中有机复合肥的施入浓度。

参考文献

[1] 陈忠萍．郁金香在新疆地区的栽培与管理．科技风，2009（13）：246～247

[2] 陈风，董小艳，涂小云．不同类型复合肥对郁金香种球复壮的影响．江苏农业科学，2011（2）：277～278

[3] 翟雷，马越，张黎霞．郁金香生长发育规律及观赏性状的调查研究．南方农业，2008（2）：5～8

[4] 曹玉梅．微量元素对郁金香切花品质的影响．青海农机推广，2008（3）：24～26

[5] 赵统利．日光温室切花郁金香高效栽培技术研究．南京农业大学（博士论文），2006

[6] 王生旭，朱东兴．保护地栽培条件下郁金香生长发育与种球复壮的研究．西北农业学报．2007（6）：178～181

[7] 康健，裴蓓，周丕生等．不同氮素水平对郁金香磷素与钾素的累积与分配的影响．上海交通大学学报（农业科学版），2005（4）：25～31